PROTEASES
Potential Role in Health and Disease

ADVANCES IN EXPERIMENTAL MEDICINE AND BIOLOGY

Recent Volumes in this Series

Volume 159
OXYGEN TRANSPORT TO TISSUE – IV
Edited by Haim I. Bicher and Duane F. Bruley

Volume 160
PORPHYRIN PHOTOSENSITIZATION
Edited by David Kessel and Thomas J. Dougherty

Volume 161
MYOCARDIAL INJURY
Edited by John J. Spitzer

Volume 162
HOST DEFENSES TO INTRACELLULAR PATHOGENS
Edited by Toby K. Eisenstein, Paul Actor, and Herman Friedman

Volume 163
FOLYL AND ANTIFOLYL POLYGLUTAMATES
Edited by I. David Goldman, Joseph R. Bertino, and Bruce A. Chabner

Volume 164
THROMBOSIS AND CARDIOVASCULAR DISEASES
Edited by Antonio Strano

Volume 165
PURINE METABOLISM IN MAN – IV
Edited by Chris H. M. M. De Bruyn, H. Anne Simmonds, and Mathias M. Müller

Volume 166
BIOLOGICAL RESPONSE MODIFIERS IN HUMAN ONCOLOGY
AND IMMUNOLOGY
Edited by Thomas Klein, Steven Specter, Herman Friedman, and Andor Szentivanyi

Volume 167
PROTEASES: Potential Role in Health and Disease
Edited by Walter H. Hörl and August Heidland

PROTEASES
Potential Role in Health and Disease

Edited by
Walter H. Hörl
and
August Heidland

University of Würzburg
Würzburg, Federal Republic of Germany

PLENUM PRESS • NEW YORK AND LONDON

Library of Congress Cataloging in Publication Data

International Symposium on Proteases: Potential Role in Health and Disease (1982:
  Würzburg, Germany)
  Proteases, potential role in health and disease.

  (Advances in experimental medicine and biology; v. 167)
  "Proceedings of the International Symposium on Proteases: Potential Role in Health
and Disease, held October 17–20, 1982, in Würzburg, Federal Republic of Germany" –
Verso t.p.
  Includes bibliographical references and index.
  1. Proteolytic enzymes – Congresses. 2. Proteolytic enzyme inhibitors – Congresses.
I. Hörl, Walter H. II. Heidland, August. III. Series. [DNLM: 1. Protease inhibitors –
Congresses. 2. Peptide hydrolases – Congresses. W1 AD559 v.167 / QU 136 I615p
1982]
QP609.P78I57   1982   599′.019256   83-19186
ISBN-13: 978-1-4615-9357-7   e-ISBN-13: 978-1-4615-9355-3
DOI: 10.1007/978-1-4615-9355-3

Proceedings of the International Symposium on Proteases: Potential Role in Health
and Disease, held October 17–20, 1982, in Würzburg, Federal Republic of Germany

© 1984 Plenum Press, New York
A Division of Plenum Publishing Corporation
233 Spring Street, New York, N.Y. 10013
Softcover reprint of the hardcover 1st edition 1984

TO OUR WIVES

Ursula Hörl
Gundula Heidland

AND TO OUR CHILDREN

PREFACE

We are pleased to present to our readers the Proceedings of
the International Symposium "Proteases: Potential Role in Health
and Disease" which was held in Würzburg (FRG) during October 17-20,
1982.

The topics discussed included those dealing with the physiology and pathophysiology of proteases and their inhibitors, the
interactions of proteases and hormones, the kallikrein-kinin system,
complement and the coagulation system, the function of proteases in
the kidney and the intestinal tract as well as the role of proteases
in lung diseases, pancreatitis, arthritis and hypercatabolic states
(multiple trauma, septicemia, acute renal failure). The papers
presented answered many questions, but raised many more concerning
the significance of proteases and their inhibitors in clinical
medicine. It was unfortunately impossible in this volume to include the extended, lively and extremely stimulating discussions
which were enjoyed by the participants during the conference.

The meeting has provided a unique framework for close interaction between scientists from various disciplines, including biochemistry, physiology, surgery, anaesthesiology, endocrinology,
hematology, pulmonology and nephrology.

We would like to express our thanks and appreciation for all
those who have stimulated, encouraged and supported us to hold
this symposium in Würzburg. This endeavor could not have been
possible without the generous financial support of the Paul-Martini-
Foundation (Mainz), Bayer AG (Leverkusen), Beiersdorf AG (Hamburg).
Bellco GmbH (Freiburg), Boehringer GmbH (Ingelheim), Boehringer
GmbH (Mannheim), Byk Gulden GmbH (Konstanz), Ciba-Geigy (Wehr),
Cyanamid GmbH (Wolfratshausen), Deutsche Abbott GmbH (Wiesbaden),
Diamed GmbH (Köln), Fresenius AG (Oberursel), Gambro KG (München),
Hoechst AG (Frankfurt), von Heyden GmbH (München), Mack (Iller-
tissen), Madaus & Co. (Köln), MSD Sharp & Dohme (München), Merck
(Darmstadt), Nordmark-Werke GmbH (Uetersen), Pfizer GmbH (Karls-
ruhe), Pfrimmer & Co. GmbH (Erlangen), Pharmacodex (München), San-

doz AG (Nürnberg), Schering AG (Berlin), Travenol GmbH (München), Wellcome GmbH (Burgwedel), Verein zur Bekämpfung der Hochdruck- und Nierenkrankheiten (Würzburg).

We also are indebted to Mrs. I. Hevendehl, Miss M. Eiring and Mrs. A. Dempwolff for their invaluable assistance and help in the organization of the meeting. It is a pleasure to acknowledge the excellent secretarial assistance of Monika Eiring for preparing the manuscripts.

Walter H. Hörl
August Heidland

CONTENTS

PHYSIOLOGY AND PATHOPHYSIOLOGY OF PROTEASES
AND THEIR INHIBITORS

Physiology and Pathophysiology of Neutral Proteinases
of Human Granulocytes 1
K. Havemann and M. Gramse

Regulation of Protease Activity 21
M. Steinbuch

Human Kininogens and Their Function in the Kallikrein-
Kinin Systems 41
W. Müller-Esterl and H. Fritz

Possible Involvement of Kinins in Muscle Energy
Metabolism 63
G. Dietze, E. Maerker, C. Lodri, R. Schifman,
M. Wicklmayr, R. Geiger, E. Fink, I. Boettger,
H. Fritz, and H. Mehnert

Structure and Function of Natural Inhibitors as
Antagonists of Proteinase Activities 73
H. Tschesche

Oxidation of Alpha-1-Proteinase Inhibitor:
Significance for Pathobiology 89
J. Travis, K. Beatty, and N. Matheson

In Vivo Significance of Kinetic Constants of
Macromolecular Proteinase Inhibitors 97
J. G. Bieth

On the Multiplicity of Cellular Elastases and their
Inefficient Control by Natural Inhibitors 111
W. Hornebeck, D. Brechemier, M. P. Jacob,
C. Frances, and L. Robert

Proteases - Proteases Inhibitors: a Local
    Cellular Information System 121
  H. Heine

                  PROTEASES AND HORMONES

Regulatory Proteolysis during Corticosteroid
    Hormone Action 129
  M. K. Agarwal

Proteases in Hormone Production and Metabolism 141
  W. A. Hsueh

Precursor Processing and Metabolism of
    Parathyroid Hormone: Regulation
    by Calcium 153
  J. A. Fischer

Processing and Degradation of Met-Enkephalin
    by Peptidase Associated with Rat Brain
    Cortical Synaptosomes 165
  W. Demmer and K. Brand

          PROTEASES IN KIDNEY AND INTESTINAL TRACT

Characterization and Clinical Significance of
    Membrane Bound Proteases from Human
    Kidney Cortex 179
  J. E. Scherberich, C. Gauhl, G. Heinert,
  W. Mondorf, and W. Schoeppe

Recent Advances in Protease Research using
    Synthetic Substrates 191
  R. Gossrau, Z. Lojda, R. E. Smith, and
  P. Sinha

Kinetic Characterization of Brush Border Membrane
    Proteases in Relationship to Mucosal
    Architecture by Section Biochemistry 209
  S. Gutschmidt, R. Hoper, and R. Gossrau

Fluorescence Detection of Proteases with AFC,
    AMC and MNA Peptides using Isoelectric
    Focusing 219
  P. Sinha, R. Gossrau, R. E. Smith, and
  Z. Lojda

PROTEASES AND BLOOD SYSTEM

Pathophysiology of the Interaction between
    Complement and Non-Complement
    Proteases 227
    U. E. Nydegger and S. Suter

Interactions between the Alternative
    Complement Pathway and Proteases
    of the Coagulation System 235
    M. D. Kazatchkine and M.-H. Jouvin

The Calcium-Dependent Neutral Protease of
    Human Blood Platelets: a Comparison
    of its Effects on the Receptors for
    von Willebrand Factor and for the
    Fc-Fragment Derived from IgG 241
    M. O. Spycher, U. E. Nydegger, and
    E. F. Luescher

Alpha-2-Plasmin Inhibitor Inactivation by
    Human Granulocyte Elastase 253
    M. Gramse, K. Havemann, and R. Egbring

Heparin and Plasma Proteinase Inhibitors:
    Influence of Heparin on the Inhibition
    of Thrombin by α_2Macroglobulin 263
    P. Lambin, F. Pochon, and M. Steinbuch

The Involvement of Plasmatic and Fibrinolytic
    Systems in Idiopathic Glomerulo-
    nephritis (GN) 273
    K. Andrassy, E. Ritz, and R. Waldherr

The Effect of Aprotinin on Platelet Function,
    Blood Coagulation and Blood Lactate
    Level in Total Hip Replacement - a
    Double Blind Clinical Trial 287
    S. Haas, R. Ketterl, A. Stemberger,
    P. Wendt, H.-M. Fritsche, H. Kienzle,
    F. Lechner, and G. Blümel

PROTEASES AND LUNG

Interaction of Granulocyte Proteases with
    Inhibitors in Pulmonary Diseases 299
    K. Ohlsson, U. Fryksmark, M. Ohlsson,
    and H. Tegner

Leukoproteinases and Pulmonary Emphysema:
    Cathepsin G and Other Chymotrypsin-
    Like Proteinases Enhance the Elasto-
    lytic Activity of Elastase on Lung
    Elastin 313
  Ch. Boudier, Ph. Laurent, and J. G. Bieth

Adult Respiratory Distress Syndrome (ARDS):
    Experimental Models with Elastase and
    Thrombin Infusion in Pigs 319
  H. Burchardi, T. Stokke, I. Hensel, H. Köstering,
    G. Rahlf, G. Schlag, H. Heine, and
    W. H. Hörl

PROTEASES AND ARTHRITIS

Interactions of Granulocyte Proteases with
    Inhibitors in Rheumatoid Arthritis 335
  L. Ekerot and K. Ohlsson

Quantitation of Human Leukocyte Elastase,
    Cathepsin G, α-2-Macroglobulin and
    α-1-Proteinase Inhibitor in Osteo-
    arthrosis and Rheumatoid Arthritis
    Synovial Fluids 345
  G. D. Virca, R. K. Mallya, M. B. Pepys,
    and H. P. Schnebli

Plasma Levels of Inhibitor Bound Leukocytic
    Elastase in Rheumatoid Arthritis
    Patients 355
  H. P. Schnebli, P. Christen, M. Jochum,
    R. K. Mallya, and M. B. Pepys

a_2M-Pasebic Assay: a Solid Phase Immuno-
    sorbent Assay to Characterize Alpha$_2$-
    Macroglobulin - Proteinase Complexes
    and the Proteinase Binding-Capacity
    of Alpha$_2$-Macroglobulin 363
  W. Borth

Role of Alpha$_2$-Macroglobulin: Proteinase
    Complexes in Pathogenesis of Inflammation:
    'F' a_2M but not 'S' a_2M Induces Synovitis
    in Rabbits after Repeated Intra-Articular
    Administration 371
  W. Borth and M. Susani

HYPERCATABOLISM

Enzyme-Linked Immunoassay for Human Granulocyte
    Elastase in Complex with α_1-Proteinase
    Inhibitor 379
    S. Neumann, N. Hennrich, G. Gunzer, and
    H. Lang

Proteinases and their Inhibitors in Septicemia
    Basic Concepts and Clinical Implications 391
    M. Jochum, K.-H. Duswald, S. Neumann,
    J. Witte, and H. Fritz

Proteolytic Activity in Patients with Hypercatabolic
    Renal Failure 405
    W. H. Hörl, R. M. Schäfer, K. Scheidhauer,
    M. Jochum, and A. Heidland

Release of Granulocyte Neutral Proteinases in
    Patients with Acute and Chronic Renal
    Failure 417
    A. Heidland, W. H. Hörl, N. Heller, H. Heine,
    S. Neumann, and E. Heidbreder

Changes in Components of the Plasma Kallikrein-
    Kinin and Fibrinolytic Systems Induced
    by a Standardized Surgical Trauma 433
    A. O. Aasen, J. Stadaas, T. E. Ruud, and
    P. Kierulf

Endotoxins and Coagulation Parameters in Patients
    with Traumatic Haemorrhagic- and Bacterio-
    toxic Shock 439
    A. Stemberger, F. Strasser, G. Blümel,
    B. v. Hundelshausen, S. Jelen, O. Schmidt,
    and G. Tempel

Studies on Pathological Plasma Proteolysis in
    Severely Burned Patients using Chromogenic
    Peptide Substrate Assays: A Preliminary
    Report 449
    T. E. Ruud, P. Kierulf, H. C. Godal, S. Aune,
    and A. O. Aasen

Changes in Components of the Plasma Protease
    Systems Related to Course and Outcome
    of Surgical Sepsis 455
    N. Smith-Erichsen, A. O. Aasen, and
    E. Amundsen

PANCREATITIS

Role of Proteases in the Development of
    Acute Pancreatitis 463
  M. Wanke

On the Potential Role of Trypsin and Trypsin
    Inhibitors in Acute Pancreatitis 477
  A. Lasson and K. Ohlsson

Studies on the Kallikrein-Kinin System in
    Plasma and Peritoneal Fluid during
    Experimental Pancreatitis 489
  T. E. Ruud, A. O. Aasen, P. Kierulf,
  J. Stadaas, and S. Aune

The Influence of the Kallikrein-Kinin System
    in the Development of the Pancreatic
    Shock 495
  H. Kortmann, E. Fink, and G. Bönner

PROTEASES AND MUSCLE FUNCTION

Proteolytic Enzymes and Enhanced Muscle
    Protein Breakdown 505
  B. Dahlmann, L. Kuehn, and H. Reinauer

Ca^2-Activated Proteinases, Protein Degradation
    and Muscular Dystrophy 519
  J. Kay

Muscle Cathepsin D Activity, and RNA, DNA and
    Protein Content in Maintenance Hemo-
    dialysis Patients 533
  G. F. Guarnieri, G. Toigo, R. Situlin,
  L. Faccini, R. Rustia, and F. Dardi

Enhanced Muscle Protein Degradation and Amino
    Acid Release from the Hemicorpus of
    Acutely Uremic Rats 545
  R. M. Flugel-Link, I. B. Salusky,
  M. R. Jones, and J. D. Kopple

Enhanced Muscle Protein Catabolism in Uremia 557
  H. R. Harter, T. A. Davis, and I. E. Karl

Catabolic Stress on Intracellular Amino Acid
        Pool 571
    P. Fürst

Rhabdomyolysis: a Clinical Entity for the
        Study of Role of Proteases 581
    S. G. Massry

INDEX 587

PHYSIOLOGY AND PATHOPHYSIOLOGY OF

NEUTRAL PROTEINASES OF HUMAN GRANULOCYTES

Klaus Havemann and Margarethe Gramse

Department of Hematology

University of Marburg, FRG

INTRODUCTION

The occurence of proteolytic enzymes in polymorphonuclear granulocytes was first demonstrated in 1888 by the famous clinician and biochemist Friedrich von Müller[1] who showed that a glycerine extract of fresh pus digests fibrin or coagulated protein at a neutral or weakly alkaline pH. Later on at the end of the last and the beginning of this century further characterization of the enzymes including their serum antiproteases was achieved by German and American scientists[2,3,4]. However, the neutral proteases then became largely forgotten as the result of the attention paid to the acid-cathepsins of the rabbit leukocyte, a convenient but somewhat misleading cell.

Advances only occured when the major granulocyte proteinases have been defined biochemically in the years from 1968 to 1975[5,6,7,8,9,10,11]. Since that time it became apparent that the neutrophil besides its role in defending microorganisms by phagocytosis and intracellular killing, is a secretory cell, which by secreting neutral proteinases and other biologically active products interferes with the regulation of the inflammatory response. Beside this benefit for the host, it also became evident, that possibly by a failure of the protective proteinase inhibitor system the granulocyte proteinases are responsible for severe tissue destruction under several clinical conditions.This short review will mainly focus on the extracellular role of the proteinases, which will be one of the main topics of this symposium.

Intracellular storage

Two distinct types of cytoplasmic granules are formed at subsequent stages of myelopoetic development: the azurophil (primary)
granules, produced in promyelocytes, and the specific (secondary)
granules formed at the myelocyte stage. The main components are
given in table 1. With the exception of a minor specific collagenase, which resides in specific granules, the neutral proteinases elastase, chymotrypsin and nonspecific collagenase are
present in primary granules together with myeloperoxidase[12,13].
Recently a third small vesicular organel has been described, containing the metalloproteinase gelatinase[14]. It is believed, that
the highly cationic neutral proteinases are neutralized within
the lysosomal granules by ionic complex formation with anionic
proteoglycans, which form a granule matrix of acid nature[15].

Table 1. Major constituents of azurophil and specific
granules of human neutrophils

	Azurophilic granules	Specific granules
Acid hydrolases	ß-Glycerophosphatase N-Acetyl-ß-Glycosamini- dase ß-Glucoronidase α-Mannosidase Cathepsin B Cathepsin D	No
Neutral pro- teinases	Elastase Chymotrypsin Non-specific collagenase	Specific collage- nase
Microbizidial enzymes	Myeloperoxidase Lysozyme	Lysozyme
Others		Lactoferrin Vit. B_{12} binding protein

In the original morphological studies of Bainton[16] it was
shown, that myeloperoxidase as the key enzyme of azurophil granules is only synthesized and packaged into the storage granules
at the promyelocyte stage as indicated by its presence in the
entire secretory apparatus, but without any further synthesis

beyond the promyelocyte. It is not known however, wether this is also true for the granulocyte proteinases. Ultrastructural cytochemical demonstration of elastase in human neutrophils employing an alaninenitrophenylester showed elastase activity in azurophil granules but also in the nuclear envelope, Golgi complex, endoplasmatic reticulum and mitochondriae suggesting either newly synthesized enzymes in both developing and mature neutrophils or a general role of elastase in protein synthesis[17,18]. Lately, granulocyte elastase and chymotrypsin was demonstrated cytochemically with alanine- and phenylalanine naphthylesters[19]. By employing these methods it has been suggested that elastase and chymotrypsin appear at a somewhat later stage of promyelocyte development then myeloperoxidase[19].

Properties of neutral proteinases

The properties of elastase like proteinase (ELP), chymotrypsin like proteinase (CLP) and of nonspecific collagenase are given in table 2[20].

Table 2. Properties of neutral proteinase from PMN

	ELP	CLP	nonspecific collagenase*
Granule content ($\mu g/10^6$ PMN)	3	1,5	4
Molecular weight	33.000	27.000	76.000
Polypeptide chains	1	1	2
Isoenzymes	3 (4)	3 (4)	2
Isoelectric point	10,8	>11	8-10 (?)
Carbohydrate content (%)	20	Ø	N.D.
pH optimum (range)	7,5-9,0	7,0-8,0	8,0-9,0
Cleavage sites	Val, Ile, Ala	Phe,Tyr, Leu	N.D.

*These data are obtained by one group of investigators and are not confirmed by others[11].

The three serine proteinases have a low substrate specificity. As a consequence a large number of natural substrates are attacked as shown in table 3.

Table 3. Natural substrates of neutral proteinases from PMN

ELP	CLP	Nonspecific collagenase
Elastin		
Proteoglycan	Proteoglycan	Proteoglycan
Fibronectin	Fibronectin	
Collagen type I,II,III,IV	Collagen type II	Collagen type I,II,III
Fibrinogen/fibrin	Fibrinogen/fibrin	Fibrinogen
Factor V*,VII,VIII,XI, XIII	Factor V,VII,VIII,XII, XIII	
Antithrombin III*		
α_2-plasmin inhibitor		
C1s, C4, C2, C3*, C5*, C9, factor B*	C1q, C1s, C4, C2, C3 C5, C9	C1s, C4, C3 C5
C1-inactivator		
IgG, IgM, Bence Jones Protein	IgG, Bence Jones Protein	
Kininogen	Angiotensinogen	
Specific collagenase		
Gelatinase	Specific collagenase* Gelatinase	

*Substrates who become transiently activated

ELP is characterized to be the only granulocyte proteinase to degrade elastin[5], whereas proteoglycan[21,22] and fibrinogen/fibrin
[23,24] are cleaved by all three proteinases. Fibronectin is
attacked by ELP[25] and CLP[26]. The three proteinases are able to
degrade different types of collagen[11,27,28,29]. It is remarkable
that the cartilage collagen type II is digested by the enzymes,
which is relative resistant to specific collagenase. ELP and CLP
attack the telopeptides of highly crosslinked structural collagen
by hydrolysis of the non helical N-terminal peptides[27]. By this,
monomeric α chains are disolved out of the fibrillar collagen.
It is therefore believed, that they play the key role in the
destruction of cartilage, the basement membranes[30] and other
connective tissue structures. Besides of fibrinogen, ELP and CLP
inactivate a variety of coagulation factors[24] with a transient
activation of factor V[24] and XIII[31]. Interestingly, also the main
inhibitors of coagulation and fibrinolysis, antithrombin III[32]
and α_2-plasmin inhibitor[33,34] are inactivated by granulocyte

elastase. Among the great number of complement components cleaved by the proteinases[35,36], C3 and C5 are extensively investigated since the enzymes are able to induce and to destroy chemotactic and opsonizing activity[37,38,39,40,41]. Immunoglobulins are degraded by ELP and CLP[42,43,44,45] leading to biologically active products[46,47]. Kininogen[48] is digested by ELP without the formation of bradykinin. The specific collagenase and gelatinase of PMN are cleaved by ELP and CLP but are transiently activated by CLP[49,50]. Recently a cleavage of angiotensinogen, leaving to biologically active angiotensin II has been reported[51].

Natural proteinase inhibitors

Because the granulocyte enzymes are released into the extra-cellular environment during inflammation the plasma- and tissue proteinase inhibitors play an important role in protecting the tissue structures. In the plasma the proteinase inhibitor α_1 proteinase inhibitor (α_1 PI; former designation α_1 antitrypsin: α_1 AT), α_1 antichymotrypsin (α_1 ACT) and α_2 macroglobulin (α_2 M) are present (table 4). The inhibitors form complexes with the proteinases in a 1:1 or in case of α_2-macroglobuline in a 1:2 molar ratio. Furthermore mucous membranes and other body fluids have low molecular weight proteinase inhibitors active against ELP and CLP[52,53,54,55].

Table 4. Natural proteinase inhibitors of granulocyte neutral proteinases

	Molecular weight	Molar plasma concentr. (µMol/l)	ELP	CLP	Nonspecif. colla- genase
α_1 proteinase inhibitor	55.000	54	90*	2*	50
α_2 macroglobulin	725.000	3	10*	86*	50*
α_1 antichymotrypsin	69.000	7	0*	12*	0*
Bronchial mucos inhibitor	10.500		+	+	Ø
Seminal plasma inhibitor	11.000		+	+	Ø
Synovial tissue inhibitor	-		+	Ø	N.D.

*Percent of binding in normal plasma

ELP has the highest affinity to α_1 PI[52] and CLP and non-specific collagenase for α_2 M[52]. The binding of the proteinases to the inhibitors in normal plasma results from the molar concentration of the inhibitors and their binding affinity to the proteinases.

All three enzymes are transferred from either α_1 PI or α_1ACT to α_2M and are still active in its complex with α_2 M against certain protein substrates[56,57,58]. Although the proteinase α_2 M complex has a rather short half life time[58], these complexes may be of particular importance for regulating the inflammatory response by a temporal generation of biologically active split products.

Under pathological conditions the plasma proteinase inhibitors have rich acces to inflammatory sites because of increased local capillar permeability. The released enzymes may therefore be rapidly inactivated providing the capacity of the inhibitor has not been saturated. However, this exhaustion of the inhibitor potential may well be operative at local sites of inflammation where very high enzyme activities are present, as for instance in abscess, and joint effusions and lavage fluids of the lung[59,60,61].

An additional important new aspect of the action of granulocyte proteinases despite of a high content of antiproteases, is the ability of active oxygen species generated from neutrophils to inactivate α_1 PI by oxidation of its methionine residues in the active center[62,63].

Degranulation of neutrophil granulocytes

The neutrophil is a secretory cell, because the content of its primary and secondary granules is frequently released after the granulocyte encounters foreign material or gets attached to appropriately coated surfaces[64,65]. This process of degranulation is accompanied by the generation and release of other inflammatory mediators such as prostaglandines, leukotrienes and active oxygen species[66,67]. A limited secretion of neutrophil proteinases seems to be of physiological importance for modulating inflammatory reactions, whereas unrestricted release may be the underlining mechanism of tissue destruction in certain diseases.

Besides by cell death extracellular release primarily occurs by two mechanisms: first by an incomplete closure of the phagosome during ingestion and second by a direct active secretion[65,68]. Both events require a receptor mediated recognition of the stimuli followed by a variety of metabolic processes finally leading to a fusion of the granule membrane with either the phagosome or the plasma membrane. Freeze etching of zymosan activated granulocytes showed a fusion of granules not only with the phagosome membrane but also with the cytoplasmic membrane[18]. The secretory process of degranulation therefore resembles secretion of other secretory tissues.

The amount of proteinases released during phagocytosis depends on the particle size and is most pronounced during frustrated phagocytosis of non ingestible surfaces, for instant basement membranes coated by IgG and complement components[64]. The secretion is initiated by soluble stimuli such as C5a, formylated methionine peptides, and endotoxin[69,70,68]. Some data indicate an active secretion of adherent neutrophils induced by these components also with an intact microtobuli apparatus not inactivated by cytochalasin B[69,70]. It is not clear, however, wether all types of granules are involved in this process.

Enzyme release during phagocytosis is observed at a very early stage when the first particles become ingested (about 15 seconds) as shown in figure 1.

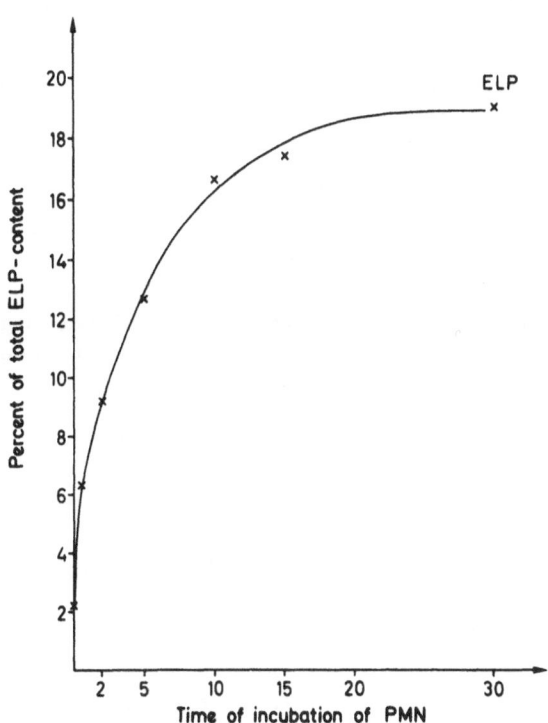

Figure 1. Time dependent release of granulocyte elastase following stimulation with zymosan.

Cytochemical determination of intracellular elastase and chymotrypsin showed a considerable partial depletion in patients with septicemia, pneumonia, Crohn´s disease, solid tumors, and acute leukemia[19]. In addition, granulocyte content of myeloperoxidase and neutral proteinase in inflammatory joint fluids and other body fluids is decreased in parallel to an extracellular increase of inhibitor bound enzymes or free proteinases.

However, systematic investigations comparing intra- and extracellular enzyme levels, inhibitor potential and protein substrate degradation are still lacking.

General role of PMN neutral proteinases in inflammatory reactions

Incubation of plasma with ELP and CLP in concentrations far below the saturation of plasma proteinase inhibitors induced considerable chemotactic activity for neutrophils (table 5)[40]. This chemotactic activity is due to C5a and eventually also to C3a[35] and fibrinopeptides[71]. These and other chemotactic peptides again induce release of PMN granule content, thus leading to enhanced chemotaxis. Locally increased chemoatractants such as C5a lead to a margination and attachment of granulocytes on vascular membranes or other cell surfaces[72].

Table 5. Possible regulatory effects of granulocyte neutral proteinases (ELP, CLP) on inflammatory and immune reactions

Substrate	Reactionproduct	Effect
C5	C5a	Activation of chemotaxis, PMN attachment on surfaces and enzyme release
C3	C3b	Opsonisation of microorganisms; activation of phagocytosis by PMN
Fibrin/ fibrinogen	Fibrinopeptides	Activation of chemotaxis, inhibition of fibrin polymerisation
IgG	Fc, Fab	Activation of oxidative metabolism of PMN and macrophages
Lymphocytes	Activation	Potentiation of proliferative response to antigen and mitogen
Macrophages	Activation	Inhibition of chemotaxis; increase of spreading, phagocytosis and lymphocyte activation
Platelets	Aggregation	Serotonine release
Mast cells, basophils	Histamine,platelet activating factor	Vasodilatation, platelet and PMN aggregation

Phagocytosis may be enhanced by granulocyte enzymes, too. So C3b can be formed by complement activation, but also as a direct effect of granulocyte elastase, leading to an amplification of phagocytosis[35,73]. In addition, the affinity of Fc receptors for immune complexes may be increased directly by granulocyte elastase and chymotrypsin[74].

However, both ELP and CLP inactivate preformed chemotactic and phagocytosis enhancing activity as well as Fc receptor binding capacity[40,75]. Thus, it is apparent that granulocyte enzymes can effect the formation and degradation of biologically active substances.

Both, the classical pathway of fibrinolysis by plasmin as well as the so called alternate pathway by PMN proteinases[76] induce the formation of fibrinogen/fibrinopeptides, which are strong chemoattractants for neutrophils[71]. In contrast to plasmin, granulocyte elastase, chymotrypsin and nonspecific collagenase induce an unrestricted cleavage of fibrinogen[23,24,77], but the split products inhibit fibrinpolymerisation similar to plasmin derived fragments[76,77]. Indirectly, also plasmin fibrinolysis is enhanced by ELP by the inactivation of the α_2-plasmin inhibitor[34].

Immunoglobulin G which during attachment of opsonized particles and immune complexes is in close contact with the granulocyte membrane via its Fc receptor, is cleaved by granulocyte elastase in Fab and Fc fragments[44,47], similar but not identical to those induced by papain (figure 2)[47]. Both elastase fragments but not the corresponding papain split products activate the oxidative metabolism of PMN[47] as shown in figure 3 and also induce granulocyte enzyme release.

In vitro incubation of lymphocytes with ELP and CLP leads to a marked potentiation of antigen and mitogen induced lymphocyte proliferation[78], suggesting a helper or initiator function of PMN neutral proteinases on the local T cell proliferative response. Enhancement of B cell maturation and function was shown by an increase of IgG synthesis in presence of leukocyte proteinases[79]. Furthermore, an activation of monocytes such as inhibition of monocyte migration and chemotaxis and increase of phagocytosis has been demonstrated[75].

All these cellular effects are shared with other proteolytic enzymes and may involve a direct proteolytic attack on the cell membrane.

CLP increases immune complex mediated platelet aggregation and serotonine release by an ADP dependent mechanism[80].

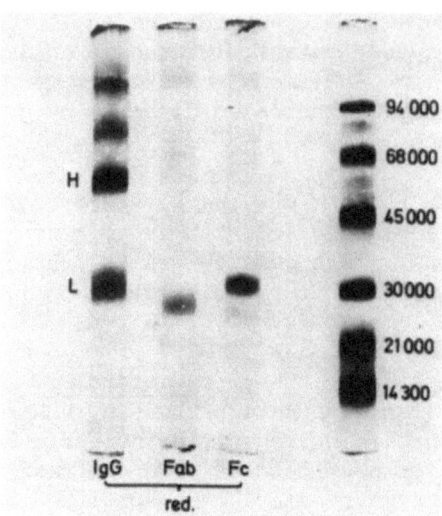

Figure 2. Polyacrylamide gel electrophoresis (7,5% containing 0,1%
          SDS) of mercaptoethanol-reduced IgG, ELP-Fab, and ELP-
          Fc (from the left to the right). The right-hand gel
          shows marker proteins from molecular weights of 14.300
          to 94.000 Da.

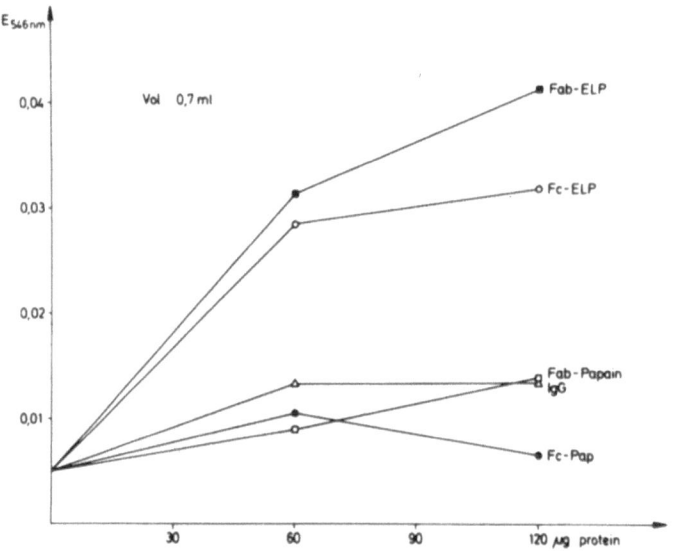

Figure 3. Comparison of the NBT reduction of papain-derived
          and ELP-derived Fab and Fc fragments; IgG = undi-
          gested IgG.

The effect of granulocyte proteinases on the <u>vascular system</u> is not yet fully established. The purified enzymes do not induce kinin from kininogen[48] and the ELP derived C3a and C5a cleavage products do not act as anaphylatoxins, since the C-terminal parts do not have an arginine residue[35]. Similar, in an animal model bradykinin induced vascular permeability was not dependent on neutrophils[81]. However, in the same animal system C5a or vaso-intestinal peptide induced vascular permeability in presence of prostaglandin E_2 was completely dependent on circulating granulo-cytes. An indirect action of the cationic proteins on <u>mast cells</u> and <u>basophils</u> with release of histamine and platelet activating has been suggested[82].

It has to be pointed out, however, that almost all of these data are obtained by in vitro experiments. Therefore, the in vivo relevance of these findings is so long uncertain until realistic animal models will allow a definite coordination of these pheno-menons with the inflammation enhancing and depressing control mechanisms of the body.

Pathogenic role of granulocyte neutral proteinase in specific disease states

In <u>septicemia</u> severe coagulation defects with decrease of fibrinogen, factor V, VIII and factor XIII have been demonstrated[83]. These coagulation defects are usually explained as a conse-quence of disseminated intravascular coagulation induced by throm-boplastic material[84]. However, disseminated thrombus formation in septicemia is extremely rare and no clear cut evidence exists that heparin is of any help[85].

On the other hand the coagulation disturbances in septicemia areparallelled by increased levels of endotoxin, a degranulation of neutrophils and an increase of extracellular granulocyte e-lastase[85,86]. In addition incubation of plasma with ELP and CLP as well as injection of the enzymes or of endotoxin into green monkeys showed similar coagulation defects as in the patients[87]. It can therefore be assumed that the PMN proteinases are respon-sible for these coagulation disorders, although the exact mecha-nism, for instance,how the enzymes can excert their effect in the presence of the serum antiproteases,is unknown.

<u>Acute respiratory distress syndrome</u> seems to be induced by an intravascular aggregation of polymorphs which then accumulate in the lung capillaries[88,89]. It has been suggested that granu-locyte elastase which is present in high amounts in the bronchio-alveolar lavage fluid is responsible for the alteration of the vascular basement membrane and the cleavage of surfactant pro-

teins, of elastin, collagen and other structural elements of the
pulmonary tissues [61,90,91,92].

Table 6. Possible pathophysiological role of neutral proteinases from PMN

Disease	Symptoms	Possible aethiology	Substrates
Septicemia	Coagulation defects	Enzyme release by endotoxin, immune complexes	Fibrinogen, factor XIII and other coagulation factors and inhibitors
ARDS-syndrome (polytrauma, septicemia)	Fluid lung (capillar injury, alveolitis)	Accumulation of aggregated PMN in the lung	Lung connective tissue + basement membrane (collagen, elastin, proteoglycan, fibronectin)
Pulmonary emphysema	Destruction of elastin containing structures	Genetic α_1PI deficiency; inactivation of α_1PI by tobacco smoke	Lung connective tissue
Rheumatoid arthritis	Synovitis, destruction of joint tissue	IgG altered by the enzymes $\rightarrow$ formation of rheumatoid factor IgG - immune-complexes $\rightarrow$ enzyme release	Joint connective tissue (collagen, proteoglycan)
Gout	Synovitis, destruction of joint tissue	Enzyme release by phagocytosis of urate cristalls	Joint connective tissue (collagen, proteoglycan)
Immune complex nephritis	Nephrotic syndrome, renal failure	Local accumulation of immune complexes $\rightarrow$ glomerular injury by PMN enzyme release	Glomerular basement membrane

Furthermore, good evidence exists that pulmonary emphysema
in genetic α_1 PI deficiency or α_1 PI inactivation by tobacco
smoke[94] is caused by the action of granulocyte elastase.

In rheumatoid arthritis or in gout where proteinase release
is either induced by immune complexes or by phagocytosis of urate
cristalls, the enzymes probably play a major role in the

destruction of joint connective tissue[22,27,28,29,95,96]. The same may be true for immune complex nephritis, where the PMN stick to the immune complex coated endothelial cells on the basement membrane of glomeruli and induce their destruction by the mechanism of reversed endocytosis[30,97].

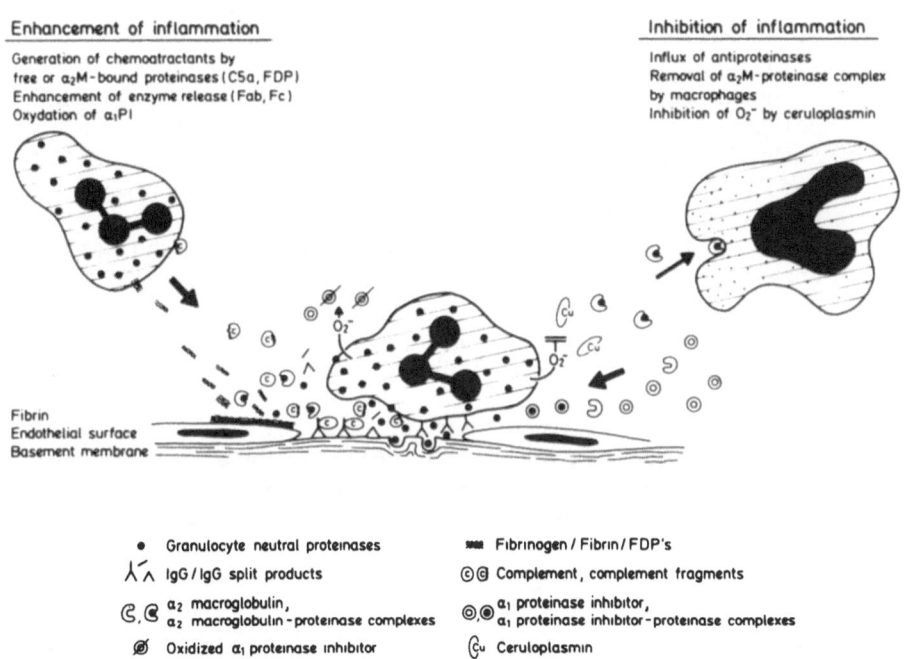

Enhancement of inflammation

Generation of chemoatractants by
free or α_2M-bound proteinases (C5a, FDP)
Enhancement of enzyme release (Fab, Fc)
Oxydation of α_1PI

Inhibition of inflammation

Influx of antiproteinases
Removal of α_2M-proteinase complex
by macrophages
Inhibition of O_2^- by ceruloplasmin

Fibrin
Endothelial surface
Basement membrane

• Granulocyte neutral proteinases	⊷ Fibrinogen / Fibrin / FDP's
⋏⋏ IgG / IgG split products	ⓒⓒ Complement, complement fragments
ⓒ,ⓒ α_2 macroglobulin, α_2 macroglobulin-proteinase complexes	◎,◎ α_1 proteinase inhibitor, α_1 proteinase inhibitor-proteinase complexes
⌀ Oxidized α_1 proteinase inhibitor	ⓒu Ceruloplasmin

Figure 4. Possible interaction of granulocyte neutral proteinases with the inflammatory response

In summary, polymorphonuclear leukocytes by exocytosis of their neutral proteinases, take part in the regulation of the inflammatory response and are producing severe tissue destruction. In both instances, the proteinases have to overcome the well equipped antiprotease potential of the body.

Enhancement of inflammation by proteinase generated active split products may be due to local exhaustion of the inhibitor capacity, inactivation of the inhibitors by granulocyte derived oxygen products or enzymatically active proteinase α_2 macroglobulin complexes. On the other hand, inhibition of inflammation may be achieved by an influx of anti-proteinases, a clearance of the α_2M-proteinase-complex by macrophages or by an inhibition of O_2^- via ceruloplasmin. Hopefully, new results will enable us to draw a more realistic picture of the regulation of inflammation by granulocyte proteinases in the near future.

REFERENCES

1. F. Müller, Apparently unpublished observations cited by: H. Kossel, "Beiträge zur Lehre vom Auswurf", Zeitschrift Klin. Med., XIII: 149 (1888).

2. G. Jochmann und E. Müller, Über proteolytische Fermentwirkung der Leukozyten, Münchner Med. Wschr. 26 (1906).

3. G. Jochmann und E. Müller, Weitere Ergebnisse unserer Methode und Nachweis proteolytischer Fermentwirkung, Münchner Med. Wschr. 41 (1906).

4. E. L. Opie, Intracellular digestion: The enzymes and anti-enzymes concerned, Physiological Reviews 2:522 (1922).

5. A. Janoff, and J. Scherer, Mediators of inflammation in leukocyte lysosomes. IX. Elastinolytic activity in granules of human polymorphonuclear leukocytes, J. exp. Med. 128:1137 (1968).

6. G. S. Lazarus, J. R. Daniels, R.S. Braun, H. A. Bladen, and H. M. Fullmer, Degradation of collagen by a human granulocyte collagenolytic system, J. Clin. Invest. 47:2622 (1968).

7. W. Schmidt, and K. Havemann, Isolation of elastase-like and chymotrypsin-like neutral proteases from human granulocytes, Hoppe-Seyler's Z. Physiol. Chem. 355:1077 (1974).

8. A. C. Gerber, J. H. Carson, and B. Hadorn, Partial purification and characterization of a chymotrypsin-like enzyme from human neutrophil leukocytes, Biochim. Biophys. Acta 364:103 (1974).

9. R. Rindler, F. Schmalzl und H. Braunsteiner, Isolierung und Charakterisierung der chymotrypsinähnlichen Protease aus neutrophilen Granulozyten des Menschen, Schweiz. Med. Wschr. 104:132 (1974).

10. K. Ohlsson, and I. Olsson, The neutral proteases of human granulocytes. Isolation and partial characterization of granulocyte elastases, Eur. J. Biochem. 42:519 (1974).

11. K. Ohlsson, and I. Olsson, The neutral proteases of human granulocytes. Isolation and partial characterization of two granu-

locyte collagenases, Eur. J. Biochem. 36:473 (1973).

12. B. Dewald, R. Rindler-Ludwig, U. Bretz, and M. Baggiolini, Sub-cellular localization and heterogeneity of neutral proteases in neutrophilic polymorphonuclear leukocytes, J. exp. Med. 141:709 (1975).

13. K. Ohlsson, J. Ohlsson, and J. K. Spitznagel, Localization of chymotrypsin-like cationic protein, collagenase and elastase in azurophil granules of human neutrophilic polymorphonuclear leukocytes, Hoppe-Seyler's Z. Physiol. Chem. 358:361 (1977).

14. B. Dewald, U. Bretz, and M. Baggiolini, Release of gelatinase from a novel secretory compartment of human neutrophils, J. Clin. Invest. 70:518 (1982).

15. J. Ohlsson, The intracellular transport of glycosaminoglycans in human leukocytes, Exp. Cell. Res. 54:318 (1969).

16. D. F. Bainton, J. L. Ullyot, and M. G. Farquhar, The development of neutrophilic polymorphonuclear leukocytes in bone marrow, J. exp. Med. 134:907 (1971).

17. J. M. Clark, B. M. Aiken, D. W. Vaughan, and H. M. Kagan, Ultra-structural localization of elastase-like enzymes in human neutrophils, J. Histochem. Cytochem. 28:90 (1980).

18. J. Poensgen, G. Adler, M. Gramse, and K. Havemann, Phagocytosis and enzyme release from stimulated neutrophils, (Abstract), Blut 45:177 (1982).

19. K. Havemann, M. Gramse, and W. D. Gassel, Cytochemical determination of granulocyte elastase and chymotrypsin in human myeloid cells and its application in aquired deficiency states and diagnosis of myeloid leukemia, Klin. Wschr. 61:49 (1983).

20. K. Havemann, and A. Janoff, Neutral proteases of human polymorphonuclear leukocytes, eds. Urban & Schwarzenberg, Baltimore-München (1978).

21. C. J. Malemud, and A. Janoff, Identification of neutral proteases in human neutrophil granules that degrade articular cartilage proteoglycan, Arthritis Rheum. 18:361 (1975).

22. A. Janoff, At least three human neutrophil lysosomal proteases are capable of degrading joint connective tissue, Ann. NY Acad. Sci. 256:402 (1976).

23. K. Ohlsson, Purification and properties of granulocyte colla-genase and elastase, in: "Neutral Proteases ..." (ref. 20), p. 89 (1978).

24. W. Schmidt, R. Egbring, and K. Havemann, Effect of elastase-like and chymotrypsin-like neutral proteases from human gra-nulocytes on isolated clotting factors, Thromb. Res. 6:315 (1975).

25. J. A. McDonald, and D. G. Kelley, Degradation of fibronectin by leukocyte elastase, J. Biol. Chem. 255:8848 (1980).

26. T. Vartio, H. Seppä, and A. Vaheri, Susceptibility of soluble
     and matrix fibronectins to degradation by tissue proteinases,
     mast cell chymase and cathepsin G, J. Biol. Chem. 256:471
     (1981).

27. P. M. Starkey, A. J. Barrett, and M. C. Burleigh, The degradat-
     ion of articular collagen by neutrophil proteinases, Bio-
     chim. Biophys. Acta 483:386 (1977).

28. J. E. Gadek, G. A. Fells, D. G. Wright, and R. G. Crystal, Hu-
     man neutrophil elastase functions as a type III collagen
     collagenase, Biochem. Biophys. Res. Com. 95:1815 (1981).

29. C. L. Mainardi, S. N. Dixit, and A. H. Kany, Degradation of type
     IV (basement membrane) collagen by a proteinase isolated
     from human polymorphonuclear leukocyte granules, J. Biol.
     Chem. 255:5435 (1980).

30. M. Davies, A. J. Barrett, J. Travis, E. Sanders, and G. A. Coles,
     The degradation of human glomerular basement membrane with
     purified lysosomal proteinases: Evidence for the pathogenic
     role of the polymorphonuclear leukocyte in glomerulonephritis,
     Clin. Sci. Mol. Med. 54:233 (1978).

31. P. Henriksson, I. M. Nilsson, K. Ohlsson, and P. Stenberg, Gra-
     nulocyte elastase activation and degradation of factor XIII,
     Thromb. Res. 18:343 (1980).

32. M. Jochum, S. Lander, N. Heimburger, and H. Fritz, Effect of
     human granulocyte elastase on isolated human antithrombin
     III, Hoppe-Seyler's Z. Physiol. Chem. 362:103 (1981).

33. H. G. Klingemann, R. Egbring, M. Holst, M. Gramse, and K. Have-
     mann, Digestion of α_2-plasmin inhibitor by neutral proteases
     from human leukocytes, Thromb. Res. 24:479 (1981).

34. M. S. Brower, and P. C. Harpel, Proteolytic cleavage and inacti-
     vation of α_2-plasmin inhibitor and C1 inactivator by human
     polymorphonuclear leukocyte elastase, J. Biol. Chem. 257:
     9849 (1982).

35. P. Venge, and J. Olsson, Cationic proteins of human granulo-
     cytes VI. Effect on the complement system and mediation of
     chemotactic activity, J. Immunol. 115:1505 (1975).

36. P. Venge, Polymorphonuclear leukocyte proteases and their effects
     on complement components and neutrophil function, in: "Neu-
     tral Proteases ..." (ref. 20), p. 264 (1978).

37. J. C. Taylor, I. P. Crawford, and T. E. Hugli, Limited degra-
     dation of the third component (C3) of human complement by
     human leukocyte elastase (HLE): Partial characterization of
     C3 fragments, Biochemistry 16:3390 (1977).

38. F. W. Orr, J. Varani, D. L. Kreutzer, R. M. Senior, and D. A.
     Ward, Digestion of the fifth component of complement by leu-
     kocyte enzymes, Am. J. Pathol. 94:75 (1979).

39. U. Johnsson, K. Ohlsson, and I. Olsson, Effects of granulocyte
     neutral proteases on complement components, Scand. J. Immunol.
     5:421 (1976).

40. C. Löffler, Human granulocyte elastase and chymotrypsin. Ge-
     neration of chemotactic activity in the presence of serum
     antiproteases, in: "Neutral proteases ..." (ref. 20), p. 292
     (1978).
41. U. Hadding, C. Löffler, M. Gramse and K. Havemann, Influence
     of elastase-like protease on guinea pig C3, factor B of the
     properdin system and the tumor cell line EL 4, in "Neutral
     Proteases ..." (ref. 20), p. 287 (1978).
42. A. Solomon, W. Schmidt, and K. Havemann, Bence Jones Proteins
     and light chains of immunoglobulins. XIII. Effect of elastase-
     like and chymotrypsin-like neutral proteases derived from
     human granulocytes on Bence Jones Proteins, J. Immunol. 117:
     1010 (1976).
43. A. Solomon, M. Gramse, and K. Havemann, Proteolytic cleavage
     of IgG molecules by neutral proteases of polymorphonuclear
     leukocytes, Eur. J. Immunol. 8:782 (1978).
44. J. D. Folds, H. E. Prince, and J. K. Spitznagel,Limited cleavage
     of human immunoglobulins by elastase of human neutrophil po-
     lymorphonuclear granulocytes, Lab. Invest. 39:313 (1978).
45. A. Baici, M. Knöpfel, K.Fehr, and A. Böni, Cleavage of human
     IgM with human lysosomal elastase, Immunol. Letters 2:47
     (1980).
46. H. E. Prince, J. D. Folds, and J. K. Spitznagel, Interaction of
     human polymorphonuclear leukocyte (PMN) elastase with human
     IgM. Production of a factor enhancing PMN migration in vitro,
     Immunol. Commun. 9:23 (1980).
47. G. Kolb, H. Köppler, M. Gramse, and K. Havemann, Cleavage of
     IgG by elastase-like protease (ELP) of human polymorphonuclear
     leukocytes: Isolation and characterization of Fab and Fc
     fragments and low-molecular-weight peptides. Stimulation of
     granulocyte function by ELP-derived Fab and Fc fragments,
     Immunobiology 161:507 (1982).
48. H. Fritz, Failure to detect intrinsic kininogenase activity in
     PMN elastase, in: "Neutral Protease ..." (ref. 20), p. 261
     (1978).
49. G. Murphy, U. Bretz, M. Baggiolini, and J. J. Reynolds, The
     latent collagenase and gelatinase of human polymorphonuclear
     neutrophil leukocytes, Biochem. J. 192:517 (1980).
50. H. W. Macartney, and H. Tschesche, Latent collagenase from hu-
     man polymorphonuclear leukocytes and activation to collage-
     nase by removal of an inhibitor, FEBS Letters 119:327 (1980).
51. M. G. Tonnesen, M. S. Klempner, K. F. Austen, and B. U. Wintroub,
     Identification of a human neutrophil angiotensin II-genera-
     ting protease as cathepsin G, J. Clin. Invest. 69:25 (1982).
52. K. Ohlsson and M. Delshammer, Interactions between granulocyte
     elastase and collagenase and the plasma proteinase inhibi-
     tors in vitro and in vivo, in: "Dynamics of Connective Tissue
     Macromolecules," P. M. C. Burleigh, A. R. Poole, eds.,
     North-Holland Publishing Company (1975).

53. H. Tegner, and K. Ohlsson, Localization of a low-molecular-
     weight protease inhibitor to tracheal and maxillary sinus
     mucosa, Hoppe-Seyler's Z. Physiol. Chem. 358:425 (1977).
54. H. Schiessler, M. Arnhold, K. Ohlsson, and H. Fritz, Inhibitors
     of acrosin and granulocyte proteinases from human genital
     tract secretions, Hoppe-Seyler's Z. Physiol. Chem. 357:1251
     (1976).
55. M. E. Englert, M. J. Landes, J. E. Birnbaum, A. L. Oronsky, and
     S. S. Kerwar, Studies on a leukocyte elastase inhibitor pre-
     sent in the culture medium of inflamed synovial tissue, Bio-
     chem. Biophys. Res. Commun. 96:498 (1980).
56. P. C. Harpel, and M. W. Mosesson, Degradation of human fibrino-
     gen by plasma $\alpha_2$macroglobulin-enzyme complexes, J. Clin. In-
     vest. 52:2175 (1973).
57. P. Venge, I. Olsson, and H. Odeberg, Cationic proteins of human
     granulocytes. V. Interaction with plasma protease inhibitors,
     Scand. J. Clin. Lab. Invest. 35:737 (1975).
58. K. Ohlsson, and C. B. Laurell, The disappearance of enzyme-inhi-
     bitor complexes from the circulation of man, Clin. Sci. Mol.
     Med. 51:87 (1976).
59. K. Ohlsson, and H. Tegner, Granulocyte elastase and plasma pro-
     tease inhibitors in purulent sputum, Europ. J. Clin. Invest.
     5:221 (1975).
60. J. Saklatvala, and A. J. Barrett, Identification of proteinases
     in rheumatoid synovium. Detection of leukocyte elastase,
     cathepsin G, and another serine proteinase, Biochim. Biophys.
     Acta 615:167 (1980).
61. C. T. Lee, A. M. Fein, M. Lippmann, H. Holtzmann, P. Rumbel, and
     G. Weinbaum, Elastolytic activity in pulmonary lavage fluid
     from patients with adult respiratory-distress syndrome, New
     Engl. J. Med. 304:192 (1981).
62. N. R. Matheson, P. W. Wong, and J. Travis, Enzymatic inactivat-
     ion of human $\alpha_1$proteinase inhibitor by neutrophil myeloper-
     oxidase, Biochem. Biophys. Res. Commun. 88:402 (1979).
63. A. B. Cohen, The effect in vivo and in vitro of oxidative damage
     to purified $\alpha_1$antitrypsin and to the enzyme inhibiting ac-
     tivity of plasma, Am. Rev. Respir. Dis. 119:953 (1979).
64. P. M. Henson, The immunologic release of constituents from neu-
     trophil leukocytes. I. The role of antibody and complement
     on nonphagocytosable surfaces or phagocytosable particles,
     J. Immunol. 107:1535 (1971).
65. G. Weissmann, R. B. Zurier, P. J. Spieler, and J. M. Goldstein,
     Mechanisms of lysosomal enzyme release from leukocyte ex-
     posed to immune complexes and other particles, J. exp. Med.
     134:1495 (1971).
66. I. Olsson, T. Olofsson, P. Venge, and I. Winquist, Release of
     biologically active compounds from neutrophil and eosinophil
     leukocyte granules, Monographs in Allergy 17:130 (1981).
67. G. Weissmann, J. E. Smolen, and H. M. Korchak, Release of in-

flammatory mediators from stimulated neutrophils, New Engl. J. Med. 303:27 (1980).

68. W. Schmidt, Differential release of elastase and chymotrypsin from polymorphonuclear leukocytes, in: "Neutral Proteases...," (ref. 20), p. 77 (1978).

69. I. Goldstein, S. Hoffstein, J. Gallin, and G. Weissmann, Mechanisms of lysosomal enzyme release from human leukocytes: Microtubule assembly and membrane fusion induced by a component of complement, Proc. Nat. Acad. Sci. (Wash.) 70:2916 (1973).

70. H. J. Showell, R.J. Freer, S.H. Zigmond, E.Schiffmann, S. Aswani-kumar, B. Corcoran, E.L. Becker, The structure-activity-relations of synthetic peptides as chemotactic factors and inducers of lysosomal enzyme secretion for neutrophils, J. exp. Med. 143:1154 (1976).

71. A. B. Kay, D. S. Pepper, and R. McKenzie, The identification of fibrinopeptide B as a chemotactic agent derived from human fibrinogen, Brit. J. Haemotol. 27:669 (1974).

72. M. Tonnesen, Personal communication (C5a attachment) (1982).

73. I. I.Gigli,and R.A.Nelson, Complement dependent immune phagocytosis. I.Requirements for C1,C4,C2,C3, Exp.Cell Res.51:45 (1968).

74. R. Hallgren, and P. Venge, Cationic proteins of human granulocytes: Enhancement of phagocytosis by staphylococcus protein A-IgG complexes, Inflammation 1:237 (1976).

75. M. E. Schmidt, S. D. Douglas, P. Quie, R. D. Nelson, W. Schmidt, and K. Havemann, Effect of neutral granulocyte proteases on human immunocompetent cells: Action of elastase-like protease and chymotrypsin-like protease on mononuclear phagocytes, in: "Neutral proteases...," (ref. 20), p. 298 (1978).

76. E. F. Plow, and T. S. Edgington, An alternative pathway for fibrinolysis. I. The cleavage of fibrinogen by leukocytic proteases at physiologic pH, J. Clin. Invest. 56:30 (1975).

77. M. Gramse, C. Bingenheimer, W. Schmidt,R.Egbring,and K.Havemann, Degradation products of fibrinogen by elastase-like neutral protease from human granulocytes.Characterization and effects on blood coagulation in vitro,J.Clin.Invest. 61:1027 (1978).

78. K. Havemann, W. Schmidt, U. Bogdanm, and M. Gramse, Effect of polymorphonuclear granulocyte proteases on immunocompetent cells, in: "Neutral Proteases...," (ref. 20), p. 306 (1978).

79. K. Ky, and M. Ziff, Enhancement of in vitro immunoglobulin synthesis of human lymphocytes by lysosomal enzymes from polymorphonuclear leukocytes, Clin. exp. Immunol. 27:254 (1977).

80. R. Hallgren, and P. Venge, Cationic proteins of human granulocytes. Effects on human platelet aggregation and serotonine release, Inflammation 1:359 (1976).

81. T. J. Williams, Personal communication (1982).

82. G. Camussi, J. M. Menzia-Huerta, and J. Benveniste, Release of platelet activating factor and histamine. I. Effect of immune complexes, complement and neutrophils on human and rabbit mastocytes and basophils, Immunology 33:523 (1978).

83. R. Egbring, W. Schmidt, G. Fuchs, and K. Havemann, Demonstration
     of granulocyte proteases in plasma of patients with acute
     leukemia and septicemia with coagulation defects, Blood 49:
     219 (1977).

84. P. M. Neame, J. G. Kelton, J. R. Walker, I. O. Stewart, H. L.
     Nossel, and J. Hirsh, Thrombocytopenia in septicemia: the
     role of disseminated intravascular coagulation, Blood 56:88
     (1980).

85. C. Merskey, Defibrination syndrome or...? Blood 41:599 (1973).

86. K. H. Duswald, M. Jochum und H. Fritz, Neue Erkenntnisse zur
     Pathobiochemie der Sepsis nach abdominal-chirurgischen Opera-
     tionen, in: "Chirurgisches Forum 82 für experimentelle und
     klinische Forschung," S. Weller, ed., Springer (1982).

87. R. Egbring and K. Havemann, Possible role of polymorphonuclear
     granulocyte proteases in blood coagulation, in: "Neutral
     Proteases..." (ref. 20), p. 442 (1978).

88. P. R. Craddock, D. Hammerschmidt, J. G. White, A. P. Dalmasso,
     and H. S. Jacob, Complement (C5a) induced granulocyte aggre-
     gation in vitro, J. Clin. Invest. 60:260 (1977).

89. P. R. Craddock, J. Fehr, A. P. Dalmasso, K. L. Brigham, and H. S.
     Jacob, Hemodialysis leukopenia and pulmonary vascular leuko-
     stasis resulting from complement activation by dialyzer cello-
     phane membranes, J. Clin. Invest. 59:879 (1977).

90. C. G. Cochrane, and B. A. Aikin, Polymorphonuclear leukocytes in
     immunologic reactions: The destruction of vascular basement
     membrane in vivo and in vitro, J. exp. Med. 124:733 (1966).

91. A. Janoff, Mediators of tissue damage in leukocyte lysosomes
     X. Further studies on human granulocyte elastase, Lab. Invest.
     22:228 (1970).

92. A. Janoff, R. A. Sandhaus, V. D. Hospelhorn, and R. Rosenberg,
     Digestion of lung proteins by human leukocyte granules in
     vitro, Proc. Soc. Exp. Biol. Med. 140:516 (1972).

93. M. Galdstone, A. L. Davis, and A. Janoff, Familial variation of
     leukocyte lysosomal protease and serum α_1 antitrypsin as de-
     terminants in chronic obstructive pulmonary disease, Am. J.
     Resp. Dis. 108:1020 (1974).

94. A. Janoff, and A. Carp, Possible mechanism of emphysema in ci-
     garette smokers: Cigarette smoke condensate suppresses pro-
     teinase inhibitors in vitro, Am. Rev. Resp. Dis. 116:65 (1977).

95. A. L. Oronsky, L. Ignarro, and R. J. Perper, Release of carti-
     lage mucopolysaccharide-degrading neutral protease from human
     leukocytes, J. exp. Med. 138:461 (1972).

96. R. A. Turner, G. B. Counts, W. J. Treadway, D. A. Holt, and C.
     A. Agudelo, Rheumatoid factor and monosodium urate crystal-
     neutrophil interactions in gouty inflammation, Inflammation
     5:353 (1982).

97. E. Sanders, G. A. Coles, and M. Davies, Lysosomal enzymes in hu-
     man urine: Evidence for polymorphonuclear leukocyte protein-
     ase involvement in the pathogenesis of human glomerulonephri-
     tis, Clin. Sci. Mol. Med. 54:667 (1978).

REGULATION OF PROTEASE ACTIVITY

Marion Steinbuch

Centre National de Transfusion Sanguine

Paris, France

INTRODUCTION

Proteases play a determinant role in many biological processes. In blood coagulation several enzymes are interrelated to form a cascade of reactions achieving amplification of the initial trigger reaction. Generally trigger mechanisms show a great diversity including negatively charged surfaces (Hageman Factor dependent pathways), antigen/antibody complexes (activation of the classical complement pathway), polysaccharides and lipopolysaccharides (alternate complement pathway) or specific enzymes such as urokinase or tissue activator in the case of fibrinolysis.

When certain proteolytic enzymes belonging to a complex biological system were isolated and studied in vitro, it became evident that they were not as specific as their known physiological role would presume. The same was observed when isolated anti-proteases were mixed with various purified enzymes. Although these in vitro experiments considerably enlarged our knowledge, inducing new ideas and concepts, they promoted the erroneous notion of broad specificity of certain enzymes and their inhibitors. In contrast, a relatively low affinity shown by other purified enzymes was considerably enhanced by the addition of a specific co-factor. These amplification proteins play an important role in the regulation of proteolytic activi-

ty and in many cases protect the enzyme from inhibition
by anti-proteases. Positive and negative feedback are
additional features of the regulatory mechanism. Finally,
one must consider the inter-relationship between several
biological systems and the potential inhibition of
activated proteases by antiproteases. Some aspects of
these reactions, essentially within plasma biological
systems, will be described in more detail below.

REGULATION BY CO-FACTORS

        Co-factors play an important role in the clotting
system. The complex interaction between factor XII, XI
and prekallikrein (PK) in the contact phase of blood
coagulation requires High Molecular Weight Kininogen
(HMWK) as cofactor. PK and Factor XI are found in
plasma as bimolecular complexes with HMWK (51). This
protein mediates the interaction of Hageman factor
substrates (PK and Factor XI) with limitating surfaces
and facilitates their activation. Kallikrein (KK)
releases the vasoactive nonapeptide bradykinin from
HMWK, but the kinin -free protein retains its accelerating effect on the interaction between activated
Hageman factor, PK and factor XI (73). Hageman factor
initiates clotting by the intrinsic pathway and
activates the kallikrein-kinin system. Fibrinolysis
may then be triggered by KK and activated factor XI
(49), HMWK being an important co-factor in these pathways.
        The other two co-factors of the clotting system,
factors VIII and V need activation for the expression
of their activity in the complexes formed with the
corresponding enzymes and phospholipides. A protease
inserted in such a complex has a different activity
and specificity than a free enzyme in solution. The
functional importance of cofactors is best demonstrated
by the clinical consequences of the absence or
abnormality of factor VIII in hemophilia A. Activation
of both co-factors implies a feed-back mechanism (see
below) and anti-coagulation is induced by protein C,
a vitamin K dependent serine protease inactivating
specifically factor V and VIII (83,82).
        Activity modulation by non-enzymatic co-factors
is also seen in both the complement and the fibrinolytic systems. Factor H (or β1H) is an important
regulatory component of the alternative complement
pathway acting as a co-factor of the enzyme C3b-INA
(59,84) and causing active disassembly of the

alternative C3/C5 convertase (57,39) whereas another
co-factor, properdin, functions as a stabilizer of the
same enzyme (21).

Plasminogen contains lysine-binding sites which
are important for the interaction with fibrin and α_2-
antiplasmin (86,54). Recently, a protein interacting
with the high affinity lysine binding site of plasmin(ogen)
has been isolated (47). This protein proved to be
identical with the histidine-rich glycoprotein. Association of half of the plasma plasminogen with this protein
results in a reduction of plasminogen available for
interaction with fibrin and in the same way interferes
with the α_2-anti-plasmin/plasmin interaction. The full
significance of this anti-fibrinolytic "control" protein
is still under close investigation.

Feed back mechanisms

Another feature of protease control are the feedback mechanisms amplifying certain proteolytic activities
or, on the contrary, achieving inactivation of important
factors. Thus in PK deficient plasma contact activation
is very slow. Indeed, a small amount of KK initially
formed by the action of Hageman Factor proves to be a
potent activator of the remaining factor XII. Whereas
KK can directly activate factor VII of the extrinsic
system (63), the latter in turn is able to activate
factor IX from the intrinsic system (56). Thrombin participates in positive and negative feedback. Thus, it
activates on the one side factor VIII and V, but on the
other side also protein C, the enzyme inactivating specifically both these factors (74,40). In association with
tissue factor, Factor VIIa activates factor X in the
extrinsic system but factor Xa can in turn activate
factor VII. The activation of the "anti-coagulation"
enzyme protein C is enhanced when thrombin binds to a
specific cell surface co-factor protein :thrombomodulin.
The enzyme specificity of thrombin becomes drastically
modified. The resulting complex retains less than 1 %
of the clotting activity and a similarly reduced capacity
for the cleavage of factor V, whereas activation of
protein C is considerably enhanced (19).

Feedback is also seen in the complement system.
Water may be sufficient to hydrolyse slowly the
thioester of C3 able to react with factor B. The latter,
in the newly formed complex, may then be activated by
factor D generating the first molecule of metastable

C3b which initiates the positive feedback reaction (58).
Feedback thus constitutes the driving force of the
alternative pathway.

Interrelationship between different biological systems

There are many interrelationships between the
different systems but only a few will be mentioned here.
Activated Hageman factor initiates activation of clotting,
the KK-kinin system, fibrinolysis and the classical pathway
of complement (25). Most interesting connections exist
between the kallikrein and the renin system causing
respectively hypo- or hypertension (17). Prorenin can be
activated to renin by kallikrein. Kininase II, a peptidase
inactivating bradykinin is identical with the enzyme
called angiotensin-converting enzyme giving rise to the
hypertensive peptide angiotensin. Kininase I, or N-
carboxpeptidase also inactivates the hypotensive nonapeptide bradykinin and is identical with the anaphylatoxin
inactivator, an enzyme inactivating, the vasoactive
peptides C3a and C5a generated during complement activation (7).

Plasmin can activate renin as it activates Hageman
factor, and fibrinolysis can be stimulated by the
anticoagulant enzyme protein C.

The influence of cell surface receptors for various
enzymatic and non enzymatic factors can only be
mentioned to underline the complexity of the regulative
mechanisms.

THE PROTEASE INHIBITORS

It is obvious that the anti-proteases are only
part of the regulatory functions. The understanding
of their biological role has been facilitated by the
clinical manifestations observed in patients with a
given antiprotease deficiency concerning for instance
antithrombin III (AT III), α_1-antitrypsin (α_1-AT),
C1-inactivator (C1-INA) or α_2-antiplasmin (α_2-AP).

α_1-antitrypsin (α_1-AT)

This inhibitor is also named α_1-protease
inhibitor (α_1-AP) in function to its broad specificity
including granulocytal enzymes and specially granulocyte
elastase (32). α_1-AT reacting in a 1:1 molar concentration with a serie of serine-proteases is the most
important anti-protease due to its plasma concentration.

α_1-AT is the support of a serum group system called Pi system (20). This polymorphism is encoded by two independant alleles at a single locus resulting in autosomal codominant inheritance. The variants of particular interest are the common Z and S mutants associated with plasma deficiency. Population surveys have shown that the S and Z variants are essentially found in individuals of European descent (19). About 110000 (12) Northern Europeans will have the PiZZ or PiSZ genotypes linked with severe inhibitor deficiency.

α_1-AT deficiency has been linked with different diseases but lung and liver disease must be specially mentioned. Early emphysema develops in individuals with severe α_1-AT deficiency due to elastin destruction by neutrophil and namely alveolar macrophage elastase. Both types of cells invade the lung on smoking and smoking PiZ subjects have a shorter survival ($\simeq$ 20 years) than the corresponding non-smokers (44). α_1-AT is inactivated by oxidizing agents which might be present in cigarette smoke (33). Another possibility of inactivation of the critical methionin residue in the active center is oxidation by leucocyte myeloperoxydase (80). Thus α_1-AT deficiency or chronic local inactivation results in protease/antiprotease unbalance achieving tissue destruction in various inflammatory diseases. It must be mentioned that only 40% of the inhibitor remain in the vascular space. Recently, the first attempts of successful replacement therapy with α_1-AT concentrates have been published (24). The patients were considerably improved and the protease (elastase) antiprotease unbalance reversed within the alveolar structure.

The above mentioned liver disease implies other features of ZZ deficiency concerning the secretion by the hepatocyte. Accumulation of intracellular amorphous aggregations of α_1-AT can be seen microscopically. The liver disease is of variable severity and in some cases fatal juvenile cirrhosis may develop.

Although the major function of α_1-AT has been pointed out its inhibitory activity against thrombin, plasma, KK, factor XI, plasmin, urokinase, akrosin, bacterial and fungal proteases has also been described. The observed inhibition is sometimes very slow and may thus have no real physiological significance. However, one has to keep in mind that α_1-AT is an acute phase protein and high α_1-AT levels have been associated with venous thromboembolism (22). In these authors view the mechanism could be explained by excessive neutral protease

inhibition inducing defective granulocyte induced pathway
Indeed, the alternative fibrinolytic pathway (61) seems
namely to be associated with leucocyte elastase whose
release during blood coagulation has been recently
reported (60).

Antithrombin III (AT III)

Another important plasma antiprotease is AT III.
This protein shows very special features as it is a
progressive inhibitor of thrombin and several other
serine proteases in the absence of heparin but its
reaction velocity is dramatically increased in the
presence of certain species of glycosamino-glycurenan-
sulfates. In the absence and presence of heparin 1:1
molar complexes are formed. When labeled thrombin is
added to defibrinated plasma in the absence of heparin,
the enzyme will be found associated with several
antiproteases : AT III, α_2-macroglobulin and even
α_1-AT (37,76). In the presence of heparin the events
are completely modified, thrombin interacting with
AT III almost instantaneously revealing the role of this
protein as heparin-cofactor (1). Whereas the anticoagulant activity of heparin became tentatively defined
as its AT III interaction capacity the biochemical
mechanism remained largely unknown. Only 20-30 % of
commercial heparin preparations contain anticoagulant
molecules.
Two methods were most frequently used to distinguish between heparin fractions having anticoagulant
activity and those having none (or only a small
activity) : on the one hand, gel-filtration, on the
other hand affinity chromatography on AT III linked
sepharose. Both techniques were also associated by some
authors.
Höök et al (30) first demonstrated that affinity
chromatography achieved the separation of high activity
from low activity species. This technique was also used
by Andersson et al (2) who then studied the molecular
size of the separated fractions. Below 9000 daltons
a decrease in specific activity was observed.
For some time the M.W. dependency of high
activity vs low activity became the most important
feature. Jordan et al. were however able to separate
highly active from low affinity species while starting
from relatively low molecular (6000 - 8000 daltons)
heparin (35). The former had 360 the latter only

4 u/mg. Developing further these studies Rosenberg
and Lam (54) identified a tetra-saccharide sequence
in sufficient amount to be present in each molecule of
active fractions whereas relatively inactive fractions
contain only a small percentage. The identification of
the reactive site in the mucopolysaccharide molecule
strengthened the view of a conformational transition within
the AT III molecule following the interaction with
heparin (35).
    The formation of a heparin-thrombin complex as an
initial event became then a new working hypothesis
studied by analysis of the kinetic data, by gel chromatography and by using chemically modified AT III and/or
thrombin. Specially the results obtained with surface-
bound heparin were in favour of an enzyme/heparin
reaction preceding the enzyme/inhibitor complex formation.
    Thrombin binds with high affinity to heparin
regardless whether the mucopolysaccharide belongs to the
high (anticoagulant) or low activity species. Jordan et al
(36) showed that factor IIa combines with 2 molecules
of heparin whereas the ratio is 1:1 for IXa and lower
affinity was observed for Xa and plasmin. Prothrombin is
unable to bind to heparin.
    The non discrimination of thrombin between active
and relatively inactive heparin became naturally the
main argument against the theory of heparin/thrombin
interaction as initial event of the inactivation process.
Although these theoretical discussions may continue,
the critical role of AT III as heparin-cofactor is out
of doubt. The single chain of AT III is cleaved during
interaction with the proteases (6) and covalent bonds
may from subsequently (48). Furthermore heparin has only
a reduced affinity for complexes and becomes free for
recycling with other molecules. The complexes are taken
up in the hepatocytes possibly by specific receptor
sites (68).
    New laboratory research work favors the isolation
of low molecular heparin most efficient in the inhibition
of Xa by AT III (14). However, in a recent study such an
oligosaccharide was only partially effective in preventing
thrombosis (77).
    Recurrent thrombosis is developing in patients
with congenital or acquired AT III deficiency when its
level is dropping below 50 %. During high risk periods
(such as delivery, eclampsia, extensive surgery) these
individuals may have benefit from substitution therapy
now entering in more generalized use (8,9,5) ;

40 u/kg/day seems to be an efficient dosage.
        AT III concentration is less than initially
presumed (78), but two new antithrombins have been
recently isolated. One of them has been called Heparin
cofactor II as heparin accelerates the reaction velocity.
Its concentration corresponds to about 90 mg/l and the
inhibitor is more efficient vs thrombin than vs Xa (78).
Another antithrombin has been called Antithrombin BM
the letters standing "for moderate binding". The
protein needs a higher amount of heparin for full thrombin
inhibitor function. It has no progressive activity in the
absence of heparin. The authors presume that about 40 %
of the antithrombin activity of human blood may be
attributed to AT BM (88).

α_2-macroglobulin (α_2-M)

        This high molecular weight protein (720 000 daltons)
has also very special properties. It is composed of 2
subunits (360 000 daltons) held together by non-covalent
forces. Each subunit is composed by 2 apparently identical
polypeptide chains (180 000) linked by disulfide bridges.
Proteolytic enzymes of low molecular weight (<28000)
interact in a 2:1 molar proportion whereas enzymes of
higher molecular weight interact in a 1:1 proportion
although 2 identical binding sites exist. The binding
is accompanied by cleavage of the α_2-M chains leaving
90 000 dalton fragments (27). SDS electrophoresis
reveals the solid (covalent) attachment of the enzymes
to these fragments (71). Whereas α_2-M has a very broad
specificity, it interacts with these enzymes leaving their
active sites free. Thus the complexes still have
enzymatic activities but only vs small molecules such as
synthetic substrates due to steric hindrance. Certain
primary amines can inactivate α_2-M (72). The underlying
mechanism corresponds to nucleophilic cleavage of the
thioester bonds of the inhibitor with the concurrent
formation of an amide with the glutamic acid residue
of the cleaved thioester (75,69). Amine treated α_2-M
has a faster electrophoretic mobility in sieving gels
than native α_2-M (72) as does α_2-M having reacted
with proteases. Barrett et al (4) studying further this
phenomenon proposed the term of "slow" and "fast" forms
of the α_2-M. The functional modification of α_2-M by
primary amines and the elucidation of its molecular
features encountered considerable interest. It has been
shown that α_2-M undergoes the same molecular

rearrangement whether it is exposed to primary amines or
to proteolytic enzymes. The clearance of both forms of
α_2-M showed the same velocity and they bound equally
to macrophage receptor sites (38,31) or fibroblasts.
Receptor-mediated endocytosis of α_2-M-trypsin complexes
could be inhibited by a monoclonal antibody to a neoantigen appearing as well on α_2-M complexed with
trypsin - as on methylamine treated α_2-M (50).

Binding of anhydrotrypsin to α_2-M has been
reported (81,66) revealing the possibility of "anomalous"
binding e.a. non covalent binding. These complexes do
not withstand the dissociating effect of SDS. Similar
anomalous complexes were observed with lysine modified
trypsin (87) and between modified α_2-M and proteases
(46,43). Despite the non physiological conditions of
these findings, it might be mentioned that disfunctioning
of modified α_2-M is one of the working hypotheses
concerning the pathology of Cystic Fibrosis.

α_2-M is an acute phase protein like α_1-AT.
Deficiency states of α_2-M are extremely rare but
considerable elevation is seen in nephrosis and
analbuminemia where part of the onchotic pressure,
normally due to albumin, is supplied by α_2-M.

C1-Inactivator (C1-INA)

This α_2-globulin containing about 35 % carbohydrates with a specially important part of sialic acid
($\simeq$ 15 %) inhibits not only the proteases of the C1
complex namely C1s and C1r but also the proteases of
the contact phase of coagulation and plasmin. However,
C1-INA is the exclusive inhibitor of C1s and C1r.

Spontaneous autoactivation of C1 at 37°C is
inhibited by C1-INA as is the activation by non-immune
activators (DNA, heparin) whereas that induced by immune
complexes is not so (89). The activated enzymes form a
complexe C1s-C1r-C1-INA (1:1:2), the 2 proteases being
associated by non covalent bonds. An ester link between
a serine residue at the active site and a non identified-
COOH in the inhibitor is postulated by Chesne et al. (13)
as the complexe can be dissociated by hydroxylamine.
Dissociable complexes seem to exist normally between
factors of the contact phase of coagulation and C1-INA
(18). Indeed when plasma is adsorbed at 4°C onto anti-
C1-INA significant amounts of kallikrein, and variable
amounts of PTA (Factor XI) are removed. Cold activation
by Hageman factor seems responsible for the observed
phenomena.

Congenital deficiency in C1-INA underlies hereditary angioneurotic edema. This disease is transmitted as an autosomal dominant. In 15 per cent of the cases normal or elevated levels of a nonfunctional (dismorphic) C1-INA are found. When the above described adsorption experiments are performed in these CRM + - plasmas no dropping of KK, HMWK or PTA is seen showing that this dismorphic material is unable to react normally. Activation of the kinin-forming cascade may occur during traumatic episodes of these patients but C1s is also activated (eventually by the HF pathway), C4 and C2 being depleted during attacks in these patients. A split product of C_2 (C2b) a kinin-like substance has also been related to the swelling episodes.

C1-inactivator concentrates are now at the disposal of the clinician and allowed to reverse promptly severe edematic attacks without any side effects (23,15). These concentrates prevented also serious hypotension following injection of albumin solutions containing prekallikrein activator (PKA) (70). Indeed, when ^{125}I-KK is added to plasma 52 % of the enzyme is found associated with C1-INA whereas 48 % appear complexed by α_2-M (67). Concentrates of the latter inhibitor have also been prepared several years ago.

α_2-antiplasmin (AP)

The discovery and isolation of this main inhibitor of plasmin was relatively late as α_2M was considered to be the most important inhibitor of plasmin. The protein is an α_2-globulin of 70 000 daltons containing approximately 13 % carbohydrates. Its concentration is insufficient ($\simeq$ 85 mg/l) to assure the inhibition of all available plasmin (16,53,55). However, the interaction of plasmin with AP is so fast that fibrinolysis starts only after considerable consumption of the inhibitor occurred. Inhibition by α_2-M is thus needed when fibrinolysis is increasing.

Plasmin/AP form stoechiometric 1:1 complexes. Interaction occurs between the light chain of plasmin and the inhibitor. A low molecular weight peptide being split concomitantly. A stable bond is formed concerning the hydroxy group of serine in the active site of plasmin and a-COOH group probably in terminal position of the inhibitor. The inactive complexes may circulate during many hours in contrast to the observations made with α_2-M (85).The interaction mechanism with the

lysine binding sites of plasmin(ogen) and the competition
with Histidine-rich glycoprotein has already been mentioned.

Synthetic fibrinolytic agents like flufenamic acid
derivatives inactivate AP (62,52). Similar effects are
observed on C1-inactivator (52,26). These examples show
best the "control" function of the inhibitors as their
inactivation induces respectively fibrinolysis or
activation of the classical complement pathway.

α_1-antichymotrypsin (α_1-ACHY)

This inhibitor has a similar molecular weight
(69 000 daltons) than the preceding AP but contains
more carbohydrates ($\simeq$ 25 %), the plasma concentration
is rather high : $\simeq$ 480 mg/l. Besides chymotrypsin,
leukocyte cathepsin G is also inhibited (79). Thus
α_1, AT and α_1-ACHY control together the tissue
destroying leukocytic enzymes : elastase and collagenase
for the former and cathepsin G for the latter. α_2-M
although able to inhibit these enzymes, remains mostly
in the intravascular space. Both inhibitors are acute
phase reactants but Laurell reported that α_1-ACHY
increases faster in the initial phase of inflammation
than α_1-AT (45) and seems to become selectively
concentrated in the bronchial lumen of patients with
chronic bronchitis (65).Beside their function in the
inflammatory process leading through excessive activity
to tissue destruction, the leucocyte proteases interact
with many plasma proteins activating enzymes on the one
side and inactivating inhibitors on the other side such
as AP,C1-INA (11) and AT III (34). These proteases are
thus desturbing the equilibrium of almost all biological
plasma systems at different levels.

Inter- α-trypsin inhibitor ($1\alpha 1$)

This inhibitor is still poorly understood. While
interacting strongly with trypsin and moderately with
chymotrypsin it is very easily degraded by various
proteolytic enzymes such as plasmin (41) but also
thrombin etc... The intact inhibitor has a molecular
weight of 180 000, the first break-down product shows a
molecular weight of 130 000. The high molecular split-
products are obtained with trace amounts of plasmin but
long digestion with a high enzyme/inhibitor ratio is
necessary to obtain a 18 000 dalton fragment (42) which

contains the trypsin inhibiting site. This split product
is similar to those obtained by Hochstrasser et al.(28)
and Bretzel et al. (10) after digestion of 1α1 with
trypsin or pancreatic kallikrein. Hochstrasser et al. observed acid stable split products of 1α1 in plasma (29)
and similar compounds in various biological fluids.
Although immunological relationships were found between
low molecular inhibitors of biological fluids and 1$^{\iota}$1,
there is no evidence that these compounds were split products of the plasma protein. 1α1 trypsin complexes dissociate in the presence of α_2-M (3) showing thus that
1α1 does not bind covalently trypsin. Similarily, we
found that plasmin was inhibited by a 1α1 derived fragment only when a chromogenic substrate was used but not
in a lysis test. Thus degradation by proteolytic enzymes
and formation of dissociable complexes characterize this
curious inhibitor whose biological role is still poorly
understood.

In conclusion, regulation of protease activity implies complexe mechanism. Antiproteases are the ultimate
control step. Their efficiency depends largely on the
accessibility of the enzyme i.e. whether the enzyme is
free or associated with other components or even inserted
in a surface structure.

REFERENCES

1. U. Abildgaard, Inhibition of the thrombin-fibrinogen
        reaction by heparin in the absence of cofactor.
        Inhibition of the thrombin-fibrinogen reaction
        by heparin and purified cofactor, Scand. J. Haemat.
        5:432 and 440 (1968).
2. L. Q. Andersson, T. W. Barrowcliffe, E. Holmer, E. A.
        Johnson, and G. Söderström, Molecular weight de-
        pendency of the heparin potentiated inhibition of
        thrombin and active factor X, Thromb. Res. 15:531
        (1979).
3. M. Aubry, and J. Bieth, A kinetic study of the in-
        hibition of human and bovine trypsins and chymo-
        trypsins by the inter-α-inhibitor from human plasma,
        Biochim. Biophys. Acta 438:221 (1976).
4. A. J. Barrett, M. A. Brown, and C. A. Sayers, The
        electrophoretically "Slow" and "Fast" forms of the
        α_2-macroglobulin molecule, Biochem.J.181:401 (1979).
5. B. Blauhut, S. Necek, H. Vinazzer, and H. Bergmann,
        Substitution therapy with an antithrombin III con-
        centrate in shock and DIC, Thromb. Res. 27:271
        (1982).

6. I. Björk, C. M. Jackson, H. Jörnvall, and W. J.
    Salsgiver, The active site of antithrombin. Re-
    lease of the same proteolytically cleaved form of
    the inhibitor from complexes with factor IXa,
    factor Xa, and thrombin, J. Biol. Chem. 257:2406
    (1982).

7. V. A. Bokisch, and H. J. Müller-Eberhard, Anaphyla-
    toxin inactivator of human plasma: its isolation
    and characterization as a carboxypeptidase, J.
    Clin. Invest. 49:2427 (1970).

8. P. Brandt, J. Jespersen, and G. Gregersen, Postpartum
    haemolytic-uraemic syndrom successfully treated
    with antithrombin III, Brit. Med. J. 280:6212
    (1980).

9. P. Brandt, Observations during the treatment of anti-
    thrombin III deficient women with heparin and an-
    tithrombin concentrate during pregnancy, partu-
    rition, and abortion, Thromb. Res. 22:15 (1981).

10. G. Bretzel, and K. Hochstrasser, Liberation of an
    acid stable proteinase inhibitor from the human
    inter-α-trypsin inhibitor by the action of kalli-
    krein,Hoppe-Seyler's Z.Physiol.Chem. 357:487 (1976).

11. M. S. Brower, and P. C. Harpel, Proteolytic cleavage
    and inactivation of α_2-plasmin inhibitor and C1-
    inactivator by human polymorphonuclear leukocyte
    elastase, J. Biol. Chem. 257:9849 (1981).

12. R. W. Carrell, J. O. Jeppsson, C. B. Laurell, S. O.
    Brennan, M. C. Owen, L. Vaughan, and D. R. Bos-
    well, Structure and variation of human α_1-anti-
    trypsin, Nature 298:329 (1982).

13. S. Chesne, C. L. Villiers, G. L. Arlaud, M. B.
    Lacroix, and M. G. Colomb, Fluid-phase interaction
    of C$\overline{1}$-inhibitor (C$\overline{1}$-Inh) and the subcomponents
    C$\overline{1}$r and C$\overline{1}$s of the first component of complement,
    C$\overline{1}$, Bioch. J. 201:61 (1982).

14. J. Choay, J. C. Lormeau, M. Petitou, P. Sinay, and
    J. Fareed, Structural studies on a biologically
    active hexasaccharide obtained from heparin, Ann.
    Nat.Acad.Sci. 370:644 (1981).

15. M. Cicardi, L. Bergamaschini, B. Marasini, G. Boccas-
    sini, A. Tucci, and A. Agostoni, Hereditary angio-
    edema: an appraisal of 104 cases, Amer. J. Med.
    Sci. 284:2 (1982).

16. D. Collen, Identification and some properties of a
    New Fast-reacting plasmin inhibitor in human plas-
    ma, Eur. J. Biochem. 69:209 (1976).

17. F. H. M. Derkx, B. N. Bouma, H. L. TanTjiong, A. J.
    Man in'T Veld, J. H. B. de Bruyn, and G. J. Wen-
    ting, The plasma kallikrein-renin connection,
    Supplementum to the archives Intern. de Pharmaco-
    dynamie et de Thérapie, 105 (1980).

18. V. H. Donaldson, and R. A. Harrisson, Complexes bet-
     ween C1-inhibitor, kallikrein, high molecular
     weight kininogen, plasma thromboplastin antecedent,
     and plasmin in normal human plasma and hereditary
     angioneurotic edema plasmas containing dysmorphic
     C1-inhibitors: role of cold activation, Blood 60:
     121 (1982).
19. C. T. Esmon, N. L. Esmon, and K. W. Harris, Complex
     formation between thrombin and thrombomodunin inhi-
     bits both thrombin-catalized fibrin formation and
     factor V activation, J. Biol.Chem. 257:7944 (1982).
20. M. K. Fagerhol, The Pi-system. Genetic variants of
     serum α_1-antitrypsin, Series Haematologica 1:153
     (1968).
21. D. T. Fearon, and K. F. Austen, Properdin: binding
     to C3b and stabilization of the C3b-dependent C3-
     convertase, J. Exp. Med. 142:856 (1975).
22. R. C. Franz, W. J. C. Coetzee, and R. Anderson,
     Venous thrombo-embolism and raised alpha-1-anti-
     trypsin levels: A possible causal relationship
     between excessive neutral protease inhibition and
     defective granulocyte-induced fibrinolysis, SA
     Medical J. 59:661 (1981).
23. J. E. Gadek, S. W. Hosea, J. A. Gelfand, M. M. Frank,
     Replacement therapy in hereditary angioedema.
     Successful treatment of acute episodes of angio-
     edema with partly purified C1-inhibitor, New Eng.
     J. Med. 302:542 (1980).
24. J. E. Gadek, H. G. Klein, P. V. Holland, and R. G.
     Crystal, Replacement therapy of alpha-1-antitryp-
     sin deficiency. Reversal of protease-antiprotease
     imbalance within the alveolar structures of PiZ
     subjects, J. Clin. Invest. 68:1158 (1981).
25. B. Ghebrehiwet, M. Silverberg, and S. P. Kaplan, Acti-
     vation of the classical pathway of complement by
     Hageman factor fragments, J. Exp. Med. 153:665(1981)
26. P. C. Giclas, Effect of plicatic acid on human serum
     complement includes interference with C1-inhibitor
     function, J. Immunol. 129:168 (1982).
27. P. C. Harpel, Studies on human plasma α_2-macroglobu-
     lin-enzyme interactions. Evidence for proteolytic
     modification of the subunit chain structure, J.
     Exp. Med. 138:508 (1973).
28. K. Hochstrasser, H. Feuth und O. Steiner, Zur Cha-
     rakterisierung der säurestabilen Proteaseninhibi-
     toren aus Humanplasma, Hoppe Seyler's Z. Physiol.
     Chem. 354:927 (1973).

29. K. Hochstrasser, J. Niebel, H. Feuth und K. Lempart,
     Über Abbauprodukte des Inter-alpha-Trypsininhibi-
     tors im Serum. I. Der Inter-alpha-Trypsininhibitor
     als Prekursor des säurestabilen Serum-Trypsin-In-
     hibitors. II. Säurelabile Derivate des Inter-alpha-
     Trypsininhibitors, Klin. Wochenschr. 55:337 (1977).
30. M. Höök,K.Björk,J.Hopwood,U.Lindahl,Anticoagulant ac-
     tivity of heparin of high activity and low activi-
     ty heparin-species by affinity chromatography on
     immobilized anti-thrombin, FEBS Lett. 66:90 (1976).
31. M. J. Imber, and S. V. Pizzo, Clearance and binding
     of two electrophoretic "Fast" forms of human α_2-
     macroglobulin, J. Biol. Chem. 256:8134 (1981).
32. A. Janoff, Inhibition of human granulocytes elastase
     by serum α_1-antitrypsin, Am. J. Resp. Dis. 105:
     121 (1972).
33. A. Janoff, H. Carp, and D. K. Lee, Cigarette smoke
     inhalation decreases α_1-antitrypsin activity in
     rat lung, Science 206:1313 (1979).
34. M. Jochum, S. Lander, N. Heimburger, and H. Fritz,
     Effect of human granulocytic elastase on isolated
     human antithrombin III, Hoppe-Seyler's Z. Physiol.
     Chem. 362:103 (1981).
35. R. E. Jordan, D. Beeler, and R. D. Rosenberg,Fraction-
     ation of low molecular weight heparin species and
     their interaction with antithrombin, J. Biol.
     Chem. 254:2902 (1979).
36. R. E. Jordan, G. M. Oosta, W. T. Gardner, and R. D.
     Rosenberg, The binding of low molecular weight
     heparin to hemostatic enzymes, J. Biol. Chem.
     255:10073 (1980).
37. F. Josso, D. Benamon-Djiane, J. M. Lavergne, C.
     Weilland, and M. Steinbuch, Investigations of
     plasma antithrombin activity using ^{131}I labelled
     human thrombin, 7th Congr. Clin. Chem. Geneva,
     M. Roth ed., Karger 2:46 (1970).
38. J. Kaplan, and M. L. Nielsen, Analysis of macrophage
     surface receptors. I. Binding of α-macroglobulin-
     protease complexes to rabbit alveolar macrophages.
     II. Internalization of α_2-macroglobulin-trypsin
     complexes by rabbit alveolar macrophages, J. Biol.
     Chem. 254:7323 and 7329 (1979).
39. M. D. Kazatchkine, D. T. Fearon, and K. F. Austen,
     Human alternative complement pathway: membrane-
     associated sialic acid regulates the competition
     between B and B1H for cell-bound C3b, J. Immunol.
     122:75 (1979).

40. W. Kisiel, W. M. Canfield, L. H. Ericsson, and E. W. Davie, Anticoagulant properties of bovine plasma protein suite. C following activation by thrombin, Biochemistry 16:5824 (1977).
41. P. Lambin, R. Audran, and M. Steinbuch, Etude des fragments de l'inter-alpha-trypsine inhibiteur obtenus par digestion plasminique, Pathol. Biol. Suppl. 25:31 (1977).
42. P. Lambin, J. M. Fine, and M. Steinbuch, Inhibition of plasmin by a small molecular weight inhibitor derived from human inter-alpha-trypsin inhibitor, Thromb. Res. 13:563 (1978).
43. P. Lambin, F. Pochon, J. M. Fine, and M. Steinbuch, α_2-M macroglobulin/thrombin interaction in the presence of hydroxylamine, in press (1983).
44. C. Larsson, Natural history and life expectancy in severe alpha$_1$-antitrypsin deficiency, PiZ. Intermediate alpha$_1$-antitrypsin deficiency, Pi M-, Acta Med. Scand. 204:345 and 353 (1978).
45. C. B. Laurell, Comparison of alpha$_1$-antitrypsin and alpha$_2$-macroglobulin, in: "Pulmonary Emphysema and Proteolysis," C. Mittman, ed., Academic Press, New York-London (1972).
46. F. van Leuven, P. Marynen, J. J. Cassiman, and H. van den Berghe, Relation of internal thioesters to conformational change and receptor recognition site in α_2-macroglobulin complexes, Biochem. J. 203:405 (1982).
47. H. R. Lijnen, M. Hoylaerts, and D. Collen, Isolation and characterization of a human protein with affinity for the lysine-binding sites in plasminogen. Role in the regulation of fibrinolysis and identification as histidine-rich-glycoprotein, J. Biol. Chem. 255:10214 (1980).
48. M. O. Longas, and T. H. Finlay, The covalent nature of the human antithrombin III-thrombin bond, Biochem. J. 189:481 (1980).
49. R. J. Mandle, and A. P. Kaplan, Hageman-factor dependent fibrinolysis: generation of fibrinolytic activity by the interaction of human activated factor XI and plasminogen, Blood 54:850 (1979).
50. P. Marynen, F. van Leuven, J. J. Cassiman, and H. van den Berghe, A monoclonal antibody to a neo-antigen on α_2-macroglobulin complexes inhibits receptor-mediated endocytosis, J. Immunol. 127:1782 (1981).
51. H. L. Meier, J. V. Pierce, R. W. Colman, and A. P. Kaplan, Activation and function of human Hageman factor. The role of high molecular weight kininogen and prekallikrein, J. Clin. Invest.60:18(1977).

52.   L. A. Miles, J. P. Burnier, M. S. Verlander, M. Good-
      man, A. J. Kleiss, and J. H. Griffin, Inactivation
      of purified human α_2-antiplasmin and purified
      human C1̄-inhibitor by synthetic fibrinolytic
      agents, Blood 57:1015 (1981).

53.   M. Moroi, and N. Aoki, Isolation and characterization
      of alpha$_2$-plasmin inhibitor from human plasma. A
      novel proteinase inhibitor which inhibits activa-
      tor-induced clot lysis, J. Biol. Chem. 251:5956
      (1976).

54.   M. Moroi, and N. Aoki, Inhibition of plasminogen
      binding to fibrin by α_2-plasmin inhibitor, Thromb.
      Res. 10:851 (1977).

55.   S. Müllertz, and I. Clemmensen, The primary inhibitor
      of plasmin in human plasma, Biochem.J.159:545 (1977).

56.   B. Osterud, and S. I. Rapaport, Activation of ^{125}I-
      factor IX and ^{125}I-factor X: Effect of tissue
      factor and factor VII, factor Xa and thrombin,
      Scand. J. Haematol. 24:213 (1980).

57.   M. K. Pangburn, and H. J. Müller-Eberhard, Complement
      C3-convertase: cell surface restriction of β1H
      control and generation of restriction on neura-
      minidase-treated cells, Proc. Nat. Acad. Sci. 75:
      2416 (1978).

58.   M. K. Pangburn, and H. J. Müller-Eberhard, Relation
      of a putative thioester bond in C3 to activation
      of the alternative pathway and the binding of C3b
      to biological targets of complement, J. Exp. Med.
      152:1102 (1980).

59.   M. K. Pangburn, R. D. Schreiber, and H. J. Müller-
      Eberhard, Human complement C3b-inactivator: iso-
      lation, characterization and demonstration of an
      absolute requirement for the serum protein β-1H
      for cleavage of C3b and C4b in solution, J. Exp.
      Med. 146:257 (1977).

60.   E. F. Plow, Leukocyte elastase release during blood
      coagulation, J. Clin. Invest. 69:564 (1982).

61.   E. F. Plow, and T. S. Edgington, An alternative path-
      way for fibrinolysis. I. The cleavage of fibrino-
      gen by leukocyte proteases at physiologic pH, J.
      Clin. Invest. 56:30 (1975).

62.   M. B. Prandini, B. Wiman, M. Samama, and D. Collen,
      Effects of the synthetic fibrinolytic agents
      ortho-thymotic acid and S-1623 on the reaction
      between human plasmin and antiplasmin, Thromb.
      Res. 13:165 (1978).

63.   R. Radcliffe, A. Bagdasarian, R. W. Colman, and Y.
      Nemerson, Activation of factor VII by Hageman
      factor fragments, Blood 50:611 (1977).

64. R. D. Rosenberg, and L. Lam, Correlation between
     structure and function of heparin, Proc. Nat.
     Acad. Sci. 76:1218 (1979).
65. H. C. Ryley, and T. D. Brogan, Quantitative immuno-
     electrophoretic analysis of the plasma proteins
     in the sol phase of sputum from patients with
     chronic bronchitis, J. Clin. Path. 26:852 (1973).
66. C. A. Sayers, and A. J. Barett, Binding of anhydro-
     trypsin to α_2-macroglobulin, Biochem.J.(Mol. As-
     pects) 189:255 (1980).
67. M. Schapira, C. F. Scott, and R. W. Colman, Contri-
     bution of plasma protease inhibitors to the in-
     activation of kallikrein in plasma, J. Clin. In-
     vest 69:462 (1982).
68. M. A. Shifman, and S. V. Pizzo, The "in vivo" meta-
     bolism of antithrombin III and antithrombin III
     complexes, J. Biol. Chem. 257:3243 (1981).
69. L. Sottrup-Jensen, T. E. Petersen, and S. Magnusson,
     A thiol-ester in α_2-macroglobulin cleaved during
     proteinase complex formation, FEBS Lett. 121:
     275 (1980).
70. P. van der Starre, D. Sinclair, J. Damen, and H.
     Brummelhuis, Inhibition of the hypotensive effect
     of plasma protein solutions by C_1-esterase inhibi-
     tor, J. Thor. Cardiovasc. Surg. 79:738 (1980).
71. M. Steinbuch, R. Audran, P. Lambin and J. M. Fine,
     Subunit structure of human alpha-2-macroglobulin.
     Its relationships with its biological activity,
     in: "The Protides of the Biological Fluids," H.
     Peters, ed., Pergamon Press, Oxford,New York (1976).
72. M. Steinbuch, L. Pejaudier, M. Quentin, and V. Martin,
     Molecular alteration of α_2-macroglobulin by ali-
     phatic amines,Biochim.Biophys.Acta 154:228 (1968).
73. T. Sugo, N. Ikari, H. Kato, S. Iwanaga, and S. Fujii,
     Functional sites of bovine high molecular weight
     kininogen as a cofactor in kaolin-mediated activ-
     ation of factor XII (Hageman factor), Biochemistry
     19:3215 (1980).
74. K. Suzuki, B. Dahlbäck, and J. Stenflo, Thrombin-ca-
     talysed activation of human coagulation factor V,
     J. Biol. Chem. 257:6556 (1982).
75. R. P. Swenson, and J. B. Howard, Characterization of
     alkylamine-sensitive site in α_2-macroglobulin,
     Proc. Nat. Acad. Sci. 76:4313 (1979).
76. A. Takada, T. Koide, and Y. Takada, Interaction of
     thrombin with antithrombin III and α_2-macroglobu-
     lin in the plasma, Thromb. Res. 16:59 (1979).

77. D. P. Thomas, R. E. Merton, T. W. Barrowcliffe, and U. Lindhal, Effects of heparin oligosaccarides with high affinity for antithrombin III in experimental venous thrombosis, Thromb. Haemost. 47:244 (1982).

78. D. M. Tollefsen, D. W. Majerus, and M. K. Blank, Heparin cofactor II, J. Biol. Chem. 257:2162 (1982).

79. J. Travis, R. Baugh, P. J. Giles, D. Johnson, J. Bowen and C. F. Reilly, Human leukocyte elastase and cathepsin G: Isolation, characterization and interaction with plasma proteinase inhibitors, in:"Neutral Proteases of Human Polymorphonuclear Leukocytes," K. Havemann, A. Janoff, eds., Urban & Schwarzenberg, Baltimore/Munich (1978).

80. J. Travis, N. Matheson, D. Johnson, and K. Beatty, Human alpha-1-proteinase inhibitor and human alpha-1-antichymotrypsin: properties and mechanism studies, in: "The Chemistry and Physiology of the Human Plasma Proteins," D. H. Bing, ed., Pergamon Press, Oxford-New York (1978).

81. D. Tsuru, K. Kado, K. Fujiwara, M. Tomimatsu, and K. Ogita, Interaction of porcine α_2-macroglobulin with chemically modified proteinases, J. Biochem. 83:1345 (1978).

82. G. Vehar, and E. W. Davies, Preparation and properties of bovine factor VIII (antihemophilic factor), Biochemistry 19:401 (1980).

83. F. J. Walker, P. W. Sexton, and C. T. Esmon, The inhibition of blood coagulation by activated protein C through the selective inactivation of activated factor V, Biochim. Biophys. Acta 571:333 (1979).

84. K. Whaley, and S. Ruddy, Modulation of the alternative complement pathway by 1H globulin, J. Exp. Med. 144:1147 (1976).

85. B. Wiman, and D. Collen, On the mechanism of the reaction between plasmin and α_2-antiplasmin. Kinetics and structural changes, in: "The Physiological Inhibitors of Coagulation and Fibrinolysis," D. Collen, B. Wiman, M. Verstraete, eds., Biomedical Press, Elsevier, Holland (1979).

86. B. Wiman, and P. Wallen, The specific interaction between plasminogen and fibrin. A physiological role of the lysine binding site in plasminogen, Thromb. Res. 10:213 (1977).

87. K. Wu, D. Wang, and R. D. Feinman, Inhibition of proteases by α_2-macroglobulin, J. Biol. Chem. 256:10409 (1981).

88. P. Wunderwald, W. J. Schrenk, and H. Port, Anti-thrombin BM from human plasma: an antithrombin binding moderately to heparin, Thromb. Res. 25: 177 (1982).
89. R. J. Ziccardi, A new role for C$\overline{1}$-inhibitor in homeostasis: control of activation of the first component of human complement, J. Immunol. 128: 2505 (1982).

HUMAN KININOGENS AND THEIR FUNCTION IN THE KALLIKREIN-KININ SYSTEMS

Werner Müller-Esterl and Hans Fritz

Abteilung für Klinische Chemie und Klinische Biochemie in der Chirurgischen Klinik Innenstadt der Universität München Nussbaumstrasse 20, D-8000 München 2, FRG

THE KALLIKREIN-KININ SYSTEMS

Kallikrein-kinin systems are present in plasma, tissues, and secretions of mammalian species (for recent reviews, see [1,2]. Five major constituents have been recognized so far which establish the kinin-forming pathway by their specific interactions (Fig. 1):

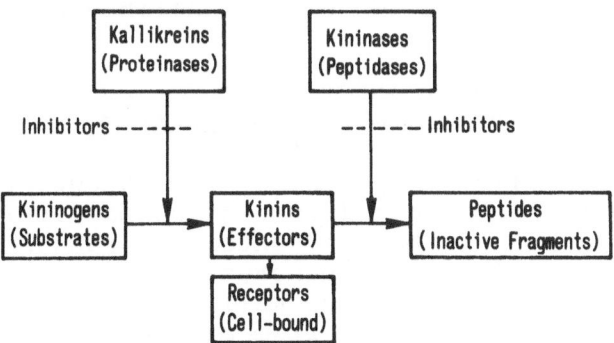

Fig. 1 The kallikrein-kinin systems. Diagrammatic representation of the major constituents establishing the kinin-forming pathway. Specific intercorrelations among the components of the systems are indicated.

(1) Kallikreins or kininogenases are proteolytic enzymes
    of the serine type of proteinases which trigger the
    systems by limited proteolysis of their natural sub-
    strates, the kininogens, thereby releasing kinins
    [3,4,5,6].
(2) Kininogens are synthesized and secreted by the liver.
    They are single-chain (glyco-)proteins comigrating
    with the α_2-fraction of human plasma proteins and re-
    present the high molecular weight precursors of the
    kinins [7,8,9].
(3) Kinins are oligopeptides composed of 9 to 11 amino
    acids which have been early discovered as the termi-
    nal effectors of the kallikrein-kinin systems [10].
    They exert at least part of their effects via second
    messengers, e.g. prostaglandins [11].
(4) Kinin receptors mediate the physiological effects of
    the kinins. The receptors have been postulated to be
    exposed on the luminal side of the target cells of
    kinin action [12].
(5) Kininases establish the second set of regulatory
    components of the kinin-generating systems. They are
    carboxy- and dipeptidylpeptidases which effectively
    limit the efficacy of the kinins with respect to time
    and space by sequential degradation of the kinins in-
    to inactive fragments [13,14].
Finally, the enzymic activity of both types of proteases,
i.e. kallikreins and kininases, is efficiently regulated
by specific inhibitors and inactivators.

According to their distribution, two principal kinin-
forming systems have been classified: the plasma kallikrein-kinin system and the tissue kallikrein-kinin system.

THE TISSUE KALLIKREIN-LMW KININOGEN-KALLIDIN SYSTEM

        Tissue kallikreins [5,6] of M_r ~ 30 000 are present
in major secretory glands (parotid, submandibular gland,
pancreas), in kidney, testis, intestine, and secretions
such as urine, saliva, and pancreatic juice. The preferential substrate of tissue kallikrein is LMW kininogen [15] which is excreted from plasma and transported
into the interstitial fluid and secretions. Alternatively, tissue kallikrein may encounter LMW*)kininogen in
_____
*Abbreviations used: LMW, low molecular weight; HMW,
 high molecular weight; H-chain, heavy chain; L-chain,
 light chain; HRP, histidine-rich peptide; SDS, sodium
 dodecyl sulfate; ELISA, enzyme-linked immunosorbent
 assay.

the plasma due to incretion of the secretory proteinase
into the blood [16]. Limited proteolysis of LMW kininogen
by tissue kallikrein results in the liberation of the
decapeptide kallidin (Lys-bradykinin) which is converted
by aminopeptidases to bradykinin. A variety of (patho)
physiological functions have been postulated for the
tissue kallikrein-LMW kininogen-kallidin systems, e.g.
regulation of salivary secretion, increase of vascular
permeability [10], modulation of intestinal resorption
[17], regulation of blood pressure [18], mediation of inflammatory and allergic reactions [19], and stimulation
of cellular glucose uptake [20]; however, a unifying concept of the physiological role of the tissue kallikrein-
kinin system is still lacking.

THE PLASMA KALLIKREIN-HMW KININOGEN-BRADYKININ SYSTEM

    In contrast, the physiological role of the plasma
kallikrein-kinin system in the contact phase of the intrinsic blood coagulation cascade is well established
[21] (Fig. 2). Following blood vessel trauma by endo-

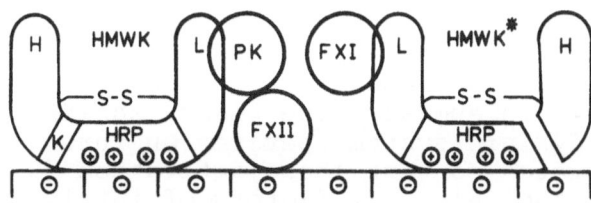

Fig. 2   Contact phase of the intrinsic blood coagulation
          cascade. HMW kininogen complexed with plasma
          pro-kallikrein and/or factor XI (plasma thrombo-
          plastin antecedent) is anchored by the histidine-
          rich peptide (HRP) to the surface of endothelial
          lesions in close vicinity to factor XII (Hageman
          factor). Subsequent reciprocal activation yields
          active Hageman factor (F XIIa) which in turn ac-
          tivates F XI thus triggering the coagulation cas-
          cade. Active plasma kallikrein splits HMW kinino-
          gen to form bradykinin (K) and the kinin-free
          HMW kininogen (HMW-K*) with a disulfide-bridged
          heavy chain (H) and light chain (L).

genous noxa, negatively charged surfaces underlying the
endothelial cells are exposed which rapidly bind factor
XII (Hageman factor). Simultaneously, HMW kininogen cir-

culating in plasma in a complex with plasma prokallikrein
[22,23] and/or factor XI (plasma thromboplastin antecedent) [24] is attached to the endothelial lesions in
proximity to F XII [8,25]. Binding of HMW kininogen is
mediated by a highly positively charged segment of the
kininogen molecule which is unusually rich in histidine
and thus is termed "histidine-rich peptide" (HRP) [26]
(cf. Fig. 2). Hence, a proper assembly of the initiator
factors of intrinsic blood coagulation at the actual site
of injury is accomplished. Subsequently, a reciprocal activation of F XII and plasma prokallikrein proceeds with
the resultant formation of activates Hageman factor
(F XIIa) [27] which in turn activated F XI thus triggering the coagulation cascade [25].

Plasma kallikrein serves two major functions in the
concerted action of contact activation:(i) augmentation
of F XII activation [28] and (ii) limited proteolysis
of HMW kininogen under the release of the nonapeptide
bradykinin. Interestingly, the nicked, kinin-free HMW
kininogen is an even more efficient cofactor of the contact phase activation than native kininogen: in vitro
studies with bovine HMW kininogen have demonstrated that
the procoagulant activity of the kinin-free kininogen is
increased by the factor of 2.2 compared to the parent molecule [29]. Once bradykinin has been liberated from its
precursor protein it may support local tissue repair due
to increase in transmembrane permeability of the endothelium [10]. In summary, the plasma kallikrein-HMW kininogen-bradykinin system operates on three levels of
the orchestrated contact phase activation [30], i.e.
assembly of the initiator factors of intrinsic blood
coagulation at the actual site of injury, activation of
the proenzyme(s) and cofactor(s) of the clotting system,
and change of capillary permeability to facilitate local
tissue repair.

Beyond its role in the initiation of intrinsic blood
clotting, activated plasma kallikrein has been implicated in the activation of several other regulatory systems: activation of prorenin thus triggering the renin-
angiotensin system which counteracts the hypotensive
effect of the kinins by liberation of angiotensin II
[31]; activation of plasminogen thus starting the fibrinolytic cascade [32]; activation of C1r, C1s [33], C3
convertase [34] and C5 [35] thus initiating the complement system via the classical and alternative pathway.
These findings provide some important links of the kallikrein-kinin system to other plasma-borne regulatory systems [36]; however, the physiological relevance of these
interactions is still in question and deserves further
experimental studies.

Elucidation of the (putative) biological roles of HMW kininogen and LMW kininogen has prompted numerous investigators to study mammalian kininogens in detail (for recent reviews, see [7,8,9]). Much of our present knowledge about mammalian kininogens has been established from the work of Kato and coworkers on bovine kininogens [37]. This report focuses on the isolation and characterization of human kininogens.

ISOLATION OF HUMAN KININOGENS

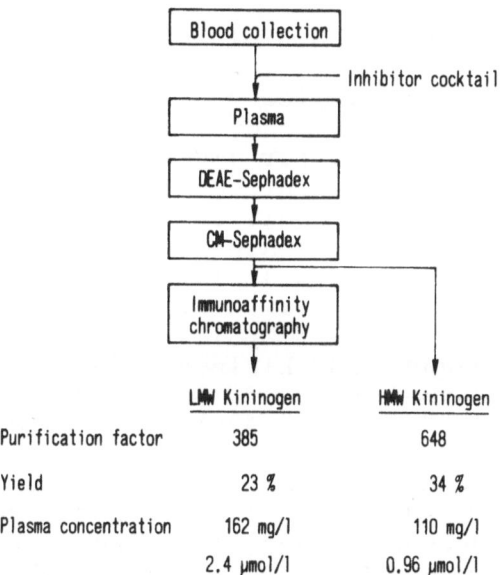

Fig. 3 Protocol used for the isolation of kininogens from human plasma. HMW kininogen and LMW kininogen are isolated from freshly drawn plasma in the presence of 300 µg/ml polybrene, 75 µg/ml soybean trypsin inhibitor, 6 µg/ml (400 KIU/ml) trasylol, 6 mM EDTA, 1 mM benzamidine·HCl (final concentrations given). Total yields are determined by the recovery of [125J]-kininogen added in trace amounts to the plasma prior to separation. Plasma concentrations of the kininogens are calculated on the basis of the given yields. Using the given protocol, a simultaneous purification of the two kininogens to apparent homogeneity and with high yields is feasible [38,39].

Simultaneous purification of both types of kininogens from freshly prepared human plasma is accomplished
by a two step isolation procedure [38] in the presence
of inhibitors such as diisopropylfluorophosphate, trasylol, soybean trypsin inhibitor, polybrene, and EDTA. Two
consecutive ion exchange chromatography steps on DEAE-
and CM-Sephadex yield homogeneous HMW kininogen with a
total recovery of 34 % and a purification factor of 650
(Fig. 3). Isolation of pure LMW kininogen requires an
additional immunoaffinity step on immobilized anti-LMW
kininogen-IgG with a maximum recovery of 23 % and a purification factor of 380 [39]. Quantitative removal of HMW
kininogen prior to the immunoaffinity step is a prerequisite for the isolation of pure LMW kininogen, as major
part of the anti-LMW kininogen IgG immobilized on Sephadex will readily cross-react with both types of kininogens.

CHARACTERIZATION OF HUMAN KININOGENS

The specific activities of the resultant kininogen
preparations are 11.2 µg bradykinin/mg for LMW kininogen
and 7.4 µg bradykinin/mg for HMW kininogen indicating
that 80 % (72 %) of HMW kininogen (LMW kininogen) is
present in the native single-chain form, while 20 %
(28 %) is present in the kinin-free two-chain form due
to limited proteolysis and kinin release. These results
are confirmed by SDS electrophoresis of the kininogens
displaying a major band of M_r 114 000 (68 000) for the
native HMW kininogen (LMW kininogen) and minor bands representing the heavy chain of M_r 63 000 (62 000) and the
light chain of M_r 58 000 (5 400). Additional molecular
parameters determined for human kininogens (pI, E_{280},
carbohydrate composition, N-terminal amino acid) are compiled in Table 1. The amino acid composition for human
HMW kininogen, the heavy and light chain of HMW kininogen, human LMW kininogen, and the light chain of LMW
kininogen are presented in Table 2.
An interesting feature of purified kininogens is
their tendency to form larger self-aggregates in the absence of dissociating agents thus suggesting multiple
forms of kininogens [38,39]. Besides the protomer, reasonable amounts of dimer, tetramer up to the decamer
were detected in gel filtration and polyacrylamide electrophoresis in the absence of SDS (cf. Table 3). Oligo-
mer formation by human kininogens may well reflect their
ability to bind other plasma proteins such as plasma prokallikrein and factor XI, which have been demonstrated to
form binary complexes with HMW kininogen [22,23,24]. Removal of the associated plasma proteins from kininogens

Table 1. Properties of human kininogens.

Properties	LMW kininogen	HMW kininogen
M_r native molecule	68 000	114 000
L–chain	5 400	58 000
H–chain	62 000	63 000
pI native molecule	4.45; 4.75; 4.9	4.45; 4.5; 4.65
L–chain	5.8	4.4; 4.6; 4.7; 4.9; 5.0
H–chain	not determined	3.85
carbohydrate content (mol/mol protein)	galactose.............3 mannose...............4 N–acetylglucosamine ....3 N–acetylneuraminic acid 3	galactose...... ....34 mannose................13 N–acetylglucosamine.....12 N–acetylgalactosamine...12 N–acetylneuraminic acid 29
specific activity	11.2 µg BK/mg protein	7.4 µg BK/mg protein
N–terminal amino acid	blocked	blocked (pyroglutamic acid) [50]
$E_{280\ nm}^{1\%}$	7.8	7.3

Table 2. Amino acid composition (residues/100 residues) of human HMW kininogen (native protein), the heavy and light chain of HMW kininogen, LMW kininogen and the light chain of LMW kininogen.

	HMW Kininogen[1]			LMW Kininogen[1] [39]	
	native	H-chain	L-chain	native	L-chain
Asx	11.6	11.4	10.8	10.4	0
Thr[2]	8.3	8.4	7.4	7.5	5.16
Ser[2]	7.3	7.0	7.7	7.5	18.4
Glx	12.7	12.2	11.4	12.4	21
Pro	6.1	6.2	8.6	6.2	7.9
Gly	7.8	6.2	10.7	5.7	7.9
Ala	5.0	6.1	3.7	6.5	7.9
1/2Cys	2.1	2.0	1.3	3.4	2.6
Val	3.4	5.9	3.1	6.4	5.3
Met	0.9	0.7	1.3	0.7	0
Ile	5.1	5.1	4.5	4.4	5.3
Leu	6.6	6.1	5.8	7.0	2.6
Tyr	2.4	4.7	1.6	3.7	2.6
Phe	4.3	4.5	3.1	4.5	0
His	5.2	2.5	7.4	1.7	2.6
Lys	8.1	7.7	8.6	7.4	7.9
Arg	3.4	3.3	3.0	3.9	10.5
Trp[3]	n.d.[4]	n.d.[4]	n.d.[4]	>0.5	n.d.[4]

[1] calculated from the nearest integer of the experimental values

[2] extrapolated to zero time

[3] determined by the method of Edelhoch [51]

[4] n.d., not determined

Table 3. Oligomerisation of human kininogens. Molecular weights were determined on a 4 - 30 % (w/v) polyacrylamide gradient gel in the absence of dissociating agents [39].

Oligomer	LMW Kininogen	HMW Kininogen
Monomer	78 000	140 000
Dimer	185 000	350 000
Tetramer	305 000	540 000
Hexamer	450 000	n.d.
Octamer	580 000	n.d.
Decamer	780 000	n.d.

n.d. = not detected

in the course of the isolation procedure may uncover competent protein binding sites on the kininogens, thus giving rise to an extensive intermolecular self-association
of the purified kininogens. Oligomerisation is completely
reversible in the presence of dissociating agents, e.g.
SDS and urea.

FRAGMENTATION BY LIMITED PROTEOLYSIS

Limited proteolysis of human kininogens by tissue
kallikrein (from porcine pancreas) results in the liberation of kinin and a concomitant generation of primary
degradation product consisting of two disulfide bridged
protein chains [40,41]. In the case of HMW kininogen,
two fragments of similar size are formed with M_r of 63 000
(heavy chain) and 58 000 (light chain). In contrast, proteolysis of human kininogen with plasma kallikrein generates a predominant light chain fragment of M_r 45 000 [40]
indicating that the light chain has been further processed. Separation of the light and heavy chain is
achieved by ion exchange chromatography on SP-Sephadex
following reduction and carboxymethylation. SDS analysis
of the isolated fragments is presented in Fig. 4. Sequence analysis of the 58 000 Dalton light chain reveals that
tissue kallikrein splits correctly at the corboxyterminal
side of the kinin segment and that the complete aminoter-
minus of the light chain is preserved during proteolysis
with tissue kallikrein [42].

In an analogous experiment, limited proteolysis of
LMW kininogen by tissue kallikrein [39] generates a
62 000 Dalton heavy chain and a light chain of apparent
M_r 5 400. Following reduction and carboxymethylation of
the nicked LMW kininogen, the light chain is isolated by
high performance liquid chromatography on a reversed
phase column*); it is eluted from the gel matrix in a
sharp peak. SDS electrophoresis with increasing amounts
of L-chain indicates that the fragment is essentially
pure (Fig. 5a). Similar results are obtained from isoelectric focusing of the isolated light chain (Fig. 5b).
Amino-terminal sequence analysis demonstrates that the
isolated light chain of human LMW kininogen is the primary split product of the kallikrein action. It is obtained without further proteolytic processing of the N-
terminus as deduced from its homology to the light chains
of human HMW kininogen and bovine LMW kininogen.

---

*)Müller-Esterl, W., Rauth, G., Lottspeich, F. and Hen-
   schen, A., manuscript in preparation

From these data an overall structure for human kininogens can be derived which comprises three functional
and structural domains (Fig. 6): the aminoterminal heavy
chain, which is indistinguishable by immunological criteria for both types of kininogens; the consecutive kinin
segment, which is identical for both kininogens; and the
carboxyterminal light chain, which is only rudimentary
in LMW kininogen compared to HMW kininogen. Notably, the
procoagulant activity of HMW kininogen resides on the
histidine-rich section of its L-chain [43], and LMW kininogen devoid of the histidine-rich peptide is not involved in the contact phase activation [43].

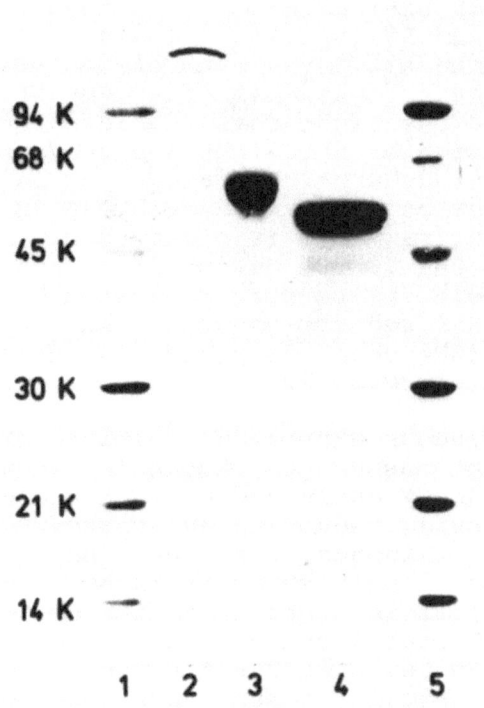

Fig. 4. SDS electrophoresis of HMW kininogen and
its major fragments. Discontinuous SDS electrophoresis was run in a 10 to 15 % (w/v) polyacrylamide gradient gel according to Laemmli [46].
Line 1,5 marker proteins (M_r indicated), 5 µg each.
Line 2, HMW kininogen, 20 µg. Lane 3, heavy chain,
20 µg. Lane 4, light chain 20 µg. All samples were
reduced with 0.32 mM dithiothreitol (5 min, 100°C)
prior to electrophoresis.

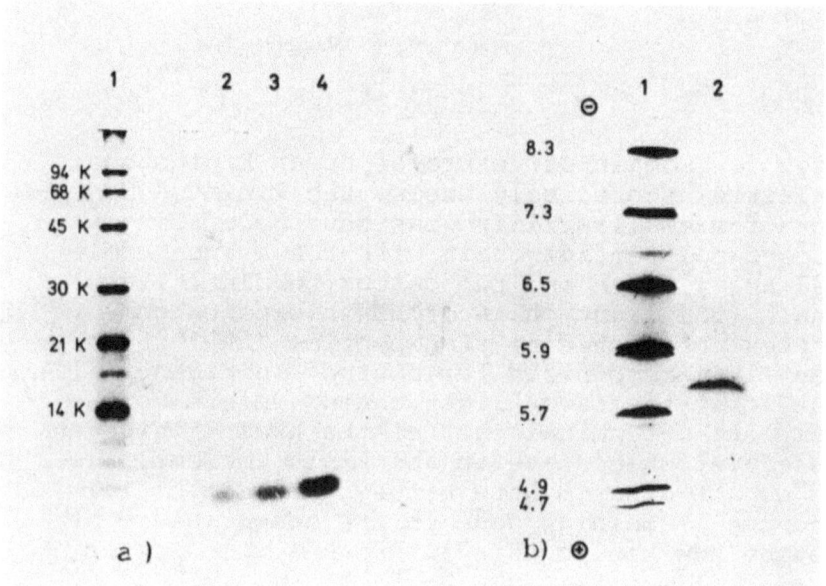

Fig. 5. Electrophoretic analysis of the LMW kininogen light chain.
a. Discontinuous SDS electrophoresis in a 10 to 15 % (w/v) polyacrylamide gradient gel according to Laemmli [46]. Lane 1, marker proteins (M_r indicated), 5 µg each. Lane 2 to 4, increasing amounts of the light chain (5, 10, 20 µg).
b. Isoelectric focusing in a 5 % (w/v) polyacrylamide gel according to Vesterberg [47]. Lane 1, marker proteins (pI indicated). Lane 2, 10 µg light chain. A pI of 5.8 was calculated for the LMW kininogen light chain from the calibration curve.

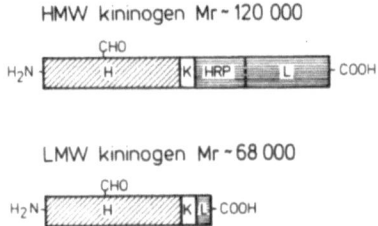

Fig. 6. Domain structure of human kininogens.
Kallikreins precisely excise the kinin segment
out of mammalian kininogens thus generating a
nicked polypeptide chain with the amino-terminal heavy chain and the carboxyterminal light
chain. The light chain of HMW kininogen comprises the histidine-rich peptide (HRP) which
was first recognized for bovine kininogens [7].
Note that the term "light chain" refers to the
complete C-terminal part of the human kininogen
molecule consecutive to the kinin segment, i.e.
we do not discriminate between a fragment 1.2
and the L-chain as done in the nomenclature of
bovine HMW kininogen [7].

PRODUCTION OF KININOGEN-DIRECTED ANTIBODIES

        Production of specific antibodies which selectively
recognize and discriminate between the two types of kininogens is hampered by the fact that IgG directed
against the heavy chains of either of the kininogens
yield a pattern of total immunological identity [23].
Two strategies may be elected to reach the final goal
of production of monospecific (polyclonal) antibodies
(Fig. 7):
(i) selection of monospecific IgG by stepwise immunoadsorption of antisera raised against HMW kininogen on
immobilized kininogens, and
(II) production of specific antibodies by immunization
with the specific fragments (L-chain) of the kininogens
(Fig. 7).

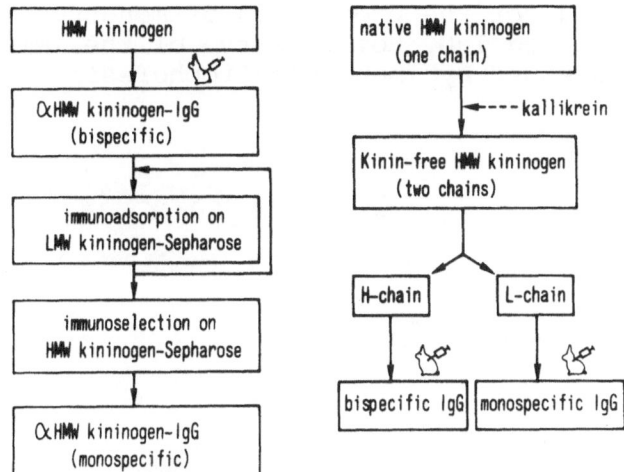

Fig. 7. Production of monospecific antibodies directed against human kininogens.
Production of anti-kininogen directed IgG was persued on two alternative routes: (i) immunization with HMW kininogen followed by immunoadsorption of cross-reacting antibodies (directed against the respective heavy chains) by repetitive immunoadsorption on immobilized LMW kininogen; final immunoselection of the unbound antibodies on HMW kininogen-Sepharose yields monospecific antibodies directed against native HMW kininogen (αHMW kininogen-IgG). (ii) Fragmentation of HMW kininogen by limited proteolysis with tissue kallikrein and subsequent isolation of the primary split products (H-chain, L-chain) followed by production of antisera against the L-chain; immunoselection of the resultant hyperimmunesera on L-chain-Sepharose thus yields monospecific antibodies directed against the carboxymethylated light chain of HMW kininogen.

        Monospecific antibodies directed against HMW kininogen are generated by removal of the cross-reacting (bispecific) antibodies via repetitive immunoadsorption of the hyperimmunesera on LMW kininogen-Sepharose, and a final immunoselection of the remaining monospecific antibodies on HMW kininogen-Sepharose. The analogous ra-

tionale is used for the production of monospecific LMW
kininogen-directed IgG. Homogeneity of the immunoselected IgG is assessed by immunoelectrophoresis (Fig. 8);

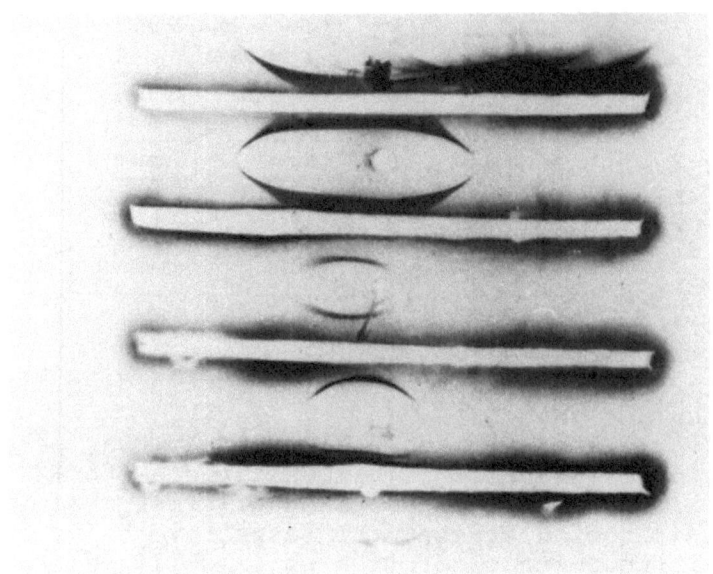

Fig. 8. Immunoelectrophoresis of antibodies directed against human kininogens.
Immunoelectrophoresis was carried out in a 1 % (w/v)
agar gel according to Graber and Williams [48]. The
wells (from bottom to top) were filled with rabbit
hyperimmuneserum against human HMW kininogen, the
IgG fraction of the anti-HMW kininogen antisera obtained by the method of Steinbuch [44], immunoselected IgG directed against human HMW kininogen, and
immunoselected antibodies directed against human
LMW kininogen (5 µl each). The graves were filled
with goat antiserum against total rabbit serum.
Formation of single precipitin bands by the immunoselected IgG indicated that the antibodies prepared
according to scheme A of Fig. 7 are largely homogeneous and free of contaminating plasma proteins.

for comparison, total rabbit antiserum and the total IgG
fraction obtained by the octanoic precipitation method
of Steinbuch [44] are simultaneouly run. Obviously, the
immunoselected antibodies (Fig. 8, lower panels) are
largely homogeneous and free of contaminating plasma
proteins. Precipitation of the competent antigen is
probed by electroimmunoassay. A linear relationship between the amount of antigen applied and the rocket height

is evident (Fig. 9). Specifity of the IgG fractions produced by either of the two approaches (Fig. 7) is tested
by enzyme-linked immunosorbent assay (ELISA).

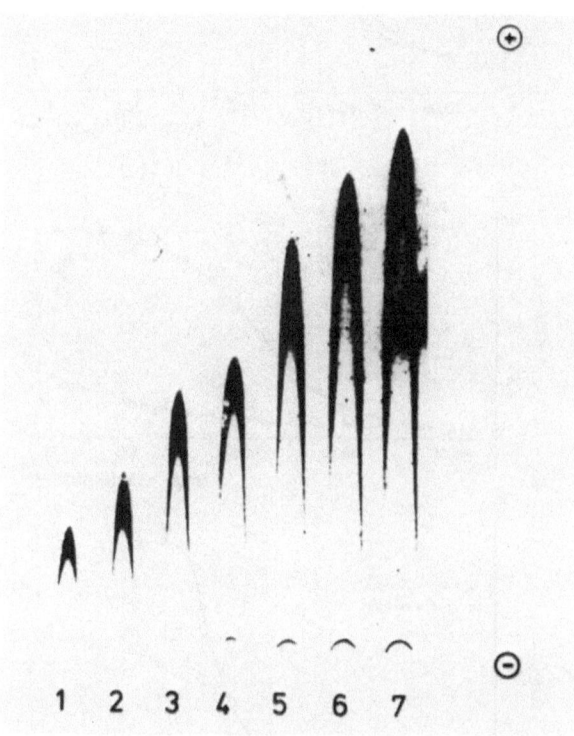

Fig. 9.   Electroimmunoassay of HMW kininogen.
Rocket electrophoresis was run in a 1 % (w/v)
agar gel including 0.5 % /w/v) antiserum against
HMW kininogen according to the method of Laurell
[49]. The wells were filled with increasing amounts
of HMW kininogen: 1, 0.2 µg; 2, 0.32 µg; 3, 0.48 µg;
4, 1.15 µg; 5, 1.76 µg; 6, 2.64 µg; 7, 4.0 µg each
per 5 µl. A linear correlation between the rocket
height and the amount of kininogen applied is obvious.

DEVELOPMENT OF SPECIFIC IMMUNOASSAYS

     Specificity of the immunoselected antibodies is
evaluated by their differential response to HMW and LMW
kininogen in the ELISA system. Titer plates are coated

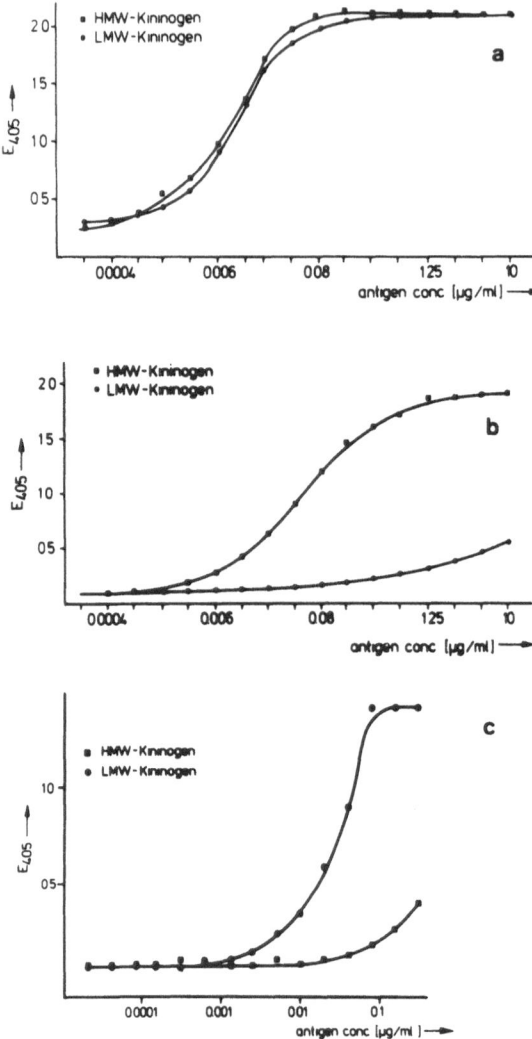

Fig. 10. Differential immunoassays for human kininogens.
a. Titer plates coated with anti-H-chain IgG
b. Titer plates coated with immunoselected antibodies
   specific for HMW kininogen
c. Titer plates coated with immunoselected antibodies
   specific for LMW kininogen
Largely congruent curves for HMW and LMW kininogen were
found with bispecific antibodies (a), while application
of immunoselected antibodies resulted in differential
curves (b,c). Quantitative analyses of the specific immunoassay revealed that ⩾ 90 % of the total IgG fractions
were monospecific for HMW kininogen (b) and LMW kininogen
(c), respectively.

with anti-kininogen IgG, then HMW kininogen or LMW kininogen is applied, and the sandwich is topped by specific
anti-kininogen directed IgG conjugated to horse radish
peroxidase [45]. Finally, the enzymic activity residing
in each cavity of the titer plate is tested with chromogenic substrate ABTS (2,2'-azino-di(3-ethylbenzthiazoline
sulfonic acid) in the presence of H_2O_2. Expectedly, application of H-chain directed antibodies yields largely
congruent curves for HMW and LMW kininogen (Fig. 10a) indicating that the coating antibodies reacts equally well
with both types of kininogens. Differential curves, however, are obtained for immunoselected anti-HMW kininogen-
IgG (Fig. 10b) which induce a full response with their
competent antigen (HMW kininogen), but react poorly with
LMW kininogen. From the inflection points of the reactions curves we calculate that the majority (> 95 %) of
the immunoselected immunoglobulins is monospecific for
HMW kininogen. Similar results are accomplished with
anti-LMW kininogen-directed IgG (Fig. 10c) thus suggesting that major part (> 90 %) of the total IgG fraction
is monospecific for LMW kininogen. Hence, antibodies of
high selectivity are now available to set up quantitative enzyme-linked immunosorbent assays. Present studies
are aimed at the validation of these test systems to
meet the final goal of selectively detecting and quanti-
tating kininogens in human plasma, secretions, and related body fluids in health and disease.

EPILOGUE

     Although our knowledge about the structure and func-
tion of human kininogens has rapidly expanded during the
past decade, some intriguing questions are still unsolved
and deserve further experimental efforts:
(1) Are the heavy chains of human kininogens identical,
    and if so, what is the organization of the gene(s)
    coding for kininogens?
(2) What is the size and the structure of the primary
    translation products of human kininogens, and are
    there precursors such as pre-pro-kininogens?
(3) What is the nature of the cells which synthesize,
    process, and secrete kininogens, and are their any
    extra-hepatic sites of kininogen biosynthesis?
(4) What is the turnover rate for human kininogens, and
    by which route are they eliminated?
(5) Are there any pathophysiological conditions with
    concomitant changes of the kininogen plasma concen-
    trations, and what is the role of the kininogens in
    the etiology of diseases related to a malfunction
    or deficiency of the kinin-forming pathway?

ACKNOWLEDGEMENT

    This work was supported by grants of the Deutsche
Forschungsgemeinschaft, Sonderforschungsbereich 51 (B/25)
and the Fond der Deutschen Chemischen Industrie (H.F.).
The authors thank Mrs. H. Gerstenberger for expertise
secretarial help. An extended version of this paper will
be published elsewhere (W. Müller-Esterl, G. Rauth, H.
Fritz, F. Lottspeich and A. Henschen, Human Kininogens,
in: Kininogenases, Kallikrein VI, G. L. Haberland, J. W.
Rohen, H. Fritz and P. Huber (eds.), Schattauer Verlag,
Stuttgart (1983)).

REFERENCES

1. R. Geiger, The kallikrein-kinin systems, in: "Pro-
     teinases and their inhibitors," V. Turk and L.
     Vitale, eds., Mladinska knjiga-Pergamon Press,
     Ljubljana/Oxford (1981).
2. E. G. Erdös, ed., Bradykinin, kallidin and kalli-
     krein. Handbook of experimental pharmacology,
     vol. XXV. Springer-Verlag, Berlin (1979).
3. M. Schachter, Kallikreins (kininogenases) a group
     of serine proteases with bioregulatory actions,
     Pharmacol. Rev. 31:1 (1980).
4. R. L. Heimark, and E. W. Davie, Bovine and human
     plasma kallikrein, Methods Enzymol. 80:157 (1981).
5. F. Fiedler, E. Fink, H. Tschesche, and H. Fritz,
     Porcine glandular kallikreins, Methods Enzymol.
     80:493 (1981).
6. R. Geiger, and H. Fritz, Human urinary kallikrein,
     Methods Enzymol. 80:466 (1981).
7. H. Kato, S. Iwanaga, and S. Nagasawa, HMW and LMW
     kininogen, Methods Enzymol. 80:172 (1981).
8. S. Iwanaga, H. Kato, T. Sugo, N. Ikari, N. Hashimo-
     to and S. Fujii, The kallikrein-kinin-system: a
     functional role of plasma kallikrein and kinino-
     gen in blood coagulation, in: "Biological funct-
     ions of proteinases," H. Holzer and H. Tschesche,
     eds., Springer-Verlag, Berlin (1980).
9. W. Müller-Esterl, B. Dittmann, H. Fritz, F. Lott-
     speich, and A. Henschen, Structural aspects of
     human kininogens, Adv. Exp. Med. Biol. 156: in
     press (1982).
10. A. R. Johnson, Effects of kinins on organ systems,
     l.c. [2], p. 357 (1979).
11. A. Nasjletti, and K. Malik, The renal kallikrein-
     kinin and prostaglandin systems interaction,
     Ann. Rev. Physiol. 43:597 (1981).

12. C. E. Odya, and T. I. Goodfriend, Bradykinin receptors, l.c. [2], p. 287 (1979).
13. T. H. Plummer, and E. G. Erdös, Human plasma carboxypeptidase N, Methods Enzymol. 80:442 (1981).
14. T. A. Stewart, J. A. Weare, and E. G. Erdös, Human peptidyl dipeptidase (converting enzyme, kininase II), Methods Enzymol. 80:450 (1981).
15. M. Komiya, H. Kato, and T. Suzuki, Bovine plasma kininogens. Structural comparison of high molecular weight and low molecular weight kininogens, J. Biochem. (Tokyo) 76:833 (1974).
16. E. Fink, R. Geiger, J. Witte, S. Biedermann, J. Seifert and H. Fritz, Biochemical, pharmacological, and functional aspects of glandular kallikreins, in: "Enzymatic Release of Vasoactive Peptides," F. Gross, and G. Vogel, eds., Raven Press, New York (1980).
17. C. Moriwaki, H. Fujimori and Y. Toyono, Intestinal kallikrein and the influence of kinin on the intestinal transport, in: "Kininogenases IV," G. L. Haberland, J. W. Rohen and T. Suzuki, eds., Schattauer Verlag, Stuttgart (1977).
18. A. Overlack, K. O. Stumpe, W. Zywzock, C. Ressel, and F. Druck, Defect of kallikrein-kinin system in essential hypertension and reduction of blood pressure by orally given kallikrein, Adv. Exp. Med. Biol. 120B:539 (1979).
19. W. Müller-Esterl and H. Fritz, Kallikreins, kinins and allergy, in: "New Trends in Allergy," J. Ring and G. Burg, eds., Springer Verlag, Berlin (1981).
20. M. Wicklmayer, G. Dietze, L. Mayer, T. Böttger, and J. Grunst, Evidence for an involvement of kinin liberation in the priming action of insulin on glucose uptake into skeletal muscle, FEBS Lett. 98:61 (1979).
21. K. Kurachi, I. Ohkubo, R. L. Heimark, K. Fujikawa, and E. W. Davie, Initiation of intrinsic blood coagulation, Adv. Exp. Med. Biol. 156: in press (1982).
22. R. J. Mandle, R. W. Colman, and A. P. Kaplan, Identification of prekallikrein and high molecular weight kininogen as a complex in human plasma, Proc. Natl. Acad. Sci. USA 73:4179 (1976).
23. D. M. Kerbiriou, N. Bouma, and J. H. Griffin, Immunochemical studies of human high molecular weight kininogen and of its complexes with plasma prekallikrein or kallikrein, J. Biol. Chem. 255:3952 (1980).

24. R. C. Wiggins, B. N. Bouma, C. G. Cochrane, and J.H.
     Griffin, Role of high molecular weight kininogen
     in surface-binding and activation of coagulation
     factor XI and prekallikrein, Proc. Natl. Acad.
     Sci. USA 74:4636 (1977).
25. J. H. Griffin, and C. G. Cochrane, Mechanism for the
     involvement of high molecular weight kininogen in
     surface-dependent reactions of Hageman factor,
     Proc. Natl. Acad. Sci. USA 73:2554 (1976).
26. Y. N. Han, M. Komiya, S. Iwanaga, and T. Suzuki, Stu-
     dies of the primary structure of bovine HMW kini-
     nogen. Amino acid sequence of a fragment ("histi-
     dine-rich peptide") released by plasma kallikrein,
     J. Biochem. (Tokyo) 77:55 (1975).
27. J. T. Dunn, M. Silverberg, and A. P. Kaplan, The
     cleavage and formation of activated Hageman factor
     by auto-digestion and by kallikrein, J. Biol.
     Chem. 257:1779 (1982).
28. T. Sugo, A. Hamaguchi, T. Shimida, H. Kato, and S.
     Iwanaga, Mechanism of surface-mediated activation
     of bovine F XII and plasma kallikrein, J. Biochem.
     (Tokyo) 92:689 (1982).
29. A. G. Scicli, R. Waldmann, J. A. Guimaraes, G. Scicli,
     O. A. Carretero, H. Kato, Y. N. Han, and S. Iwana-
     ga, Relationship between structure and correcting
     activity of bovine high molecular weight kininogen
     upon the clotting time of Fitzgerland-trait plasma,
     J. Exp. Med. 149:847 (1979).
30. A. P. Kaplan, M. Silverberg, J. T. Dunn, and G. Mil-
     ler, Mechanism for Hageman factor activation and
     role of high molecular weight kininogen as a coa-
     gulation cofactor, Ann. N. Y. Acad. Sci. 370:254
     (1981).
31. K. W. Rumpf, K. Becker, U. Kreusch, S. Schmidt, R.
     Vetter, and F. Scheler, Evidence for a role of
     plasma kallikrein in the activation of prorenin,
     Nature 283:482 (1980).
32. R. J. Mandle, and A. P. Kaplan, Hageman factor sub-
     strates. Human plasma prekallikrein: mechanism
     of activation by Hageman factor and participation
     in Hageman-factor dependent fibrinolysis, J. Biol.
     Chem. 252:6097 (1977).
33. N. R. Cooper, L. A. Miles, and J. H. Griffin, Acti-
     vation of purified C1, the first component of
     complement, by purified plasma kallikrein and
     plasmin, Thromb. Haemostasis 42:251 (1979).
34. R. G. DiScipio, The activation of the alternative

pathway of C3 convertase by human plasma kallikrein, Immunology 45:587 (1982).

35. R. G. Wiggins, P. C. Giclas, and P. M. Henson, Chemo-
tactiv activity generated from the fifth component
of complement by plasma kallikrein of the rabbit,
J. Exp. Med. 153:1391 (1981).

36. A. P. Kaplan, B. Ghebrehiwet, M. Silverberg, and J.
E. Sealey, The intrinsic coagulation-kinin pathway, complement cascades, plasma renin-angiotensin
system, and their interrelationships, CRC Crit.
Rev. Immunol. 3:75 (1981).

37. H. Kato, Y. N. Han, S. Iwanaga, N. Hashimoto, T. Sugo, S. Jujii and T. Suzuki, Mammalian plasma kininogens: their structure and function, in: "Kininogenases IV," G. L. Haberland, J. W. Rohen,
and T. Suzuki, eds., Schattauer-Verlag, Stuttgart
(1977).

38. B. Dittmann, A. Steger, R. Wimmer, and H. Fritz, A
convenient large-scale preparation of high molecular weight kininogen from human plasma, Hoppe-
Seyler's Z. Physiol. Chem. 362:919 (1981).

39. W. Müller-Esterl, M. Vohle-Timmermann, B. Boos, and
B. Dittmann, Purification and properties of human
low molecular weight kininogen, Biochim. Biophys.
Acta 706:145 (1982).

40. D. M. Kerbiriou, and J. H. Griffin, Human high molecular weight kininogen, J. Biol. Chem. 254:12020
(1979).

41. S. Schiffman, C. Mannhalter, and K. Tyner, Human high
molecular weight kininogen. Effects of cleavage
by kallikrein on protein structure and procoagulant activity, J. Biol. Chem. 255:6433 (1980).

42. B. Dittmann, F. Lottspeich, A. Henschen and H. Fritz,
Structural and functional aspects of human HMW ki-
ninogen, in: "Protides of biological fluids," H.
Peeters, ed., Pergamon Press, New York (1981).

43. T. Sugo, N. Ikari, H. Kato, S. Iwanaga, and S. Fujii,
Functional sites of bovine high molecular weight
kininogen as a cofactor in koalin-mediated activation of factor XII (Hageman factor), Biochemistry 19:3215 (1980).

44. M. Steinbuch, and R. Audran, The isolation of IgG
from mammalian sera with the aid of caprylic acid.
Arch. Biochem. Biophys. 134:279 (1969).

45. A. Voller, Heterogeneous enzyme-immunoassay and their
applications, in:"Enzyme-immunoassay," E. T.
Maggio, ed., CRC Press, Boca Raton (1980).

46. U. K. Laemmli, Cleavage of structural proteins during
     assembly of the head of bacteriophage T4, _Nature_
     227:680 (1970).
47. O. Vesterberg, Isoelectric focusing of proteins,
     Methods Enzymol. 22:389 (1971).
48. P. Graber, and C. A. Williams, Méthode permettant l'
     étude conjugée des propriétés électrophorétic et
     immunochimiques d'un mélange de proteines. Appli-
     cation du sérum sanguin, _Biochim. Biophys. Acta_
     10:193 (1953).
49. C. B. Laurell, Quantitative estimation of proteins
     by electrophoresis in agarose gels containing an-
     tibodies, _Anal. Biochem_. 15:45 (1966).
50. K. Mori, W. Sakamoto, and S. Nagasawa, Studies on
     human HMW kininogen. III. Cleavage of HMW kinin-
     ogen by the action of human salivary kallikrein,
     J. Biochem. (Tokyo) 90:503 (1981).
51. H. Edelhoch, Spectroscopic determination of trypto-
     phan and tyrosine in proteins, _Biochemistry_ 6:
     1948 (1967).

POSSIBLE INVOLVEMENT OF KININS IN MUSCLE ENERGY METABOLISM

G. Dietze, E. Maerker, C. Lodri, R. Schifman, M. Wicklmayr,
R. Geiger, E. Fink, I. Boettger, H. Fritz, and H. Mehnert

III. Med.Dept. (Metabolism and Endocrinology)Schwabing
City Hospital and Diabetes Research Unit, Dept.Clin.
Biochem. Munich University School of Medicine, Ludwig-
Maximilians-University, Munich

It is a well-known phenomenon since the end of the last century that muscle shortening as it is initiated in the forearm by mov-
ing one finger is accompanied by accelerated blood flow through the
contracting forearm-muscle[1] and by increased glucose uptake into
the working tissue[2], although cardiac output and the arterial concentrations of insulin and glucose do not change. That means that
these phenomenons are controlled locally.

What we know so far of this metabolic adaptation may be summarized as follows:
Energy for fibre shortening is provided by breakdown of ATP
which may be regenerated by creatine phosphate. However, the concentrations of these energy rich compounds are falling in muscle
tissue corresponding to the work load.

That their levels decrease slowly is due to the continuous
substrate utilization of the working tissue which gains ATP. Since
at the beginning, the external supply with substrates and oxygen
is missing, the working tissue will use that substrate for ATP-ge-
neration which is stored intracellulary as glycogen and may be utilized without oxygen, namely glucose. Since under anaerobic conditions one mole of glucose yields only three moles of ATP, the activity of glycogenolysis and, thus, of its key-enzyme phospho-fructokinase has to be accelerated 2000 fold[3]. This acceleration of phospho-
fructokinase can very nicely be shown in skeletal muscle in man:
The substrate of the enzyme namely fructose-6-phosphate falls
and its product namely fructose-di-phosphate rises[4].

As Sir Hans Krebs has pointed out in one of his last reviews on the control of this enzyme[5], the remarkable acceleration of its capacity during work cannot entirely be explained by the activators we know so far e.g. cyclic-AMP, inorganic phosphate and AMP.

If the energy necessary for contraction may not be provided by accelerated glycogenolysis and accelerated anaerobic glycolysis, a "signal" will be generated which induces an enlargement of the capillary net providing oxygen and glucose for oxidative energy yield gaining 35 moles of ATP more per mole of glucose as compared to anaerobic conditions. That means that the acceleration of capillary blood flow is absolutely indispensable for a greater energy yield and, thus, for the continuation of muscle performance.

Many naturally occuring compounds have been considered as mediators of functional vasodilation[6]. Some authors have attempted to ascertain a possible role also for kinins[7]. They were suggested to be suitable because they exhibited many of the phenomena that may be observed during muscle exercise at much lower concentrations than those known from other factors under discussion[8].

The potential role of kinins was further underlined by findings on a more molecular level: Frey showed in 1930, that the kallikrein-induced liberation of bradykinin from kininogen is increased by lowering the blood-pH as occurs during muscle work[9]. That kinin-degradation is reduced simultaneously has been observed by Edery and Lewis in 1962[10]. Thus, it was not surprising that Carretero and coworkers measured increased kinin release from working skeletal muscle in man[11] and this was furthermore strengthened by the finding that the level of the kinin precursor, namely kininogen, was reduced simultaneously in the deep venous blood of the working muscle tissue[12].

In order to prove whether kinins would be involved in the adaptation of muscle metabolism during muscle work, we decided to block the liberation of kinins by the application of a protease inhibitor during a well-defined muscular performance: If kinin liberation would play a role, the acceleration of blood flow and of muscle glucose uptake should be prevented (Fig. 1): When kinin liberation was inhibited during the infusion of aprotinin(R), blood flow, glucose and oxygen uptake were significantly reduced[13]. If, however, in another group synthetic bradykinin was applied in addition to aprotinin(R), the normal response could be reestablished again. The same results have been obtained under ischemia in the forearm: The normal adaptation of blood flow and metabolism was abolished by the inhibitor and normalized again by bradykinin[14].

These results together with the findings indicated above strengthened the view that kinins might very probably be involved in the "signal", which is responsible for the adaptation of energy metabolism in the working muscle tissue. Since prostaglandin-release is also increased from working muscle and since functional vasodilation could be prevented by indomethacine[R] pretreatment[15] corresponding to the reduction of prostaglandin release, kinins seem to exhibit their action via these second messengers. Thereby, kinins will rather play a role as tissue hormones, carrying the signal from the working muscle tissue to the precapillary sphincters for the enlargement of the capillary net, and prostaglandins will rather act as cellular mediators. However, it will further be an open question which product out of the anaerobic metabolism will work as the signal which induces kinin liberation.

This notion was underlined by the findings that synthetic bradykinin at small concentrations (6×10^{-10} mol/l) was able to increase the capillary blood flow and the glucose uptake also in the resting forearm muscle (Fig. 2)[15]. This insulin-like activity of physiological concentrations of bradykinin could be shown not only on carbohydrate metabolism but also on amino acid metabolism[17]. There were two ways how kinins might reveal their metabolic actions: (1) either they were working via insulin (a) by improvement of insulin-receptor binding due to the enlargement of the capillary net yielding new receptor sites, (b) by an involvement of kinins in the action of insulin, or (2) they were working separately from insulin. Since papaverine exhibited accelerated capillary blood flow as bradykinin but revealed no metabolic action, the way under (1,a) was not very probable. However, the way under (1,b) has become attractive, recently, when evidence has been presented that a plasma membrane protease is activated by insulin, generating a peptide with an apparent molecular weight of 1500 - 2500, capable of modulating the activity of target enzymes of insulin such as glycogen-synthetase and pyruvate-dehydrogenase in muscle and fat cell preparations [18 - 20]. These new data were underlined by the finding that trypsin which is well-known to release kinins from kininogen and exhibits insulin-like activity on glucose uptake into muscle tissue[21], was able to release the same peptide as insulin. It is furthermore interesting that the liberation of the peptide induced by insulin and by trypsin could be inhibited by aprotinin[(R)20]. That these results obtained in the in vitro experiment might be relevant for the in vivo situation could be suggested from in vivo experiments: Aprotinin[(R)] was able to reduce the effect of insulin on glucose uptake into the human forearm muscle by 50 % [22] (Fig.3).

In order to test the third possibility, namely that kinins might have an action independent of insulin, the following experiments have been performed in the isolated perfused rat heart: We used a modification of the technique introduced by Bleehen and Fischer[23]. The main differences were concerning the preperfusion periode which was longer to assure complete insulin-washout and

the glucose concentration which was lower as compared to[23]. The
tissue-levels of the energy-rich phosphates were constant during
sixty minute reperfusion which guaranteed stable conditions in
energy metabolism (Fig. 4). This was also underlined by steady concentrations of glycogen, glucose and lactate and also by persistent
rates of glucose uptake and lactate release.

Using a similar concentration of bradykinin as in the in vivo
experiment namely 6×10^{-10}Mol/l, we yielded the following results:
While insulin stimulated glucose uptake by 100 % and the lactate
release by 400 %, bradykinin, in contrast to what we would have
expected from the in vivo experiments, reduced the uptake of glucose and the release of lactate in the absence of insulin by about
50 %. Assuming the same release of glucose from glycogen, the reduced uptake of glucose had to be interpreted as a reduction of the
rate of glycolysis yielding smaller energy for cardiac performance.
However, there was no reduction of cardiac performance as proven by
normal frequency and contraction.

This discrepancy could be explained when we determined the myocardial substrate stores as glycogen and triglycerides and also the
intermediates of glycolysis (Fig. 5). Thereby, it got evident that
bradykinin exhibits not only significant reduction of the myocardial glycogen content but also of the intermediates in front of the
key-enzyme phospho-fructokinase, the substrate of which namely fruc-
tose-(6)-phosphate was reduced by about 50 % and its product namely
fructose-di-phosphate was increased by about 80 %, yielding a typical cross-over plot at this point of the glycolytic pathway.

From these findings obtained without insulin, one may suggest
kinins to have also an independent action on carbohydrate metabolism, namely an accelerating effect on glycolysis via the activation of its key-enzyme. Whether the accelerated glycogenolysis is
due to the effect of bradykinin on glycolysis leading indirectly
via reduction of glucose-6-phosphate to an acceleration of the
rate of glycogenolysis, or is due to a direct effect on glycogen
metabolism, cannot be differentiated so far. Accordingly, bradykinin seems just to mimick the well-known metabolic adaptation as
it occurs initially during muscle work and as described above. The
glucose necessary for the same energy yield is rather taken from
the intracellular stores than from outside the cell. At first
sight one would suggest from these findings that kinins have an
opposite effect in the in vitro system as compared to the in vivo
situation. However, if one considers that insulin is acting preferably on glucose transport and on glycogen synthesis and bradykinin is working preferably on glycolysis, it is not surprising
that the latter acts synergistically.

In summary then: From earlier data and the data presented

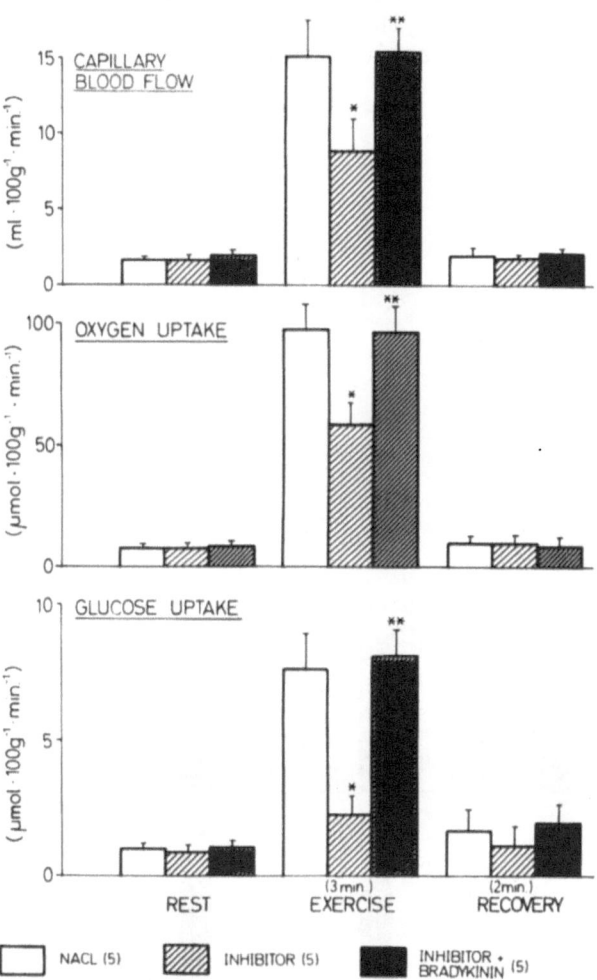

Fig. 1: Capillary blood flow, glucose and oxygen uptake in human forearm muscle during standardized forearm exercise and during recovery (from Dietze and Wicklmayr, 1977). Values are given as mean + SEM (n = 4 in each group) during the infusion of saline i.a. (empty columns), dito plus infusion of a protease inhibitor i.v. (aprotinin(R), hadged) and aprotinin(R) i.v. plus bradykinin i.a. (dark).
  * significant to saline, p < 0,05;
 ** to aprotinin(R) p < 0,025

here, we would like to suggest kinins to have a significant role in
the physiology of muscle energy metabolism, being involved in the
adaptation between energy expenditure and energy yield. Further
studies will have to prove, especially with the isolated phospho-
fructokinase whether kinins themselves or their possible mediators
namely prostaglandins are active at the enzyme level.

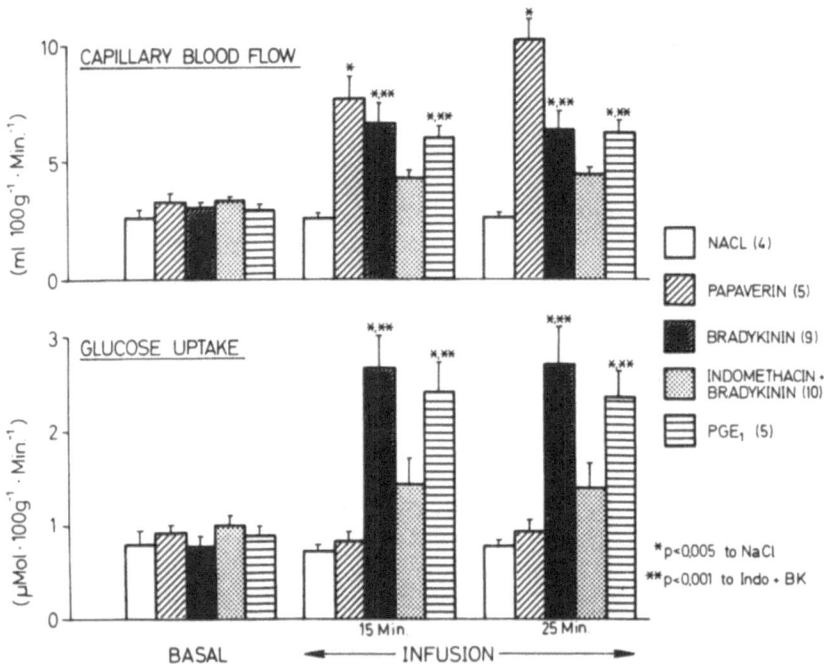

Fig. 2: Capillary blood flow and glucose uptake of the human forearm during intraarterial-infusion of the compounds indicated.
Values are given as the mean ± SEM. For details, see the text.

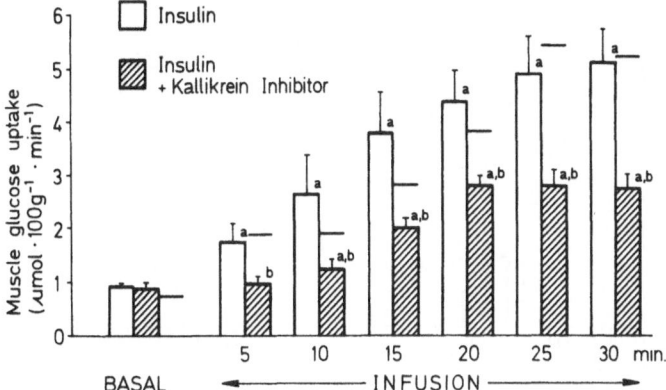

Fig. 3: Glucose uptake into the human forearm muscle and deep
venous concentrations of insulin during the arterial
infusion of insulin and of insulin + aprotinin(R)
(Dietze et al., 1978b). Values are indicated as the
mean ± SEM of 6 (insulin; empty columns) and of 5
(insulin + aprotinin) subjects, [a]signigicant basal
(p < 0,05 - 0,005); [b]significant to insulin (p < 0,05 -
0,005).

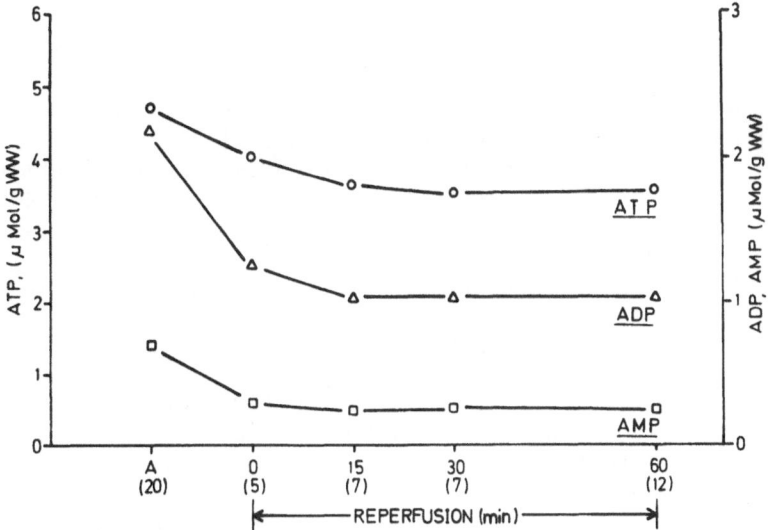

Fig. 4: ATP-, ADP- and AMP-content of the isolated perfused rat
heart after excision (A), preperfusion (O) and 15,30 and
60 minutes reperfusion. Values are indicated as the mean,
Numbers in brackets. A: means after excision in Nembutal(R),
O: after 10 min. or preperfusion.

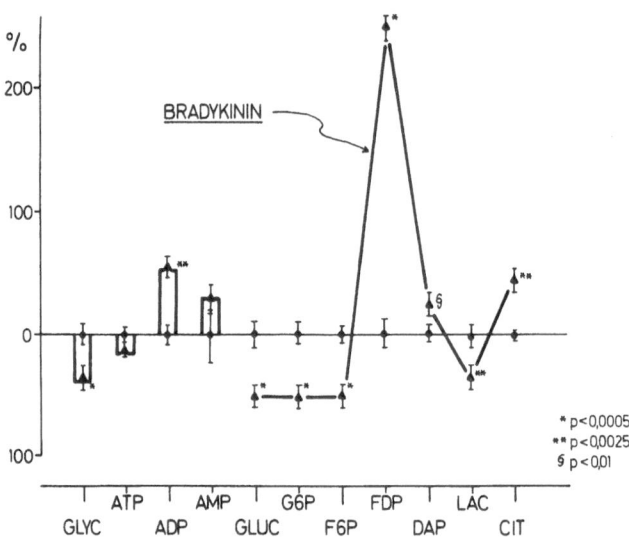

Fig. 5: Myocardial content of glycogen (GLYC) of the energy rich
         phosphates (ATP etc.) and glycolytic intermediates (G6P,
         F6P etc.) during the perfusion with bradykinin (BK).
         The values (n = 20) give the mean difference from the
         controls (n = 20) represented by the seroline. Signifi-
         cant changes are indicated by *p < 0.0005, **p < 0.0025,
         §p < 0.01.

REFERENCES

1. W. H. Gaskell, The changes of the blood-stress in muscles through
      stimulation of their nerves, J. Anat. 11:360 (1877).
2. M. A. Chaveau, and M. Kaufman, Expériences pour la détermination
      du coefficient de l'activité nutritive et respiratoire des
      muscles en repos et en travail, Compt. Rend. Acad. Sci. 104:
      1126 (1887).
3. E. A. Newsholme, Regulation of carbohydrate metabolism in muscle:
      Identification of regulatory enzymes, in: "Regulation in
      Metabolism," E. A. Newsholme, G. A. Start, eds., J. Wiley and
      Sons, London (1973).
4. S. Bergström, and L. A. Carlson, Inhibitory action of prosta-
      glandin E 1 on the mobilization of free fatty acids and
      glycerol from human adipose tissue in vitro, Acta Physiol.
      Scand. 63:195 (1965).
5. H. A. Krebs, The Pasteur effect and the relations between res-
      piration and fermentation, Essays Biochem. 8:1 (1972).
6. O. Hudlicka, "Muscle blood flow: Its relation to muscle meta-
      bolism and function," Swets and Zeitlinger, Amsterdam (1973).
7. S. M. Hilton, and G. P. Lewis, The relationship between glandu-
      lar activity, bradykinin formation and functional vasodilat-
      ion in the submandibular salivary gland, J. Physiol. 134:471
      (1956).

8. R. Zacest, and M. L. Mashford, Blood bradykinin levels in the human, Aust. J. Exp. Biol. Med. Sci. 45:89 (1967).
9. E. K. Frey, Zur Deutung der reaktiven Hyperämie, Arch. Klin. Chir. 162:334 (1930).
10. H. Edery, and G. P. Lewis, Kinin-forming activity and histamine in lymph after tissue injury, J. Physiol. 169:568 (1963).
11. O. Carretero, A. Nasjletti, and J. C. Fasciolo, The kinin content of human blood at rest and during vasodilation, Experientia 21:141 (1965).
12. K. Scharnagel und K. Greef, Änderungen des Kininogengehalts des Plasmas bei Hypoxie und lokaler Ischämie, Naunyn-Schmiedebergs Arch. Pharmacol. 253:81 (1966).
13. G. Dietze, and M. Wicklmayr, Evidence for a participation of the kallikrein-kinin system in the regulation of muscle metabolism during muscular work, FEBS. Lett. 74:205 (1977a).
14. G. Dietze, M. Wicklmayr, and L. Mayer, Evidence for a participation of the kallikrein-kinin system in muscle metabolism during hypoxia, Hoppe-Seyler's Z. Physiol. Chem. 358:633 (1977).
15. A. Kilbom, and A. Wennmalm, Endogenous prostaglandins as local regulators for blood flow in man: effect of indomethacin on reactive and functional hyperemia, J. Physiol. 257:109 (1976).
16. G. Dietze, M. Wicklmayr, L. Mayer, I. Boettger, and H. J. Funcke, Bradykinin and human forearm metabolism: inhibition of endogenous prostaglandin synthesis, Hoppe-Seyler's Z. Physiol. Chem. 359:369 (1978a).
17. R. Schifman, M. Wicklmayr, I. Boettger, and G. Dietze, Insulin-like activity of bradykinin on amino acid balances across the human forearm, Hoppe-Seyler's Z. Physiol. Chem. 361: 1193 (1980).
18. H. Hidaka, B. V. Howard, F. C. Kosmakos, R. M. Fields, J. W. Craig, P. H. Bennet, and J. Larner, Insulin stimulation of glycogen synthetase in cultured human diploid fibroblasts, Diabetes 29:806 (1980).
19. F. L. Kiechle, N. Kotagal, D. A. Popp, and L. Jarett, Isolation from rat adipocytes of a chemical mediator for insulin activation of pyruvate dehydrogenase, Diabetes 29:852 (1980).
20. M. P. Czech, and J. N. Fain, Antagonism of insulin action on glucose metabolism in white fat cells by dexamethasone, Endocrinology 91:224 (1981).
21. P. Rieser, and C. H. Rieser, Enzymatic hypoglycemia in alloxan diabetic rats, Proc. Soc. Exp. Biol. Med. 116:669 (1964).
22. G. Dietze, M. Wicklmayr, I. Boettger, and L. Mayer, Inhibition of insulin action on glucose uptake into skeletal muscle by a kallikrein-trypsin inhibitor, Hoppe-Seyler's Z. Physiol. Chem. 359:1209 (1978b).
23. N. M. Bleehen, and R. B. Fischer, The action of insulin in the isolated rat heart, J. Physiol. 123:260 (1954).

STRUCTURE AND FUNCTION OF NATURAL INHIBITORS

AS ANTAGONISTS OF PROTEINASE ACTIVITIES

Harald Tschesche

University of Bielefeld
Faculty of Chemistry
Universitätsstraße
D-4800 Bielefeld

INTRODUCTION

Enzymic activities are controlled and regulated in nature by various processes which can be subheaded under the notions given in Table 1. For proteolytic enzymes the processes 1-5 are known to be valid while the other processes have so far only been established for other enzyme classes.

Table 1. Regulation of enzyme activities.

1. De novo Biosynthesis
2. Zymogen Activation
3. Competitive Inhibitors
4. Disulfide-thiol Interchange in Complexes
5. α_2-Macroglobulin (Thioestertrap)
6. Intramolecular S-S-Reduction
7. Allostery
8. Covalent Modification (Phosphorylation)

The control of proteolytic activities by proteinase inhibitors can follow either one of the three different mechanisms of action, notions 3-5, Table 1. Thus, the inhibitors will be grouped according to their mechanism of action regardless of molecular weight, source of appearance, or protein properties. Examples of each of these types of inhibitors will be given and briefly discussed. The important group of plasma inhibitors will be classified into the type of competitive inhibitors from which the low molecular weight inhibitors have so far been the most excessively studied and well characterized class of proteinase-inhibitors. Most emphasis will be given to the newly discovered type of inhi-

bitors regulating proteolytic activity by a disulfide-thiol interchange reaction with the proteinase depending on the redox potential of the medium.

COMPETITIVE INHIBITORS

    The large number of blood plasma inhibitors, Table 2, belong
to the mechanistic class of competitive inhibitors. Only α_2-macro-
globulin (see below) and β_1-anticollagenase (see below) follow a
different mechanism of action. The α_1-proteinase inhibitor (formerly α_1-antitrypsin), the most predominant plasma inhibitor , is the
physiological antagonist of the leukocyte elastase. The reactive
site residue is methionine which can be oxidized under the action
of peroxidases (or cigarette smoke) which then· drastically changes
the kinetics of the reaction (this book, other contributions).

Table 2.   Survey of the protease inhibitors from human plasma.

Name	Concentration (mg/100 ml)	Mr	Amino acid content %	Carbo- hydrate content %	Chain compo- sition (number)
α_1-antitrypsin	290 + 45	54,000	86	12	1
α_1-antichymo- trypsin	49 $\pm$ 7	69,000	73	25	1
inter-α-trypsin inhibitor	50	160,000	90	8	
antithrombin III	24 $\pm$ 2	65,000	85	13	1
C_1-inactivator	24 $\pm$ 3	104,000	65	35	1
α_2-antiplasmin	7 $\pm$ 1	70,000	87	13	1
inhibitor of plas- minogen activation		80,000			
α_2-macroglobulin	260 $\pm$ 70	725,000	92	8	4
β_1-anticollagenase		30,000	95	2	1
α_2-thiol proteinase inhibitor		90,000			

    The principle of the inhibitory action by this type of inhi-
bitor is the strong substrate-like association of the inhibitor
protein with the enzyme at the substrate binding site, thus mask-
ing the enzyme, Fig. 1.

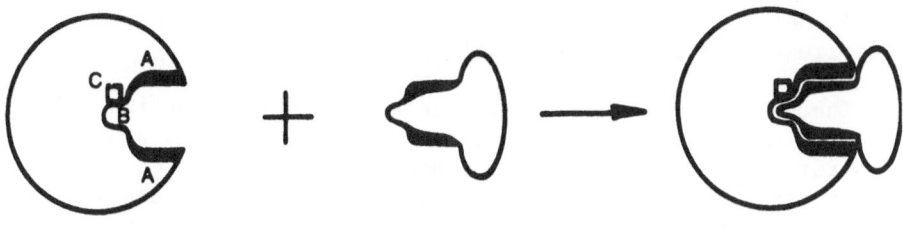

Active enzyme   **Inhibitor**   **Inactivated enzyme**

Fig. 1.   Association of enzyme and its inhibitor at the catalyti-
          cally active site of the enzyme, leading to inhibition. A,
          Contact area and substrate binding site; B, specificity
          pocket accommodating the inhibitor reactive site residue
          P_1; C, catalytically active site residue, i.e., serine in
          serine hydrolases.

At least ten protein classes are known to date from various animal
and plant sources which follow this general principle, Table 3.

Table 3.   Families of Protein Proteinase Inhibitors of Serine
           Proteinases with Low Molecular Weight.

I.	Bovine pancreatic trypsin inhibitor (Kunitz) family
II.	Pancreatic secretory trypsin inhibitor (Kazal) family
III.	Streptomyces subtilisin inhibitor family
IV.	Soybean trypsin inhibitor (Kunitz) family
V.	Soybean proteinase inhibitor (Bowman-Birk) family
VI.	Potato I inhibitor family
VII.	Potato II inhibitor family
VIII.	Ascaris trypsin inhibitor family
IX.	Other families

   The mechanism of this association during which a substrate-like
cleavage can occur at the reactive site peptide bond-P_1-P_1'-
(Schechter-Berger notation [1]) has been the subject of detailed
investigations on several low-molecular weight inhibitors (I) from
various tissues. The inhibitor with the cleaved peptide bond (modified inhibitor I*) still exhibits inhibitory activity. Thus, the
general mechanism of enzyme inhibitor interaction follows the minimal mechanism:

$$E + I \rightleftharpoons L \rightleftharpoons C \rightleftharpoons X \rightleftharpoons L^* \rightleftharpoons E + I^*$$

Fig. 2.   I = virgin inhibitor
          I*= modified inhibitor
          L, L* = non covalent Complexes
          X = Intermediate
          C = stable Enzyme-Inhibitor Complex
          E = Enzyme

Several books [2-6] and review articles [7-10]have appeared that
summarize the status of this field.

   Though the reversibility of the association according to this
mechanism has not yet been demonstrated for the plasma proteinase
inhibitors numerous indirect evidence exists that indicates a
similar behaviour. A very strong indication that e.g. the inhibitory action excerted by the inter-α-trypsin inhibitor from human
plasma follows the general competitive minimal mechanism comes from
its structure, Fig. 2. The amino acid sequence analysis has revealed that the inter-α-trypsin inhibitor contains two domains that are
homologous in sequence to the bovine trypsin inhibitor (Kunitz,
BPTI)[11], for which this mechanism has been shown to be valid[12].

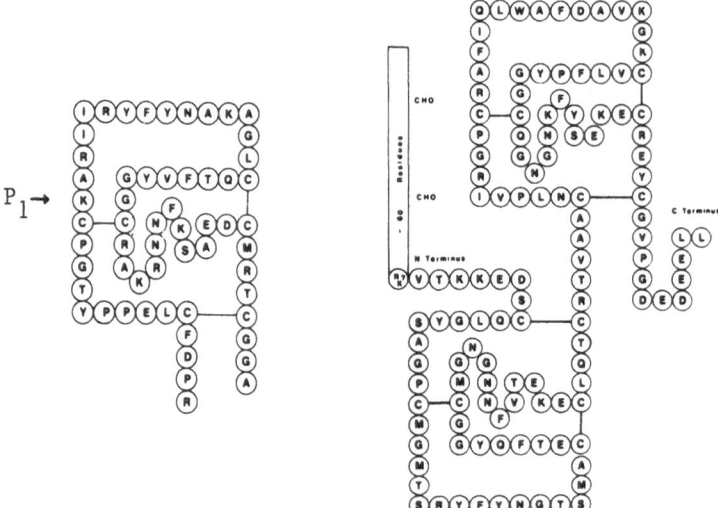

Fig. 3.   Covalent structures of the two domain inter-α-trypsin in-
          hibitor fragment and the bovine inhibitor (Kunitz), shown
          on left according to [11], P_1 = Lys 15 reactive site.

   The inhibitory specificity of a particular proteinase inhibitor is obviously determined by

1) its surface subsite contacts complementary to the enzyme's substrate binding region and

2) dominated by the type of P_1 reactive site residue.

This has been demonstrated by detailed studies on the inhibitors
specificities within several classes of homologous inhibitors having P_1 natural reactive site mutations, e.g. in the classes of Ka-
zal and ovoinhibitors [3,7,8,9,12] or Kunitz-inhibitors and avian
ovomucoids [3,7,8,9,13]. This has been further evidenced by single
amino acid replacements of the P_1 residue either by means of enzymatic mutation [14,15] or "chemical mutation" [16] of the reactive
site residue P_1.

Such replacements have been carried out in our laboratory on the example of the bovine trypsin inhibitor (Kunitz, BPTI). The experiments have been described [15,16] and the results confirmed the expectations that significant specificity changes could be obtained within the same class of competitive serine proteinase inhibitors by just a single amino acid replacement at the P_1 position. The P_1 positions at which "mutations" have been performed are indicated in Fig. 3. The primary inhibitory specificity as determined by the type P_1 residue clearly correlates with the splitting specificity of the inhibited proteinase. The apparent changes in primary inhibitory specificity are roughly presented in Table 4. Inhibitors with Lys, or Arg at the P_1 position are trypsin inhibitors, those with Tyr, Phe, Trp, (or Leu) are good chymotrypsin inhibitors, while inhibitors with Ala and Val in the P_1 position are specific for pancreatic and leukocytic elastase respectively. A detailed study on the correlation of amino acid sequence with inhibitor activity and specificity of protein inhibitors of serine proteinases is under way in the laboratory of Laskowski, Jr. [18].

Table 4.   Inhibitory Specificity of P_1 Substituted Bovine Inhibitor (Kunitz).

Reactive Site	Inhibited Proteinase			
Residue P_1	Trypsin	Chymotrypsin	Pancreas-	PMN-Elastase
Lys	+++	++	o	(+)
Arg	+++	++	o	o
Trp	++	+++	o	o
Phe	++	+++	o	o
Leu	+	+++	+++	+++
Met	+	+++	+	+
Val	+	+	+	+++
Ala	+	+	+	+
Gly	+++	++	o	o

Inhibition: +++, strong; ++, medium; +, little; o, not detectable.

α_2-MACROGLOBULIN INHIBITOR

The inhibition of proteinases from all four mechanstic classes, i.e. serine, thiol, aspartic and metallo enzymes, has been observed with α_2-macroglobulin. A characteristic feature of α_2-macroglobulin is that binding of the proteinase onto the inhibitor protein abolishes the characteristic proteolytic features of the enzyme against high molecular weight substrates but still allows the enzyme to hydrolyse low molecular weight (peptide) substrates.

To explain this peculiar feature the entrapment-hypothesis had been proposed [19] involving a subunit cage-like structure. This concept has been developed further and the complete amino acid sequence of one of the four identical subunits of the quarternary structure has been resolved. This includes the interesting bait region of the sequence, Fig. 4 [20]. It contains all the cleavage sites for the primary proteolytic attack at the α_2-macroglobulin molecule by any of the different types of proteinases (in which almost all have been shown to lead to proteolytic cleavage). The model proposed by the Danish group involves four steps for the interaction between enzyme and inhibitor, Fig. 5 [21].

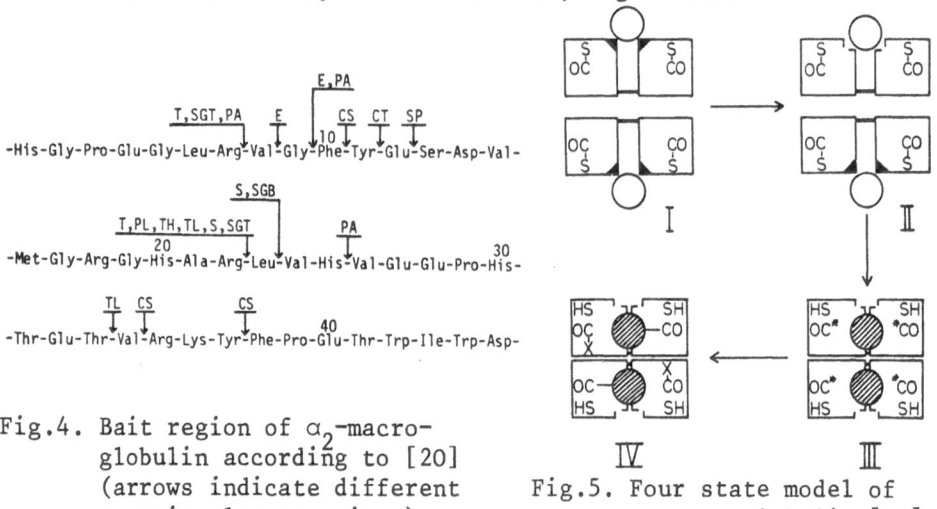

                                                        E,PA
        T,SGT,PA   E   CS CT SP
-His-Gly-Pro-Glu-Gly-Leu-Arg-Val-Gly-Phe-Tyr-Glu-Ser-Asp-Val-
                                    10
                S,SGB
    T,PL,TH,TL,S,SGT   PA
            20   30
-Met-Gly-Arg-Gly-His-Ala-Arg-Leu-Val-His-Val-Glu-Glu-Pro-His-

    TL CS   CS
                        40
-Thr-Glu-Thr-Val-Arg-Lys-Tyr-Phe-Pro-Glu-Thr-Trp-Ile-Trp-Asp-

Fig.4. Bait region of α_2-macro-
       globulin according to [20]
       (arrows indicate different
       enzymic cleavage sites).

Fig.5. Four state model of
       α_2-macroglobulin [21].

State   I. Noncovalent binding of the enzyme(s) and formation
            of the initial α_2-macroglobulin complex (binding
            mode I).
State   II. Cleavage of two "bait" region sequences by one
            trypsin molecule.
State III. A "nascent" α_2-macroglobulin inhibitor is formed
            which can entrap two trypsin molecules due to con-
            formational change. The intrachinar thiol esters
            become cleaved resulting in appearance of free
            SH-groups.
State   IV. The acyl groups react with part of the entrapped
            trypsin molecules to form covalent α_2-macroglobu-
            lin trypsin complexes.

The final complex formed from the proteinase and α_2-macroglo-
bulin is (with very few exceptions) a covalent complex, in which
the enzyme is linked via a peripheral surface side chain to the
γ-carboxylate of the accepting γ-glutamyl thiol ester in α_2-macro-
globulin [21,22].

It is interesting to note that the sequence containing the reactive γ-glutamyl thiol ester in α_2-macroglobulin is highly homologous to a sequence in the α-chain of the complement component C 3 containing a similar reactive γ-glutamyl thiol ester site [23].

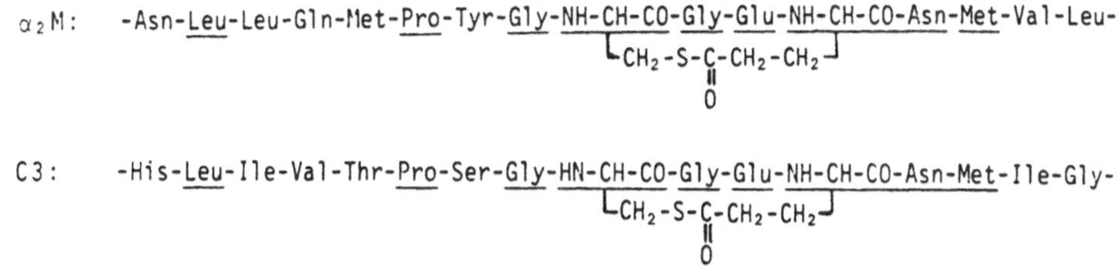
α2M: -Asn-Leu-Leu-Gln-Met-Pro-Tyr-Gly-NH-CH-CO-Gly-Glu-NH-CH-CO-Asn-Met-Val-Leu-
CH2-S-C-CH2-CH2
O
C3: -His-Leu-Ile-Val-Thr-Pro-Ser-Gly-HN-CH-CO-Gly-Glu-NH-CH-CO-Asn-Met-Ile-Gly-
CH2-S-C-CH2-CH2
O

Fig. 6. Partial sequences with the γ-glutamyl thiol ester of α_2-macroglobulin (upper row) and complement component C 3 (lower row) according to [23].

INHIBITORS ACTING BY THIOL-DISULFIDE INTERCHANGE

Latent Collagenase is an Enzyme-Inhibitor-Complex.

Recently we have demonstrated that the latent collagenase from human leukocytes is an enzyme-inhibitor complex [24], which is activatable to yield active enzyme and a collagenase inhibitor by a variety of compounds, e.g. active and inactivated cathepsin G, trypsinogen, immunoglobulin G, insulin, relaxin, cystine and oxidized glutathione [25]. These compounds have been indicated to act by surface accessible disulfide bonds. Oxidized glutathione is one of the most effective activators of the latent collagenase and therefore it has been proposed, that it may function as one of the physiological activators according to a mechanism schematically given in Fig. 7.

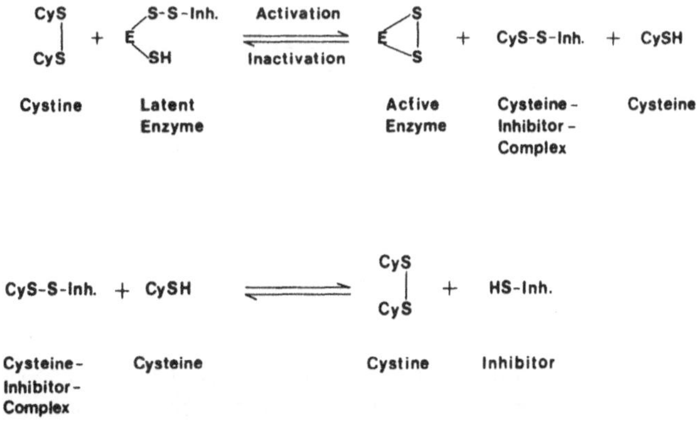
CyS
CyS
S-S-Inh.
E
SH
Activation
Inactivation
S
E
S
CyS-S-Inh.
CySH
Cystine
Latent Enzyme
Active Enzyme
Cysteine-Inhibitor-Complex
Cysteine
CyS-S-Inh.
CySH
CyS
CyS
HS-Inh.
Cysteine-Inhibitor-Complex
Cysteine
Cystine
Inhibitor

Fig. 7. Schematic representation of the mixed disulfide complex of enzyme and inhibitor (latent enzyme) and its activation by disulfide-thiol interchange reaction with cystine [24].

This general non proteolytic activation mechanism is supported
by the findings that:
1. The molecular weight of M_r 91,000 for the latent enzyme is
   reduced to M_r 64,000 for the active enzyme by release of a
   M_r 25,000 inhibitor [24-27].
2. A free SH-group on the inhibitor is essential for inhibitory activity [24,25].
3. A 1:1 stoichiometric complex is formed between enzyme and
   inhibitor as indicated by
      a) titration experiments
      b) amino acid composition of the complex and
      c) SDS-gel electrophoresis experiments [26].
4. 179-203-di-S-carboxymethyl-trypsinogen is incapable of activating the latent enzyme [24].
5. Carboxyamidomethylation of the latent enzyme prevents activation of the latent enzyme [26].

ACTIVATION COUPLED TO THE GLUTATHIONE CYCLE

The optimal activation of the latent leukocyte collagenase by
oxidized glutathione can be demonstrated in vitro [25]. Generation
of collagenolytic activity against type I collagen is straight
forward dependent on the amount of oxidized glutathione [25],Fig.8.

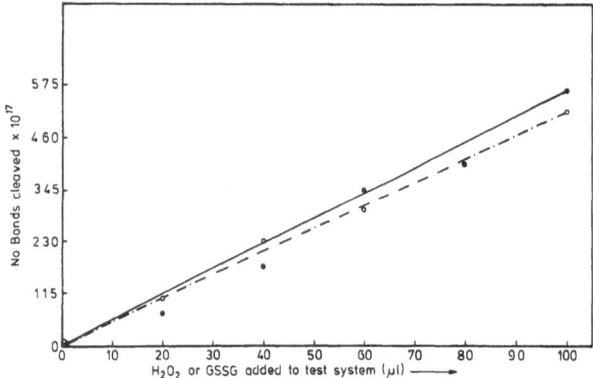

Fig. 8. Activation of latent collagenase via the "peroxidase/hy-
         drogen peroxide" system with increasing amounts of hydro-
         gen peroxide, or GSSG according to [25].

The activation can be coupled to the in vitro generation of oxidized glutathione from the reduced form by glutathione peroxidase and
hydrogen peroxide as the oxidizing agent, Fig. 8. The presence of
the peroxidase is a prerequisite, but glutathione peroxidase can be
substituted by any other peroxidase, e.g. lactoperoxidase, horse
radish peroxidase or myeloperoxidase, which is a major constituent
of the polymorphonuclear leukocyte comprising 5 % of the dry weight
of the cell [28]. Other peroxidases may also be employed. Instead
of adding hydrogen peroxide to the system, it may be generated from
glucose and glucose oxidase in vitro [25].

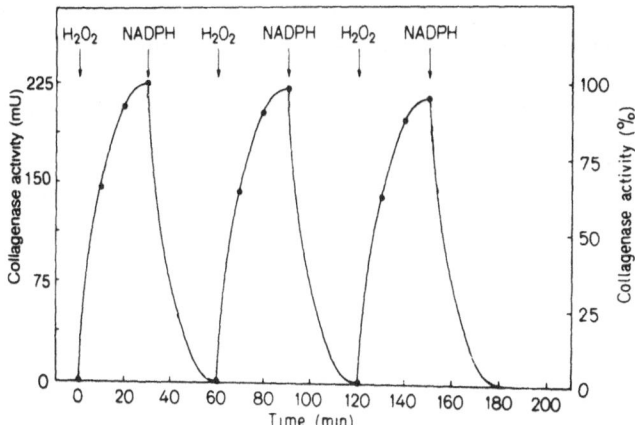

Fig. 9. Activation and inactivation of latent collagenase via H_2O_2/
NADPH/glutathione peroxidase/glutathione reductase system
according to [25].

The activation via addition of oxidizing equivalents in the
presence of a peroxidase can be inhibited by poisining the peroxidase with sodium azide, or by including an excess of glutathione
reductase or catalase in the system [25]. The activation can be reversed by adding excess amounts of reducing equivalents such as
NADH or NADPH to the system. This provides regulatory possibilities
for a system comprising of latent collagenase and the physiological
concentrations in the resting leukocyte of glutathione peroxidase,
glutathione reductase and reduced and oxidized glutathione [25].
Addition of oxidizing equivalents (H_2O_2) activates this system,
while addition of reducing equivalents (NADPH) abolishes the proteolytic activity [25], Fig.10. The process of activation and inactivation may be reversed and repeated several times, Fig.10. The

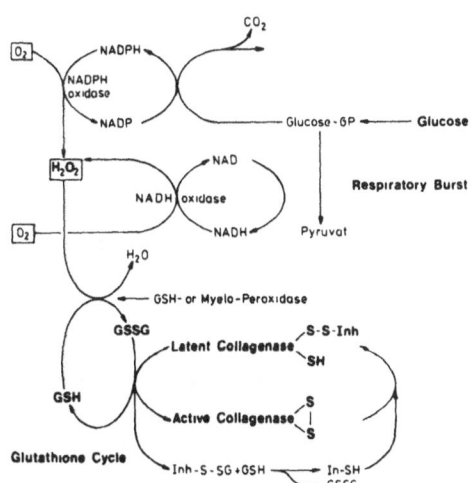

Fig.10. Scheme of the hypothetical regulation
of human leukocyte
collagenase via the
redox potential of
the glutathione cy-
cle linked to glu-
cose and oxygen
metabolism of the
cell according to
[25].

proteolytic activity may be regulated via the redox potential in
the system, the redox state of the glutathione cycle [25], Fig. 10.

It is well established that phagocytosis by polymorphonuclear
leukocytes is associated with the production of hydrogen peroxide,
superoxide and other oxygen radicals [29-32]. This process is coupled to an increase in oxygen metabolism and to an increased glucose
catabolism through the hexose monophosphate (HMP) shunt. The overall process is known as the "respiratory burst", in which oxidizing
equivalents are produced in excess to that required for microbici-
dal function. A shift of the ratio NADPH/NADP from 9 to approximately 3 in phagocitizing leukocytes was determined [31] and the level of reduced glutathione decreased at least by 25 - 30 % in 10 -
15 min [32]. The glutathione cycle in leukocytes is an integral
link between the phagocytosis-associated respiratory burst and the
increase in hexose monophosphate shunt activity. Thus, it seems
highly plausible that activation of the latent enzyme either in the
phagosome or in the extracellular space (several mechanisms for enzyme release are known) might be linked in vivo to the phagocytizing activity of the cell and finally to its glucose metabolism.

SIMILAR SYSTEMS IN OTHER TISSUES

A collagenase inhibitor has been described in human plasma
[33]. The inhibitor has been isolated to apparent homogeneity and
characterized as a protein of M_r 30,000 exhibiting a different
amino acid composition than the leukocyte collagenase inhibitor
[34]. This plasma collagenase inhibitor also contains a free sulfhydryl group essential for inhibitory activity. The same mechanism
for inactivation of human leukocyte collagenase by this inhibitor
has been demonstrated [34]. The inactivated (latent) enzyme is subject to reactivation by the same activators or activating H_2O_2/per-
oxidase/glutathione system as described above for the latent leukocyte collagenase. The plasma collagenase inhibitor seems to act as
the plasma antagonist for neutralizing leukocyte collagenase. No
transfer of collagenase from the inhibitor complex to α_2-macroglo-
bulin was observed [35].

These observations are good indications that latent collagenases from other tissues, such as e.g. fibroblasts, synovial cells
and others operate by a similar system, i.e. that the enzyme's la-
tency is due to a disulfide-linked inhibitor. It was indeed demonstrated that tumor tissue produces an inhibitor for pancreatic
trypsin which operates by the thiol/disulfide interchange mechanism described [36].

Another system, presently under investigation in our laboratory, which seems to function by the same disulfide interchange
mechanism between an inhibitor and the active enzyme seems to be

the gelatinase-inhibitor system in human leukocytes. This system
is located in another different compartment of the cell. While the
leukocyte collagenase-inhibitor system is obtained from secondary
(specific) granules the gelatinase is obtained from tertiary granules, a type of secretory organelle [37].

The inhibition of proteolytic enzymes by inhibitors with a
surface exposed sulfhydryl group reacting with an accessible disulfide bond of the enzyme in a thiol/disulfide interchange mechanism
seems to be a general phenomenon of which the first example is described here with the latent collagenase [24-27]. This implies the
possibility of some of these systems being regulated by the effective redox potential present in the cell.

POSSIBLE SIGNIFICANCE IN PATHOLOGICAL PROCESSES

A number of pathophysiological states are known which are due
to an imbalance between the amounts of active proteinase released
and the availability of the inhibiting antagonist to block the deleterious action of the enzyme.

The most well known disease in this respect is pulmonary emphysema in which the leukocyte elastase is not sufficiently blocked
by α_1-proteinase inhibitor [38]. This situation is discussed in
other chapters of this book. Degradation of arterial elastin by
neutrophils may be involved in the pathogenesis of arteritis.
Attack upon cartilage proteoglycan by elastase and in a concerted
action by collagenases could play a role in certain inflammatory
joint disorders such as arthritides, gout, and the accumulation of
chemotactic immune complexes in joints.

Our limited investigations of synovial fluids from normal subjects and from patients who have undergone surgery of the hip
joint due to rheumatoid arthritis revealed that all contained latent and varying amounts of active collagenase (Tschesche et al.,
in preparation). Changes in the content of glutathione in erythrocytes from arthritic patients have been reported [39]. The decrease in serum sulfhydryl groups in rheumatoid arthritis has been
noticed [40]. An unbalanced redox state with high levels of NADPH
was observed in rheumatoid cells [41,42]. This could be an indication that perturbations of the redox potential within the cells or
in the interstital space are of causal significance for the deleterious activation of some latent enzymes such as collagenase and
gelatinase. Perturbation of cell membranes by leukocyte elastase
may contribute to cocarcinogenic sequela of recurrent inflammation.
The production of imflammatory kinins seems to be the result of a
specific leukocyte kininogenase (kallikrein) which is not identical to tissue or plasma kallikrein [43].

Understanding of the control and regulatory phenomena of proteinase activation under pathophysiological conditions would enable us to develop suitable therapeutical approaches.

ACKNOWLEDGEMENT

The author is indepted to generous support of his work by the Stiftung Volkswagenwerk (Hannover, FRG), by the Deutsche Forschungsgemeinschaft (Bonn-Bad Godesberg, FRG), and the Fonds der Chemischen Industrie (Frankfurt, FRG).

REFERENCES

1. I. Schechter, and A. Berger, On the size of the active site in proteases. I. Papain, Biochem. Biophys. Res. Commun. 27:157 (1967).
2. R. Vogel, I. Trautschold und E. Werle, "Natürliche Proteinasen-Inhibitoren," Thieme Verlag, Stuttgart (1966).
3. H. Fritz,and H. Tschesche, eds., "Proceedings of the International Research Conference on Proteinase," de Gruyter, Berlin-New York (1971).
4. H. Fritz, H. Tschesche, L. J. Greene,and E. Truscheit, eds., "Proteinase Inhibitors,"-Bayer Symposium V, Springer Verlag, Berlin - Heidelberg - New York (1974).
5. E. Reich, D. B. Rifkin,and E. Shaw, eds., "Proteases and Biological Control," Cold Spring Harbor Conf. on Cell Proliferation, Vol. 2, Cold Spring Harbor Laboratory, Cold Spring Harbor (1975).
6. H. Holzer,and H. Tschesche, eds., "Biological Functions of Proteinases," Springer Verlag, Berlin - Heidelberg - New York (1979).
7. M. Laskowski,Jr.and R.W. Sealock, Protein-proteinase-inhibitors - Molecular aspects, in: "The Enzymes, Vol III," Academic Press, New York (1971).
8. H. Tschesche, Biochemie natürlicher Proteinase-Inhibitoren, Angew. Chem. 86:21; Angew. Chem. Internat. Ed. 13:10 (1974).
9. M. Laskowski,Jr., and I. Kato, Protein inhibitors of proteinases, Annu. Rev. Biochemistry 49:593 (1980).
10. H. Tschesche, Proteolytic enzyme inhibitors, in: "Medicinal Chemistry Advances,"De Las Heras, F.G., Vega, S., eds., Pergamon Press, Oxford (1981).
11. E. Wachter, and K. Hochstrasser, Kunitz-type proteinase inhibitors derived by limited proteolysis of the inter-α-trypsin inhibitor, III, Z. Physiol. Chem. 360:1305 (1979).
12. U. Quast, J. Engel, E. Steffen, H. Tschesche, and S. Kupfer, Stopped-flow kinetics of the resynthesis of the reactive site peptile bond in kallikrein inhibitor (Kunitz) by beta-trypsin, Biochemistry 17:1675 (1978).

13. H. Tschesche, S. Kupfer, R. Klauser, E. Fink, and H. Fritz, Structure, biochemistry and comparative aspects of mammalian seminal plasma acrosin inhibitors, in: "Protides Biol. Fluids, 23 Colloqu.," H. Peeters, ed., Pergamon Press, Oxford-New York (1976).

14. I. Kato, W. J. Lohr, and M. Laskowski, Jr., Evolution of Avian Ovomncoids, in: "Proc. 11th FEBS Meeting Regulatory Proteolytic Enzymes and their Inhibitors," S. Magnusson et al., eds., Pergamon Press, Oxford (1978).

15. R. W. Sealock, and M. Laskowski, Jr., Enzymatic replacement of the arginyl by a lysyl residue in the reactive site of soybean trypsin inhibitor, Biochemistry 8:3703 (1969).

16. H. Jering, and H. Tschesche, Replacement of lysine by arginine, phenylalanine and tryptophan in the reactive site of the bovine trypsin-kallikrein inhibitor (Kunitz) and change of the inhibitory properties, Eur. J. Biochem. 61:453 (1976).

17. H. R. Wenzel und H. Tschesche, "Chemische Mutation" durch Aminosäureaustausch im reaktiven Zentrum eines Proteinase-Inhibitors und Änderung seiner Hemmspezifität, Angew. Chem. 93:292; Angew. Chem. Internat. Ed. 20:295 (1980).

18. M. Laskowski, Jr., M. W. Empie, I. Kato, W. J. Kohr, W. Ardelt, W. C. Bogard, Jr., E. Weber, E. Papamokos, W. Bode, and R. Huber, Correlation of amino acid sequence with inhibitor activity and specify of protein inhibitors of serine proteinases, in: "Structural and Functional Aspects of Enzyme Catalysis," H. Eggerer, R. Huber, eds., Springer Verlag, Berlin - Heidelberg - New York (1981).

19. A. J. Barrett, and P. M. Starkey, The interaction of α_2-macroglobulin with proteinases. Characteristics and specificity of the reaction, and a hypothesis concerning its molecular mechanism, Biochem. J. 133:709 (1973)

20. B. Mortensen, L. Sottrup-Jensen, M. F. Hansen, T. E. Petersen, and St. Magnusson, Primary and secondary cleavage sites in the bait region of α_2-macroglobulin, FEBS letters 135:295 (1981).

21. L. Sottrup-Jensen, T. E. Petersen, and St. Magnusson, Mechanisms of proteinase complex formation with α_2-macroglobulin, FEBS letters, 128:127 (1981).

22. L. Sottrup-Jensen, T. E. Petersen, and St. Magnusson, A thiol ester in α_2-macroglobulin cleaved during proteinase complex formation, FEBS letters 121:275 (1980).

23. L. Sottrup-Jensen, F. M. Hansen, S. B. Mortensen, T. E. Petersen, and St. Magnusson, Sequence location of the reactive thiol ester in human α_2-macroglobulin, FEBS letters 123:145 (1981).

24. H. W. Macarney, and H. Tschesche, Latent collagenase from human polymorphonuclear leukocytes and activation to collagenase by removal of an inhibitor, FEBS letters 119:327 (1980).

25. H. Tschesche, and H. W. Macartney, A new principle of regulation of enzymic activity, Eur. J. Biochem. 120:183 (1981).

26. H. W. Macartney, and H. Tschesche, Latent and active human polymorphonuclear leukocyte collagenases: Isolation, purification and characterization, Eur. J. Biochem. 130:71 (1983).

27. H. W. Macartney, and H. Tschesche, The collagenase inhibitor from human polymorphonuclear leukocytes. Isolation, purification and characterization, Eur. J. Biochem. 130:79 (1983).

28. J. Schultz, and K. Kaminker, Myeloperoxidase of the leucocyte of normal human blood. I. Content and localization, Arch. Biochem. Biophys. 96:465 (1962).

29. B. M. Babior, R. S. Kipnes, and J. T. Cumutte, Biological defense mechanisms. The production by leukocytes of superoxide, a potential bacterial agent, J. Clin. Invest. 52:741 (1973).

30. G. Y. N. Iyer, M. F. Islam, and I. H. Quastel, Biochemical aspects of phagocytosis, Nature 192:535 (1961).

31. M. Zatti, and F. Rossi, Early changes of hexose monophosphate pathway activity and of NADPH oxidation in phagocytizing leucocytes, Biochim. Biophys. Acta 99:557 (1965).

32. S. Klebanoff,and R. A. Clark, "The Neutrophil," North-Holland Publ. Comp. Amsterdam – New York – Oxford (1978).

33. D. E. Woolley, D. R. Robert, and I. M. Evanson, Small molecular weight β_1 serum-protein which specifically inhibits human collagenases, Nature 261:325 (1976).

34. H. W. Macartney, and H. Tschesche, Characterisation of β_1-anticollagenase from human plasma and its reaction with polymorphonuclear leukocyte collagenase by disulfide/thiol interchange, Eur. J. Biochem. 130:85 (1983).

35. H. W. Macartney, and H. Tschesche, Interaction of β_1-anticollagenase from human plasma with collagenases from various tissues and competition with α_2-macroglobulin, Eur. J. Biochem. 130:93 (1983).

36. F. S. Steven, V. Podrazky, and S. Itzhaki, Evidence for the presence of a trypsin inhibitor within rabbit and mouse tumour cells, Biochim. Biophys. Acta 483:211 (1977).

37. B. Dewald, U. Bretz, and M. Baggiolini, Release of gelatinase from a novel secretory. Compartment of human neutrophils, J. Clin. Invest. 70:518 (1982).

38. C. B. Laurell, and S. Eriksson, The electrophoretic α_1-globulin pattern of serum in α_1-antitrypsin deficiency, Scand. J. Clin. Invest. 15:132 (1963).

39. E. Munthe, E. Kass,and E. Jellum, Glutathione in erythrocytes: a parameter of change in disease activity and response to drugs in rheumatoid arthritis, in: "Inflammation: Mechanism and Treatment," Proc. 4th Int. Meet. Future Trends Inflammation – London 1980, D. A. Willoughby, J. Giround, eds., MTP Press, Lancaster (1980).

40. M. Haataja, Serum sulphydryl levels in rheumatoid patients
     treated with gold thiomalate and pluicillamine, <u>Scand</u>. J.
     <u>Rheumatol</u>. 4:7 (1975).
41. J. Chayen, L. Bitensky, R. G. Butcher, and L. W. Poulter, Redox
     control of lysosomes in human synovia, <u>Nature</u> 222:281 (1969).
42. J. Chayen, L. Bitensky, R. G. Butcher, and B. Cashman, The
     effect of experimentally induced redox changes on human
     rheumatoid and non-rheumatoid synovial tissue in vitro,
     <u>Beitr. Pathol</u>. 149:127 (1973).
43. U. Lüpke, W. Rautenberg, and H. Tschesche, Kininogenase from
     human polymorphonuclear leukocytes, <u>in</u>: "Recent Progr.
     Kinins," Internat. Conf. Kinin 81, Munich, H. Fritz, H.
     Dietze, F. Fiedler, G. L. Haberland, eds., Birkhäuser Ver-
     lag, Basel - Boston - Stuttgart (1982).

OXIDATION OF ALPHA-1-PROTEINASE INHIBITOR: SIGNIFICANCE FOR PATHOBIOLOGY

James Travis, Keith Beatty, and Nancy Matheson

Department of Biochemistry
University of Georgia
Athens, Georgia 30602

The control of proteolytic enzyme activities in blood and other tissues is exerted, primarily, by nine plasma proteins (1). These inhibitors, which represent more than 10% of the total protein in plasma, have a broad spectrum of regulatory functions in controlling coagulation, fibrinolysis, complement activation, and connective tissue turnover.

One of the most thoroughly studied of these plasma inhibitors is alpha-1-proteinase inhibitor (α-1-PI), a protein which has been implicated in controlling elastin turnover. This inhibitor has been well characterized and found to have the following characteristics: 1. It is a glycoprotein containing 394 amino acid residues and whose sequence has been determined (2). 2. It has three carbohydrate side chains of variable composition (3). 3. Isoforms of the protein have been found which are due primarily to aberrations in carbohydrate structure but also to changes in amino acid sequence (3,4). 4. The inhibitor has a methionyl-seryl residue at its reactive site and interaction at this position is most rapid with neutrophil elastase (5,6). 5. Oxidation of the reactive site methionine either chemically or enzymatically reduces the rate of interaction with neutrophil elastase (6-8). 6. Deficiencies in the inhibitor levels in plasma due to improper synthesis result in the development of familial emphysema (9).

Based on the above data it has been suggested that oxidation of α-1-PI might result in a pseudo genetic defect by reducing active inhibitor levels and thereby allowing elastolysis of connective tissue and subsequent development of lung disease. This report summarizes data currently available in support of this hypothesis.

KINETIC INVESTIGATIONS

The second order association rates (k_{ass}) of α-1-PI with several proteinases have been measured in order to determine the relative importance of the suggested roles of this inhibitor in controlling various proteolytic activities. As can be seen in Table 1, neutrophil elastase reacts most rapidly, with the highest k_{ass} ever reported for a proteinase-proteinase inhibitor system, about 15 times slower than the diffusion-controlled limit for the rate of collision of two proteins. This k_{ass} is significantly reduced when the inhibitor is oxidized, either chemically with N-chlorosuccinimide (7) or enzymatically with neutrophil myelo-peroxidase (8). However, with the exception of porcine pancreatic elastase, which does not react at all with α-1-PI$_{ox}$, most of the proteinases form complexes, albeit at a slower rate.

Table 1.   Effect of Oxidation on Alpha-1-Proteinase Inhibitor Association Rates[*]

Enzyme	Alpha-1-PI	Alpha-1-PI (ox)
Human Neutrophil Elastase	6.5×10^7	3.0×10^4
Human Chymotrypsin I	5.4×10^6	1.0×10^6
Human Neutrophil Cathepsin G	4.1×10^5	6.4×10^2
Porcine Pancreatic Elastase	1.0×10^5	zero
Human Cationic Trypsin	1.1×10^4	3.0×10^3
Human Plasmin	1.0×10^2	zero
Human Thrombin	4.8×10^1	zero

[*]$M^{-1} sec^{-1}$, at $37°$, pH 8.0

In order for α-1-PI to carry out its suggested physiological function of controlling neutrophil elastase the inhibitor must be able to react rapidly enough with this enzyme to prevent degradation of elastin, particularly in the lung. From the values of k_{ass} and the normal concentration of inhibitor in plasma (130 mg/100 ml), the half-time of association ($t_{\frac{1}{2}}$) in vivo can be approximated and for this system is 0.61 msec (10). Such a rapid $t_{\frac{1}{2}}$ strongly indicates the physiological role of α-1-PI in controlling this enzyme, its importance in protecting lung tissue, and the deleterious results when genetic deficiencies of the inhibitor occur. Furthermore, when the inhibitor is oxidized an increase in $t_{1/2}$ to 1330 msec occurs which is more than 2000-fold longer than that with the native inhibitor.

It is obvious from the above results that while complex formation is significantly retarded when the inhibitor is oxidized, it does occur. The question then remains as to the stability of such a complex. By adding alpha-2-macroglobulin ($\alpha_2 M$) to α-1-PI-proteinase complexes it has been found that some dissociation does occur, both in the forward reaction to give inactive inhibitor and free enzyme but also by reversing the original reaction to give native inhibitor and free enzyme. In fact, by trapping released enzyme in an $\alpha_2 M$ complex it is possible to obtain dissociation rates for both α-1-PI-elastase complexes and α-1-PI$_{ox}$-elastase complexes.

The results obtained indicate a doubling in the dissociation rate of α-1-PI$_{ox}$-elastase complexes relative to the native inhibitor and a K_{eq} for the interactions of 3.29×10^{-14}M for native α-1-PI vs 1.55×10^{-10}M for α-1-PI$_{ox}$. Thus, complete oxidation of α-1-PI would be expected to result in a 4000-fold increase in free elastase. To be fair, and although this difference appears to indicate a large increase in the amount of free elastase if α-1-PI is oxidized, the amount of free enzyme which would be expected is still vanishingly small (1.5×10^{-10}M). However, these calculations do not take into account the presence of a substrate for elastase, such as elastin, a major structural component of the lungs. Furthermore, since this protein is insoluble, no kinetic data is available on its interactions with elastase in order to examine any competitive effects which this protein might show.

EFFECT OF OXIDATION ON ELASTIN PROTECTION

The comparative protective effects of α-1-PI and α-1-PI$_{ox}$ for elastin degradation can be measured by preincubation of native or oxidized inhibitor, enzyme, and elastin, in varying combinations and then measuring elastin hydrolysis. The data obtained should then give a comparison as to the protective capacity of α-1-PI$_{ox}$, relative to that of native α-1-PI.

As shown in Figure 1A, either pre-incubation of native inhibitor with elastase or addition to an elastase-elastin mixture markedly reduced elastolysis (experiments 3 and 4) relative to the control without inhibitor (experiment 1), whereas α-1-PI$_{ox}$ showed only slight protection, even after ten minutes pre-incubation (experiment 2). Some elastolysis still occurred in those samples containing native inhibitor, presumably because

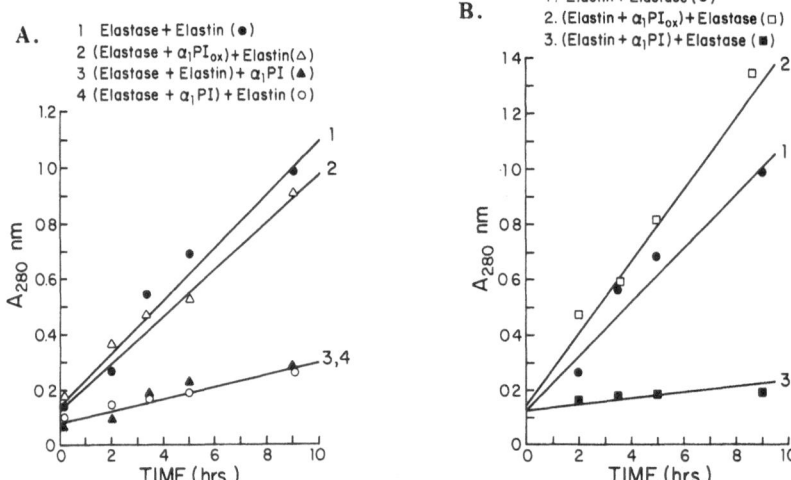

Figure 1.   The effect of native or oxidized α-1-PI on elastin
            solubilization by neutrophil elastase. Enzyme, native
            or oxidized α-1-P, and human lung elastin, were incu-
            bated in varying combinations as shown, inhibitor and
            enzyme being in equal concentration. After various time
            intervals, aliquots were removed, centrifuged, and the
            filtrates read at 280nm. A. Preincubation of elastase
            with inhibitor. B. Preincubation of elastin with in-
            hibitor.

elastin effectively competes for elastase slowly released from complexes. Similar results were obtained when the elastin was preincubated with either α-1-PI or α-1-PI$_{ox}$ and then enzyme added (Figure 1B). In fact, it appears that α-1-PI$_{ox}$ actually caused an increase in elastolysis relative to the control (experiments 1 and 2) while α-1-PI was an effective inhibitor (experiment 3). Clearly, α-1-PI$_{ox}$ offers little or no protection to elastin degradation.

CIGARETTE SMOKE AND ALPHA-1-PI OXIDATION

It has been clearly shown that oxidizing agents present in cigarette smoke are capable of inactivating α-1-PI (11,12). Furthermore, α-1-PI$_{ox}$ has also been detected in the lung lavage fluid of smokers (13), as well as in rheumatoid synovial fluid (14), the former occurring presumably because of chemical and enzymatic oxidation and the latter to enzymatic oxidation alone. Since the oxidized form of the inhibitor cannot inactivate porcine pancreatic elastase (Table 1), but still complexes pancreatic trypsin, an assay has been set up to determine the presence of α-1-PI$_{ox}$ in blood (15). If no α-1-PI$_{ox}$ is present the ratio of trypsin inhibitory activity to elastase inhibitory activity should be very nearly 1.0. However, any α-1-PI$_{ox}$ in the sample would alter the balance, showing an increased trypsin inhibitory activity: elastase inhibitory activity ratio. Indeed, when a sample of normal plasma was submitted to chemical oxidation this ratio approached infinity with over 99% of the inhibitor being oxidized.

When the serum of normal individuals was compared to those who had just finished smoking it was found that the ratio had shifted for the latter group and that an average of 23% α-1-PI was present in an oxidized form. Such results are compatible with other data indicating that at least 60% of the α-1-PI in lung lavage fluid of smokers is inactive towards pancreatic elastase and is oxidized (13). Presumably, the time elapsed after smoking, prior to either serum withdrawl or lung lavage, is important in these determinations because of dilution of the α-1-PI$_{ox}$ with native inhibitor in the whole body pool and, perhaps, because of increased inhibitor synthesis or oxidized inhibitor uptake.

DISCUSSION

From the foregoing data it can be concluded that 1), oxidation of α-1-PI markedly affects its rate of interaction with its target enzyme, neutrophil elastase; 2), α-1-PI$_{ox}$ shows little protection of lung elastin against degradation by neutrophil elastase; 3), chemical and biological oxidants, in particular those in cigarette smoke and neutrophils, rapidly oxidize α-1-PI, and the latter can be detected in serum, bronchial lavage fluids, and rheumatoid synovial fluid. Based on these results it would seem logical to conclude that in pulmonary emphysema development associated with smoking, a "pseudo-genetic defect" is created whereby serum inhibitor levels remain normal or slightly elevated while inhibitory activity is markedly decreased. This latter effect lowers the protective capacity normally served by native α-1-PI against enzymes released from phagocytic cells, attracted to this organ in response to the inhaled foreign materials in the smoke. As a result, both elastolytic and oxidative enzymes are released, either through phagocytosis or cell death, with the ultimate result being abnormal elastin degradation (elastase) and continued oxidation of fresh α-1-PI diffusing into the lung (myeloperoxidase). Without interruption of this process by cessation of inhalation of foreign matter (quitting smoking; inhalation of cleaner air), the ultimate result is reduction of normal lung function by this combination of oxidation and proteolysis.

ACKNOWLEDGEMENTS

This research was supported, in part, by grants from the National Heart, Lung, and Blood Institutes, and The Council for Tobacco Research-U.S.A.

REFERENCES

1. J. Travis and G. S. Salvesen, Human plasma proteinase inhibitors, Ann. Rev. Biochem., in press (1983).
2. K. Kurachi, T. Chandra, S. J. F. Degen, T. T. White, T. L. Marchiano, S. L. C. Woo, and E. W. Davie, Clonine and sequence of cDNA coding for α_1-antitrypsin, Proc. Natl. Acad. Sci. U.S.A. 78:6826 (1981).
3. L. Vaughan and R. Carrell, α_1-antitrypsin isoforms: Structural basis of microheterogeneity, Biochemistry International 2:461 (1981).
4. J. O. Jeppson, Amino acid substitution Glu-Lys in α_1 antitrypsin PiZ, FEBS Letters 65:195 (1976).
5. D. A. Johnson and J. Travis, Structural evidence for methionine at the reactive site of human α_1-proteinase inhibitor, J. Biol. Chem. 253:7142 (1978).
6. K. Beatty, J. Bieth, and J. Travis, Kinetics of association of serine proteinases with native and oxidized α_1-proteinase inhibitor and α_1-antichymotrypsin, J. Biol. Chem. 255:3931 (1980).
7. D. A. Johnson and J. Travis, The oxidative inactivation of human α_1-proteinase inhibitor. Further evidence for methionine at the reactive center, J. Biol. Chem. 254:4022 (1979).
8. N. R. Matheson, P. S. Wong, and J. Travis, Enzymatic inactivation of human α_1-proteinase inhibitor by neutrophil myeloperoxidase, Biochem. Biophys. Res. Commun. 88:402 (1979).

9.   C. B. Laurell and S. Eriksson, The electrophoretic α_1-globulin pattern of serum in α_1-antitrypsin deficiency, Scand. J. Clin. Lab. Invest. 15:132 (1963).

10. J. G. Bieth, Pathophysiological interpretation of kinetic constants of protease inhibitors, Bull. Europeen de Physiopath. Respiratoire 15:Suppl 183 (1980).

11. A. Janoff, H. Carp, D. K. Lee, and R. T. Drew, Cigarette smoke inhalation decreases α_1-antitrypsin activity in rat lung, Science 206:1313 (1979).

12. J. E. Gadek, G. A. Fells, and R. G. Crystal, Cigarette smoking induces functional antiprotease deficiency in the lower respiratory tract of humans, Science 206:1315 (1979).

13. H. Carp, F. Miller, J. R. Hoidal, and A. Janoff, Potential mechanism of emphysema: α_1-proteinase inhibitor recovered from lungs of cigarette smokers contains oxidized methionine and has decreased elastase inhibitory capacity, Proc. Natl. Acad. Sci. U.S.A. 79:2041 (1982).

14. P. S. Wong and J. Travis, Isolation and properties of oxidized α_1-proteinase inhibitor from human rheumatoid synovial fluid. Biochem. Biophys. Res. Commun. 96:1449 (1980).

15. K. Beatty, P. Robertie, R. M. Senior, and J. Travis, Determination of oxidized α_1-proteinase inhibitor in serum. J. Lab. Clin. Med. 100:186 (1982).

IN VIVO SIGNIFICANCE OF KINETIC CONSTANTS OF

MACROMOLECULAR PROTEINASE INHIBITORS

Joseph G. Bieth

INSERM - Unité 237 - Faculté de Pharmacie
Université Louis Pasteur
67048 Strasbourg (France)

INTRODUCTION

Many physiological and pathological processes are controlled by proteinases whose activity is itself controlled by macromolecular inhibitors. In recent years a growing number of investigators have been interested in searching for novel inhibitors in biological fluids or tissues. It is commonly thought that if such inhibitors are active in vitro, they will also be active in vivo. This is not always the case. To decide whether an inhibitor plays a physiological role or not, two sets of data are required : the in vivo inhibitor concentration and the kinetic parameters describing the proteinase/inhibitor interaction. This paper describes briefly these parameters, their measurement and their use for defining the role of proteinase inhibitors in health and disease.

DEFINITION OF KINETIC PARAMETERS

The number and the nature of the macroscopic kinetic constants describing the enzyme/inhibitor interaction depend upon the "nature" of the inhibition process. Many inhibitors (I) react with proteinases (E) to form a reversible complex (E·I) according to scheme 1 :

$$E + I \underset{k_{off}}{\overset{k_{on}}{\rightleftharpoons}} E \cdot I \qquad (1)$$

where k_{on} and k_{off} are the rate constants for complex formation and dissociation respectively. Such enzyme/inhibitor systems are also characterized by K_i the equilibrium dissociation constant of the complex (i.e. the inhibition constant). The three constants are related : Ki = koff/kon (2).

Scheme 1 is certainly an oversimplification of the reaction pathway of most reversible inhibitors[1,2]. The above three ma-croscopic parameters are nevertheless valuable constants for in-ferring the physiological role of the inhibitors.

Most of the plasma proteinase inhibitors undergo peptide bond hydrolysis upon reaction with proteinases. The resultant comple-xes are highly stable to denaturants[3]. Such inhibitors may be con-sidered as "irreversible" ones and are characterized by kon only.

The inhibition is said to be temporary if active proteinase can escape spontaneously from the complex due to the breakdown of the inhibitor[1]. The most simple and general scheme is as follows :

$$\text{E + I} \underset{koff}{\overset{kon}{\rightleftharpoons}} \text{E·I} \xrightarrow{k} \text{E + inhibitor products (3)}$$

DETERMINATION OF KINETIC PARAMETERS

It is usually easy to measure kinetic parameters. Theoretical background, some practical points and useful references will be given below.

Association rate constant kon

For an irreversible inhibitor, the rate of association $- \dfrac{d(E)}{dt}$ is given by :

$$- \frac{d(E)}{dt} = \text{kon (E)(I)} \qquad (4)$$

whereas for a reversible one it is given by :

$$- \frac{d(E)}{dt} = \text{kon (E)(I)} - \text{koff(EI)} \qquad (5)$$

If koff is not too high (say koff $\leqslant 10^{-3}$ sec^{-1}) eq. 4 may be used even for reversible inhibitors (at least during the initial stages of the inhibition process). With equimolar concentrations of enzyme and inhibitor, eq. 4 integrates into :

$$\frac{1}{(E)} = \frac{1}{(E°)} + \text{kon.t} \qquad (6)$$

where (E°) and (E) are the enzyme concentrations at time zero and at any time t. (E) values are easily measured with substrate. kon will then be determined from a plot of $1/(E)$ vs.t (e.g.ref. 4,5).

If koff is too high, the integrated form of eq. 5 must be used[6]. This equation is somewhat complex and needs computer assistance. It yields both kon and koff values. On the other hand, kon may be determined by competition experiments between two inhibitors and an enzyme[5].

Dissociation rate constant koff
_____

In order to measure koff, equilibrium 1 must be shifted towards the left by one of the following means :
    - extensive dilution of the complex
    - trapping of free enzyme by a substrate[5], a synthetic irreversible inhibitor[7], another macromolecular proteinase inhibitor[8], α_2-macroglobulin[9] which forms enzymatically active complexes with most proteinases
    - trapping of free inhibitor with another proteinase[10]

Dissociation is then monitored by measuring the time-dependent appearence of (E) or (I) using direct or indirect enzymatic methods. Its rate is given by :

$$ -\frac{d(EI)}{dt} = koff\ (EI) \qquad (7) $$

which integrates into :

$$ \ln \frac{(EI)}{(EI^\circ)} = -\ koff.t \qquad (8) $$

an equation from which koff may be determined graphically (e.g. ref. 5).

Equilibrium dissociation constant Ki
_____

This is determined by measuring the enzymatic activity of mixtures formed of constant amounts of proteinase and increasing amounts of inhibitor. Equilibrium must be attained before addition of substrate. If the inhibition curve is a straight line, Ki cannot be determined directly from the data. If the curve is concaved, Ki may be determined either directly from the curve by the new method of Dixon[12] or by a replot[11] of the data in accordance with eq. 9 :

$$ \frac{(I^\circ)}{1-a} = \frac{Ki}{a} + (E^\circ) \qquad (9) $$

where a is the fractional activity (rate in presence of inhibitor/
rate in its absence). These two methods may however, yield Ki
values with large errors. It is safer[13] to use a computer-assisted
nonlinear regression method based for instance on eq. 10

$$a = 1 - \frac{(E°)+(I°)+Ki - \left[(E°)+(I°)+Ki)^2-4(E°)(I°)\right]^{\frac{1}{2}}}{2(E°)} \qquad (10)$$

which yields Ki and its standard deviation. The Ki value used for
the first interation may be obtained by one of the above graphical
methods or by using the following formula[11] :

$$Ki = \frac{a}{1-a}(I°) - a(E°) \qquad (11)$$

For reasons given elsewhere[11], the familiar Lineweaver-Burk or
Dixon plots are useless for tight-binding inhibitors.

     If there is competition between the inhibitor and the substra-
te used to measure inhibition, the inhibition constant determined
by the above methods in an apparent one (Ki(app)) and is related
to the true Ki through eq. 12 :

$$Ki(app) = Ki (1 + (S°)/Km) \qquad (12)$$

Km must therefore be known in this case. Competition between
inhibitor and substrate may be diagnosed by methods outlined
elswhere[11,14].

     The breakdown constant k of temporary inhibitors (eq. 3)
may be determined by measuring the enzymatic activity of proteinase and inhibitor mixtures incubated during variable intervals of
time and plotting the data on a semi-logarithmic scale.

IN VIVO SIGNIFICANCE OF THE KINETIC PARAMETERS

     For beeing efficient in vivo, a macromolecular inhibitor must
react very fast with proteinases which are liberated accidentally
or physiologically. On the other hand, the binding of the proteinase to the inhibitor must be irreversible or at least pseudoirreversible. In this section we shall attempt to correlate these two
parameters (i.e. fastness of inhibition and stability of complexes)
with the in vitro measurable kinetic parameters.

The delay time of inhibition

We call "delay time of inhibition d(t)" the time required for almost complete inhibition of a proteinase in vivo. We shall demonstrate that d(t) is related to kon and (I°), the in vivo inhibitor concentration. Since (I°) is higher than the concentration of liberated proteinase (otherwise the inhibitor does not play any role), the association of E and I is pseudofirst-order with a rate constant of $\mathrm{kon(I°)}$ $(\mathrm{E} \xrightarrow{\mathrm{kon(I°)}} \mathrm{E \cdot I.})$. The half-life time of inhibition will be :

$$t_{\frac{1}{2}} = 0.693/\mathrm{kon(I°)}. \qquad (13)$$

After seven half-life times the association will be more than 99 % complete (i.e. $d(t) \simeq 7\, t_{\frac{1}{2}}$). Hence :

$$d(t) \simeq 5/\mathrm{kon(I°)} \qquad (14)$$

so that d(t) may be calculated knowing kon and (I°). Stricktly speaking, eq. 14 is valid only for irreversible inhibitors (e.g. α_1-proteinase inhibitor). It may however also be used for reversible inhibitors whith have a pseudo-irreversible behavior in vivo (i.e. inhibitors for which $(I°)/\mathrm{Ki} \gg 100$ see also the following section).

Table 1 lists the delay times for the inhibition of some proteinases by plasma α_1-proteinase inhibitor. It can be seen that this inhibitor reacts extremely fast with leukocyte elastase. Of the two trypsins present in human pancreas which may be massively liberated into blood during acute pancreatitis, only the anionic one is inhibited reasonably fast. Cationic trypsin may reach targets like prothrombin or fibrinogen before beeing inhibited. α_1-proteinase inhibitor is not a physiological inhibitor of the important blood coagulation proteinases plasmin or thrombin, contrary to general opinion.

The above calculations do not take into account the rate-depressing effect of physiological substrates. Since both inhibitors and substrates are proteins, it is likeky that they compete for the binding of proteinases. Theoretically this effect may easily be taken into account[16]. Let (S°) and Km be the in vivo substrate concentration and Michaelis constant respectively. It may be shown that :

$$\mathrm{kon'} = \mathrm{kon}/(1 + (S°)/\mathrm{Km}) \qquad (15)$$

Table 1. Physiological function of human plasma α_1-protei-
          nase Inhibitor as inferred from the measurement
          of its Association rate constant (kon) which
          various proteinases

Proteinases	kon $(M^{-1}s^{-1})^a$	delay time d(t)	physiological function
Leukocyte elastase	6×10^7	2 msec	certain
Pancreatic anionic trypsin	7×10^4	2 sec	likely
Pancreatic cationic trypsin	1×10^4	12 sec	unlikely ?
Plasmin	190	12 min	unlikely
Thrombin	48	42 min	unlikely

^{a}data from Beatty et al.[15]

where kon' is the association rate constant in the presence of a
substrate. Hence d'(t), the delay time of inhibition in the presence of substrate is given by :

$$d'(t) \simeq \frac{5 \ (1 + (S°)/Km)}{kon \ (I°)} \qquad (16)$$

Of course, if several substrates compete with the inhibitor, their
effects are "additive" $(1 + (S_1^°)/Km_1 + (S_2^°)/Km_2 + \ldots)$. Competition is neglectible if $(S°) \ll Km$.

     From a practical point of view, the substrate effect is diffi-
cult to quantitate precisely because proteinases are essentially
active on any protein. This problem may be circumvented by determining association kinetics in the presence of the biological
substrates (i.e. measuring kon' directly).

The partition of a proteinase between two inhibitors

When a proteinase is liberated into a biological fluid it may encounter more than one inhibitor. The question of how it will partition between these inhibitors may be partially answered if kon values are known. Let I_1 and I_2 be two inhibitors which react with the same enzyme E :

$$E + \quad \begin{array}{l} I_1 \xrightarrow{\quad kon_1 \quad} E \cdot I_1 \\[2em] I_2 \xrightarrow{\quad kon_2 \quad} E \cdot I_2 \end{array} \qquad (15)$$

it may be shown that[5] :

$$\frac{kon_1}{kon_2} = \frac{\log((I_1^o)/(I_1^o) - (E \cdot I_1))}{\log((I_2^o)/(I_2^o) - (E \cdot I_2))} \qquad (16)$$

hence, if the two association rate constants as well as the two in vivo inhibitor concentrations (I_1^o) and (I_2^o) are known, the ratio of the two complexes $((E \cdot I_1)/(E \cdot I_2))$ may be calculated.

Table 2 shows how these calculations were used to find out the partition of leukocyte elastase between α_1-proteinase inhibitor and bronchial mucus inhibitor.

Stricktly speaking, eq. 16 is valid only for irreversible inhibitors. If the two inhibitors are reversible, eq. 16 describes the initial partition of enzyme between them. As time elapses, exchange will occur until thermodynamic equilibrium will be reached. The ultimate partition of enzyme will then be a function of the equilibrium dissociation constants $(Ki_1$ and $Ki_2)$ of the two complexes[5,18].

On the other hand, if one inhibitor is irreversible and the other one is reversible as in the example shown in table 2, the partition may be described by eq. 16 provided that the concentration of the reversible inhibitor is much higher than that of the irreversible one.

The stability of protease-inhibitor complexes

To be efficient, a macromolecular inhibitor does not only have to react fast : it must also form stable complexes with

Table 2.   Partition of leukocyte elastase between α_1-pro-
           teinase inhibitor and bronchial inhibitor in bron-
           chial secretions

	kon $(M^{-1}s^{-1})^a$	Concentration[b] (μM)	Percent elastase bound
α_1-proteinase inhibitor	6.0×10^7	0.1	30
bronchial inhibitor	1.2×10^7	1.4	70

[a] from references 8 and 15

[b] in bronchial lavage fluid[17]. The concentrations in bronchial secretions do not need to be known since the partition depends only
upon the relative concentrations of the two inhibitors.

proteinases. The term "stability" may be viewed in three different
ways :

(i) the bulk of the proteinase molecules must be bound to the
     inhibitor once the association process is terminated.

(ii) the E·I complexe should not undergo dissociation by macromo-
      lecules (substrates) present in the biological medium.

(iii) the E·I complex should not undergo spontaneous degradation
       with liberation of active proteinase.

     Let us first consider point (i). This requirement is obviously
fulfilled in the case of irreversible inhibitors. By contrast,
reversible ones are efficient only if they exhibit a "pseudoirreversible behavior". This behavior depends upon the ratio $(I°)/Ki$.
It may be shown that[11] for $(I°)/Ki \geqslant 100$ there will be virtually
no free enzyme present in an enzyme + inhibitor mixture (i.e. there
will be pseudoirreversible inhibition). Table 3 lists some illus-
trative examples. Both inter-α-inhibitor and bronchial inhibitor
form reversible complexes with proteinases. Two of these complexes
have a pseudoirreversible behavior.

     Let us now analyze point (ii). Only reversible inhibitors may
undergo kinetic dissociation by substrates. The question whether
dissociation may occur in vivo depends both upon the value of koff
and upon the presence of potential dissociation agents in the biological medium. Since the latter are difficult to know, dissocia-

Table 3.   In vivo pseudoirreversible behavior of
reversible inhibitors

Inhibitor	Concentration (μM)	Proteinase	Ki (M)	(I°)/Ki
Inter-α-inhi-bitor	3.1[a]	Trypsin[c]	1.2×10^{-8}	260
Inter-α-inhi-bitor	3.1[a]	Chymotrypsin[c]	1.3×10^{-7}	24
Bronchial inhibitor	14.0[b]	Elastase[d]	1.2×10^{-11}	10^6

[a] in plasma[19]

[b] in bronchial mucus[17] assuming that the mucus has been diluted ten-times during lavage

[c] human pancreatic proteinases[9]

[d] human leukocyte elastase[8]

tion is, a priori, difficult to predict. Limiting cases may, however, be considered. We know that if substrate-induced dissociation occurs indeed, its half-life time ($t_{\frac{1}{2}}$) is related to koff through the following equation :

$$t_{\frac{1}{2}} = 0.693/koff \qquad (17)$$

Let us then assume that the minimal time during which the complex remains essentially undissociated (i.e. the time during which virtually no dissociation occurs) is equal to $t_{\frac{1}{2}}/10$ and let us call this time the "stability time" :

$$\text{stability time} \simeq 0.07/koff \qquad (18)$$

According to this formula, the stability time varies from 7 seconds to about 2 hours when koff varies from 10^{-2} to 10^{-5} s^{-1}. For instance, the stability time of the trypsin-intera-trypsin inhibitor complex (koff = 1.7×10^{-3} s^{-1} ref. 9) is ca 40 seconds. Therefore, this complex which we have considered as pseudoirreversible on the basis of its (I°)/Ki value (table 3), is theoretically subject to fast dissociation in the presence of a strong dissociating macromolecule. Whether or not it will dissociate in vivo, depends upon the presence and the affinity of such dissociating macromolecules (e.g. one or several substrates of the proteinase) :

$$E + I \rightleftharpoons E \cdot I, \text{ dissociation constant : Ki}$$

+

S

$$E \cdot S, \text{ dissociation constant : Ks}$$

Efficient dissociation occurs only if (S°)/Ks > (I°)/Ki. If (S°)/Ks is unknown, the best way to solve the problem, is to incubate the E·I complex with the physiological substrate S and to measure the breakdown of the latter as a function of time. If no substrate breakdown occurs during a prolonged period of time, E·I may be considered as stable in the presence of substrate. We would suggest that such an experiment is necessary only if koff > 10^{-4} s^{-1} i.e. if the theoretical stability time is lower than ca. 10 minutes.

Point (iii) concerns temporary inhibitors. In that case, the in vitro measured hydrolysis constant k (see eq. 3) may be used directly to infer the in vivo stability of E·I. Using the same reasonning as before we may define a stability time equal to ca. 0.07/k. By contrast with the kinetic dissociation process (point ii) the stability time of temporary inhibitors does not depend upon the presence of surrounding substrate molecules.

DISCUSSION

The present paper should be considered as a preliminary attempt to correlate in vitro determined kinetic parameters to in vivo function of proteinase inhibitors in health and disease. One of our goals was to emphasize that demonstration of inhibition in the test tube is by far not sufficient to claim that an inhibitor plays a physiological function in the living organism. We hope that at least this minimal message will reach many physiological chemists. We also hope that the theoretical and practical points outlined in this paper will help those who are interested in the physiological function of new proteinase inhibitors.

Let us now attempt to summarize the salient content of this dissertation. Say you have discovered and isolated a new proteinase inhibitor I which blocks the enzymatic activity of a proteinase E in vitro. You measure the pertinent kinetic parameters as well as the in vivo concentration (I°).

(i) the inhibition is irreversible : you measure kon and you calculate d(t). If d(t) is in the millisecond range you may safely conclude that the inhibition is fast enough in vivo even in the presence of substrates. If not, try to measure kon' in the presence of physiological substrates of E and calculate d'(t). If d'(t) ≤ 1 second, the inhibition may still be considered as fast enough in vivo. Test then the E·I system for temporary inhibition. If k is zero or very low, the inhibitor may be considered as efficient in vivo.

(ii) the inhibition is reversible : you measure kon, koff (and Ki). Again you calculate d(t) or d'(t). If these parameters are satisfactory, you must investigate the "stability" of E·I in vivo. If (I°)/Ki < 100, the inhibitor is inefficient even if d(t) is satisfactory. If (I°)/Ki ≫ 100, calculate the stability time using koff. If this time is > 10 minutes, the inhibitor is efficient. If not, investigate the stability in the presence of macromolecular ligands of the enzyme (e.g. substrates). Of course, the E·I complex has also to be tested for temporary inhibition.

In this report natural proteinase inhibitors have been considered as efficient only if they react "fast" with proteinases and if the complexes they form with enzymes are irreversible or pseudoirreversible. This is of course an oversimplification which was used for the sake of brevity. We are aware that (i) some inhibition processes have to be "slow" to be efficient (e.g. inhibition of some coagulation proteinases) (ii) the phenomenon of temporary inhibition is of prime importance for pancreatic proteinases.

REFERENCES

1. M. Laskowski Jr. and R.W. Sealock, Protein proteinase inhibitors. Molecular aspects, in : The Enzymes vol. III 3rd ed. Academic Press, New York, (1971).
2. U. Quast, J. Engel, H. Heumann, G. Krauze and E. Steffen, Kinetics of the interaction of bovine pancreatic trypsin inhibitor (Kunitz) with α-chymotrypsin, Biochemistry 13 : 2512 (1974).
3. D.A. Johnson and J. Travis, Human alpha-1-proteinase inhibitor mechanism of action : evidence for activation by limited proteolysis, Biochem. Biophys. Res. Commun. 72 : 33 (1976).

4. J.F. Meyer, J. Bieth and P. Métais, On the inhibition of elastase by serum. Some distinguishing properties of α_1-antitrypsin
   and α_2-macroglobulin, Clin. Chim. Acta 62 : 43 (1975).
5. J.P. Vincent and M. Lazdunski, Trypsin-pancreatic trypsin inhibitor association. Dynamics of the interaction and role of disulfide bridges, Biochemistry 16 : 2967 (1972).
6. K.J. Laidler, "The chemical kinetics of enzyme action", Oxford
   University Press, London (1958).
7. J.C. Zahnley and J.G. Davis, Determination of trypsin.inhibitor
   complex dissociation by use of the active site titrant, p-nitrophenyl p'-guanidinobenzoate, Biochemistry 9 : 1428 (1970).
8. F. Gauthier, U. Fryksmark, K. Ohlsson and J.G. Bieth, Kinetics
   of the inhibition of leukocyte elastase by the bronchial inhibitor, Biochim. Biophys. Acta 700 : 178 (1982).
9. M. Aubry and J. Bieth, A kinetic study of the inhibition of human and bovine trypsin and chymotrypsin by the inter-α-inhibitor
   from human plasma, Biochim. Biophys. Acta 438 : 221 (1976).
10. H. Ako, R.J. Foster and C.A. Ryan, Mechanism of action of natu-
    rally occuring proteinase inhibitors. Studies with anhydrotrypsin
    and anhydrochymotrypsin purified by affinity chromatography,
    Biochemistry 13 : 132 (1974).
11. J. Bieth, Some kinetic consequences of the tight binding of
    protein-proteinase inhibitors to proteolytic enzymes and their
    application to the determination of dissociation constants, in
    Proteinase Inhibitors H. Fritz, H. Tschesche, L. Greene and
    E. Trutscheit, eds. Springer, Berlin (1974).
12. M. Dixon, The graphical determination of Km and Ki, Biochem.
    J. 129 : 197 (1972).
13. W.R. Greco and M.T. Hakala, Evaluation of methods for estimating
    the dissociation constant of tight binding enzyme inhibitors,
    J. Biol. Chem. 254 : 12104 (1979).
14. J.G. Bieth, Pathophysiological interpretation of kinetic cons-
    tants of protease inhibitors, Clin. Resp. Physiol. 16, suppl.
    183 (1980).
15. K. Beatty, J. Bieth and J. Travis, Kinetics of association of
    serine proteinases with native and oxidized α_1-proteinase inhi-
    bitor and with α_1-antichymotrypsin, J. Biol. Chem. 255 : 3931
    (1980).
16. W.X Tian and C-L Tsou, Determination of the rate constant of
    enzyme modification by measuring the substrate reaction in the
    presence of the modifier, Biochemistry 21 : 1028 (1982).
17. K. Ohlsson, The low molecular weight proteinase inhibitor in
    bronchial mucus, in :"Le lavage broncho-alvéolaire chez l'homme",
    G. Biserte, J. Chrétien and C. Voisin, eds, Editions INSERM,
    Paris (1979).
18. J. Bieth, M. Aubry and J. Travis, The interaction of human ca-
    tionic trypsin and chymotrypsin II with human serum inhibitors,
    in : "Proteinase inhibitors", H. Fritz, H. Tschesche, L. Greene
    and E. Trutscheit, eds. Springer, Berlin (1974).

19.N.Heimburger, H. Haupt and H.G. Schwick, Proteinase inhibitors of human plasma, in : Proc. Int. Res. Conf. Pro. Inh., H. Fritz and H. Tschesche, eds. Walter de Gruyter, Berlin, (1971).

ON THE MULTIPLICITY OF CELLULAR ELASTASES AND THEIR INEFFICIENT

CONTROL BY NATURAL INHIBITORS

W. Hornebeck, D. Brechemier, M.P. Jacob,
C. Frances and L. Robert

Labo. de Biochimie du Tissu Conjonctif, GR CNRS n° 40
Faculté de Médecine, 8 rue du Gén. Sarrail, 94010 CRETEIL

Several elastic tissue diseases are mainly characterized by the conspicuous fragmentation and loss of orientation of the elastic fibers and it is now generally accepted that the degradation of elastin plays a primordial role in the development of human diseases as emphysema and arteriosclerosis[1,2]. This generalized age-dependent lysis of elastin, which is accelerated under pathological circumstances, has been attributed to the proteolytic action of endopeptidases designated as elastases[3]. Such an example is the specific degradation of elastin in lung tissue which is considered as a necessary requirement for the experimental induction of emphysema. The severity of the induced lesion is directly related to the elastolytic efficiency of the intratracheally infused protease[4].

The first complexity in understanding the mechanisms of elastin degradation occuring during a given pathology resides in the multiplicity of cellular elastases which can be directly or secondarily involved in the process. Elastases have been previously isolated and characterized from several cell types (leucocytes[5], platelets[6], macrophages[7]) and in addition we will report here that cells of mesenchymal origin (aorta smooth muscle cells, human skin fibroblasts) also contain similar types of proteases. There is ample experimental evidence to indicate that the quantitative variation and secretion of these different neutral proteases are under the control of several different mechanisms. It could be speculated that each enzyme, in addition to its capacity to hydrolyze elastin, may have other biological functions[9].

The elastolytic activity of almost all these endopeptidases

is controlled by protease inhibitors present in physiological
fluids as serum for instance[10]. Nevertheless, there are conditions where the main plasma elastase inhibitor e.g. α_1-antipro-
tease (α_1-antitrypsin), do not provide a complete protection of
elastic fibers against elastolysis. Genetic deficiency of this
inhibitor often leads to the appearance of emphysematous lesions[11].
The modification of the inhibitory site of this molecule by oxydi-
zing agents induces a considerable decrease of the functionnal
activity of the inhibitor[12]. Furthermore, several proteases unaffected by the α_1-antiprotease inhibitor, including for instance
cathepsin B[13] and macrophage elastase[14], are capable to degrade
the molecule rendering it inactive to further inhibit other
proteases.

We have shown already that, even if stoecchiometric quantities
of elastase were injected in puppies (which were entirely complexed by circulating blood plasma inhibitors as shown in previous
in vitro experiments), in vivo attack of elastic fibers in the
lung and in aorta could still be demonstrated[15]. The results of
these experiments suggested the existence of escape mechanisms
which would enable elastolytic proteases to remain active even in
the presence of an excess of biological inhibitors[16].

Elastases as well as other pancreatic enzymes, could be liberated into the bloodstream after pancreatic injury and elevated
amounts of immunoreactive serum elastase (of pancreatic origin)
have been quantitated following experimentally induced pancreatitis[17]. Immunologically reactive circulating human leucocyte elastase levels also varied considerably in patients presenting chronic obstructive lung disorders[18].

We also presented evidence that human serum does inhibit activity against synthetic substrates considered as specific for elastases and also towards soluble forms of elastin[19]. Part of the
whole activity is associated with lipoproteins in human plasma[20].

Finally, major emphasis should be led on the fact that the
level of elastases secreted by cells may well exceed in the
extracellular space the normal amount of elastase inhibitors. In
this case, elastases will adsorb preferentially on to elastic fibers. Recent experiments showed us that in its adsorbed state,
the protease is partly refractory to the inhibition by the α_1-anti-
protéase[21].

All the above mentioned facts suggested the incomplete protection of elastic fibers against elastolysis by naturally occuring inhibitors. Inhibitors of elastase of microbial[22] or synthetic origin[23] have been used as a therapeutic approach mainly in
the treatment of emphysema induced by intratracheal instillation
of elastase[22]. The multiplicity of cellular elastases involved in
elastic degradation renders quite difficult such an approach. We

will emphasize on alternatives in controlling elastin degradation as occuring during pathological circumstances.

AORTA SMOOTH MUSCLE CELLS ELASTASE-TYPE PROTEASES

In 1978, we demonstrated that the extractible elastase-type activity from human arterial wall increases in a statistically significant manner with the age of the individual and the degree of atheroma[24]. Parallely it was found that the amount of reticulated elastin as determined by the quantitation of desmosines per tissue dry weight decreases significantly in pathological aortas. Subsequently, a neutral serine protease possessing elastase-type activity was purified from human arterial wall by affinity chromatography and it could be differentiated from other elastases by enzymatical and immunological criteriae[25]. On this basis, we postulated the endogenous origin of such an enzyme. Indeed, elastase activities could be demonstrated in Triton X-100 extracts of aorta smooth muscle cells originated from several species[26].

Elastase activities on Succinoyl tri-alanine paranitroanilide and ^{3}H-insoluble ligamentum nuchae elastin have been determined in Triton X-100 extracts and serum free culture medium of rat aorta smooth muscle cells. Most of the enzymic activity (representing 2 ng equivalent of porcine pancreatic elastase on Suc (Ala)$_3$NA and 38 ng equivalents of porcine pancreatic elastase on ^{3}H-insoluble elastin per 10^6 cells, 8th passage) was recovered intracellularly. Addition of Trypsin to serum free culture medium did not result in the appearance of detectable amounts of latent enzyme. Elastase activity increases within the cell up to 48 hrs of culture and this increase could be inhibited by cycloheximide. Interestingly the levels of elastase-type activities quantitated on Triton X-100 extracts of the cells increased exponentially with cell passages. This variation in enzymic activity could be closely related to the exponential enhancement of elastase activities extractible from human aortas as a function of the age of the individuals[27].

Porcine aorta smooth muscle cells were used to determine the subcellular localization of elastase activities. The cells were therefore subjected to analytical subcellular fractionation on sucrose density gradients. The distribution of elastase-type activity towards Suc(Ala)$_3$ NA and insoluble elastin was similar to that of 5' nucleotidase activity, a plasma membrane marker enzyme and was distinct from those of the marker enzymes for other organelles. When the cell homogenate was treated with digitonin, the modal equilibrium density of the plasma membrane was increased from 1.13 to 1.19 g/ml. The elastase-type activities showed similar increases in equilibrium density, thus confirming their plasma membrane localization[28].

The neutral protease(s) was partially purified by ion exchange chromatography on a carboxymethyl cellulose column[9]. The inhibition profile of this enzyme(s) as quantified on $Suc(Ala)_3NA$ and ^3H ligamentum nuchae insoluble elastin was determined and found to justify classification as a serine protease. In addition, the rate of hydrolysis of $Suc(Ala)_3NA$ was decreased by α_1-proteinase inhibitor, α_2-macroglobulin and by $Ac(Ala)_2 Pro Val CH_2Cl$. It does not hydrolyze BANA, BINA and type I collagen from rat tail tendon.

HUMAN SKIN FIBROBLAST ELASTASE-TYPE PROTEASE

Succinoyl-tri-alanine paranitroanilide, a specific synthetic substrate of elastases, could be also hydrolyzed by triton X-100 extracts of human skin fibroblasts at near neutral pH^{26}. The neutral endopeptidase has been partially purified (86 fold) by ion exchange chromatography followed by affinity chromatography using an AH Sepharose $(Ala)_3$ column[29]. In its partially purified form, the protease appeared to be a metallo-endopeptidase as evidenced by its inhibitory profile. Structural glycoproteins microfibrils (sgp mf) isolated from porcine aorta are extensively degraded by the neutral protease. The enzyme could also hydrolyze, but to a lesser extent, insoluble elastins from human or porcine aortas but was totally inactive towards ligamentum nuchae insoluble elastin. Its capacity to degrade skin elastic fibers has been assessed by a morphometric analysis of human skin tissue sections (appropriately stained) prior and after the action of the purified enzyme. It was observed that human skin elastase-type protease was nearly devoid of proteolytic action on elastic fibers but could, in contrast, rapidly hydrolyze both elaunin and oxytalan fibers of human dermis. A similar protease has been isolated from human vulva fibroblasts[30]. In addition, when partially purified protease was injected intradermally to young rabbits, it produced appreciable degradation of the elastic fiber system.

ELASTASE-TYPE ACTIVITY OF HUMAN SERUM

Despite the presence of high levels of elastase inhibitors elastase-type activities could be detected in human serum both on $Suc(Ala)_3NA$ and radiolabelled elastin peptides (kappa-elastin) covalently coupled to agarose beads as substrates[19]. Both types of activities could be partly abolished by serine active-site titrants and partly by neutral chelating agents. Only the combination of both types of inhibitors could completely abolish the enzy-

mic activity of human serum on both substrates. These findings indicated the presence of at least two different types of elastases
(serine and metallo-proteases) in human serum. In recent experiments elastase(s) levels were compared in control healthy non-smo-
king individuals and in patients presenting clinical symptoms of
atherosclerosis or/and suffering from chronic obstructive lung
diseases. Sera from the control group had average serum elastase
level equal to 78.1 ng/ml on Suc(Ala)$_3$ NA and 688.8 ng/ml on soluble elastin peptides as substrate. No statistically significant
differences were observed in elastase(s) activities in the sera of
individuals presenting clinical symptoms of atherosclerosis as
compared to the controls. In contrary, the sera of patients suffering from chronic obstructive lung diseases contained higher quantities of elastase-type activities : 237.2 ng/ml on Suc(Ala)$_3$ NA
and 1096 ng/ml on soluble elastin respectively[19]. It was also
reported[18], that the average serum concentration of leukocyte elastase determined by immunochemical methods, was higher in the sera
of patients suffering from chronic obstructive bronchitis than in
age matched control subjects.

     In attempts to identify the protease(s) responsible for the
elastase-type activity of human serum, it was fractionated by
gel permeation chromatography. Most of the activity hydrolyzing
the above mentioned elastase substrates was found associated
with fractions of high molecular weight. Some of the active fractions also contained lipoproteins. We could confirm that isolated
lipoproteins of all classes (HDL, LDL, VLDL) do exhibit elastase-
type activity as previously reported[20]. A metallo-protease hydrolyzing Suc(Ala)$_3$ NA and soluble elastin peptides was further isolated from apoHDL by Sepharose-elastin chromatography in close association with apoprotein A$_1$[20]. Quite recently, we have shown that
incubation of apoHDL at physiological temperature in presence of
calcium ions at neutral pH resulted in an appreciable degradation
of apoproteins (mainly of apo A$_1$). The inhibitory profile of this
apo A$_1$ degradation closely resembles those of the isolated apolipoprotein bound metallo-elastase-type protease. It is tempting
to speculate on a possible biological function of this enzyme in
the metabolism of apolipoprotein A$_1$[31].

ELASTIN-ELASTASE INTERACTION AND THEIR IMPORTANCE IN CONTROLLING
ELASTIN DEGRADATION BY NATURALLY OCCURING INHIBITORS

     Several investigators, including ourselves, pointed out that
the first and main step in the degradation of insoluble elastin
catalyzed by elastases consisted in a rapid adsorption of the
enzyme on to its substrate[32,33]. This interaction has an electrostatic nature and involves the free non amidated carboxylic groups
of the substrate[34]. The unique structural properties of elastin
render quite difficult to study such substrate enzyme interactions
in a qualitative and quantitative way and several important ques-

tions remained unanswered. It was already shown that elastins originated from various sources or purified according to different isolation procedures presented different susceptibilities to the degradation by a given elastase[35], but the reason for such variations remained obscure. In this respect only a few informations are available concerning the relationship between the conformational state of the substrate and the catalytic efficiency of elastases. Although, it is highly probable that different elastases would bind to the polymeric substrate at different locations and the resulting solubilized peptides might exhibit different biological functions. We already mentioned that, once bound to elastin, elastase molecules could not freely diffuse from its substrate[33]. Unfortunately, due to the insoluble character of the substrate, the elastin-elastase association constants could not yet be determined. Particularly relevant to pathological problems would be the determination of the partition coefficient of free elastase with elastin and α_1-proteinase inhibitor.

We recently studied the differences in susceptibility of elastase against its inhibitors, in solution or after adsorption to its substrate elastin[21]. Low molecular mass inhibitors including an irreversible active site titrant (MeO Suc(Ala)$_2$ Pro Val CH$_2$ cl) and a peptide inhibitor from leeches (Eglin Mr = 8,100) inactivated leukocyte elastase nearly completely regardless of whether or not the enzyme was adsorbed to its substrate elastin prior to the addition of the inhibitor. In contrast, we found that elastase preadsorbed to its substrate was much less sensitive (50 %) to the inhibitory effect of α_1-proteinase inhibitor than the free enzyme.

CONCLUDING REMARKS

Active site titrant inhibitors as well as tight binding inhibitors of microbial origin have been used more or less successfully to inhibit the development of emphysematous lesions induced by intratracheal instillation of elastase in experimental animal models[22,23]. The use of α_1-antiproteinase inhibitor has also been suggested and utilized as substitution therapy in case of a genetic deficiency of this inhibitor. Besides leukocyte elastase, other elastases could also play a major role in the degradation of elastic fibers in lung diseases and also in several other pathological conditions. The proper design and use of a synthetic inhibitor implies the knowledge of the type of protease involved in the lysis of elastin. As cellular elastases may exhibit different specificities (cf. leukocyte elastase and macrophage elastase[36]), their control would necessitate the administration of different synthetic inhibitors. Another drawback in the use of synthetic inhibitors in case of a deficiency of natural inhibitors resides in the fact that cellular elastases may well possess other important metabolic functions than degrading elastin.

In our opinion, elastolysis by cellular elastases could be alternatively controlled by the "a priori" protection of elastic fibers. Several substances are known to interact with insoluble elastin rendering it more or less susceptible to the degradation by elastases. Calcium ions and lipids bind to elastin in a synergestic fashion and further increase the rate of the solubilization of the polymer by a fixed amount of elastase. Other macromolecules such as structural glycoproteins (microfibrils) presenting an affinity for elastin could have a protective action increasing the non-productive catalytic sites of the elastic fiber[37].

ACKNOWLEDGEMENTS : This work was supported by CNRS (GR n° 40), INSERM (ATP n° 825012) and by the Conseil Scientifique of the University Paris-Val-de-Marne.

REFERENCES

1. Anon. The pathogenesis of pulmonary emphysema, Lancet, 1:743 (1980).
2. B. Robert, L. Robert and A.M. Robert, Elastine, élastase et artériosclérose, Pathol.Biol. 363:455 (1972).
3. J. Bieth, Elastases : structure, function and pathological role. In :"Frontiers of Matrix Biology", L. Robert, ed., Karger,Basel (1978).
4. J. Bignon and H. de Crémoux, Pathological and pathogenetic aspects of chronic obstructive lung disease, Bull. Europ. Physiopath. Resp. 16:13 (1980).

5. K. Ohlsson and I. Olsson, The neutral proteases of human granulocytes. Isolation and partial characterization of granulocyte elastases, Eur.J.Biochem. 42:519 (1974).
6. W. Hornebeck and Y. Legrand, Possible role of elastase-like enzymes isolated from human blood platelets and arterial wall in arteriosclerosis. In:"Frontiers of Matrix Biology", L.Robert, ed., Karger, Basel (1980).
7. H. de Cremoux, W. Hornebeck, M.C. Jaurand, J. Bignon and L. Robert, Partial characterization of an elastase-like enzyme secreted by human and monkey alveolar macrophages. J.Pathol. 125:172 (1978).
8. P.M. Starkey, Elastase and cathepsin G ; the serine proteinases of human neutrophil leukocytes and spleen. In :"Proteinases in Mammalian Cells and Tissues", A.J.Barrett, ed., North Holland, Amsterdam (1977).
9. W. Hornebeck, D. Brechemier, M.C. Bourdillon and L. Robert, Isolation and partial characterization of an elastase-like protease from rat aorta smooth muscle cells. Possible role in the regulation of elastin biosynthesis. Conn.Tiss.Res. 8:245 (1981).
10. N. Heimburger und H. Haupt. Zur Spezifität der Antiproteinasen des Humanplasmas für Elastase. Klin.Wochenschr. 44:1196 (1966).

11. J.O. Morse, Alpha$_1$ antitrypsin deficiency, New Engl.J.Med. 299: 1099 (19.78).

12. D. Johnson and J. Travis, The oxidative inactivation of human alpha$_1$ proteinase inhibitor by thiol proteinases, Biochem.J. 163:639 (1977).

13. D. Johnson and J. Travis, The oxidative inactivation of human alpha$_1$ proteinase inhibitor. Further evidence for methionine at the ractive center, J.Biol.Chem. 254:4022 (1979).

14. M.J. Banda, E.J. Clark and Z. Werb, Limited proteolysis by macrophage elastase inactivates human alpha$_1$ proteinase inhibitor, J. Exp.Med. 152:1563 (1980).

15. G.M. Turino, W. Hornebeck and B. Robert. In vivo effect of pancreatic elastase. Studies on the serum inhibitors, Proc.Soc. Exp.Biol.Med. 146:712 (1974).

16. L. Robert, G. Bellon and W. Hornebeck, Characterization of different elastases ; their possible role in the genesis of emphysema, Bull.Europ.Physiopath.Resp. 16:199 (1980).

17. K. Sakate, Y.Sukchung, T.Yoshimoto, K. Umeyama, T. Ohyama and K. Katayama, Radioimmunoreactive serum elastase levels and histological changes during experimental pancreatitis in rats, Arch.Surg. 117:777 (1982).

18. R.A.Stockley, K. Ohlsson, Serum studies of leukocyte elastase in acute and chronic lung diseases, Thorax 37:114 (1982).

19. W. Hornebeck, J.P.Potazman, H. de Crémoux, G. Bellon and L. Robert, Elastase-type activity of human serum ; its variation in chronic obstructive lung diseases and atherosclerosis, Lab.Invest. (Submitted) (1982).

20. M.P. Jacob, G. Bellon, L. Robert, W. Hornebeck, M. Ayrault-Jarrier, J. Burdin and J. Polonovsky, Elastase-type activity associated with high density lipoproteins in human serum, Biochem.Biophys.Res.Comm. 103:311 (1981).

21. W.Hornebeck and H.P. Schnebli, Effect of different elastase inhibitors on leukocyte elastase preadsorbed to elastin, Hoppe-Seyler's Z.Physiol.Chem. 363:455 (1982).

22. A. Yoshida and S.Yu, In vivo suppression of elastase emphysema in hamsters by alpha$_1$ antitrypsin or elastatinal, Fed.Proc. 38:1205 (1979).

23. P.J. Stone, E.C. Lucey, J.D. Calore, J.C. Powers, G.L. Snider and C. Franzblau, The moderation of elastase induced emphysema in hamsters by post-treatment with a chloromethylketone elastase inhibitor, Fed.Proc. 119:364 (1979).

24. W. Hornebeck, J.J. Adnet and L. Robert, Age-dependent variation of elastin and elastase in aorta and human breast cancer. Exp. Gerontol. 13:293 (1978).

25. G. Bellon, T. Ooyama, W. Hornebeck and L. Robert, Isolation and partial characterization of an elastase-like protease from human arterial wall by lima bean trypsin inhibitor affinity chromatography, Artery 7:290 (1980).

26. M.C. Bourdillon, D. Brechemier, N. Blaes, J.C. Derouette, W. Hornebeck and L. Robert, Elastase-like enzymes in skin fibroblasts and rat aorta smooth muscle cells, Cell.Biol.Intern.Rep. 4:313 (1980).

27. D. Brechemier, L. Robert and W. Hornebeck, Identification of elastase-type proteases in rat aorta smooth muscle cells, Biochem.J. (submitted) (1982).

28. D.S. Leake, W.Hornebeck, D.Brechemier and T.J. Peeters, Elastase-like activities of arterial smooth muscle cells in culture are located on the plasma membrane, Clinical Science 63:73 (1982).

29. M. Szendroï, H. Bakala, G. Meimon, C. Frances, L. Robert and W. Hornebeck, Isolation and partial characterization of a neutral protease hydrolyzing Suc(Ala)$_3$NA in human skin fibroblasts ; morphometric analysis of its potential to degrade human skin elastic fibers, Conn.Tiss.Res. (Submitted) (1982).

30. G. Godeau, C. Frances, W. Hornebeck, D. Brechemier and L.Robert, Isolation and partial characterization of an elastase-type protease in human vulva fibroblasts :its possible involvement in vulvar elastic tissue destruction in patients with LSA, J.Invest.Dermatol. 78:270 (1982).

31. P. Codogno, J. Burdin, M. Ayrault-Jarrier and W. Hornebeck. Degradation of apolipoproteins A$_1$ by bound elastase-type protease. B.B.R.C. (Submitted) (1982).

32. A. Gertler, The non-specific electrostatic nature of the adsorption of elastase and other basic proteins on elastin, Eur.J. Biochem. 20:541 (1971).

33. B. Robert, W. Hornebeck and L. Robert, Cinétique hétérogène de l'interaction élastine-élastase, Biochimie 56:239 (1974).

34. H.M. Kagan and R.M. Lerch, Amidated carboxyl groups in elastin, Biochem.Biophys.Acta 434:223 (1976).

35. R.E.Jordan, N. Hewitt, W. Lewis, H. Kagan and C. Franzblau, Regulation of elastase-catalysed hydrolysis of insoluble elastin by synthetic and naturally occuring hydrophobic ligands, Biochemistry 13:3497 (1974).

36. M.S. Banda and Z. Werb, Mouse macrophage elastase : purification and characterization as a metalloproteinase, Biochem J. 193:589 (1981).

37. W. Hornebeck, On the degradation of elastic fibers by elastases In : "Colloques Internationaux du CNRS : Biochimie des Tissus Conjonctifs normaux et pathologiques", Robert A.M. and Robert L. ed., CNRS Paris (1980).

PROTEASES - PROTEASES INHIBITORS:

A LOCAL CELLULAR INFORMATION SYSTEM

Hartmut Heine

Institute of Anatomy

Koellikerstraße 6, D-8700 Würzburg

INTRODUCTION

Proteases are enzymes which cleave peptides. It is generally accepted that proteases are involved in the destruction of proteins and peptide conjugates within the cells and the extracellular space (review Lojda, 1982). In clinical practice the role of the proteases is generally considered as destructive in nature. This evidently bases on a misunderstanding. Proteases are intracellularly stored in primary lysosomes. Substances taken up by the cell via phagocytosis or endocytosis are destructed proteolytically within secondary lysosomes. An intracellular opening of the lysosomes causes a proteolytic destruction of the cell; a phenomenon which induced immediately after its first description the name "suicide bags" for lysosomes (review de Duve and Wattiaux, 1966).

It is well known that lysosomes are involved in inflammation by their cell and tissue destructive potency of their proteases playing an important role in the clotting and lysis system. Deleterious effects of deficiencies of those proteases have been observed. Situations of protease inhibitors deficiency have been evaluated e. g. deficiency of C1-inhibitor involved in the angioneurotic edema or deficiency of α_1-antitrypsin concomitant with the development of lung emphysema.

Up to now an acceptable concept referring to the functional feedback mechanisms between proteases and their inhibitors at the cellular level is not available. As a consequence I would like to find a common denominator.

THE CONDUCTANCE FOR EXTRA-INTRACELLULAR INFORMATIONS

Activation and inhibition represent a space and time related principle of changes in the formation and preservation of organ tissues. Biological systems as energetically open systems are characterized by a steady state; they are consequently far away from a thermodynamically stable balance. To preserve their cellular stability a steady flux of informations between extra- and intracellular space is necessary. The cell membrane acts as a mediator and filter for information. Its asymmetric structure causes an imbalance of the ion distribution between the intra- and extracellular space which by itself allows the existence of potentials and the activation of enzymes. Changes in the membrane potential induce generally changes in the cellular metabolism. Nicolson and Singer (1972) have shown the cell membrane to represent a fluid mosaic in which the glycoproteins can move laterally in a lipid phase. Recent studies suggest that also lipid molecules form domains in membranes especially in close vicinity to the glycoproteins (for literature see Karnovsky et al., 1982).

Bennett and coworkers (1963) have recognized for the first time that oligosaccharide chains are fixed to proteins and lipids of the cell membrane being directed to the extracellular space. It has been shown that the glycocalyx determines essential cell properties (for instance surface antigenicity, cell contact, phagocytosis and exocytosis; review Spicer et al., 1981). Changes of the cell glycocalyx are combined with changes in the cell function as it is observed in tumor cells. The glycocalyx connects cell funtions with the extracellular matrix (fibre proteins, glycoproteins, proteoglycans). The proteoglycans (PG) participate to a major part in the extracellular flow of informations (Heine and Schaeg, 1979). Because of their negative charge PG are able to immobilize water, to exchange ions (mono- against bivalent cations) and to form widespread polymerisats. PG therefore guarantee an isoionic, isoosmotic and isotonic condition within the matrix. PG influence the polymerisation of collagen and elastin; they are part of the glycocalyx in form of heparansulfate molecules and they influence the cell surface functions and the extracellular distribution of certain extremely protease sensitive glycoproteins such as fibronectin, laminin, chondronectin (as far as they are identified) (review Comper and Laurent, 1978; Skosey et al., 1981) (Fig. 1).

The PG of the extracellular space are synthetized exclusively by fibroblasts and myocytes of the vessel wall whereas heparansulfate of the cell surface can be synthetized apparently by any cell type (review Pickrell, 1981). The physiological breakdown of the PG is started by serine proteases of macrophages and leukocytes which cleave the sugar component from the protein backbone.

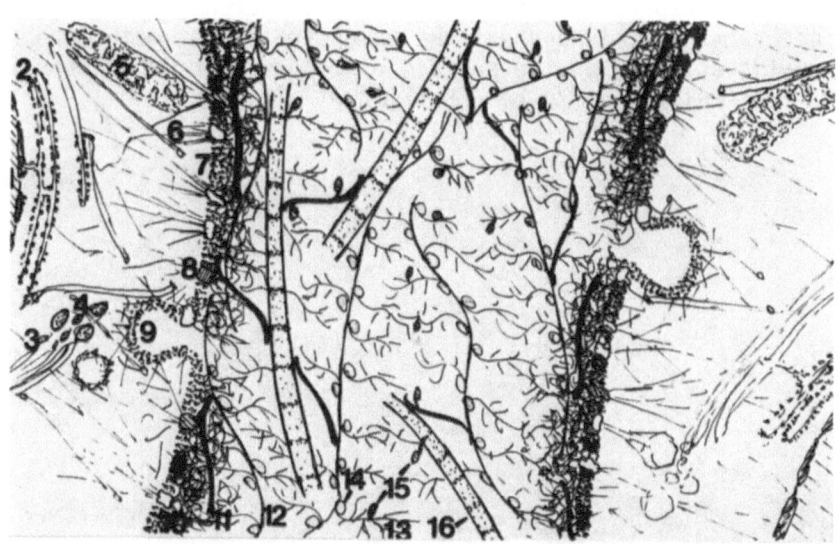

Fig. 1. Structures of the intercellular information pathway. 1 - 10
Cell components. (1) Cell nucleus,(2) endoplasmic reticulum, (3)
Golgi-apparatus, (4) lysosomes, (5) mitochondrion, (6) filaments
of the cytoskeleton, (7) lipids and glycolipids of the cell membrane, (8) integrative proteins of the cell membrane (in contact
to fibronectin, 11), (9) endocytotic (or phagocytotic) invaginat-
ion of the cell membrane, (10) glycocalyx. 10 - 16 Components of
the extracellular space. (10) Oligo-Saccarid chains of the glycoproteins of the glycocalyx, (11) fibronectin, (12) hyaluronic acid,
(13) proteoglycan, (14) link protein of the proteoglycan to hyaluronic acid, (15) protease-protease inhibitor complex, (16) collagen.

EXTRA-INTRACELLULAR ·INFORMATION FLOW MODULATED BY PROTEASE SENSI-
TIVE GLYCOPROTEINS OF THE EXTRACELLULAR SPACE

     Any change of the PG is followed by changes of the closely
located cell glycocalyx. These induce transformations in the cell
membrane by activation of the second messengers (Ca^{2+},calmodulin,
cAMP, cGMP). By that way the genetic material in the cell nucleus
is reached finally via cytoplasmatic enzyme cascades. As a consequence changes of translation and transcription of the genetic material take place according to the actual demands followed by an
adapted synthetic activity of the cells. Morphologically,proteolyt-
ic changes at the cell surface cause in vitro as well as in vivo
distinct changes of the plasmamembrane in form of various protrusions vesicula buds and invaginations shaped like coated vesicles
(Fig. 2; Huet and Herzberg, 1973). This may be considered as the

morphological expression of functionally different areas in the
cell membrane. In this context the demonstration of membrane-bound
proteolytic activity on the surface of mononuclear leukocytes may
be of great relevancy (Zucker-Franklin et al., 1981).

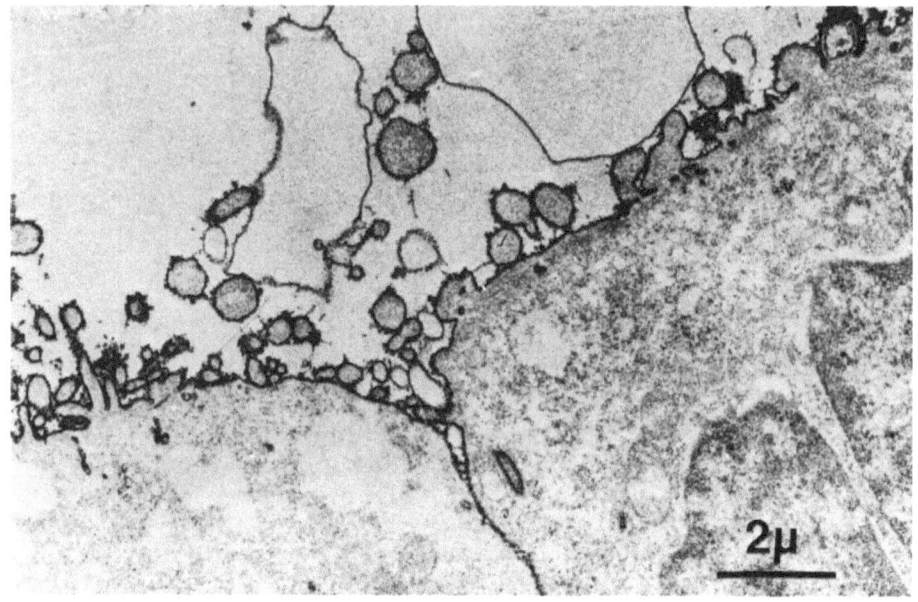

Fig. 2. Surface of cultured human skin epithelium cells (NCTC
strain 2544; Flow Lab., U.K.) after trypsin incubation (0.005 %,
30 minutes at 20 °C). Various protrusions and blebs of the plasma
membranes. Glycocalyx was marked with ruthenium red according to the
method of Luft (1971).

        The glycoproteins of the cell surface play a major role in the
extra-intracellular flow of informations. In this connection the
extreme protease sensitive fibronectin (FN) is best known with
regard to its function. In the extracellular space FN is found in
dimeric form in the intravasal space in monomeric form as cold insoluble protein (literature Erickson et al., 1981). At the cell
surface FN is coupled on certain integrative membrane proteins
which are interconnected with the filament system of the cytoskeleton on the inner parts of the membrane. There is evidence that FN
is directly transmembraneously coupled with contractile proteins of
the cytoskeleton (Singer, 1982). By that a direct transfer of information is given from the extra- to the intracellular space.
Different to that tumor cells have less FN at their surface besides
that a transmembraneous coupling on cytoplasmic filaments is also

missing (Singer, 1982). At least for fibroblasts it is shown that
a part of heparin (or heparansulfate) and fibronectin bound in the
glycocalyx participates in the regulation of the synthesis of collagen (Pickrell, 1981). For the intercellular flow of information it
may be important that FN synthetized in the extracellular matrix
by fibroblasts is associated with different types of collagen, procollagen and PG. Fibronectin is also bound on different native proteins with collagen-like sequences as acetylcholine or C1q which is
the first component of the complement activation after binding on
antigen-antibody complexes (Burrell and Hill, 1981). In the extracellular space FN couples on collagen where it is also attacked
by collagenase (Pickrell, 1981). FN may therefore be considered as
a protease sensitive matrix determining protein. The flow of information between the cells via closely located glycocalyxes or
via intercalated matrix areas is apparently managed to a good deal
by FN and its binding sites at the PG, collagen, procollagen, plasma membrane proteins or via transmembraneous coupling to the cytoskeleton. Of real influence on spreading and modulation of informations may be the extraordinary plasmin sensitivity of FN. The
intra - extravasal plasmin concentration depends on the catecholamine level. An increase of the catecholamine concentration elicites
an augemented synthesis of plasminogen activators by various cell-
types especially the capillary endothelium. Mast cells counter-re-
gulate these processes with their granula-bound spectrum of biologically active substances. Such substances contain highly active
plasmin inhibitors of the Kunitz-type (Aprotinin; Fritz et al.,
1979) and also heparin as activator of antithrombin III. Plasmin
inhibitors of the aprotinin-type are part of the low molecular
weight inhibitors which easily penetrate the capillary barrier in
both directions. This is possible since vasoactive substances are
released with the mast cell granula. These inhibitors have at the
same time a selective binding capacity with rescoproteins and proteoglycans (Heine and Schaeg, 1979). Besides that,they have an inhibitory effect on decoupling of the glycosaminoglycans from their
protein backbone induced by serin proteases. In addition,a steady
level of protease-protease inhibitors necessary for the intercellular flow of informations is performed by amoeboid movable macrophages and leukocytes (review Page et al., 1978). As in vitro
studies have shown they release a wide spectrum of proteases which
mostly are inhibited by heparin-antithrombin III. By that matter
mast cells apparently control macrophages (Seljelid et al., 1980).
Mast cells are found not in their close neighborhood but also mast
cells are neutrop towards terminal vegetative axons and are seen
always in close vicinty to the capillaries. Polymorphonuclear leukocytes participate in those cellular informational recoupling besides their immunogenic competence in the special way of releasing
wide spectrum inhibitors such as α_2-macroglobulins (Stimson and
Blackstock, 1975).

These evidently complicated mechanisms may be reduced to a quite simple common basis. The in vitro experiment shows that changes of the oxygen tension (PO_2) causes lysosomal enzyme release from human peripheral blood leukocytes (Skosey et al., 1981). In vivo a lowered tissue PO_2 (phenomenon of hidden acidosis) also stimulates phagocytotic cells to release lysosomal enzymes which themselves provoke the mast cells. The resulting changes in the intercellular matrix finally induce cellular reactions just characterized which regulate on the actual level of metabolic demands.

REFERENCES

Bennett, H. S., 1963, Morphological aspects of extracellular poly-saccarides, J. Histochem. Cytochem., 11:14.
Burrell, R., and Hill, J.O., 1976, The effect of respiratory immunization on cell-mediated immune effector cells of the lung, Clin. Exp. Immunol., 24:116.
Comper, W., and Laurent, T., 1978, Physiological function of connective tissue polysaccarides, Physiol. Rev., 58:255.
De Duve, C., and Wattiaux, R., 1966, Functions of lysosomes, Ann. Rev. Physiol., 28:435.
Erickson, H. P., Carell, N., and McDonagh, J., 1981, Fibronectin molecule visualized in electron microscopy: A long, thin, flexible strand, J. Cell Biol., 91:673.
Fritz, H., Kruck, J., Rüsse, I., and Liebich, H. G., 1979, Immunfluorescence studies indicate that the basic trypsin-Kallikrein-inhibitor of bovine organs (Trasylol) originates from mast cells, Hoppe-Seyler's Z. Physiol. Chem., 360:437.
Heine, H., Schaeg, G., 1979, Informationssteuerung in der vegetativen Peripherie, Z. Hautkr., 54:590.
Huet, Ch., and Herzberg, M., 1973, Effects of enzymes and EDTA on ruthenium red and concanavalin A labeling of the cell surface, J. Ultrastruct. Res., 42:186.
Karnovsky, M. J., Kleinfeld, A. M., Hoover, R. L., and Klausner, R. D., 1982, Concept of lipid domains in membranes, J. Cell Biol., 94:1.
Lojda, Z., 1981, Proteinase in Pathology. Usefulness of histochemical methods, J. Histochem. Cytochem., 29:481.
Luft, J. H., 1971, Ruthenium red and violet. II.Fine structural localization in animal tissues, Anat. Rec., 197:369.
Page, R. C., Davies, Ph., and Allison, A.C., 1978, The macrophage as a secretory cell, Int. Rev. Cytol., 52:119.
Pickrell, J. A., 1981, Interaction of mesenchymal cells with the extracellular matrix-A speculativemodel, in: "Lung Connective Tissue: Location, Metabolism, and Response to Injury," J. A. Pickrell, ed., CRC Press, Boca Raton.
Seljelid, R., Bäckström, G., and Lindahl, U., 1980, Proteinase activity in macrophage cultures, Exp. Cell. Res., 129:478.

Singer, I. I., 1982, Association of fibronectin and vinculin with focal contacts and stress fibers in stationary hamster fibroblasts, J. Cell Biol., 92:398.

Skosey, J., Chow, D. C., Nusinow, S., May, J., Gestautas, V., and Niwa, Y., 1981, Effect of oxygen tension on human peripheral blood leukocytes: Lysosomal enzyme release and metabolic responses during phagocytosis, J. Cell Biol., 88: 358.

Spicer, S. S., Baron, D. A., Sato, A., and Schulte, B. A., 1981, Variability of cell surface glycoconjugates - Relation to differences in cell function, J. Histochem. Cytochem., 29: 994.

Stimson, W. H., and Blackstock, J. C., 1975, Synthesis of a pregnancy-associated α-macroglobulin by human leucocytes, Experientia, 31:371.

Zucker-Franklin, D., Lavie, G., and Franklin, E.C., 1981, Demonstration of membrane-bound proteolytic activity on the surface of mononuclear leukocytes, J. Histochem. Cytochem., 29:451.

REGULATORY PROTEOLYSIS DURING CORTICOSTEROID HORMONE ACTION

M. K. Agarwal[1]

Centre National de la Recherche Scientifique and
Laboratoire de Physio-Hormono-Récepterologie,
Université Pierre et Marie Curie, Paris 75006

INTRODUCTION

Proteases are involved at various stages in the mechanism of hormone action in the mammal. Proteolytic enzymes are causal to the genesis of physiologically active hormones from prepro- and pro- polypeptides. In a subsequent step, the role of proteases in the action of cyclases, and possibly in the coupling mechanism between hormone receptors and cyclases, too, has been known for some time. More recently, regulatory proteolysis has been implicated in the genesis of receptors for various hormonal steroids.

The initial step in the mechanism of corticosteroid hormone action is said to consist of the association of the steroid molecule with its specific intracellular receptor. All five major groups of steroids (androgens, estrogens, progestagens, glucocorticoids and mineralocorticoids) are believed to bind with high affinity to a cytoplasmic protein in the appropriate target tissue. Translocation of the steroid-receptor complex to nuclear acceptor sites initiates organ specific physiological modulation (for recent reviews see 1-4).

In most instances, the study of hormone-receptor binding is limited to attempts designed to assess the association-dissociation constants of the complex formed between the cytosol vector and a steroid molecule alone or in presence of an agonist or antagonist of choice. Analysis by Scatchard plots of such results is based upon the widely accepted dogma that the receptor protein is a

---

[1]All correspondance should be addressed to:
15 rue de l'Ecole de Médecine, 75270 Paris Cédex 06, France.

simple, unitary entity whose physiological activity is based solely
upon Mass action. Thus, increase and decrease in the quantity of a
steroid bound to high affinity receptor sites is directly equated
with positive and negative modulation, respectively, of the biological activity of the test material.

As has been shown in several other biological systems, it is
entirely conceivable that "proreceptor", "isoreceptor", "activator",
and "inhibitor" proteins may actually be present in cell cytoplasm.
Thus, competition studies alone can not distinguish between the
possibility where an agonist may actually alter binding of an
active material to the same protein species versus saturation of
another population having similar affinity constants. The same is
true when an antagonist is considered especially since several,
biologically active, steroids exhibit both agonist and antagonist
activity, depending upon the animal species and the organ. To some
extent, difficulty in technical separation of steroid-receptor
complexes impaired conceptual development in this regard (1-5).

Our work on steroid receptors dates back to 1969 when this
field was just beginning to emerge and was looked upon with some
suspicion. At that time we had already established very well
defined chromatographic procedures capable of resolving high affinity, low capacity hormone-ligand complexes from low affinity,
high capacity hormone-transcortin or hormone-albumin binding (6,7).
Analysis of glucocorticoid responsive target tissues, primarily
liver, by these procedures had revealed that the receptor resolves
into a polymorphic, heterogeneous entity of which only a given
subpopulation may be saturated depending upon the nature of the
target organ and the steroid (6,7). This work has been extended
to other organs in recent years in our laboratory and we have
studied a whole array of agonists and antagonists (1-5).

What is the nature of this multiplicity? Among other explanations is the possibility that endogenous proteases could modify
the tissue receptor during preparation of cytosol. This derives
support from the observation that proteases and their inhibitors
can influence hormone vector binding in vitro (8-10).

The purpose of this review is to summarize the nature of
receptor multiplicity and to analyze the influence of various
proteolytic enzymes, and their inhibitors, on the conformation of
the tissue receptor for various classes of hormones.

Over the past years the work in our laboratory has been aided by
grants from: Centre National de la Recherche Scientifique (AI 03
1917); Institut National de la Santé et de la Recherche Médicale
(CRL 76.5.001.4) and the D.G.R.S.T. (IMB 75.7.0744; BFM 76.7.0725).

STUDIES ON THE CONFORMATION OF STEROID HORMONE RECEPTORS

In Table 1 a summary is provided of our results on the conformation of various steroid receptor subpopulations. The GR_1 component of GR is revealed more readily with synthetic steroids (dexamethasone, triamcinolone acetonide); GR_2 has hitherto been detected only in the liver, whereas GR_4 has been observed in all tissues thus far tested. Both GR_2 and GR_4 are labelled exclusively with natural steroids such as corticosterone and GR_3 with only synthetic molecules (1-7).

This sort of polymorphism is even more evident with MR where various grades of agonists and antagonists are more readily available than with GR. Thus, MR_1 and MR_2 were optimally labelled with aldosterone in rat kidney whereas deoxycorticosterone filled the MR_4 component that coeluted with CBG, albumin and GR_4 but the binding of 18-hydroxy-deoxycorticosterone was limited exclusively to the MR_1 moiety. Thus, the relative physiological action in vivo in the order: aldosterone > DOC > 18-OH-DOC seems to be expressed via preferential saturation of the MR_2, MR_4 and MR_1 components, respectively. Even more recently it has been shown (12,13) that progestagens (progesterone, R-5020), known to antagonize the action of mineralocorticoids in vivo (3,4), saturate the MR_4 entity almost exclusively for which the name Protoreceptor was proposed (12). In addition, the relative abundance of MR_2 is dependent upon the species; rat kidney possesses a lot more MR_2 whereas human kidney has little if any MR_2 activity (14).

More recently liver ER and AR were analyzed in our laboratory in this context (11). Whereas tamoxifen exerts potent antiestrogenic action in ER dependent breast cancer, it was largely without effect on rat liver gluconeogenesis which could be dramatically diminished by a number of estrogens and androgens (11). Although estradiol was preferentially bound to an ER_4 component that coeluted with CBG from DE-52 columns, ^3H-tamoxifen labelled the ER_3 moiety that was clearly distinct from transcortin. Similarly, testosterone was bound to the AR_4 component but R-1881 was eluted in the AR_3 region. Estradiol and testosterone also labelled the ER_1 and AR_1 components albeit with lesser affinity than for ER_4 and AR_4, respectively.

All these data have been summarized in Table 1. It has previously been shown that coelution with CBG is clearly distinct from transcortin itself but these arguments will not be repeated here again (1-5, 11-14).

Glucocorticoid Receptor = GR; Mineralocorticoid Receptor = MR; Androgen Receptor = AR; Estrogen Receptor = ER; Corticosteroid Binding Globulin = CBG = Transcortin. DOC = Deoxycorticosterone; 18-OH-DOC = 18-hydroxy-deoxycorticosterone.

Table 1. Selected Hydrodynamic Properties of Various
Steroid Receptor Components.

Component[a]	Steroid Specificity	Organ Distribution	M Phosphate[b]	Conductance[c] (miliSiemens)
GR_1	Synthetic > Natural	All tested	0.001 M	0.18 0.2
GR_2	Cortico-sterone	Liver only	0.02 M	5.0
GR_3	Synthetic only	All tested	0.04 M	5.8
GR_4	Natural only	All tested	0.06 M	9.2
MR_1	18-OH-DOC	Kidney	0.001 M	0.2
MR_2	Aldosterone > TA	Kidney, Heart?	0.006 M	2.3
$MR_4 (=MR_3)$	DOC Progesterone R-5020	Kidney	0.06 M	9.2
AR_1	Testosterone	Liver	0.001 M	
AR_3	R-1881	Liver	0.04 M	
AR_4	Testosterone	Liver	0.06 M	
ER_1	Estradiol	Liver	0.001 M	
ER_3	Tamoxifen	Liver	0.04 M	
ER_4	Estradiol	Liver	0.06 M	
Transcortin	Natural corticoids	Serum, Tissue???	0.06 M	9.8
Albumin	All tested	Plasma	0.06 M	8.9

DOC = Deoxycorticosterone; TA = Triamcinolone acetonide.

[a]Abbreviations have been explained in the text.

[b]Elution molarities from DEAE-cellulose-52 column (0.001-0.2 M PO_4)

[c]The peak of the steroid-receptor complex from the DE-52 column
was brought to 20°C for determination of specific conductivity.
For further details see (1-7, 11-14).

INFLUENCE OF PROTEASES AND INHIBITORS ON HORMONE-RECEPTOR BINDING

Data in Table 2 show the binding of five different hormones to their respective tissue vectors. Both papain and trypsin were effective in decreasing the binding of ^{3}H-cortisol to liver high affinity receptors but α-chymotrypsin was definitely weaker in this respect and pepsin was totally without effect (not shown). Papain was also very potent with liver androgen, but not estrogen, receptors and the binding decreased effectively. Surprisingly, trypsin actually increased the binding of both sex steroids to their liver vectors, contrary to the results obtained with cortisol, and α-chymotrypsin decreased estradiol binding but increased the binding of testosterone in the liver. Kidney bound ^{3}H-aldosterone was decreased in a dose dependent manner with trypsin and α-chymotrypsin, equally, whereas the latter was less potent in the liver in this respect when cortisol was used as the hormone. Papain was the most effective protease in diminishing kidney MR saturation and pepsin was without effect. The binding of progesterone in the kidney was diminished in the order: α-chymotrypsin < trypsin < papain, and pepsin was ineffective (not shown). Collectively, these results show that the availability of protease substrate on different receptors is not the same; the fact that progesterone binding was not affected in the same manner as aldosterone binding in the kidney would furthermore argue for receptor heterogeneity.

Table 2. The Effect of Various Proteases on the Binding of Different Steroid Hormones to their Cellular Receptor.

	Hormone concentration was 10^{-7} M ^{3}H-Steroid				
Enzyme Conc	Cortisol	Estradiol	Testosterone	Aldosterone	Progesterone
	--- Liver -------------------------------				
None	1697	2448	1857	1261	4106
Papain					
100 μg	640	1165	1199	555	1678
500 μg	264	1011	908	206	-----
2000 μg	84	1024	443	12	----
Trypsin					
100 μg	710	9769	3565	671	2823
200 μg	518	10903	3782	541	1243
2000 μg	358	9067	3012	349	1020
α-Chymotrypsin					
100 μg	1121	3260	3293	670	2390
200 μg	967	275	3613	633	2191
2000 μg	724	---	3545	381	1695

In other studies, the influence of trypsin was totally reversed by lima bean trypsin inhibitor on a mole per mole basis. This confirms the fact that the effect of protease was due to the enzyme itself and not related to the theoretical presence of a contaminant.

Attention was next directed to the possibility that endogenous proteases could be responsible for receptor processing and its eventual conformation during chromatography. Organ cytosol was therefore incubated with the steroid of choice alone or in presence of 10 mg of a protease inhibitor. Data in Table 3 show that neither trypsin inhibitor alone nor ε-aminocaproic acid (3070) influenced the binding of either cortisol or aldosterone to GR and MR in the liver and the kidney, respectively; both inhibitors increased the binding of estradiol to liver ER under these conditions. Leupeptin very effectively decreased cortisol and estradiol binding to liver GR and ER, respectively, but had no effect on aldosterone binding to renal MR. Contrarily, liver testosterone binding to AR increased with leupeptin, decreased in presence of ε-aminocaproic acid, and remained unaltered by trypsin inhibitor. Thus, liver vectors for sex steroids exhibited differences not only among each other but also as compared to liver GR whereas kidney MR appeared entirely different in this respect.

Table 3. The Influence of Selected Protease Inhibitors on the Binding of Steroids Hormones to Cellular Receptors.

Inhibitor (10 mg)	Hormone Concentration = 10^{-7} M in all cases			
	Cortisol	Estradiol	Testosterone	Aldosterone
	---- Liver --------------------------			--Kidney -----------
None	246	912	563	461
3070	304	1156	119	463
Trypsin Inhibitor	215	1189	552	580
Leupeptin	47	570	884	378

MODIFICATION OF RECEPTOR STRUCTURE BY PROTEASES AND INHIBITORS

Data in Fig. 1a show the elution profile of liver cytosol - ^{3}H-corticosterone complex revealing the GR_1 and GR_2 moities of the glucocorticoid receptor and the transcortin like binder (T). When a total of 1 mg trypsin was present during the incubation of cytosol with the tritiated hormone prior to chromatography, another component (I) was made evident (Fig. 1b), in addition to the components shown in Fig. 1a without trypsin. With 5 mg trypsin, only a portion of the peak in the T region could be observed with concomitant increase in peaks in the low ionic prewash, and with total abolition of peaks eluted in the intermediate concentrations of the phosphate gradient (such as GR_2). Under conditions of total proteolysis (incubation with trypsin at 20°C prior to chromatography) no bound radioactivity could be eluted (not shown). It therefore appears unlikely that the multiplicity of peaks may be the expression of endogenous proteolysis under normal conditions. In other experiments, papain had a similar effect but did not induce additional peaks (such as I with trypsin). Thus, the component I can not be said to represent protease-hormone complex especially since trypsin[3]H-corticosterone mixtures eluted in the void volume of the DE-52 column.

Data in Fig. 1c show that the inhibitor specific for trypsin did not alter the elution profile of liver GR saturated with corticosterone. Similar results were obtained when 3070 was used in place of trypsin inhibitor (Fig. 1d). These results establish that endogenous proteolysis does not appear to be responsible for the multiplicity of peaks observed with liver GR.

All results in tables 2 and 3 are expressed as CPM/mg protein. Further details have been published before (2, 5, 15). Briefly, 0.5 ml cytosol was incubated in presence of the indicated tritiated steroid in presence of either the protease (table 2) or the inhibitor (table 3). All determinations were in triplicate and each experiment was repeated thrice. All results were corrected for non-specific binding.

Data in Fig. 2a show the MR_1, MR_2 and GR like entities when kidney cytosol bound ^{3}H-aldosterone was fractionated on DE-52 columns. All these components still persisted in presence of 100 mg of trypsin inhibitor during organ homogenization (Fig. 2b), although the MR_2 component appeared to elute into two entities. In other experiments, 3070 was without effect under these conditions (not shown). Data in Fig. 2c show that 5mg papain effectively decreased the quantity of bound aldosterone but trypsin was less

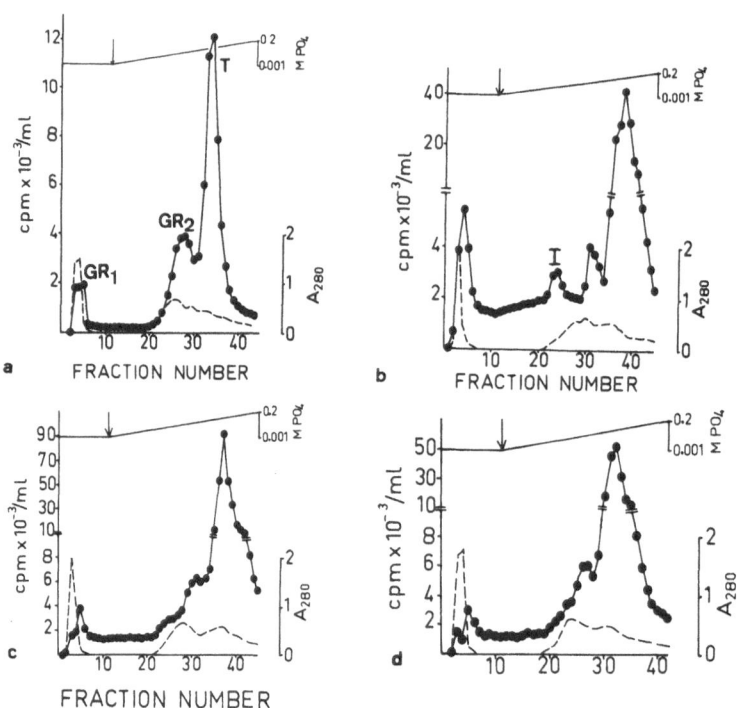

Fig. 1. Influence of Proteases and Inhibitors on Liver GR Profile
        from DEAE-52 columns.
4 ml liver cytosol was incubated in presence of 10^{-7} M ^{3}H-corti-
costerone and 0 (a), or 1 (b) mg trypsin; 100 mg trypsin inhibitor
(c) or 3070 (d) was added just prior to organ homogenization in
order to inhibit endogenous proteolysis and cytosol was thereafter
incubated with tritiated corticosterone. After 1 h in presence of
the radioactive hormone, free steroid was removed by charcoal
treatment and the cytosol loaded onto the DEAE-cellulose-52
columns equilibrated with 0.001 M phosphate buffer, pH 7.5. After
an initial prewash with this buffer, protein was eluted by a
gradient between 0.001 and 0.2 M phosphate, pH 7.5. Fractions
were processed for radioactivity determination and absorbance.
For further details see earlier publications (1-7, 11-15).
-------- A_{280}.   ●----------● ^{3}H.

effective under these conditions. Data in Fig. 2d show that serum
bound corticosterone, in presence of 5 mg trypsin, did not elute
in GR or MR positions noted above. Thus, contamination with CBG
does not appear to lead to receptor multiplicity.

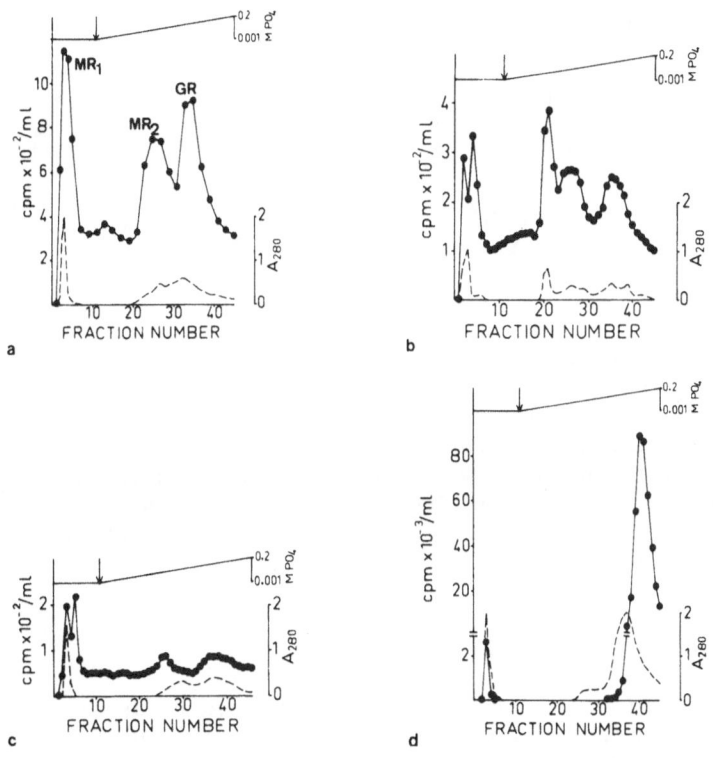

Fig. 2. Influence of Proteases and Inhibitors on Kidney Receptor
        Profile eluted from DEAE-52 Columns.
4 ml renal cytosol was incubated with 10^{-8} M of ^{3}H-aldosterone
alone (a) or in presence of 100 mg trypsin inhibitor (b), or 5 mg
papain (c). Trypsin inhibitor was added at the time of organ
homogenization whereas papain was added just prior to incubation
with the steroid. 4 ml serum was incubated with 10^{-8} M ^{3}H-cortico-
sterone and 5 mg trypsin. Chromatography was thereafter performed
as for Fig. 1. Further details have been published earlier and may
be consulted (1-7, 11-15).

Data in Fig. 3 show that limiting concentrations of α-chymotrypsin did not alter the elution profile of liver estradiol
binders (compare figs. 3a and 3b). The intermediate peak in Fig.
2b was not seen in 2a because of the relative concentrations of
the hormone (10^{-8} and 10^{-7} M respectively). With more complete
digestion (4 mg of the protease), bound radioactivity was limited
to the component in the low ionic prewash (not shown). Similar
results were obtained when testosterone was used in place of
estradiol.

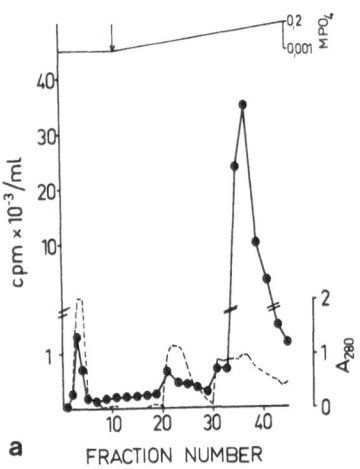
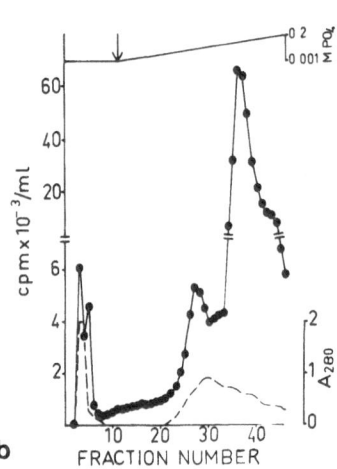

Fig. 3. Influence of Proteases on Rat Liver Sex Steroid Receptors.
4 ml liver cytosol was incubated with either 10^{-8} or 10^{-7} M of
^{3}H-estradiol (a and b respectively) 1 mg of α-chymotrypsin was
added with the steroid for fig. 3b. Chromatography was thereafter
performed on DE-52 columns as for Figs. 1 and 2. For further
details see (1-7, 11-15).

CONCLUSIONS AND PERSEPECTIVE

Experiments described here were designed to assess whether
the multiple peaks of various hormone receptors in the liver and
the kidney could be written off merely as a result of endogenous
proteolysis during organ homogenization, cytosol preparation and
subsequent chromatography. All results consistently show that
multiplicity is a property innate to the nature of the receptor
as previously described in our Multipolar Model (5). This is not
to say that endogenous proteolytic action does not have any
influence on receptor function since acceptor translocation and
transformation factors in the cytoplasm were not explored. It is,
however, a false notion to entertain when the receptor is conceived
of as a single protein unit capable of accepting all grades of
agonist and antagonist activity. This ignorance arises primarily

from the widespread use of Scatchard analysis, alone, to study the receptor function. One can not find multiplicity when one is not looking for it and when all data are faithfully interpreted to fit the preconceived notion of receptor homogeneity. Rather, concepts should be advanced to explain data.

From a totally physiological view point, too, receptor multiplicity would appear imperative for hormone activity. Thus, the affinity of DOC and of 18-OH-DOC is only 20% that of aldosterone when MR is assayed by classical methods. It therefore implies that for 18-OH-DOC to be active in vivo, 5-10 fold greater concentrations of this hormone would be needed so that aldosterone may be replaced by it from the receptor. In reality, however, only trace amounts of 18-OH-DOC are produced and DOC production is only little more than 18-OH-DOC formation. Yet both these steroids are active in vivo. How? The results shown in Fig. 2, and published in far greater detail elsewhere (1-7, 11-16), resolve this dilemma very nicely. DOC saturates the MR_4 whereas 18-OH-DOC has special affinity for the MR_1 entity; neither would need to compete with aldosterone for MR_2 to initiate physiological action in vivo.

In more recent studies, tamoxifen has been shown to satuarte ER_3 and estradiol ER_4. This can explain the fact that tamoxifen is a potent antiestrogen in the mammary gland but a weak agonist in the liver (11).

Further support for the Multipolar Model stems from the observation that GR_3 appears several days post partum in the liver whereas GR_4 is present even at birth. Liver becomes responsive to steroid hormone modulation approximately 7-10 days post partum (17).

The manner in which proteolytic action is limited in time and space must await further experimental work. It is, however, paradoxical that the most effective proteases and inhibitors are not of mammalian origin. In fact, the role of endogenous tissue proteases has received little attention. Thus, papain and antipain along with leupeptins are widely used with great efficacy but macroglobulins, somatomedins and lysosomal enzymes appear to have only minor effect. Activation of proteases during various diseases in the human may offer a good opportunity to delineate regulatory proteolysis during hormone action.

REFERENCES

1. M. K. Agarwal (ed) Multiple Molecular Forms of Steroid Hormone Receptors, Elsevier/North Holland (1977).
2. M. K. Agarwal (ed) Proteases and Hormones, Elsevier (1979).
3. M. K. Agarwal (ed) Antihormones, Elsevier (1979).

4.   M. K. Agarwal (ed) Hormone Antagonists, Walter de Gruyter,
     Berlin, New York (1982).
5.   M. K. Agarwal, Physical characterization of cytoplasmic gluco-
     and mineralo-steroid receptors, FEBS Letts. 85:1 (1978).
6.   R. S. Snart, N. N. Sanyal, and M. K. Agarwal, Binding of corti-
     costerone in rat liver, J. Endocrinol. 47:149 (1970).
7.   R. S. Snart, R. E. Shepherd, and M. K. Agarwal, Studies of cor-
     ticosterone binding in rat liver, Hormones 3:293 (1972).
8.   O. Wrange, and J. A. Gustafsson, Separation of the hormone and
     DNA binding sites of the hepatic glucocorticoid receptor by
     means of proteolysis, J. Biol. Chem. 253:856 (1978).
9.   C. Notides, D. E. Hamilton, and J. H. Rudolph, The action of
     a human uterine protease on the estrogen receptor, Endo-
     crinology 93:210 (1973).
10. M. R. Sherman, L. A. Pickering, F. M. Rollwagen, and L. K.
     Miller, Mero-receptors: proteolytic fragments of receptors
     containing the steroid binding sites, Fed. Proc. 37:167
     (1978).
11. M. K. Agarwal, Paradoxical nature of estrogen agonist and
     antagonist binding in rat liver, Biochem. Biophys. Res.
     Comm. 109:291 (1982).
12. M. K. Agarwal, J. Paillard, and M. Philippe, Evidence for an
     unusual multifunctional protoreceptor in hormone action,
     Experientia 36:1010 (1980).
13. M. K. Agarwal, and J. Paillard, Paradoxical nature of mineralo-
     corticoid receptor antagonism by progestins, Biochem. Bio-
     phys. Res. Comm. 89:77 (1979).
14. J. Paillard, E. Baviera, and M. K. Agarwal, Gluco- and mineralo-
     corticoid receptors in human liver and kidney, Biochem. Med.
     24:210 (1980).
15. M. K. Agarwal, and M. Philippe, The influence of various
     proteases and inhibitors on steroid hormone receptors in
     rat liver and kidney, Biochem. Med. 26:265 (1981).
16. M. K. Agarwal, Differential binding to renal receptor sub-
     populations as an explanation of mineralocorticoid agonist
     action, FEBS Letts. 67:260 (1976).
17. M. K. Agarwal, Genesis of rat liver glucocorticoid receptor
     multiplicity as a function of age, Biochem. Biophys. Res.
     Comm. 106:1412 (1982).

PROTEASES IN HORMONE PRODUCTION AND METABOLISM

Willa A. Hsueh

Section of Endocrinology, Department of Medicine
LAC/USC Medical Center
Los Angeles, California 90033

The physiologic effect of a hormone is governed by its bio-availability to receptors on the target organ. Chertow (1) has suggested that proteases can potentially regulate hormone availability by 1) intracellular conversion of precursor hormones to active hormones, 2) degradation of hormone in the cell prior to secretion, 3) facilitation of release of hormone from the cell, 4) activation or inactivation of the hormone in the circulation, and 5) degradation of hormone in target tissue. These regulatory mechanisms particularly apply to polypeptide hormones and enzymes, which are produced as larger molecular weight precursors and subsequently converted by limited proteolysis to their secreted forms.

Utilizing subcellular fractionation and pulse-labeling techniques (2,3), investigators have been successful in uncovering the cellular processes involved in hormone biosynthesis and secretion. General aspects of hormone biosynthesis are illustrated in Figure 1, although variations in the biosynthetic process exist for each endocrine system. "Preprohormone" is the primary translation product of messenger-RNA in the endocrine cell. The "pre" sequence, usually consisting of about 15-30 amino acids residing at the N-terminal, allows the ribosome to attach to the microsomal membrane so that the forming hormone can penetrate the membrane of the endoplasmic reticulum. Following penetration, a "clipase" located on the microsomal membrane immediately cleaves the "pre" segment leaving the "prohormone".

Examples of polypeptide prohormones are listed in Table I. The "pro" sequence can reside at the N-terminal as in the case of pro-PTH, at the C-terminal as in the case of progastrin, or in the middle, i.e., C-peptide of proinsulin. In general prohormones are

141

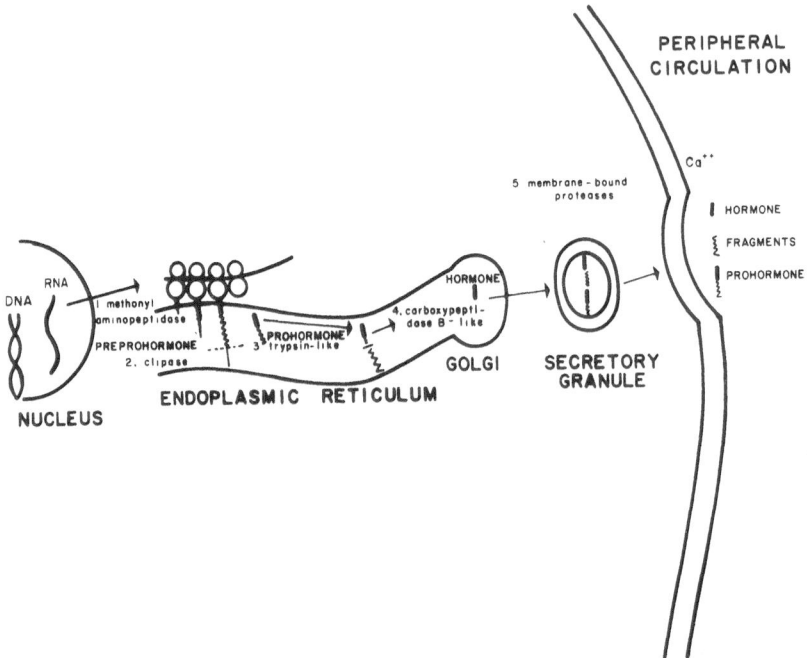

Figure 1:   Schematic representation of hormone biosynthesis. Pos-
            sible proteases involved in hormone processing are
            numbered. (See text for details).

biologically inactive, although several fragments can have biologic
activity, i.e., fragments of proACTH and endorphin. Chertow (1) has
postulated that the profragment may facilitate transport and processing of the hormone through the endoplasmic reticulum, may protect the secreted form from degradation, or enable the secreted form
to assume a conformation which allows for biologic activity.

     Following transport to the Golgi area, the prohormone is pack-
aged into secretory granules. These vesicles contain enzymes that
cleave the prohormone resulting in formation of the secreted form of
the hormone. Examples of these lysosomal enzymes are listed in
Table II. The particular enzyme involved in conversion of prohormone to secreted hormone may be specific to individual hormone systems. However, the prominent lysosomal enzymes of the cathepsin
family (B,D,R, etc.)are implicated to be responsible for physiologic conversion of prohormone to active hormone for an increasing
number of endocrine systems (4-8).

Table 1. Proteases Involved in Peptide Hormone Biosynthesis

Endopeptidases
Trypsin
Cathepsin B
Cathepsin D
Plasmin
Glandular Kallikrein
Exopeptidases
Carboxypeptidase B
Aminopeptidases

Table 2. Examples of ProHORMONES

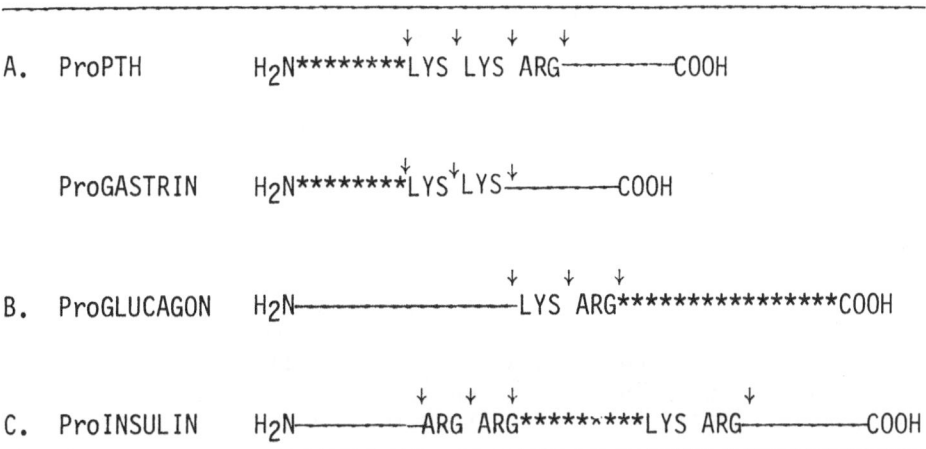

Prosequence of the molecule represented by ********.
Arrows represent sites of cleavage.

The packaged hormone is released by exocytosis. Conceivably proteolytic enzymes could enhance release of the hormone by acting on the membrane of the secretory granule or on the cell membrane.

If hormone release is suppressed, increased intracellular degradation or "crinophagy" of the hormone occurs. In this process crinophagic vacuoles digest secretory granules, and this has been observed by electron microscopy of several types of endocrine cells, e.g., β cells of the pancreas, parathyroid gland cells (1). Crinophagy may be an important physiologic mechanism regulating the amount of hormone available for cellular secretion, but the mecha-

nisms which regulate crinophagy are unclear. However, an increase
in crinophagic vacuolar activity is seen in tissue in which hormone
secretion has been inhibited. For example, bovine parathyroid tissue contains an increased number of phagocytic and crinophagic vacuoles in the presence of high calcium concentration, which inhibits
parathyroid hormone secretion (9). The enzymes which are involved
in crinophagy are not defined, but presumably they are lysosomal proteases.

Extracellular degradation of hormone occurs in blood, in target
organs, and in some nontarget organs such as liver and kidney. Multiple endopeptidases and exopeptidases appear to be involved in degradation. Some hormones have specific degradating enzymes such as
"TRHase", "LH-RHase", "somatostatinase", etc. (1).

Insulin provides a model of polypeptide hormone biosynthesis
and metabolism (10). In the β-cell of the pancreas, preproinsulin
is the major ribosomal product. Within a minute of entering the endoplasmic reticulum, preproinsulin is converted to proinsulin. Within the next 10-20 minutes proinsulin is transported through the endoplasmic reticulum and in the Golgi area is converted to insulin and
C-peptide. During the next 20 minutes, insulin, C-peptide, and some
proinsulin are packaged into secretory granules. Upon stimulation,
the secretory granule releases insulin and C-peptide in a 1:1 molar
ratio. Less than 10% of the total radioimmunoassayable insulin released is proinsulin. The structure of proinsulin is shown in Figure 2. C-peptide appears to allow correct folding of the proinsulin
molecule and alignment of the A and B chains (11). A trypsin-like
enzyme cleaves off C-peptide at arginine-glutamine and arginine-gly-
cine bonds leaving the A and B chains of insulin connected by two
disulfide bridges. The exact identity of this trypsin-like enzyme
is unknown. Candidates include trypsin, glandular kallikrein, plasmin, and cathepsin B (6,11-14). A carboxypeptidase B-like enzyme
then cleaves the two carboxyterminal arginines from the B chain,
which is necessary for the secreted insulin to have full biologic
activity (11).

Crinophagy of insulin has been demonstrated in pancreatic islets of normal rats and in rats in which insulin secretion is impaired following injection with diphenylhydantoin or diazoxide (15,
16). Crinophagy does not appear to be an acute response to suppression of insulin, for example by acute hypoglycemia, but is evident
only after prolonged suppression (17). C-peptide may affect crinophagy, as it inhibits intracellular degradation of insulin (18).
The liver and kidney are the major organs that metabolize insulin.
Following binding of insulin to a membrane receptor on the hepatocyte, the insulin-receptor complex is internalized in pinocytotic
vesicles. The exact correlation between receptor binding, hormone
internalization and hormone activity remains unclear and how intracellular degradation of the hormone in the target organ affects

HUMAN PROINSULIN

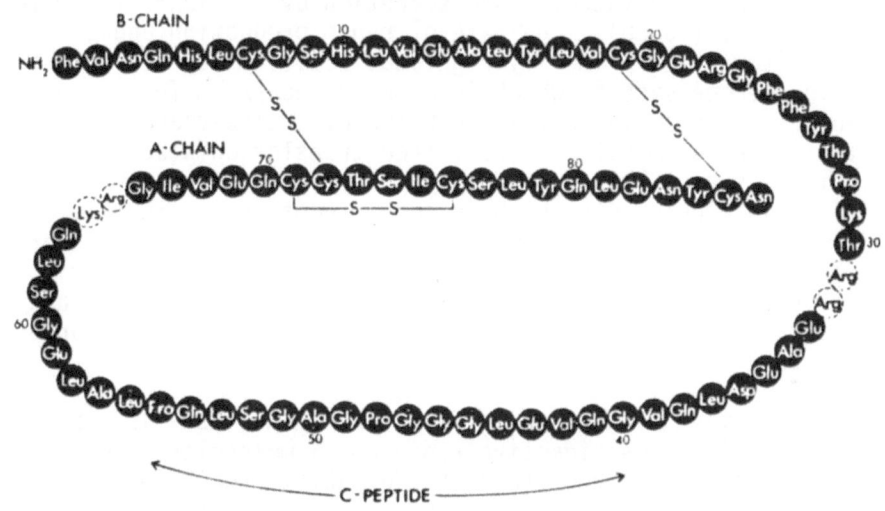

Figure 2:   Aminoacid sequence of proinsulin. Amino acid bonds
            involved in cleavage are denoted by open circles
            (Reprinted with permission of Oyer, et al J. Biol. Chem).

activity is unknown. Lysosomal proteases appear to degrade internalized insulin. These include an "insulin-specific" protease which
cleaves the B chain, a glutathione-insulin transhydrogenase which
reduces the disulfide bonds, and other endopeptidases (such as cathepsins B, D, L) which further degrade the A and B chains (19-21).

        Several clinical abnormalities stem from abnormal precursor
processing of insulin. Insulin-secreting tumors secrete large
amounts of proinsulin as well as insulin, as demonstrated by gel
filtration of plasma from patients with insulinoma (22). In fact,
the production of large amounts of proinsulin is often indicative
of a carcinomatous rather than adenomatous process. Another example
of abnormal precursor processing of insulin is the familial defect
in conversion of proinsulin to insulin (23,24). One such family
produces an abnormal proinsulin molecule. Another family appears
to have a deficiency of the proinsulin-cleaving enzyme within the
β cell. This defect results in high circulating levels of proinsulin, but rarely leads to insulin deficiency or diabetes mellitus.
Measurement of plasma levels of the "pro" sequence, or C-peptide,
have clinical utility. Levels of C-peptide reflect endogenous insulin secretion and may be particularly helpful in assessing insu-
lin-treated diabetic patients who may have circulating antibodies
to insulin or in patients suspected of surreptitious use of insulin.

Another less well understood system in which abnormalities of precursor processing have been postulated to result in clinical pathology is renin biosynthesis and secretion by the juxtaglomerular cells of the kidney (25,26). The rate of renin production and secretion into the circulation primarily regulates activity of the renin-angiotensin-aldosterone system, which is a major mechanism for blood pressure homeostasis and sodium-potassium balance. Pulse-chase studies in isolated canine glomeruli and cell-free translation studies of m-RNA from mouse submaxillary gland and kidney indicate that renin is synthesized as preprorenin which is converted to prorenin (27,28). Little information about the biochemical characterization of prorenin is available except that it appears to have a molecular weight about 8-10,000 greater than the active secreted form of the enzyme and that microsomal preparations from dog pancreas are capable of converting preprorenin to prorenin (27,29).

An inactive form of renin found in human plasma is a putative prorenin (25). Human inactive renin has a molecular weight of 50-55,000 which is larger than that of active renin 44-48,000. It can be activated in vitro by low pH and by a variety of proteases including trypsin, pepsin, plasmin, plasma and glandular kallikrein, and cathepsins B, D, and H (8,30-35). Comparison of these various techniques of activation yield quantitatively similar amounts of inactive renin in plasma (35,36). In vitro activation is generally not accompanied by an apparent reduction in molecular weight, although Luetscher et al. (8) have recently reported that treatment of renal inactive renin with cathepsins B, D, or H decreases the molecular weight of renal inactive renin to a size similar to that of active renin. They have suggested that one of the cathepsins may be a physiologic mediator of renin activation.

Inactive renin comprises 60-90% of the total renin concentration in normal human plasma and 10-50% of the total renin concentration in human renal cortical homogenate (25,37-40). Extensive biochemical characterization of semipurified preparations of renal and plasma inactive renin indirectly suggests structural homology between inactive renins from the two sources (28). In addition, Atlas et al. (37) have demonstrated the presence of inactive renin in perfusate of human cadaver kidney. We have recently demonstrated a significant gradient of inactive renin across the kidney in subjects receiving chronic treatment with converting enzyme inhibitor, which is known to stimulate inactive renin secretion (39). Taken together these data would suggest that the kidney is a major source of circulating inactive renin in man.

Abnormal precursor processing of renin may occur in two clinical situations. First, patients with renin-secreting tumors, either renal or nonrenal, have large amounts of inactive renin in the tumor and in their circulation (41,42). Atlas et al. (43) have demonstrated that the inactive renin of these tumors has similar biochemical

characteristics to renal and plasma inactive renin. They suggested
that since inactive renin is produced when extra-renal tissue is
transformed into renin-producing cells, there is tight genetic link-
age of active and inactive renin. The large amounts of inactive
renin in the tumor compared to normal plasma would be consistent
with ineffective precursor processing in tumor tissue.

Second, diabetic patients with nephropathy and hyporeninemic
hypoaldosteronism have 3-5 times the normal levels of inactive renin
in their plasma (44,45). In contrast, patients with nondiabetic
nephropathy and the syndrome have normal circulating levels of inactive renin as shown in Figure 3 (46). Upon stimulation of renin
production, the diabetics have an impaired active renin response, but
a marked rise in circulating inactive renin levels. Upon stimulation in the nondiabetics, there is a blunted rise in active renin
and no change in inactive renin. We postulate that diabetic nephropathy may specifically impair precursor processing of inactive
(pro-) renin to active renin, while other renal causes of hyporeninemic hypoaldosteronism may be characterized by impaired production
of all forms of renin. Whether an enzyme defect in diabetic hyporeninemic hypoaldosteronism accounts for abnormalities in renin activation is unknown. Renal kallikrein has been implicated to be an

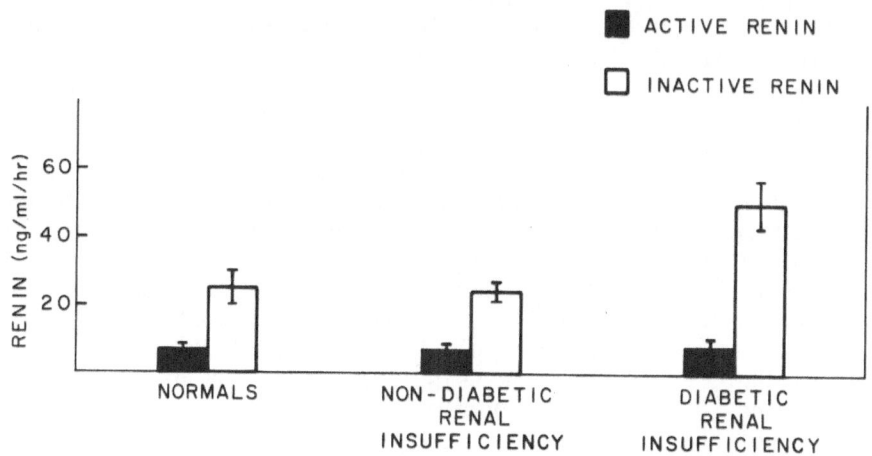

Figure 3:   Active and inactive renin concentration in blood of patients with hyperkalemia and diabetic or non-diabetic
renal insufficiency and age-matched normals (Reprinted
with permission of Hsueh, et al. Am. J. Nephrology).

in vivo activator of renin (33), and, indeed, these patients have
been demonstrated to have subnormal levels of urinary kallikrein
(47). However, subnormal levels of urinary kallikrein occur in patients with renal insufficiency and in patients with hypertension
(48) without the syndrome of hyporeninemic hypoaldosteronism, so
the significance of a renal kallikrein deficiency is unknown. Indirect evidence also suggests that prostaglandins may play a role in
conversion of inactive to active renin. Prostaglandins mediate active renin secretion in vivo (49) and some patients with hyporeninemic hypoaldosteronism have been reported to have low levels of urinary PGE_2 (50). Indeed, the prostaglandin synthetase inhibitor, indo-
methicin, has been shown to induce the syndrome of hyporeninemic hypoaldosteronism (51). Rodemann et al. (52) have suggested that PGE_2
can activate the lysosomal apparatus. Thus, one postulated mechanism
for the renin deficiency in diabetic hyporeninemic hypoaldosteronism
is that impaired prostaglandin production leads to inactivation of
lysosomal enzymes which prevents activation of renin. This could
result in a build up of inactive renin stores in the kidney and increased release of inactive renin into the circulation. Further
studies of the possible effect of prostaglandins on lysosomal proteases and on renin activation will help to evaluate this hypothesis.

REFERENCES

1. B.S. Chertow, The role of lysosomes and proteases in hormone se-
    cretion and degradation, Endo Rev., 2:137 (1981).
2. G. Palade, Intracellular aspects of the process of protein syn-
    thesis, Science, 189:347 (1975).
3. V.R. Lingappa and G. Blobel, Early events in the biosynthesis of
    secretory and membrane proteins: the signal hypothesis, Rec.
    Prog. Horm. Res. 36:451 (1980).
4. E. Barat, A. Patthy, and L. Graf, Action of cathepsin D on human
    β-lipotropin: a possible source of β-melanotropin, Proc. Natl.
    Acad. Sci. USA 76:6120 (1980).
5. R.V. Lewis, A.S. Stern, S. Kimura, J. Rossier, S. Stein, and S.
    Undenfriend, An about 50,000-Dalton protein in adrenal medulla:
    a common precursor of [Met]- and [Leu], Enkephalin. Sci. 208:
    1459 (1980).
6. R.B. Puri, K. Anjaneyulu, J.R. Kidwai, and V.K.M. Rao, In vitro
    conversion of proinsulin to insulin by cathepsin B and role
    of C-peptide, Acta Diabetol. Lat. 15:243 (1978).
7. J. Fischer, Precursor processing and metabolism of parathyroid
    hormone (PTH), this symposium.
8. J.A. Luetscher, J.W. Bialek, and G. Grislis, Human kidney cath-
    epsins B and H activate and lower the molecular weight of hu-
    man inactive renin, Clin. Expt. Hypertension (in press, 1982).
9. B.S. Chertow, D.J. Manke, G.A. Williams, G.R. Baker, G.K. Hargis,
    and R.J. Buschmann, Secretory and ultrastructural responses

of hyperfunctioning human parathyroid tissues to varying calcium concentration and vinblastine, Lab. Invest. 36:198 (1977).

10. D.F. Stemer and H.S. Tager, Biosynthesis of insulin and gluca-gon, in: "Endocrinology,", L.J. DeGroot, Grune and Stratton, New York (1979).

11. D.F. Steiner, W. Kemmler, H.S. Tager, A.H. Rubenstein, A. Lern-mark, and A. Zuhlke, Mechanisms in the biosynthesis of poly-peptide hormones, in: "Proteases and Biological Control," E. Reich, D.B. Rifkin, and E. Shaw, Cold Spring Harbor La-boratory, Col Spring Harbor (1975).

12. M.A. Virji, J.D. Vassalli, R.D. Estensen, and E. Reich, Plas-minogen activator of islets of Langerhans: modulation by glucose and correlation with insulin production, Proc. Natl. Acad. Sci. USA 77:875 (1980).

13. C.C. Yip, A bovine pancreatic enzyme catalyzing the conversion of proinsulin to insulin, Proc. Natl. Acad. Sci. USA 68:1312 (1971).

14. O. Ole-Moiyoi, G.S. Pinkus, J. Spragg, and K.F. Austen, Identi-fication of human glandular kallikrein in the β cells of the pancreas, N. Eng. J. Med. 200:1289 (1979).

15. G. Bommer, H-J. Schafer, and G. Koppel, Morphological effects of diazoxide and dephenylhydantoin on insulin secretion and biosynthesis in B cells of mice, Virchows. Arch. [Pathol. Anat.] 371:227 (1976).

16. W. Creutzfeldt, C. Cruetzfeldt, H. Frerichs, E. Perings, and R. Sickinger, The morphological substrate of the inhibition of insulin secretion by diazoxide, Horm. Metab. Res. 1:53 (1969).

17. C.L.G. Dumm, O.R. Rebolledo, and J.J. Gargliardino, Ultrastruc-tural responses of pancreatic β cells to metabolic alkalosis, Cell Tissue Res. 201:159 (1979).

18. G.A. Burghen, R.A. McKenzie, and T.M. Howarth, Evidence that C-peptide inhibits and divalent cations stimulate degrada-tion of insulin by insulin protease, Diabetes 20:46A (1981).

19. W.C. Duckworth, F.B. Stentz, M. Heinemann, and A.E. Kitabchi, Initial site of insulin cleavage by insulin protease, Proc. Natl. Acad. Sci. USA 76:635 (1979).

20. P.T. Varandi, Insulin degradation. V. Unmasking of glutathione-insulin transhydrogenase in rat liver microsomal membrane. Biochim. Biophys. Acta 304:1973 (1973).

21. H. Keilova, On the specificity and inhibition of cathepsins D and B, in: "Tissue Proteinases," A.J. Barett and J.T. Din-gle, North Holland Publishing Co., Amsterdam (1971).

22. A.H. Rubenstein, Insulin, proinsulin and C-peptide: Secretion, metabolism, regulation in health and disease, in: "Endocrin-ology," L.J. DeGroot, Grune and Stratton, New York (1979).

23. D.H. Gabbay, K. DeLuca, J.N. Fisher, Jr., M.E. Mako, and A.H. Rubenstein, Familial hyperproinsulinemia: An autosomal do-minant defect, N. Engl. J. Med. 294:911 (1976).

24. A.S. Kitabchi, Proinsulin and C-peptide: A review, Metabolism
    26:547 (1977).
25. J.E. Sealey, S.A. Atlas, and J.H. Laragh, Prorenin and other
    large molecular weight forms of renin, Endocrine Rev. 1:365
    (1980).
26. W.A. Hsueh, Inactive renin in human plasma--is it prorenin?,
    Min. Elect. Metab. 7:169 (1982).
27. W. Carlson, V. Dzau, S. Quay, J. Kreisberg, and E. Haber, Bio-
    synthesis of renin in the dog kidney: Evidence for the pres-
    ence of prorenin, Circulation 62:111 (1980).
28. K. Poulsen, J. Vuust, and T. Lund, Renin precursor from mouse
    kidney identified by cell-free translation of messenger
    RNA, Clin. Sci. 59:297 (1980).
29. V.J. Dzau, O. Lanaka, and R.E. Pratt, The nature of inactive
    renin and renin precursor and inactive renin, Clin. Exp.
    Hypertension, A4:1973 (1982).
30. R.M. Cooper, G.E. Murray, and D.H. Osmond, Trypsin-induced acti-
    vation or renin precursor in plasma of normal and anephric
    man, Circ. Res. 40:171 (1976).
31. B.J. Morris, Activation of human inactive ("pro-") renin by ca-
    thepsin D and pepsin, J. Clin. Endocrinol. Metab. 46:153
    (1978).
32. D.H. Osmond, E.K. Lo, A.Y. Loh, E.A. Zingg, and A.H. Hedlin,
    Kallikrein and plasmin as activators of inactive renin,
    Lancet 2:1375 (1978).
33. J.E. Sealey, S.A. Atlas, J.H. Laragh, N.B. Oza, and J.W. Ryan,
    Activation of prorenin-like substance in human plasma by
    trypsin and by urinary kallikrein, Hypertension 1:179 (1979).
34. J.E. Sealey, S.A. Atlas, J.H. Laragh, M. Silverberg, and A.P.
    Kaplan, Initiation of plasma prorenin activation by Hageman
    factor-dependent conversion of plasma prekallikrein to kalli-
    krein, Proc. Natl. Acad. Sci. USA 76:5914 (1979).
35. W.A. Hsueh, E.J. Carlson, and M. Israel-Hagman, Mechanism of
    acid-activation of renin: Role of kallikrein in renin acti-
    vation, Hypertension 3:I-22 (1981).
36. W.A. Hsueh, J.A. Luetscher, E.J. Carlson, and G. Grislis, In-
    active renin of high molecular weight (big renin) in normal
    human plasma, Hypertension 2:750 (1980).
37. S.A. Atlas, J.E. Sealey, B. Dharmgrongartama, T.E. Hesson, and
    J.H. Laragh, Detection and isolation of inactive, large
    molecular weight renin in human kidney and plasma, Hyper-
    tension 3:I-30 (1981).
38. J.J. Chang, M. Kisarag, H. Okamoto, and T. Inagami, Isolation
    and activation of inactive renin from human kidney and plas-
    ma: Plasma and renal inactive renin have different molecular
    weights, Hypertension 3:509 (1981).
39. W.A. Hsueh, E.J. Carlson, and V.J. Dzau, Characterization of
    renal and plasma inactive renin: Evidence for a renal source
    of circulating inactive renin, J. Clin. Invest. (in press,
    1983).

40. S.A. Atlas, J.E. Sealey, T.E. Hesson, A.P. Kaplan, J. Ménard, P. Corvol, and J.H. Laragh, Biochemical similarity of partially purified inactive renins from human plasma and kidney, Hypertension 4:II-86 (1982).
41. R.P. Day and J.A. Luetscher, Big renin: A possible prohormone in kidney and plasma of a patient with Wilms' tumor, J. Clin. Endocrinol. Metab. 38:923 (1974).
42. A. Mimran, B.J. Leckie, J.C. Fourcade, P. Beldet, A. Davratil, and P. Barjon, Blood pressure, renin-angiotensin system and urinary kallikrein in a case of juxtaglomerular cell tumor, Am. J. Med. 65:527 (1978).
43. M.C. Ruddy, S.A. Atlas, and F.G. Salerno, Hypertension associated with a renin-secreting adenocarcinoma of the pancreas, N. Engl. J. Med. 307:993 (1982).
44. A. Deleiva, A.R. Christlieb, J.C. Melby, C.A. Graham, R.P. Day, J.A. Luetscher, and P.G. Zager, Big renin and biosynthetic defect of aldosterone in diabetes mellitus, N. Engl. J. Med. 295:639 (1976).
45. W.A. Hsueh, E.J. Carlson, J.A. Luetscher, and G. Grislis, Activation and characterization of inactive big renin in plasma of patients with diabetic nephropathy and unusual active renin, J. Clin. Endocrinol. Metab. 51:535 (1980).
46. R. Goldstone, E.J. Carlson, and W.A. Hsueh, Evidence for two independent mechanisms of juxtaglomerular (JG) cell impairment in hyporeninemic hypoaldosteronism, Endocrine Society Program (1982).
47. J.A. Hahn, R.D. Zipser, A. Burg, R.A. Stone, P.R. Zia, W.A. Hsueh, and R. Horton, Studies of the renal vasoactive systems in hyporeninemic hypoaldosteronism, Prost. and Med. 6: 549 (1981).
48. J.A. Mitas, S.B. Levy, R. Holle, R. Frigon, and R.A. Stone, Urinary kallikrein activity in the hypertension of renal parenchymal disease, N. Engl. J. Med. 299:162 (1978).
49. J. Oates, R. Whorton, J. Gerkins, R. Banch, J. Hollified, and J. Frolich, The participation of prostaglandins in control of renin release, Fed. Proc. 38:72 (1979).
50. S.Y. Tan, I. Antonipillai, and P.J. Mulrow, Inactive renin and prostaglandin E_2 production in hyporeninemic hypoaldosteronism, J. Clin. Endocrinol. Metab. 51:849 (1980).
51. S.Y. Tan, R. Shapiro, R. Franco, H. Stockard, and P.J. Mulrow, Indomethacin-induced prostaglandin inhibition with hyperkalemia. A reversible cause of hyporeninemic hypoaldosteronism, Ann. Intern. Med. 90:783 (1979).
52. H.P. Rodemann, L. Waxman, and A.L. Goldberg, The stimulation of protein degradation in muscle by Ca^{2+} is mediated by prostaglandin E_2 and does not require the calcium-activated protease, J. Biol. Chem. 257:8716 (1982).

PRECURSOR PROCESSING AND METABOLISM OF PARATHYROID HORMONE:

REGULATION BY CALCIUM

Jan A. Fischer

Research Laboratory for Calcium Metabolism, Departments of Orthopedic Surgery (Balgrist) and Medicine University of Zurich, 8008 Zurich, Switzerland

INTRODUCTION

Parathyroid hormone (PTH) raises serum calcium levels through stimulation of the tubular reabsorption of calcium and of bone resorption, and of the formation of 1,25-dihydroxycholecalciferol enhancing the intestinal absorption of calcium (for references see Fischer, 1982). Conversely, the extracellular calcium concentration is the most important regulator of the biosynthesis and secretion, and possibly of the metabolism of the hormone. Indeed, biosynthesis and secretion of PTH are stimulated by low extracellular calcium concentrations (Moran et al., 1981; Sherwood et al., 1968). Biosynthetic precursors of PTH have not been detected outside of parathyroid cells. However, proportionally higher amounts of biologically inactive PTH fragments are secreted in the face of raised extracellular calcium levels (Hanley et al., 1978; Mayer et al., 1979). The metabolism of PTH, furthermore, takes place in peripheral organs such as the liver and the kidneys (Canterbury et al., 1975; Hruska et al., 1977).

PRECURSORS OF PARATHYROID HORMONE

The initial precursor of PTH is PreProPTH with 31 additional amino acid residues at the amino-terminal end of PTH-(1-84) (Fig. 1). PreProPTH was originally identified in a heterologous cell-free extract using wheat germ as the translation product of a messenger RNA isolated from calf parathyroid glands (Habener et al., 1978). The structure has subsequently been confirmed by analysis of the sequence of the desoxyribonucleic acid coding for PreProPTH (Kronenberg et al., 1979). PreProPTH is converted into ProPTH within one minute after completion of its synthesis on the rough

```
TAG CAG CTG ATG CTT TCT CAA AGT TGA GTA AAC CTG AGA AGG CTG ATA AAT TGA GCT GCT AAT ACA TTT
 1          10          20          30          40          50          60
```

```
                                            -31
                                            met met ser ala lys asp met val lys val met ile
GAA AGA AGA TTG TAT CCT AAG ACG TGT GTT AAT ATG ATG TCT GCA AAA GAC ATG GTT AAG GTA ATG ATT
 70          80          90          100         110         120         130
```

```
                                      -6                        -1 +1
val met leu ala ile cys phe leu ala arg ser asp gly lys ser val lys lys arg ala val ser glu
GTC ATG CTT GCC ATC TGT TTT CTT GCA AGA TCA GAT GGG AAG TCT GTT AAG AAG AGA GCT GTG AGT GAA
 140         150         160         170         180         190         200
```

```
                                                              +22
ile gln phe met his asn leu gly lys his leu ser ser met gly arg val glu trp leu arg lys lys leu
ATA CAG TTT ATG CAT AAC CTG GGC AAA CAT CTG AGC TCC ATG GAA AGA GTG GAA TGG CTG CGG AAA AAG CTA
 210         220         230         240         250         260         270
```

```
gln asp val his asn phe val ala leu gly ala ser ile ala tyr arg asp gly ser ser gln arg pro
CAG GAT GTG CAC AAC TTT GTT GCC CTT GGA GCT TCT ATA GCT TAC AGA GAT GGT AGT TCC CAG AGA CCT
 280         290         300         310         320         330         340
```

```
arg lys lys glu asp asn val leu val glu ser his gln lys ser leu gly glu ala asp lys ala asp
CGA AAA AAG GAA GAC AAT GTC CTG GTT GAG AGC CAT CAG AAA AGT CTT GGA GAA GCA GAC AAA GCT GAT
 350         360         370         380         390         400         410
```

```
                          +84
val asp val leu ile lys ala lys pro gln stop
GTG GAT GTA TTA ATT AAA GCT AAA CCC CAG TGA AAA CAG ATA TGA TCA GAT CA
 420         430         440         450         460         470
```

Figure 1. Nucleotide sequence of sense strand of parathyroid insert in pPTHml. Numbers above the protein sequence refer to amino acid position. -31 through -7 is the presequence, -6 through -1 the prosequence, and +1 through +84 the sequence of intact PTH (from Kronenberg et al., 1979).

endoplasmic reticulum (Fig. 2) (Habener et al., 1979).

Proparathyroid hormone has been isolated and identified with amino acid incorporation experiments (Hamilton et al., 1974). Pro-PTH consists of 90 amino acid residues with six residues still being attached to the amino-terminal end of intact PTH-(1-84). Immunoreactive ProPTH was detected in parathyroid tissue extracts, but not in the venous effluent of the parathyroid glands (Habener et al., 1974). ProPTH is converted to PTH-(1-84) at the Golgi region within 15 to 20 minutes after its first appearance in the tissue (Chu et al., 1973, Habener et al., 1979).

CLEAVAGE MECHANISMS OF BIOSYNTHETIC PRECURSORS

The conversion of PreProPTH to ProPTH occurs cotranslation-ally, and it is presumably activated by proteases localised in the endoplasmic reticulum (Table 1). Proteases involved in the cleavage of the signal sequence of PreProPTH remain to be characterized. The PreProPTH leader sequence was synthesized and shown to inhibit cell-free processing not only of PreProPTH but also of placental and pituitary prehormones (Majzoub et al., 1980). Besides enzymatic processing, precursor sequences may therefore be recognized by a

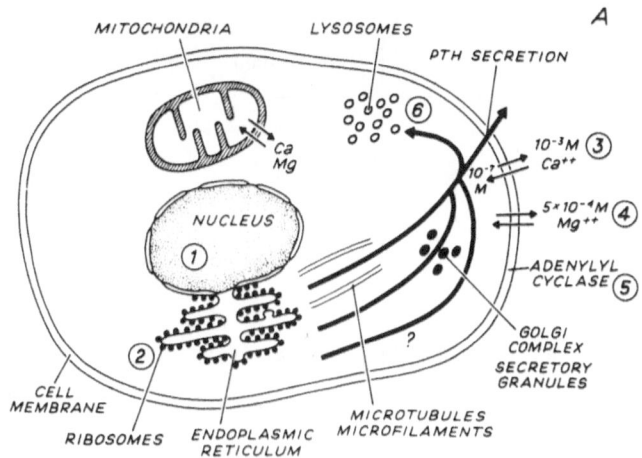

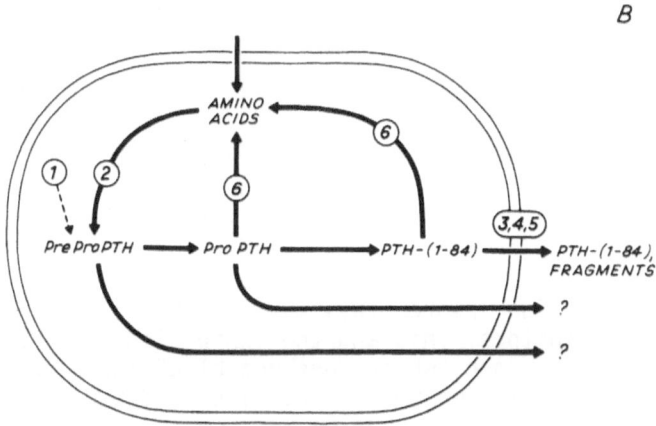

Figure 2. Biosynthesis and control points (numbers) of formation and release of PTH in parathyroid cells. A, morphology; B, biosynthetic pathways. 1, transcription 2, translation, and synthesis of PreProPTH and ProPTH. 3, regulation of PTH release by calcium, and 4, by magnesium, and magnesium requirement. 5, activation of adenylyl cyclase. 6, metabolism of PTH (after Fischer, 1982).

common binding and translocation site, or receptor in the rough endoplasmic reticulum.

Proparathyroid hormone was converted to the intact PTH-(1-84) by mild trypsinization (Goltzman et al., 1976). Digestion of the prohormone with 0.1% pancreatic trypsin resulted in the formation of biologically active intact PTH-(1-84). The nature of the con-

Table 1. Proteases in PTH biosynthesis and metabolism.

SUBSTRATE ➛ PRODUCT	PEPTIDASE	REGULATION
PREPROPTH ➛ PROPTH	?	RECEPTOR MEDIATED (?)
PROPTH ➛ PTH-(1-84)	TRYPSIN-LIKE (0.1%)	-
PROPTH, PTH-(1-84) ➛ SMALL PEPTIDES	TRYPSIN-LIKE (10%)	STIMULATED BY CALCIUM
PROPTH, PTH-(1-84) ➛ PTH-(37-84)+ SMALL PEPTIDES	CATHEPSIN B (LYSOSOMAL PROTEASE)	REQUIRES EDTA, THIOL-DEPENDENT

verting enzyme(s) in parathyroid cells remains to be determined.
The converting activity was localized to particulate fractions
of parathyroid cells (MacGregor et al., 1976; Habener et al., 1977).
The enzymatic activity had an optimal pH range between 7 and 9, and
was inhibited by EDTA and the trypsin inhibitor benzamidine, but
not by tosyl-L-lysine chloromethyl ketone, and pancreatic and soybean trypsin inhibitors. This and the inhibition by chloroquine, an
inhibitor of cathepsin B, which does not block the effects of trypsin, indicates that the enzymatic activity is different from pancreatic trypsin.

DEGRADATION OF PARATHYROID HORMONE

We have presented evidence that PTH is more rapidly destroyed
by parathyroid homogenates in the presence of 1 mM calcium as compared to 5 mM EDTA, suggesting the presence of trypsin-like activity enhanced by calcium (Fischer et al., 1972). Similarly, incubation with 10% pancreatic trypsin was shown to result in extensive fragmentation of ProPTH (Goltzman et al., 1976). To this end,
ProPTH and PTH-(1-84) were shown to be preferentially degraded in
cultured parathyroid explants under conditions of decreased parathyroid secretory activity in the presence of high as compared to
low medium calcium concentrations (Chu et al., 1973; Habener et al.,
1975). Morphological evidence suggests that degradation of PTH-(1-
84) takes place in lysosomes in a high calcium environment (Oldham
et al., 1971; Capen et al., 1971; Roth and Capen, 1974).

CLEAVAGE OF PTH-(1-84) INTO AMINO- AND CARBOXYL-TERMINAL FRAGMENTS

The biological activity of intact PTH-(1-84) resides in the amino-terminal third of the molecule (Rosenblatt, 1982). Parathyroid hormone-(1-84) is degraded into biologically active amino-terminal and biologically inactive carboxyl-terminal fragments both in parathyroid cells and in peripheral organs predominantly the liver and the kidneys. The question as to whether cleavage of PTH-(1-84) is essential for biological activation of the hormone is not entirely resolved. Intact PTH-(1-84) was shown to stimulate both renal and skeletal adenylate cyclase activity without the requirement for proteolysis (Goltzman et al., 1976; Goltzman, 1978). Martin et al. (1978), on the other hand, have presented evidence that intact PTH-(1-84) is less active in perfused bone in causing cyclic AMP production than the synthetic PTH-(1-34) fragment, suggesting that metabolism of PTH may be required for the expression of biological activity. The occurrence of bioactive and immuno-reactive amino-terminal PTH fragments has only occasionally been noted in the peripheral circulation (Canterbury et al., 1973; Goltzman et al., 1980; Fischer et al., 1974).

Following i.v. administration of intact bovine PTH-(1-84) in calves predominantly carboxyl-terminal fragments and only minimal amounts of an amino-terminal fragment have been detected in the circulation (Hunziker et al., 1977). The quantity of carboxyl-terminal fragments formed is greatly reduced in hepatectomized rats suggesting that besides the parathyroid cells the liver is an important cleavage site of PTH-(1-84) (Segre et al., 1981a). Together the liver and the kidney accounted for all but a small proportion of the metabolic clearance rate of intact PTH-(1-84) (Hruska et al., 1981).

An enzymatic activity cleaving [^{125}I]bovine PTH-(1-84) into smaller molecular weight PTH forms has been extracted from parathyroid glands, the liver and other porcine tissues (Fischer et al., 1972). This thiol-dependent protease was activated by EDTA at pH 6 and was resistant to trypsin inhibitor. The enzymatic activity, cathepsin B, has subsequently been isolated from extracts of the parathyroid glands and the liver by MacGregor et al. (1979a). Cathepsin B is probably of lysosomal origin. Both ProPTH and PTH-(1-84) are cleaved between amino acid residues 36 and 37 (MacGregor et al., 1979b). The carboxyl-terminal fragment PTH-(37-84) was not degraded further, whereas the amino-terminal fragments were digested into small non-identified peptides and amino acids. Besides intact PTH-(1-84), carboxyl- and amino-terminal fragments are secreted by the parathyroid glands (Flueck et al., 1977). Generation of these fragments of PTH both by the parathyroid glands and by peripheral organs such as the liver and the kidneys may be activated by cathepsin B. Indeed, PTH-(34-84) and -(37-84) are secreted by the parathyroid glands, but they are also generated after i.v. administration

of PTH-1-84) and during incubation of Kupffer cells of the liver
with the hormone (Morrissey et al., 1980; Segre et al., 1977, 1981b).
A predominant carboxyl-terminal fragment formed in the kidneys is
PTH-(39-84) (D'Amour et al., 1979).

The clearance of carboxyl-terminal fragments was shown to be
reduced in anephric and uremic man (Berson and Yalow, 1968; Silverman and Yalow, 1973; Freitag et al., 1978). The kidneys are probably
the principal organ of disposal of the biologically inactive fragments.

REGULATION OF THE BIOSYNTHESIS, SECRETION AND METABOLISM OF PARA-
THYROID HORMONE

The release of PTH from parathyroid cells is inversely related to the extracellular calcium concentration (Sherwood et al.,
1970; Hamilton et al., 1971; Oldham et al., 1971). The stimulation
of the secretion of PTH by low extracellular calcium is associated
with a raised incorporation of $[^3H]$-thymidine into desoxyribonucleic acids in parathyroid glands (Lee and Roth, 1975), as well as
enhanced incorporation of radioactive amino acids into ProPTH and
PTH-(1-84) (Sherwood et al., 1970; Hamilton et al., 1971; Moran et
al., 1981). Other workers have concluded that the biosynthesis of
PTH was not influenced by changes of the extracellular calcium concentration (Habener et al., 1975; Morrissey and Cohn, 1978). Chu et
al. (1973) suggested that the control of PTH biosynthesis by calcium may occur at the level of intracellular turnover and degradation of ProPTH. Conversely, Habener et al. (1975) demonstrated
that the conversion of ProPTH to PTH-(1-84) was not regulated by
the extracellular calcium concentration, but implied that PTH-(1-84)
was degraded within parathyroid cells in a calcium dependent manner.

Indeed, parathyroid glands have been shown to secrete biologically inactive carboxyl-terminal PTH fragments in proportionally
higher amounts than intact PTH-(1-84) under high calcium conditions
in the face of suppressed PTH secretion (Hanley et al., 1978; Mayer
et al., 1979). Moreover, the secretion of intact PTH-(1-84), as
estimated in the venous parathyroid effluent in man, was inhibited
within minutes in response to calcium infusions (Born et al., 1982).
The suppression of the secretion of PTH-(1-84) was accompanied by a
slower fall of the secretion of carboxyl-terminal PTH fragments.

Results concerning calcium regulated metabolism of PTH in
peripheral organs are contradictory. The metabolic clearance rate
of exogenously administered bovine PTH-(1-84) in dogs is the same
during hypocalcemia and hypercalcemia (Fox and Heath, 1981). The
rate of metabolism of PTH-(1-84) was considered to be increased
under conditions of low extracellular calcium in perfusion experiments of rat livers (Canterbury et al., 1975). Low extracellular
calcium in renal perfusion experiments was shown to, respectively

accelerate in canine kidneys (Hruska et al., 1977), or to suppress in rat kidneys (Hanano et al., 1978) the metabolism of PTH-(1-84). Moreover, the clearance of intact PTH-(1-84) across the human kidney is related to the serum calcium concentration (Oldham et al., 1978). Finally, both calcium and vitamin D metabolites enhance PTH degrading activity in rat kidney tissue (Fujita et al., 1980).

The rates of disappearance of secreted or of exogenously administered intact PTH-(1-84) are comparable and in the range of minutes (Born et al., 1982; Hunziker et al., 1977). The elimination from the circulation of secreted biologically inactive carboxyl-terminal PTH fragments in response to calcium infusions, or of comparable fragments generated after the exogenous administration of PTH-(1-84) is slower than of the intact hormone. Moreover, carboxyl-terminal PTH fragments remain in the peripheral circulation of hyperparathyroid man longer following calcium infusions than after parathyroidectomy (von Lilienfeld-Toal et al., 1977). The findings suggest that a sizeable fraction of carboxyl-terminal fragments in peripheral blood originates from the parathyroid glands, and that the secretion of both intact PTH-(1-84) and to a lesser extent of its carboxyl-terminal fragments is calcium regulated.

SUMMARY AND CONCLUSIONS

The initial biosynthetic precursor of PTH-(1-84) is PreProPTH. Within one minute of its synthesis PreProPTH is converted into ProPTH in the rough endoplasmic reticulum (Habener et al., 1979). The formation of PTH-(1-84) from ProPTH takes place at the Golgi region 15 to 20 minutes later. Both ProPTH and PTH-(1-84) are degraded into non-identified small peptides and amino acids by a calcium-dependent trypsin-like activity of lysosomal origin (Fischer et al., 1972; Habener et al., 1975). The nature of the proteases converting PreProPTH to ProPTH and of the PTH degrading activity remains to be elucidated. Besides these enzymatic activities, cathepsin B has been isolated from extracts of the parathyroid glands and of the liver (MacGregor et al., 1979a). Cathepsin B converts both ProPTH and PTH-(1-84) into biologically inactive PTH-(37-84) and non-identified small peptides from the amino-terminal region of PTH-(1-84) (MacGregor et al., 1979b). Similarly, PTH-(34-84) and -(37-84) are secreted by the parathyroid glands (Morrissey et al., 1980). Carboxyl-terminal fragments together with a presumably biologically active amino-terminal PTH fragment are also generated in peripheral cells such as Kupffer cells of the liver and renal tubular cells (Segre et al., 1981b; D'Amour et al., 1980). Preliminary evidence suggests that the cleavage of intact PTH-(1-84) into a biologically active amino-terminal fragment may be required at least in perfused canine bone for the accumulation of cyclic AMP and presumably the expression of biological activity (Martin et al., 1978). Finally, the biologically inactive carboxyl-terminal PTH fragments are predominantly degraded by the kidneys (Martin et al., 1977).

        Calcium is the most important regulator of PTH secretion.
The biosynthesis of ProPTH and the release of the intact PTH-(1-
84) are inversely related to the extracellular calcium concentration (Moran et al., 1981; Sherwood et al., 1968). In the presence
of raised extracellular calcium levels proportionally higher amounts
of biologically inactive carboxyl-terminal fragments are secreted
by the parathyroid glands besides the intact PTH-(1-84) (Hanley et
al., 1978). The results suggest that a sizeable fraction of carboxyl-terminal fragments encountered in the peripheral circulation
originates from the parathyroid glands provided the renal function
and elimination of these fragments by the kidneys is adequate. In
chronic renal failure the concentration of carboxyl-terminal fragments is greatly increased (Dambacher et al., 1979). The issue as
to whether the metabolism of PTH in peripheral organs is calcium
regulated or not remains to be resolved.

ACKNOWLEDGEMENT

        This work was supported by grant 3.813-0.81 of the Swiss
National Science Foundation.

REFERENCES

Berson, S. A., and Yalow, R. S., 1968, Immunochemical hetero-
        geneity of parathyroid hormone in plasma, J. Clin. Endo-
        crinol. Metab., 28:1037.
Born, W., Dambacher, M. A., Meyrier, A., Ardaillou, R., and
        Fischer, J. A., 1982, Parathyroid suppressibility in hyper-
        parathyroidism due to chronic renal failure: Studies with
        autotransplanted parathyroid tissue, Clin. Endocrinol.,
        17: in press.
Canterbury, J. M., Bricker, L. A., Levey, G. S., Kozlovskis, P. L.,
        Ruiz, E., Zull, J. E., and Reiss, E., 1975. Metabolism of
        bovine parathyroid hormone. Immunological and biological
        characteristics of fragments generated by liver perfusion,
        J. Clin. Invest., 55:1245.
Canterbury, J. M., Levey, G. S., and Reiss, E., 1973, Activation
        of renal cortical adenylate cyclase by circulating immuno-
        reactive parathyroid hormone fragments, J. Clin. Invest.,
        52:524.
Capen, C. C., 1971, Fine ultrastructural alterations of parathy-
        roid glands in response to experimental and spontaneous
        changes of calcium in extracellular fluids, Am. J. Med.
        50:598.
Chu, L. L. H., MacGregor, R. R., Anast, C. S., Hamilton, J. W.,
        and Cohn, D. V., 1973, Studies on the biosynthesis of rat
        parathyroid hormone and proparathyroid hormone: Adaptation
        of the parathyroid gland to dietary restriction of calcium,
        Endocrinology, 93:915.

Dambacher, M. A., Fischer, J. A., Hunziker, W. H., Born, W., Moran, J., Roth, H.-R., Delvin, E. E., and Glorieux, F. H., 1979, Distribution of circulating immunoreactive components of parathyroid hormone in normal subjects and in patients with primary and secondary hyperparathyroidism: the role of the kidney and of the serum calcium concentration, Clin. Sci., 57:435.

D'Amour, P., Segre, G. V., Roth, S. I., and Potts, J. T., Jr., 1979, Analysis of parathyroid hormone and its fragments in rat tissues. Chemical identification and microscopical localisation, J. Clin. Invest., 63:89.

Fischer, J. A., 1982, Parathyroid hormone, in: "Disorders of Mineral Metabolism", F. Bronner and J. W. Coburn, eds., Academic Press, New York, 2:271.

Fischer, J. A., Binswanger, U., and Dietrich, F. M., 1974, Immunological characterization of antibodies against a glandular extract and the synthetic amino-terminal fragments 1-12 and their use in the determination of immunoreactive hormone in human sera, J. Clin. Invest., 54:1382.

Fischer, J. A., Oldham, S. B., Sizemore, G. W., and Arnaud, C. D., 1972, Calcium-regulated parathyroid hormone peptidase, Proc. Natl. Acad. Sci. U.S.A., 69:2341.

Flueck, J. A., Di Bella, F. P., Edis, A. J., Kehrwald, J. M., and Arnaud, C. D., 1977, Immunoheterogeneity of parathyroid hormone in venous effluent serum from hyperfunctioning parathyroid glands, J. Clin. Invest., 60:1367.

Fox, J., and Heath III, H., 1981, Does plasma calcium concentration regulate metabolic clearance of parathyroid hormone?, Calcified Tissue Int., 31:321 (abstract).

Freitag, J., Martin, K. J., Hruska, K. A., Anderson, C., Conrades, M., Ladenson, J., Klahr, S., and Slatopolsky, E., 1978, Impaired parathyroid hormone metabolism in patients with chronic renal failure, New Engl. J. Med., 298:29.

Fujita, T., Uezu, A., Ota, K., Ohata, M., Fukushima, M., and Nishii, Y., 1980, Vitamin D and parathyroid hormone degradation by kidney, Contr. Nephrol., 22:51.

Goltzman, D., 1978, Examination of the requirement for metabolism of parathyroid hormone in skeletal tissue before biological action, Endocrinology, 102:1555.

Goltzman, D., Callahan, E. N., Tregear, G. W., and Potts, J. T., Jr., 1976, Conversion of proparathyroid hormone to parathyroid hormone: Studies in vitro with trypsin, Biochemistry, 15:5076.

Goltzman, D., Henderson, B., and Loveridge, N., 1980, Cytochemical bioassay of parathyroid hormone. Characteristics of the assay and analysis of circulating hormonal forms, J. Clin. Invest., 65:1309.

Goltzman, D., Peytremann, A., Callahan, E. N., Segre, G. V., and Potts, J. T., Jr., 1976, Metabolism and biological activity of parathyroid hormone in renal cortical membranes, J. Clin. Invest., 57:8.

Habener, J. F., Amherdt, M., Ravazzola, M., and Orci, L., 1979,
     Parathyroid hormone biosynthesis. Correlation of conversion
     of biosynthetic precursors with intracellular protein migra-
     tion as determined by electron microscope autoradiography,
     J. Cell Biology, 80:715.
Habener, J. F., Chang, H. T., and Potts, J. T., Jr., 1977, Enzymic
     processing of proparathyroid hormone by cell-free-extracts
     of parathyroid glands, Biochemistry, 16:3910.
Habener, J. F., Kemper, B., and Potts, J. T., Jr., 1975, Calcium-
     dependent intracellular degradation of parathyroid hormone:
     A possible mechanism for the regulation of hormone stores,
     Endocrinology, 97:431.
Habener, J. F., Rosenblatt, M., Kemper, B., Kronenberg, H. M.,
     Rich, A., and Potts, J. T., Jr., 1978, Pre-proparathyroid
     hormone: Amino acid sequence, chemical synthesis, and some
     biological studies of the precursor region, Proc. Natl. Acad.
     Sci. U.S.A., 75:2616.
Habener, J. F., Tregear, G. W., Stevens, T. D., Dee, P. C., and
     Potts, J. T., Jr., 1974, Radioimmunoassay for proparathyroid
     hormone, Endocr. Res. Commun., 1:1.
Hamilton, J. W., Niall, H. D., Jacobs, J. W., Keutmann, H. T.,
     Potts, J. T., Jr., and Cohn, D. V., 1974, The N-terminal
     amino-acid sequence of bovine proparathyroid hormone, Proc.
     Natl. Acad. Sci., U.S.A., 71:653.
Hamilton, J. W., Spierto, F. W., MacGregor, R. R., and Cohn, D. V.,
     1971, Studies on the biosynthesis in vitro of parathyroid
     hormone. II. The effect of calcium and magnesium on synthe-
     sis of parathyroid hormone isolated from bovine parathyroid
     tissue and incubation medium, J. Biol. Chem., 246:3224.
Hanano, Y., Ota, K., and Fujita, T., 1978, in: "Endocrinology of
     Calcium Metabolism", D. H. Copp and R. V. Talmage, eds.,
     Excerpta Medica, Amsterdam, p. 359 (abstract).
Hanley, D. A., Takatsuki, K., Sultan, J. M., Schneider, A. B., 1978,
     Direct release of parathyroid hormone fragments from func-
     tioning bovine parathyroid glands in vitro, J. Clin. Invest.,
     62:1247.
Hruska, K. A., Korkor, A., Martin, K., and Slatopolsky, E., 1981,
     Peripheral metabolism of intact parathyroid hormone, J. Clin.
     Invest., 67:885.
Hruska, K. A., Martin, K., Mennes, P., Greenwalt, A., Anderson, C.,
     Klahr, S., and Slatopolsky, E., 1977, Degradation of parathy-
     roid hormone and fragment production by the isolated perfused
     dog kidney, J. Clin. Invest., 60:501.
Hunziker, W., Blum, J. W., and Fischer, J. A., 1977, Plasma kinetics
     of exogenous bovine parathyroid hormone in calves, Pflügers
     Arch., 371:185
Kronenberg, H. M., McDevitt, B. E., Majzoub, J. A., Nathans, J.,
     Sharp, P. A., Potts, J. T., Jr., and Rich, A., 1979, Cloning
     and nucleotide sequence of DNA coding for bovine prepropara-

thyroid hormone, Proc. Natl. Acad. Sci. U.S.A., 76:4981.

Lee, M. J., and Roth, S. I., 1975, Effect of calcium and magnesium on deoxyribonucleic acid synthesis in rat parathyroid glands in vitro, Lab. Invest., 33:72.

Lilienfeld-Toal von, H., Lambert, P. W., Arnaud, C. D., 1977, Endogenous parathyroid hormone: Metabolism in hyperparathyroid man, in: "Program of the 59th Annual Meeting of The Endocrine Society, Chicago IL, 1977", abstract 500.

MacGregor, R. R., Chu, L. L. H., and Cohn, D. V., 1976, Conversion of proparathyroid hormone to parathyroid hormone by a particulate enzyme of the parathyroid gland, J. Biol. Chem., 251: 6711.

MacGregor, R. R., Hamilton, J. W., Kent, G. N., Shofstall, R. E., and Cohn, D. V., 1979b, The degradation of proparathormone and parathormone by parathyroid and liver cathepsin B, J. Biol. Chem. 254:4428.

MacGregor, R. R., Hamilton, J. W., Shofstall, R. E., and Cohn, D. V., 1979a, Isolation and characterization of porcine parathyroid cathepsin B, J. Biol. Chem., 254:4423.

Majzoub, J. A., Rosenblatt, M., Fennick, B., Maunus, R., Kronenberg, H. M., Potts, J. T., Jr., and Habener, J. F., 1980, Synthetic pre-proparathyroid hormone leader sequence inhibits cell-free processing of placental, parathyroid, and pituitary prehormones, J. Biol. Chem., 255:11478.

Martin, K. J., Freitag, J. J., Conrades, M. B., Hruska, K. A., Klahr, S., and Slatopolsky, E., 1978, Selective uptake of the synthetic amino terminal fragment of bovine parathyroid hormone by isolated perfused bone, J. Clin. Invest., 62:256.

Martin, K. J., Hruska, K. A., Lewis, J., Anderson, C., and Slatopolsky, E., 1977, The renal handling of parathyroid hormone. Role of peritubular uptake and glomerular filtration, J. Clin. Invest., 60:808.

Mayer, G. P., Keaton, J. A., Hurst, J. G., and Habener, J. F., 1979, Effects of plasma calcium concentration on the relative proportion of hormone and carboxyl fragments in parathyroid venous blood, Endocrinology, 104:1778.

Moran, J. R., Born, W., Tuchschmid, C. R., and Fischer, J. A., 1981, Calcium-regulated biosynthesis of the parathyroid secretory protein, proparathyroid hormone, and parathyroid hormone in dispersed bovine parathyroid cells, Endocrinology, 108:2264.

Morrissey, J. J., and Cohn, D. V., 1978, The effects of calcium and magnesium on the secretion of parathormone and parathyroid secretory protein by isolated porcine parathyroid cells, Endocrinology, 103:2081.

Morrissey, J. J., Hamilton, J. W., MacGregor, R. R., and Cohn, D. V., 1980, The secretion of parathormone fragments 34-84 and 37-84 by dispersed porcine parathyroid cells, Endocrinology, 107:164.

Oldham, S. B., Finck, E. J., and Singer, F. R., 1978, Parathyroid hormone clearance in man, Metabolism (Baltimore), 27:993.

Oldham, S. B., Fischer, J. A., Capen, C. C., Sizemore, G. W., Arnaud, C.D., 1971, Dynamics of parathyroid hormone secretion in vitro, Am. J. Med., 50:650.

Rosenblatt, M., 1982, Structure-activity relations in the calcium-regulating peptide hormones, in: "Endocrinology of Calcium Metabolism", J. A. Parsons, ed., Raven Press, New York, p. 103.

Roth, S. I., and Capen, C. C., 1974, Ultrastructural and functional correlations of the parathyroid gland, Int. Rev. Exp. Pathol., 13:161.

Segre, G. V., D'Amour, P., Hultman, A., and Potts, J. T., Jr., 1981a, Effects of hepatectomy, nephrectomy, and nephrectomy/uremia on the metabolism of parathyroid hormone in the rat, J. Clin. Invest., 67:439.

Segre, G. V., Niall, H. D., Sauer, R. T., and Potts, J. T., Jr., 1977, Edman degradation of radioiodinated parathyroid hormone: Application to sequence analysis and hormone metabolism in vivo, Biochemistry, 16:2417.

Segre, G. V., Perkins, A. S., Witters, L. A., and Potts, J. T., Jr., 1981b, Metabolism of parathyroid hormone by isolated rat Kupffer cells and hepatocytes, J. Clin. Invest., 67: 449.

Sherwood, L. M., Mayer, G. P., Ramberg, C. R., Jr., Kronfeld, D. S., Aurbach, G. D., and Potts, J. T., Jr., 1968, Regulation of parathyroid hormone secretion: proportional control by calcium, lack of effect of phosphate, Endocrinology, 83: 1043.

Sherwood, L. M., Rodman, J. S., Lundberg, W. B., 1970, Evidence for a precursor to circulating parathyroid hormone, Proc. Natl. Acad. Sci. U.S.A., 67:1631.

Silverman, R., and Yalow, R. S., 1973, Heterogeneity of parathyroid hormone. Clinical and physiologic implications, J. Clin. Invest., 52:1958.

PROCESSING AND DEGRADATION OF MET-ENKEPHALIN BY PEPTIDASES ASSO-

CIATED WITH RAT BRAIN CORTICAL SYNAPTOSOMES

Wolfgang Demmer and Karl Brand

Institute of Physiological Chemistry

University of Erlangen-Nuremberg, FRG

More than 20 peptides have been identified in neurones of the brain, spinal cord and in the periphery. There is growing evidence that some of these peptides act as neurotransmitters in the nervous system (for review see Hökfelt et al., 1980, Snyder, 1980, Konishi et al., 1981). The discovery of the enkephalins, Tyr-Gly-Gly-Phe-Met and Tyr-Gly-Gly-Phe-Leu, (Hughes et al., 1975) has led to numerous studies of their metabolism. For the regulation of the activity of these peptides the mode of processing from larger precursor proteins of peptides and of inactivation by proteolytic cleavage is of fundamental importance. Enkephalin containing polypeptides have been described in different organs such as adrenal medulla, intestine mucosa and brain (Lewis et al., 1979, Stern et al., 1981) as potential processing products of the larger pro-enkephalin precursor. The carboxyterminus of pro-enkephalin consists of the met-enkephalin-Arg[6]-Phe[7] sequence (Gubler et al., 1982, Noda et al., 1982).

A variety of enzymes isolated from whole brain (Knight and Klee, 1979) pituitary (Orlowski and Wilk, 1981), adrenal medulla (Hook et al., 1982) as well as from rabbit kidney and brain (Benuck et al., 1981) are capable of generating met-enkephalin from enkephalins with C-terminal extensions. These enzymes could possibly play an important role in the control of the activity of met-enkephalin at the sites of its action.

The synapses, loci of the process of neurotransmission between nerve cells, are potential sites for processing and degradation of peptides acting as neurotransmitters or neuromodulators. Isolated nerve endings - synaptosomes - should therefore be suitable for studying such processing and degradation reactions. We studied the de-

gradation of met-enkephalin-Arg[6]-Phe[7] to demonstrate, that synaptosomes are capable of generating met-enkephalin from a putative precursor molecule and the way in which the generated enkephalin is me-
tabolized.

MATERIAL AND METHODS

Met-enkephalin-Arg[6]-Phe[7], possible degradation intermediate
peptides, amino acids and inhibitors were obtained from SERVA Heidelberg, FRG. All other chemicals were of analytical grade. The preparation of the synaptosomes followed the method of Krueger et al.,
(1977) with minor modifications as described by Bauer and Brand (1982).
If not used subsequently the pellet was resuspended in 4 ml of a
buffer containing 124 mM NaCl, 4 mM KCl, 1.3 mM $MgSO_4$, 16 mM Na_2H
PO_4 adjusted to pH 7.4 and saturated with O_2 and stored at -20°C until use (frozen synaptosomes). The extent of contamination of the
synaptosomal preparation with free mitochondria has been followed by
measuring NADH-cytochrome c reductase as mitochondrial marker enzyme
according to Duncan and Mackler (1966). LDH activity was assayed in
the absence and presence of 0.1 % Triton-X-100 in order to assess the
integrity of the synaptosomes.

Electron microscopy

From the last pellet of the preparation of synaptosomes specimen
were removed and fixed in a 2.5 % solution of buffered glutaraldehyde
for 4 h and then refixed in a 1 % solution of osmium tetroxide, de-
hydrated and embedded in Epon. Ultrathin sections were investigated
with a Siemens electron microscope (Elmiscope 101 A).
The micrographs were taken under a magnification of x 6000 and mag-
nified to x 34000.

Incubation conditions

Determination of the pH-profile and the time dependency of the
degradation of met-enkephalin-Arg[6]-Phe[7], met-enkephalin, the tetrapeptide Gly-Gly-Phe-Met the tripeptide Gly-Gly-Phe and the dipeptide
Gly-Gly were performed by incubation freshly prepared synaptosomes
or 1 x frozen synaptosomes resp. (600 ug of protein) with 100 nmoles
of the respective peptides in 200 ul of 0,1 mol/l phosphate buffer
(pH 4.0 - 8.0), or 200 ul of 0,1 mol/l Tris-HCl buffer (pH 7.0 - 10.5)
for 10 min at 25°C. The reactions were terminated at the appropriate
times by the addition of sufficient 6,25 mol/l HCl to bring the pH
values of the samples to pH 2.0. The samples were boiled for 15 min
and made up to 2 ml by addition of H_2O and stored frozen at -20°C.
After thawing to 4°C the samples were centrifuged at 1000 x g for
10 min and the supernatants were fractionated by HPLC.

High performance liquid chromatography

The HPLC system (Kontron, Zürich) used in this study consisted of a 600/200 pump, a 250 x 4 mm, 10 um Lichrosorb RP 18 (Merck AG Darmstadt, FRG) column and a Rheodyne 7520 injector valve. The effluent was monitored at 206 nm with a UVICORD detector, equipped with an analytical HPLC-cell (LKB, Bromma Sweden). Samples (200 ul) were eluted with a 30 min linear gradient of 0-45 % Acetonitrile in 0,1 % H_3PO_4 adjusted to pH 2.2 with NaOH. Column temperature was $30^{\circ}C$, flow rate 1 ml/min.
The kinetic of the individual reactions catalyzed by the peptidase activities of rat brain cortical synaptosomes were determined from the disappearance of the substrate peak and appearance of the product peak(s) with time. With respect to the diglycine peptide the liberated amounts of glycine were determined by amino acid analyses and calculated by the integration system connected to the amino acid analyzer. They are expressed as nmol glycine per mg synaptosomal protein per minute.

Amino acid analyses

The fractionated peaks from the individual HPLC runs were pooled, dried in vacuo and hydrolyzed for 24-36 hours with 500 ul 6 N HCl containing 1 % Phenol and 0.1 % Thioglycolic acid. After evaporation under reduced pressure the samples were reconstituted and the amino acid composition of the samples was established by ophthalaldehyde fluorescence using a Biotronik LC 6000 E automatic amino acid analyzer coupled with a Biotronik System I computing system. Quantification of the individual amino acids were performed by injecting known amounts of amino acid standards.

Protein determination

Protein was measured according to the method of Lowry (1951).

RESULTS

Electron microscopy

Fig. 1 shows a typical preparation of rat brain cortical synaptosomes. As it can be seen, a large number of closed membrane structures which contain synaptic vesicles (Syn) and some which contain mitochondria too, are present. A few extrasynaptosomal mitochondria (M) and some unidentified membraneous elements are present too. From 6 g of rat cerebral cortex 60 mg of synaptosomal protein was obtained.

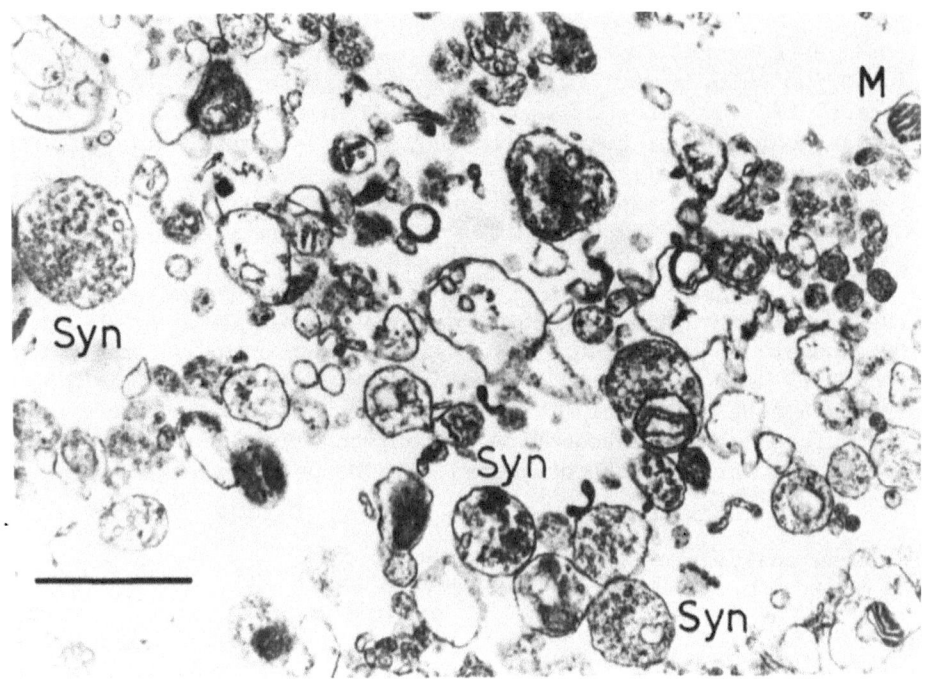

Fig. 1. Electron micrograph of the synaptosomal fraction from rat
       brain cortical cortex. See text for preparation, fixation
       etc. Several synaptosomal profiles (Syn), containing
       vesicles and mitochondria (M), bounded by an unit membrane,
       are clearly visible. The bar represents 1 um. (By courtesy
       of Dr. Drecoll-Lütjen, Department of Anatomy, University
       of Erlangen-Nuremberg).

Peptide separation by HPLC

     Fig. 2 shows a typical separation run of met-enkephalin-Arg[6]-
Phe[7], met-enkephalin and their possible degradation products by
HPLC. All standard peptides and amino acids, except Met-Arg-Phe and
Gly-Gly-Phe-Met could be resolved under the run conditions described
in Materials and Methods. Because these two peptides mentioned above
cannot occur as degradation products of met-enkephalin-Arg[6]-Phe[7] at
the same time, a further resolution was not intended.

Integrity of the synaptosomes and contamination with other cellular
structures

     To investigate the integrity of the synaptosomes prepared by
our standard procedure, the LDH activity of the preparation was

measured in the presence or absence of 0,1 % Triton X-100. The activity of the preparation without added Triton X-100 was only 7 $\pm$ 3 %
of the value obtained in the presence of 0,1 % Triton X-100. To study
the degree of contamination of the synaptosomes by extra synaptosomal
mitochondria, NADH-cytochrome c reductase activity has been measured
to be 5,2 $\pm$ 0,5 nmol x g protein^{-1}. This value is in good agreement
with that of Booth and Clark (1978) who calculated on the basis of
an activity of 4.6 nmol x g^{-1} x min^{-1} a contamination of their synaptosomal preparation with free mitochondria in the range of only 5 %.

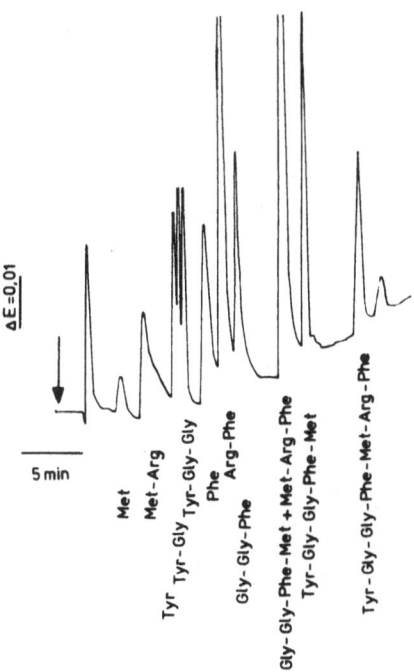

Fig. 2. Separation of methionine-enkephalin-Arg[6]-Phe[7] and possible
        degradation intermediates and products. A standard mixture
        of 2 ug of each peptide was separated by reversed phase
        HPLC on Lichrosorb RP-18. The arrow indicates injection of
        the sample and start of gradient as described in the text.

      A contribution of proteolytic enzymes from intrasynaptosomal
lysosomes to the degradation rates of the synaptic membranes could
be excluded in the following way. Synaptosomes were prepared freshly
and divided into aliquots. One aliquot was used immediately for the
peptide degradation study while an other was stored frozen at -20°C
overnight. The frozen aliquot was allowed to thaw slowly to 4°C the
following day and was then used. The results of this experiment are

shown in Fig. 3. No significant differences in the degradation pattern and the degradation rate was found between the two samples with
respect to the splitting of the Tyr-Gly peptide bond in met-enkephalin and the Gly-Phe peptide bond in the generated tetrapeptide.

pH Profiles

Fig. 4 shows the pH optima for the degradation of met-enkephalin Arg^6-Phe^7 (A), met-enkephalin (B), the tetrapeptide Gly-Gly-Phe-
Met (C) and the tripeptide Gly-Gly-Phe (D). They were all in the
range of pH 7-8. All subsequent experiments therefore were performed
in 0.1 M Tris-HCl buffer at pH 7.8.

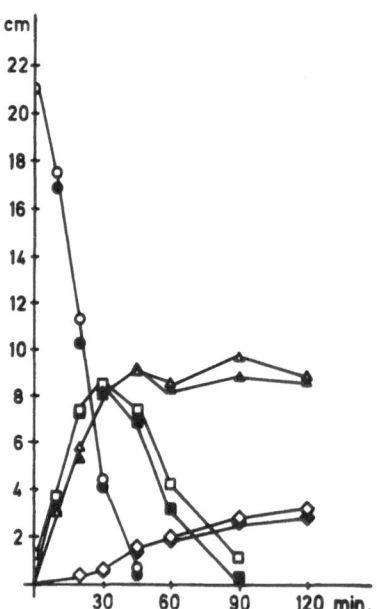

Fig. 3. Time course of the degradation of met-enkephalin by freshly
prepared (open symbols) and frozen and thawed synaptosomes
(closed symbols). The ordinate gives the relative heights
of the peaks as read directly from the recorder stripchart.
(o) Tyr-Gly-Gly-Phe-Met, (Met-enkephalin), (△) Tyr,
(□) Gly-Gly-Phe-Met, (◇) Phe.

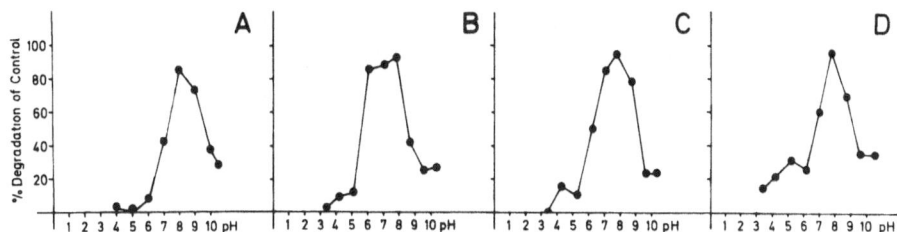

Fig. 4. pH-profiles of the cleavage of met-enkephalin-Arg^6-Phe^7 and
intermediate degradation peptides. (A) cleavage of met-
enkephalin-Arg^6-Phe^7, (B) cleavage of met-enkephalin, (C)
cleavage of Gly-Gly-Phe-Met, (D) cleavage of Gly-Gly-Phe.
Samples were analyzed by HPLC as described in the Material
and Methods section.

Kinetic of conversion of met-enkephalin-Arg^6-Phe^7 to met-enkephalin
and subsequent degradation of the generated peptides

    Fig. 5 shows the kinetic of processing of met-enkephalin (5B)
from met-enkephalin-Arg^6-Phe^7 (5A) by splitting off the C-terminal
Arg-Phe dipeptide (5B) during incubation with freshly prepared cortical synaptosomes. The next fastest step in the degradation is the
release of free tyrosine from met-enkephalin (5C) thus generating
the tetrapeptide Gly-Gly-Phe-Met (5C). The Arg^6-Phe^7 dipeptide released in the first cleavage step is degraded rapidly to the free
amino acids (5B).

    Figure 6 shows results of experiments in which the kinetic of
degradation of met-enkephalin and its possible cleavage products
Gly-Gly-Phe-Met, Gly-Gly-Phe and Gly-Gly upon incubation with synaptosomes are followed. The fastest process is the cleavage of the N-
terminal tyrosine from met-enkephalin (6A) with a half live of around
3 min. The time course for the degradation of the tetrapeptide Gly-
Gly-Phe-Met and the release of phenylalanine is shown in Fig. 6B.
Since the intermediates Gly-Gly-Phe and Phe-Met were not formed
during the incubation in detectable amounts it has to be assumed that

the cleavage occurs first at the Gly-Phe bond, generating the dipeptide Phe-Met which rapidly is cleaved and therefore does not accumulate. For comparison the kinetics of the degradation of the tripeptide Gly-Gly-Phe is given in Fig. 6C. The final step of the complete
degradation of met-enkephalin is the breakage of the Gly-Gly bond.
The kinetic of this reaction is shown in Fig. 6D. It should be noted
that the rates differ as much as 40 fold.

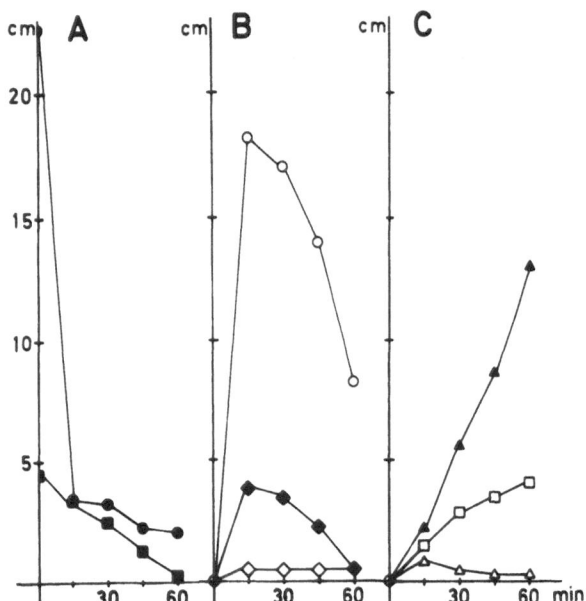

Fig. 5. Time course of the degradation of met-enkephalin-Arg^6-Phe^7
by freshly prepared synaptosomes. (A) Degradation of met-
enkephalin-Arg^6-Phe^7 (●) and an unknown impurity of the
synthetic peptide (■), (B) Generation and subsequent degradation of met-enkephalin (○), Arg-Phe (◆), (C) Liberation of Gly-Gly-Phe-Met (□) and Tyr (▲) from the generated met-enkephalin. (◇) Tyr-Gly-Gly-Phe and (△) Gly-
Gly-Phe-Met-Arg-Phe are minor compounds which do not accumulate during the incubation time.

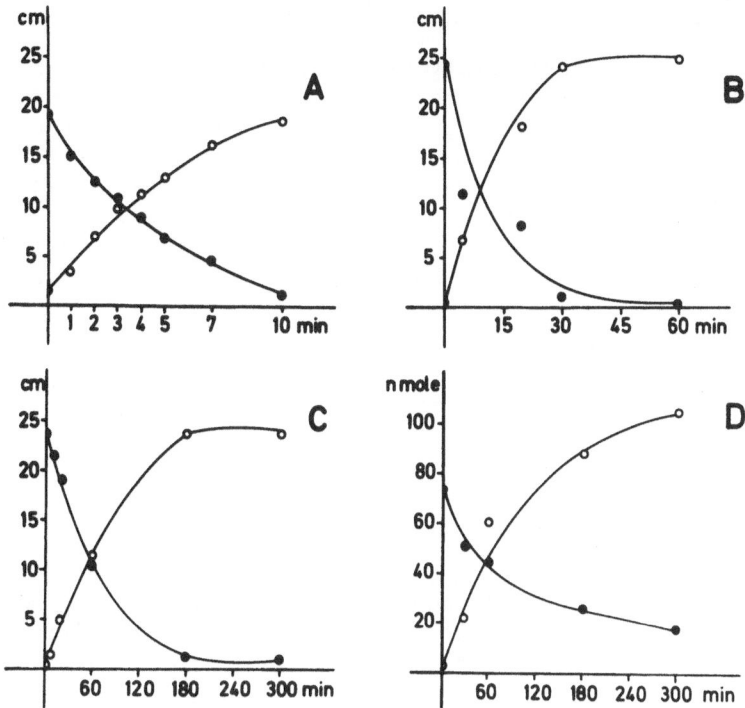

Fig. 6. Time course of the degradation of met-enkephalin and related
peptides by frozen and thawed synaptosomes. (A) Disappearance
of met-enkephalin (●) and formation of the tetrapeptide
Gly-Gly-Phe-Met (o). (B) Disappearance of the tetrapeptide
(●) and liberation of Phe (o). (C) Breakdown of the tripeptide Gly-Gly-Phe (●) and release of Phe (o). (D)
Cleavage of Gly-Gly (●) and release of Gly (o). The ordinates indicate the relative peak height except for the
cleavage of Gly-Gly where it indicates nmol.

Table 1. INHIBITION OF THE FIRST TWO CLEAVAGE STEPS OF THE DEGRA-
DATION OF MET-ENKEPHALIN BY VARIOUS INHIBITORS

INHIBITOR[a]	% INHIBITION OF CONTROL	
	STEP 1	STEP 2
Puromycin	100	0
PCMB	100	100
Captopril[b]	0	0

a: concentration of all inhibitors used was 10^{-3}M.
b: Captopril was a generous gift from Heyden Munich, FRG.

Inhibition of degradation of Met-enkephalin

In Table 1 the sensitivity of the first two cleavage reactions
in the degradation of met-enkephalin (Fig. 6A and B) towards various
inhibitors is summarized. Step one, the cleavage of the Tyr-Gly peptide bond is inhibited to 100 % by Puromycin (ID_{50} = 5 x 10^{-5}M) and
by PCMB at inhibitor concentrations of 10^{-3}. Step two, the cleavage
of the Gly^3-Phe^4 peptide bond of the tetrapeptide Gly-Gly-Phe-Met is
inhibited to 100 % by PCMB (ID_{50} = 5 x 10^{-4}M). Both reactions are
not inhibited by captopril, a known inhibitor of Angiotensin Converting Enzyme, at a concentration of 10^{-3}M.
In the presence of 10^{-3}M Puromycin the cleavage of the Gly^3-Phe^4
bond of met-enkephalin does not occur indicating that the N-terminal
Tyr has to be cleaved off before the dipeptidyl carboxypeptidase can
split the Gly^3-Phe^4 bond.

DISCUSSION

Metabolically competent synaptosomes were isolated by following
essentially the method of Krueger et al.(1977) with minor modifications (Bauer and Brand 1982). The high degree of integrity and low
contamination with free mitochondria of our preparation is comparable with synaptosomal fractions previously described by others.
e.g.: (Booth and Clark 1978, Hargittai et al. 1982).
Recent studies on enkephalin biosynthesis have led to the discovery
of proenkephalin, a polypeptide in adrenal medulla, the sequence of
which has been established by molecular cloning (Gubler et al. 1982,
Noda et al. 1982). It contains approximately six to seven met-enkephalin sequences and one leu-enkephalin sequence. The heptapeptide
met-enkephalin-Arg^6-Phe^7 represent the carboxyl terminus of pro-enkephalin in bovine adrenal medulla (Gubler et al. 1982).
It has been shown (Kojima et al. 1982), that this peptide can be converted to met-enkephalin by sequential action of trypsin and carboxypeptidase B activities. This action of trypsin-like and carboxy-
peptidase-like activities upon prohormones has been proposed to be
a common feature for their processing (Stein et al. 1981).
In order to see if this mechanism is effective also in nerve termi-
nals, we studied the kinetic of cleavage of met-enkephalin-Arg^6-Phe^7.
The products of the cleavage have been identified to be met-enkephalin and the dipeptide Arg-Phe indicating that a dipeptidyl carboxypeptidase is mainly responsible in synaptosomes for the liberation
of met-enkephalin. The fact that met-enkephalin and the dipeptide
Arg-Phe accumulates in the early period of incubation with synaptosomes shows that this processing reaction is faster than the cleavage reactions leading to degradation of met-enkephalin.
As a consequence, we studied the sequence of degradation of met-enkephalin by synaptosomes and found, that the fastest step was the
splitting of the Tyr^1-Gly^2 peptide bond by the action of an aminopeptidase which is sensitive towards puromycin and parachloromer-

curibenzoate (Table 1). A number of aminopeptidases acting on enkephalins have been described. For reviews see Schwartz (1981) and
Hersh (1982). The significance of aminopeptidases to the degradation
of endogenous enkephalins has been called in question (Schwartz 1981),
but this is however a tentative conclusion which must be documented
by further experimentation.

The generated tetrapeptide Gly-Gly-Phe-Met is then splitted at the
Gly^3-Phe^4 peptide bond (Fig. 6B) with the simultaneous release of a
free Phe residue. The mode of sequential cleavage by an carboxypeptidase activity could be excluded by the observation that the intermediate Gly-Gly-Phe does not accumulate during any stage of incubation. So it has to be assumed that the generated Phe-Met dipeptide is immediately after liberation cleaved too. For comparison
(Fig. 6C) the splitting off the C terminal Phe from Gly-Gly-Phe
occurs slower supporting the proposed sequence of splitting events.
This dipeptidyl carboxypeptidase activity is different from Angiotensin Converting Enzyme since it is not inhibited by Captopril (Table
1) and SQ 20881 (data not shown). The identity with one of the recently reviewed dipeptidyl carboxypeptidases (Hersh 1982, Schwartz
1981, Printz et al. 1982). remains to be established. The slowest
step of degradation of met-enkephalin seems to be the conversion of
the Gly^2-Gly^3 dipeptide to the free aminoacids, which is approximately 40 times slower than the aminopeptidase reaction (Fig. 6D).
A cobalt-requiring enzyme carrying out this reaction has been described by Marks (1970).

Our results clearly show that synaptosomes are capable of processing
and degrading met-enkephalin, the fastest reaction being the splitting off the dipeptide Arg^6-Phe^7 generating met-enkephalin followed
by the release of tyrosine by the action of an aminopeptidase which
concommitantly leads to inactivation.

REFERENCES

Bauer, U., and Brand, K., 1982, Carbon balance studies of glucose
    metabolism in rat cerebral cortical synaptosomes, J. Neuro-
    chem., 39:239.
Benuck, M., Berg, M. J., and Marks, N., 1981, Met-enkephalin Arg^6-
    Phe^7 metabolism: Conversion to met-enkephalin by brain and
    kidney dipeptidyl carboxypeptidases, Biochem. Biophys. Res.
    Comm.,99:630.
Booth, R. F. G., and Clark, J. B., 1978, A rapid method for the pre-
    paration of relatively pure metabolically competent synapto-
    somes from rat brain, Biochem. J., 176:365.
Duncan, H. M., and Mackler, B., 1966, Elektron transport system of
    yeast, J. Biol. Chem., 241:1964.
Gubler, U., Seeburg, P., Hoffman, B. J., Gage, L. P., and Udenfriend,
    S., 1982, Molecular cloning established proenkephalin as pre-
    cursor of enkephalin-containing peptides, Nature, 295:206.

Guyon, A., Roques, B. P., Guyon, F., Foucoult, A., Perdrisot, R., Swerts, J.-P., and Schwartz, J. Ch., 1979, Enkephalin degradation in mouse brain studied by a new H.P.L.C. method. Further evidence for the involvement of carboxydipeptidase, Life Sci., 25:1605.

Hargittai, P., Agoston, D., and Nagy, A., 1982, Comparative biochemical and biophysical studies on rat brain synaptosomes, FEBS Lett. 137:67.

Hersh, L. B., 1982, Degradation of enkephalins: The search for an enkephalinase, Mol. Cell. Biochem., 47:35.

Hökfelt, T., Johansson, O., Ljungdahl, Å., Lundberg, J. M., and Schultzberg, M., 1980, Peptidergic neurons, Nature, 284:515.

Hook, V. Y. A., Eiden, L. E., and Brownstein, M. J., 1982, A carboxypeptidase processing enzyme for enkephalin precursors, Nature, 295:341.

Hughes, J., Smith, T. W., Kosterlitz, H. W., Fothergill, L. A., Morgan, B. A., and Morris, H. R., 1975, Identification of two related pentapeptides from the brain with potent opiate agonist activity, Nature, 258:577.

Knight, M., and Klee, W. A., 1979, Enkephalin generating activity of rat brain endopeptidase, J. Biol. Chem., 254:10426.

Kojima, K., Kilpatrick, D. L., Stern, A. S., Jones, B. N., and Udenfriend, S., 1982, Proenkephalin: A general pathway for enkephalin biosynthesis in animal tissues, Arch. Biochem. Biophys., 215:638.

Konishi, S., Tsunoo, A., and Otsuka, M., 1981, Enkephalin as a transmitter for presynaptic inhibition in sympathetic ganglia, Nature, 294:80.

Krueger, B. K., Forn, J., and Greengard, P., 1977, Depolarization induced phosphorylation of specific proteins, mediated by calcium ion influx, in rat brain synaptosomes, J. Biol. Chem., 252:2764.

Lewis, R. V., Stern, A. S., Rossier, J., Stein, S., and Udenfriend, S., 1979, Putative enkephalin precursors in bovine adrenal medulla, Biochem. Biophys. Res. Comm., 89:822.

Lowry, O. H., Rosebrough, N. J., Farr, A.-L., and Randall, R. J., 1951, Protein measurement with the Folin phenol reagent, J. Biol. Chem., 193:265.

Marks, N., 1970, Peptide Hydrolases, in: "Handbook of Neurochemistry," A. Lajtha, ed., Plenum Press, New York.

Noda, M., Furutani, Y., Takahashi, H., Toyosato, M., Hirose, T., Inayama, S., Nakanishi, S., and Numa, S., 1982, Cloning and sequence analysis of cDNA for bovine adrenal preproenkephalin, Nature, 295:202.

Orlowski, M., and Wilk, S., 1981, A multicatalytical protease complex from pituitary that forms enkephalin and enkephalin containing peptides, Biochem. Biophys. Res. Comm., 101:814.

Printz, M. P., Ganten, D., Unger, T., and Phillips, M. I., 1982,
        Minireview: The brain renin angiotensin system 1982, in:
        "The Renin Angiotensin System in the Brain," D. Ganten, M.
        Printz, M. I. Phillips, and B. A. Schölkens, eds., Springer-
        Verlag, Heidelberg.
Schwartz, J.-C., 1981, Biological inactivation of enkephalins and
        the role of enkephalin-dipeptidyl-carboxypeptidase ("enkepha-
        linase") as neuropeptidase, Life Sci., 29:1715.
Snyder, S. H., 1980, Brain peptides as neurotransmitters, Science,
        209:976.
Stern, A. S., Lewis, R. V., Kimura, S., Rossier, J., Stein, S., and
        Udenfriend, S., 1980, Opioid hexapeptides and heptapeptides
        in adrenal medulla and brain possible implications on the
        biosynthesis of enkephalins, Arch. Biochem. Biophys.,
        205:606.
Stern, A. S., Jones, B. N., Shively, J. E., Stein, S., and Udenfriend,
        S., 1981, Two adrenal opioid polypeptides: Proposed inter-
        mediates in the processing of proenkephalin, Proc. Natl.
        Acad. Sci. USA, 78:1962.

CHARACTERIZATION AND CLINICAL SIGNIFICANCE OF MEMBRANE

BOUND PROTEASES FROM HUMAN KIDNEY CORTEX

J. Scherberich, C. Gauhl, G. Heinert,
W. Mondorf, and W. Schoeppe

Department of Nephrology, Department of Virology
Department of Urology
University Hospital, Frankfurt am Main, W-Germany

INTRODUCTION

Human kidney tissue contains high concentrations of various proteases; especially the epithelia of the proximal tubule are rich in
leu-(ala-) aminopeptidase (AAP) and gamma-glutamyl transpeptidase
(GGT) activity, enzymes which are bound to the outer luminal plasma membrane [1].
The physiological role of kidney brush border proteases still
remains unclear. Experimental data using rat and rabbit kidneys
suggest that renal proteases (exopeptidases) are involved in
cleavage and consequently inactivation of peptides which are filtered by the glomeruli such as glucagon, somatotropin, insulin
parathormone, angiotensin and other vasoactive components [2] .
In a second step, degradation products might then be resorbed via
endocytosis, facilitated diffusion or through an active counter co-
transport system [3,4] .
Immunochemical and ultrastructural studies carried out in our
laboratory have shown that human kidney AAP and GGT exhibit a
characteristic arrangement within the brush border membrane (BB)
of the proximal tubule [1,5] . However, under pathological conditions
e.g. in patients with kidney diseases (glomerulo-, pyelonephritis,
lupus-nephritis), during kidney allograft rejection crises and drug
induced nephrotoxicity due to antibiotics, cytostatics and x-ray
contrast media, the typical architecture of these membrane bound
constituents can be lost[6,7]. Enhanced structural imbalance of the
tubule cell might lead to exfoliation of brush border proteases
into the lumen followed by an increased urinary output of these
enzymes (= tissue-proteinuria).
In the present paper we report on further biochemical, immunological, ultrastructural and histological characterization of BB-bound

proteases from human kidney under normal and clinically defined
conditions.

Our results indicate that even in the early phase of kidney
damage the tissue as well as the urinary profile of protease acti -
vity is changed, an observation that might be of significant value
for diagnostic and prognostic purposes [8,9] .

MATERIAL AND METHODS

Plasma-membranes rich in BB-fragments were isolated from human
kidney cortex by differential centrifugation and buffer free-flow
electrophoresis as previously described [1,5] . Specific activity
of BB-marker enzymes such as AAP (E.C 3.4.4.1-) and GGT (E.C
2.3.2.2) were assayed as reported earlier [1] . For the isolation of
BB bound proteases, membranes were digested with papain for 30 min.
at 37 °C; the ratio BB-protein : papain was 2o : 1 (w/w). After
incubation, the suspension was centrifuged (30 min, 60,000 x g)
and the supernatant containing the main part of BB surface AAP and
GGT was further separated by lectin-specific affinity chromato -
graphy [1] .   In order to isolate peripheral as well as integral BB
proteins, membranes were treated in similar manner with proteinase-
K , bromelain, DOC, and triton X 100 respectively according to
standard procedures.

Antibodies against BB and BB-lectin receptors including BB-AAP
and GGT were raised in rabbits and goats. Lipoproteins of sera
were removed by treatment with aerosil; the IgG fraction was isolated by DEAE-Trisacryl-ion exchange chromatography. Antisera were
totally absorbed against normal plasma protein antibodies using
plasmaprotein- Sepharose followed by a free absorption step with
plasmaproteins. Quantitative radialimmunodiffusion, one-and two -
dimensional electroimmunoassay, immunotitration, immunospecific
affinity chromatography, and immunofluorescence microscopy was
carried out as previously decribed [1,5,10] .

For negative staining of membrane preparations, specimens were
placed on an agarose surface and a formvar carbon-coated copper
grid (3oo mesh) was laid on the drops (samples). After appropriate diffusion time, the grid was placed on a drop of PBS(1%)-
formaldehyde, washed with 0.1 M ammoniumacetate, dried and stained
with 3 % phosphotungstic acid, pH 6.8 or 2 % uranylacetate, pH 4
for 3 minutes. In addition, BB protease activity in urine samples
of patients was documented by specific reaction with anti-protease
antibody (anti-AAP) followed by ferritin labelling of the antigen
antibody complexes through ferritin-conjugated anti-rabbit IgG.
Labelled complexes were centrifuged at 12,000 x g for 10 min;
pellets were washed twice, and were then prepared for negative staining. Thoroughly dried grids were viewed in a Siemens Elmiskop 102
at a primary magnification of 100,000.

Quantitative distribution patterns of BB related AAP and GGT on kidney tissue sections were evaluated by computer assisted histo-photometry applying high-speed image processing Micro-Videomat-2 equipment, as well as an " interactive image analysis system " IBAS (Kontron-Zeiss, Oberkochen, FRG). Regions of interest were measured using various grey levels and by automatically increasing the background contrast of tissue areas which had to be discriminated for BB- proteolytic activity. Image signals of 30 grey levels (Micro-Videomat) or of 256 grey levels (IBAS) were scanned. In the case of kidney slices areas containing 100 - 200 tubules were differentiated, where $3.2 \times 10^5 - 1.28 \times 10^6$ image points (specific AAP, GGT chromogen) were measured by an electronic beam and rapidly analyzed ("Axiomate", "eye-TV-camera", "PDP-11/50 computer", "Tektronic" 4014 monitor).   In similar manner tissue sections stained for immunofluorescence were evaluated through a negative image using " Tessovar"- macroscope equipment [11] .

RESULTS AND DISCUSSION

    Plasma-membranes from human kidney cortex could be separated into two fractions applying sucrose density gradient centrifugation and free-flow electrophoresis. One fraction exhibited high specific activity of Na-K- ATPase, mainly due to the presence of basal-lateral membranes; another fraction revealing high AAP and GGT activity showed vesicles and microvillous fragments ("brush border fraction").

    Using negative staining technique the surface of BB-vesicles revealed numerous globular particles (fig. 1), which could be selectively cleaved off after limited proteolytic digestion with papain and bromelain respectively (fig. 1 c).
    Removal of these BB surface components was associated with rise of BB-AAP and GGT activity into the supernatant, where approximately 90 % of total protease activity could be solubilized without loss of enzymatic or immunological reactivity [1,5] .
    Fig.2 shows the results after differential solubilization of BB associated AAP with agents including trypsin, chymopapain, papain, and detergents. In a modification of one-dimensional rocket electro-immunoassay it was possible to demonstrate a peripheral as well as a minor intrinsic BB-portion of AAP.
    Wheat germ agglutinin, added to BB proteases, was capable of precipitating both AAP and GGT, depending on the lectin concentration chosen (fig. 3). After incubation with the lectin, substrate splitting by the enzymes was not abolished, however. In contrast to wheat germ lectin Concanavalin A did not significantly bind to GGT but to AAP, indicating differences in the pattern of surface sugars of both BB-proteases. Applying a combination of lectin-specific affinity chromatography (Con A, WGA- Sepharose) and gelfiltration (Sephadex G 200), AAP and GGT could be isolated from BB - vesicles at high purity; no additional immunoprecipitates were observed when purified enzymes were immunoprecipitated with polyspecific anti-BB antibody (fig. 4).

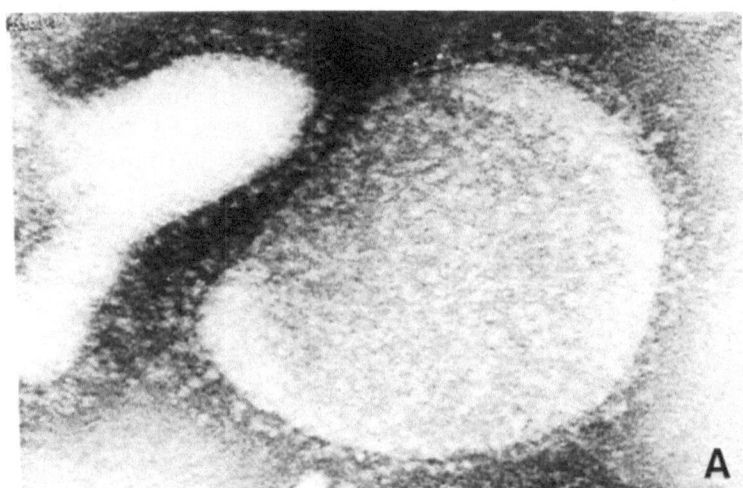

FIG. 1 : Arrangement of brush border constituents from human kidney
as revealed by negative staining. Fig. 1 A: Isolated BB vesitcles
exhibiting randomly distributed 5-7 nm globular surface particles;
magnification approximately x 100,000.
Fig. 1 B: high resolution of peripheral BB-antigens seen if fig.1A
(x 200,000); the "knobs" bound to the BB-matrix through stem-like
constituents represent 90% of total BB-ala-aminopeptidase and ~35%
of total gamma-glu-transpeptidase activity. Proteolytic treatment
with papain cleaves off BB surface proteins (proteases), the remaining membrane now appearing smooth (fig. 1C). Intrinsic part of AAP
(minor portion) and GGT, associated with BB-matrix is not sensitive
to additional proteolytic attack, but might be solubilized by detergents (DOC, triton X 100) as shown in Fig. 2.

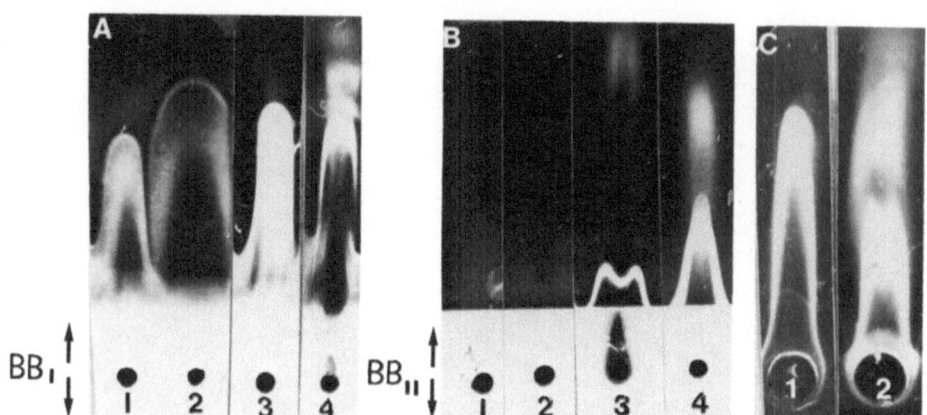

FIG. 2 : Evaluation of peripheral and integral brush border proteases AAP (& GGT) by modified fused rocket immunoeletrophoresis.
Fig. 2A, 2B : BB I = untreated brush border fragments (see Fig. 1A
and B), BB II = brush border membranes treated with papain (fig. 1C),
immobilized in 1.4% agarose gel. BB proteins (AAP) solubilized by
proteolytic agents or detergents applyed into the wells 1-4 led to
precipitation in the upper gel, containing 5% brush border antibody
(well 1 = trypsin, 2 = papain, 3 = DOC, 4 = Triton x 100).
Compared to the left gel (BB$_1$), no precipitates were obtained with
pretreated BB after trypsin and papain digestion (BB$_{II}$, well 1+2)
but after incubation with DOC and triton indicating release of an
<u>intrinsic</u> BB-AAP portion. Fig. 2C : Conventional one-dimensional
electroimmunodiffusion of supernatants after treatment of BB fragments with chymopapain (well 1) and triton X 100 (well 2), staining
for AAP (GGT).

Further studies showed that AAP was activated by 0.5 M Mg^{++} and
Co^{++} but inhibited by Mn^{++} and Zn^{++}; the Michaelis-Menten constant
of AAP was 9.5 x 1o^{-4} M (Ala- p- nitranilide, o.1 M phosphate-
buffer, pH 8.0); pH optimum 8.5; molecular weight (analytical gel-
filtration): 240,000 dalton, using SDS-PAGE : 160,000 (Mercaptoethanol added). The molecular weight of peripheral BB-GGT was determined to 126,000 dalton; ph optimum 7.5.
     BB proteases were used as standard antigens for quantifying the
excretion of kidney BB constituents under pathological conditions
(fig. 4c). Antisera raised against BB-AAP were not capable of precipitating GGT. Equivalent findings were obtained with antisera
against GGT, where AAP activity remained unchanged after incubation
with the GGT antibody (immunoinhibition, immunotitration studies).
     Tissue specificity of antisera directed against BB proteases
was evaluated by immunofluorescence microscopy (fig. 5).
Immunochemically crossreacting constituents were found in intestine
and syncytio(-cyto)-trophoblast of placental tissue. BB-antisera

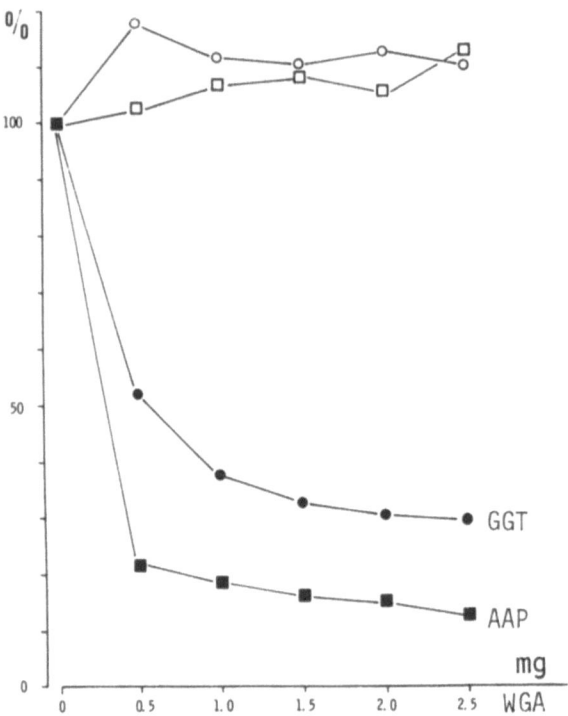

FIG. 3 : Interaction of wheat germ agglutinin (WGA) with soluble BB-AAP and GGT. Abscissa: concentration of lectin added, ordinate: AAP activity found in the supernatant (15 min, 12,000 x g), black dots. Open circles: AAP and GGT activity without centrifugation, indicating no major change of active site by the lectin binding site of the enzymes.

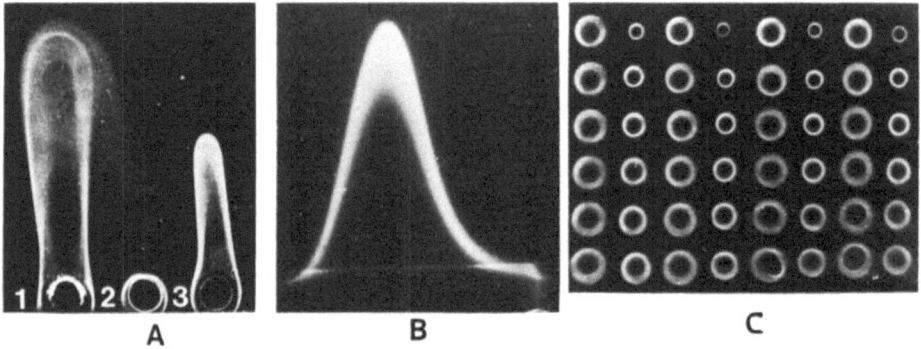

FIG. 4a,b : One/two dimensional immunoelectrophoresis: fig. 4c: radial-immunodiffusion; in the gel: 5% anti-BB antibody. Fig. 4a: 1= BB AAP after solubilization by papain, 2= eluate, 3= desorbate (glycoprotein) after WGA-affinity chromatography; Fig. 4b: purified BB-AAP. Fig. 4c: BB-AAP used as a standard for the evaluation of BB-antigens in urine (see fig. 7).

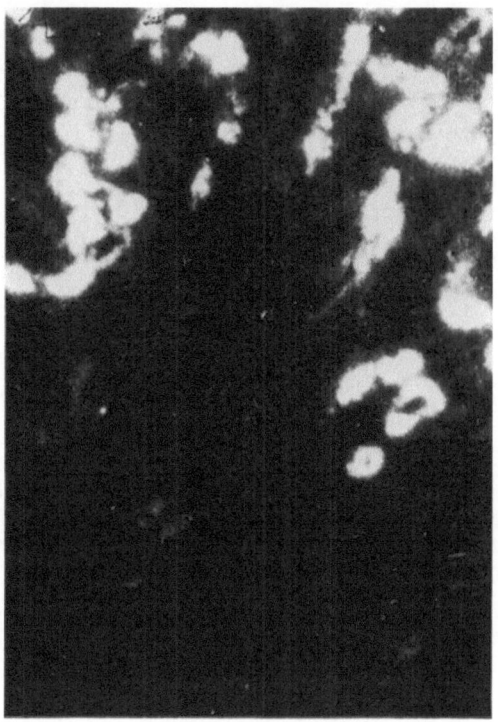

FIG. 5 : Tissue section of human kidney (7μ) incubated with an antibody raised against brush border protease alanine-aminopeptidase.
A specific immunofluorescence of luminal sites of the proximal tubule epithelia only was observed; glomeruli, distal tubule and
medulla remaining negative. Sandwich technique, FITC-anti-rabbit IgG,
mognification x 60.

were used for quantitative histophotometry of kidney sections (distribution of AAP and GGT under normal and pathological conditions, see
figures 9, lo).
A hypothetic model of the arrangement of BB-AAP and GGT within the
membrane matrix is seen in fig. 6.

CLINICAL STUDIES

   In the following study, antisera against BB-proteases were used to
monitor early kidney damage that might be associated with an increased elimination of kidney tissue-proteins in urine [6,7] . In healthy
persons (n= 44), no immunoreactive BB-AAP (= SGP antigen) or GGT
could be measured by applying sensitive counter immunoelectrophoresis,
radial immunodiffusion and onedimensional electroimmunoassay. However, a low but constant urinary excretion of AAP activity was found
in normal persons : male 2.3o2 $\pm$ o.931 U/24 hrs AAP; Cv: 4o %, n= 88;
female 1.43$\pm$ o.53 U/24 hrs AAP, Cv : 3o %, n=36. Clinical studies

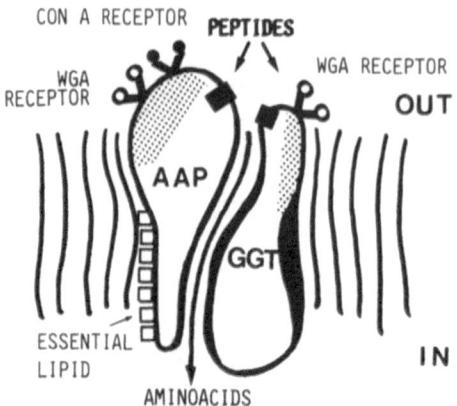

FIG. 6 : Hypothetic arrangement of BB proteases AAP and GGT; both
enzymes consist of a peripheral and an integral portion. The hydro-
philic peripheral part exhibits lectin binding sites and is sensitive
to proteolysis. In contrast to GGT, n-butanol destroys AAP activity
which is in part reversible after adding low MW lipids. AAP and GGT
might be involved in the degradation and transport (?) of peptides;
Recent data suggest that AAP and GGT are strong autoantigens.

revealed that the appearance of immunoreactive BB-proteases AAP (SGP)
and GGT in urine was always related to a clinically defined situation
of renal injury e.g. in patients with glomerulonephritis, rejection
episodes after grafting, kidney involvement in systemic diseases
(multiple myeloma, lupus erythematodes), drug induced nephrotoxicity
(antibiotics, cytostatics or x-ray contrast media, see fig. 7, lo),
and infectious diseases including malaria. A significant correlation
was found between rise of SGP-antigen and AAP-activity in urine under
pathological conditions : R_S = 0.777, 2p = 4 x10^{-5}, n = 29.
   45/49 patients with histologically diagnozed glomerulonephritis
excreted BB-proteins at an increased rate, the 4 "non-excreting"
patients all showed clinical signs of complete remission.
   High concentrations of BB-AAP and GGT were found in urine of
patients after various operations (n= 31).   In addition, protease
activity of BB origin in urine was correlated with the duration of the
operation: group I (35-85 min) : group II (95-16o min) 2p < o.o2;
group I : group III (19o-39o min) 2p < o.oo1 (Mann-Whitney test).
Furthermore, recipients of cadaver kidneys showed a significantly
higher excretion of SGP compared to recipients of kidneys from
related donors (2p < o.o1, observation period up to 2 months after
grafting). A coincidence of 84,5 % was found between significant
rise of SGP, AAP and GGT and the clinical diagnose of an acute rejection episode (total of 26 rejections evaluated).

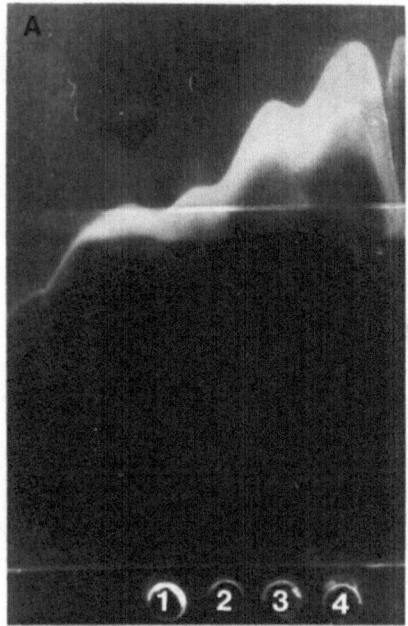
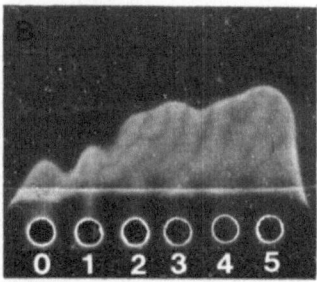
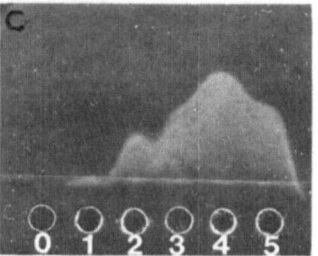

FIG. 7A : One-dimensional tandem immunoelectrodiffusion demonstrating rise of BB-AAP in urine during a acute renal graft rejection crisis; well 1-4 urine specimens of four consecutive days; upper gel: 5% anti -BB-antibody. The immunoprecipitation profil is not changed by the lower gel containing antibody against normal human plasmaproteins. FIG. 7B, 7C : Excretion of BB-AAP in two patients with solid tumors before (o) and during (1-5) treatment with potentially nephrotoxic cytostatics; 24 hrs. specimens, sample volume 10 μl.

All patients (n = 61) receiving x-ray contrast media developed a highly significant rise of BB-AAP, SGP and GGT activity in urine with a maximum between the 24th - 48th hour after administration (i.v. urography, renal angiography). Low osmol contrastmedium (Solutrast, 6oo mosmol/kg) induced a lower increment of BB proteases in urine than high osmol contrast agents (Angiografin, 15oo mosmol/kg) : rise of AAP : $2p<o.o16$, GGT $2p<o.ooo15$ (Mann-Whitney test).

Immunoreactive BB-AAP (SGP antigen) could be demonstrated in urine specimens of patients using ferritin labelled antibodies (fig. 8). Molecular weights of BB AAP and GGT in urine were lower than those of the enzymes isolated from BB fragments of human kidney (AAP: 156,ooo, GGT 93,ooo dalton); nevertheless, BB constituents of urine and from kidney vesicles were completely immunologically identical [6]. In addition no change in lectin binding (Con A, WGA) characteristics compared to the kidney BB enzymes was observed.

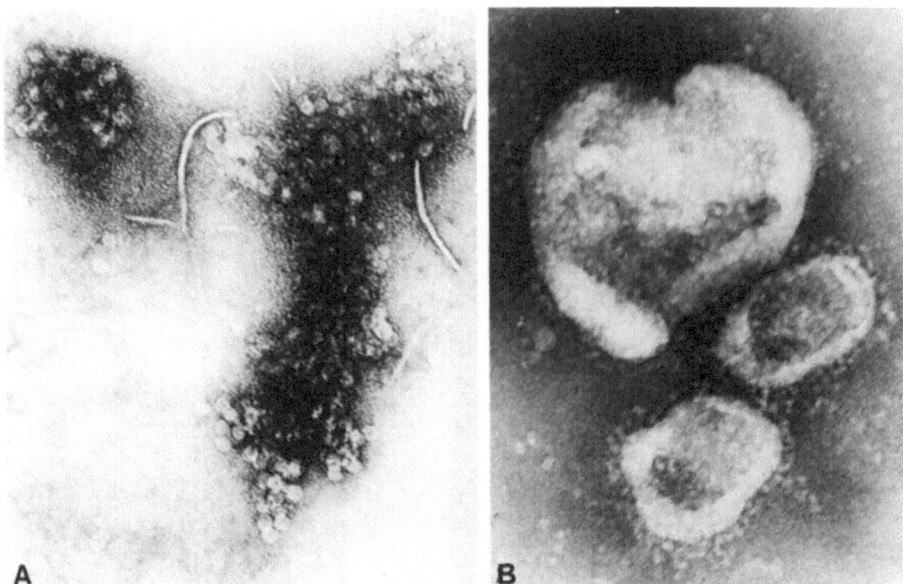

FIG. 8A : Negative staining of kidney brush border AAP in urine of
a patient during cytostatic treatment; native urine was incubated
with anti-AAP antibody, centrifuged and labelled with ferritin-
conjugated anti-rabbit IgG. The figure shows complexes tagged with
ferritin (translucent ring with central dark dot), not found in
healthy persons. FIG. 8B : Excretion of membrane vesicles with
surface particles in a patient with tubule derangement due to drug
nephrotoxicity; both figures x 250,000.

     Besides soluble BB-AAP and GGT supramolecular BB-like vesicles
were also found in urine of patients, especially in those treated
with nephrotoxic cytostatics such as cis-platinum (fig.8 B).
     In order to elucidate tissue patterns of protease activity, kidney
sections were scanned for AAP, SGP applying quantitative high speed
image analysis device technique. Compared with normal kidney, AAP
(SGP) staining profile for instance, were drastically altered (fig.
9, fig.lo). Kidneys of patients with interstitial nephropathy, hydro-
nephrosis, renal arterial stenosis, chronic allograft rejection,
amyloidosis, as well as with renal carcinoma revealed a significant
decrease of BB-AAP (SGP) and GGT activity. This indicates that
parallel to an enhanced loss of BB constituents into the urine, kidney damage might lead to a reduced ability of the tubule cell to
synthesize these membrane proteins in more severe stages of the
underlying disease. Loss of differentiated kidney proteins (proteases) in renal adenocarcinoma might be due to retro-/ dedifferentia-
tion of malignant cell clones (fig. lo). [12]

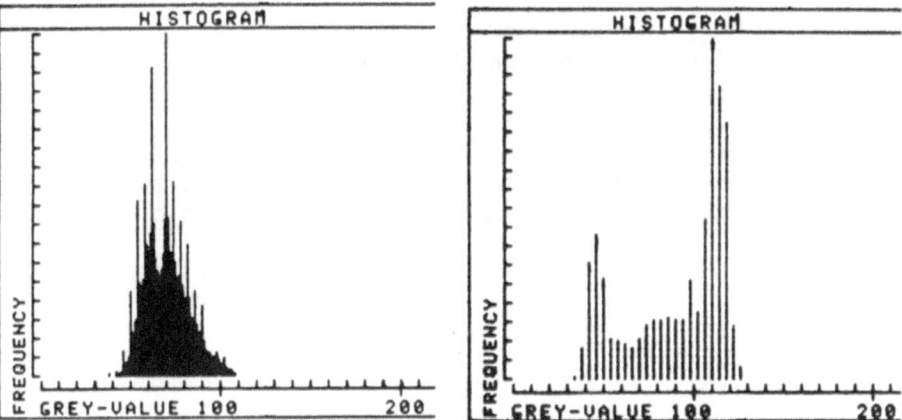

FIG. 9 : Computer assisted histograms of AAP-activity (discrimination of grey levels) in normal human kidney (left) and in a renal allograft that had to be removed due to endstage functional deterioriation (right); dramatic change of normal staining profile in the explanted kidney.

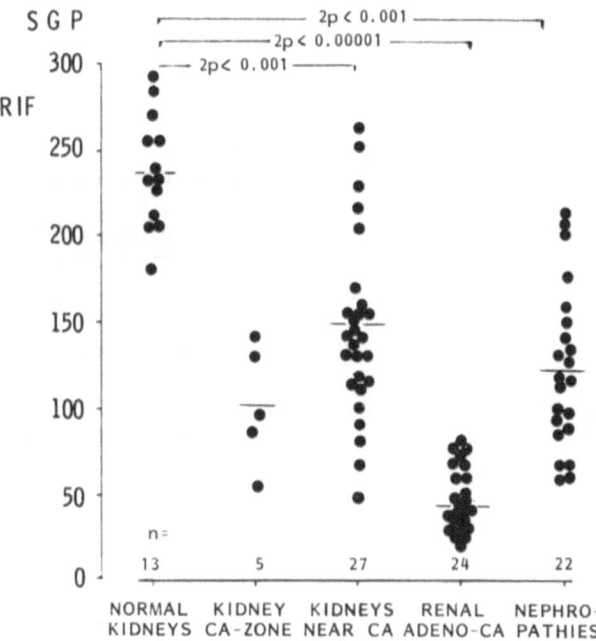

FIG. 10 : Scanning of SGP antigen (=AAP) on 7μ sections of normal, pathological kidneys and renal cancer using electronic image analysis device (Micro-Videomat-2). RIF= relative immunofluorescence; tissue sections were incubated with 1:10 diluted anti-AAP. Loss of SGP under pathological conditions.

REFERENCES

1. J. Scherberich, C. Gauhl, and W. Mondorf, Biochemical, immunological and ultrastructural characterization of brush border
   membranes from human kidney, in: "Curr. Probl. Clin. Biochem.
   8," W. Guder, U. Schmidt, eds., H. Huber Publ., Bern (1979).
2. F. A. Carone, and D. R. Peterson, Hydrolysis and transport of
   small peptides by the proximal tubule, Am. J. Physiol. 7:F 151
   (1980).
3. S. Silbernagl, Renal handling of aminoacids and oligopeptides,
   in:"Experimental and Clinical Aspects of Proteinuria," M. Wei-
   se, ed., Karger, Basel, New York (1981).
4. R. Kinne, Current concepts of renal proximal tubular function,
   Contr. Nephrol. 14:14 (1978).
5. J. E. Scherberich, R. Kinne, C. Gauhl, W. Mondorf, and W.
   Schoeppe, Isolation and characterization of basal-lateral and
   luminal plasma-membranes from the proximal tubule of human
   kidney, in: "Protides Biol. Fluids," H. Peeters, ed., Pergamon,
   Oxf. N. York (1981).
6. J. E. Scherberich, and W. Mondorf, Excretion of kidney brush
   border antigens as a quantitative indicator of tubular damage,
   in "Curr. Probl. Clin. Biochem.," Huber Publ., Bern (1979).
7. W. Mondorf, J. E. Scherberich, T. Stefanescu, P. Mitrou, and
   W. Schoeppe, The elimination of brush border membrane proteins
   in urine caused by toxic alteration of the tubular cell, Contr.
   Nephrol. 24:99 (1981).
8. W. Mondorf, J. Hendus, J. Breier, J. Scherberich, and W. Schoeppe, Tubular toxicity induced by aminoglycosides in human kidneys, in: "Nephrotoxicity," J. P. Fillastre, ed., Masson Publ.
   (1978).
9. J. E. Scherberich, W. Mondorf, and W. Schoeppe, Side effects of
   antimicrobial therapy: nephrotoxic potential of antibiotics
   (aminoglycosides, cephalosporins) as monitored by brush border
   tissue proteinuria, in: "Pyelonephritis, Urinary Tract Infection IV," Losse, Asscher, Lison, eds., Thieme, Stuttgart, New
   York (1980).
10. J. Scherberich, B. Kleemann, and W. Mondorf, Isolation of kid-
    ney brush border gamma-glutamyl transpeptidase from urine by
    specific antibody gel chromatography, Clin. Chim. Acta 93:35
    (1979).
11. G. Heinert, J. E. Scherberich, W. Mondorf, B. Kleemann, Enzyme
    histograms in normal kidneys and renal cancer by image analysis
    device, Verhdl. Dtsch. Krebsges. (Springer) 2:440 (1979).
12. J. E. Scherberich, T. Stefanescu, C. Gauhl, R. Sayre, B. Klee-
    mann,and W. Mondorf, Lectin receptors of plasma-membranes from
    human kidney and renal adenocarcinoma, in: "Prot. Biol. Fluids,"
    H. Peeters, ed., Pergamon, Oxf., New York (1979).

RECENT ADVANCES IN PROTEASE RESEARCH

USING SYNTHETIC SUBSTRATES

[1]Reinhart Gossrau, [2]Zdenek Lojda, [3]Robert E. Smith, and [1]Pranav Sinha
[1]Department of Anatomy, Free University, [2]Laboratory of Histochemistry, Charles University and [3]Lawrence Livermore Laboratories, [1]Berlin 33, FRG, [2]Prague, ČSSR and [3]Livermore, USA

INTRODUCTION

Until recently the use of synthetic peptides in protease research was rather limited; this was mainly due to the lack of suitable substrates and effective reaction principles (Pearse, 1972; Smith and van Frank, 1975; Lojda et al., 1979). However, especially during the last years substantial progress was achieved in this field; it is the aim of the present report to describe the most important advances in the use of synthetic peptide substrates.

DETECTOR GROUPS AND SUBSTRATE SYNTHESIS

The detector groups or tags are 2-naphthylamine (2NA), 4-methoxy-2-naphthylamine (MNA), 7-amino-4-methylcoumarin (AMC), 7-amino-4-trifluoromethylcoumarin (AFC; cf. Huseby and Smith 1980) and a new tag (Fig. 1). This group is already introduced in peptide synthesis. For reasons of patent law we are provisionally obliged to call it only X. All these detector groups can be linked to non-blocked or N-terminal blocked amino acids or peptides. In order to design relatively specific substrates the so-called binding site or recognition site theory is now employed for the synthesis of synthetic peptides (cf. Smith and van Frank, 1975; Huseby and Smith, 1980). An example for this principle is the 2NA peptide for the investigation of enteropeptidase. The development of the NA peptide for this protease based on the fact that the

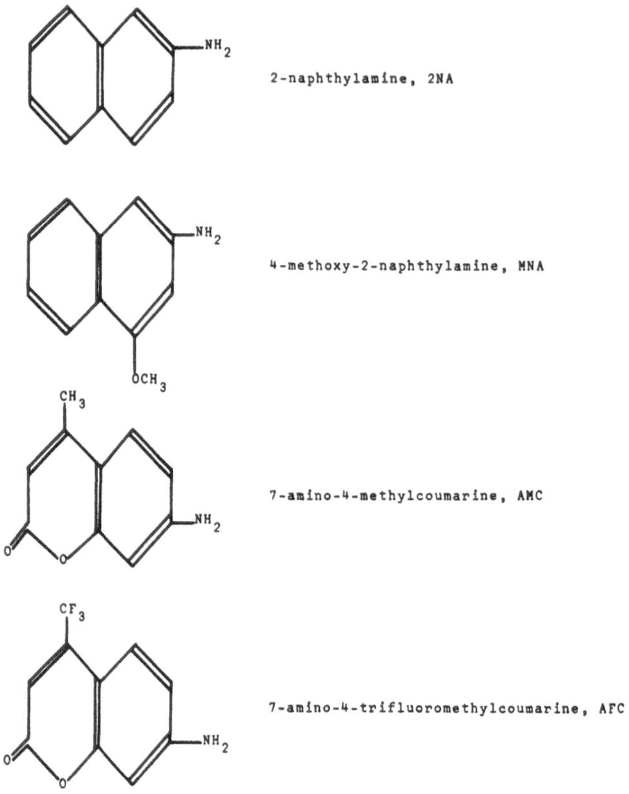

2-naphthylamine, 2NA

4-methoxy-2-naphthylamine, MNA

7-amino-4-methylcoumarine, AMC

7-amino-4-trifluoromethylcoumarine, AFC

Fig. 1. Structural formula, names and abbreviations
of detector groups for protease research

activation peptide of trypsin or the enteropeptidase
substrate has the amino acid sequence, e.g. in man of
Ala-Pro-Phe-Asp-Asp-Asp-Asp-Lys. The corresponding synthetic substrate contains most of the amino acids of the
activation peptide, i.e. Asp-Asp-Asp-Asp-Lys linked via
an amide bond to 2NA. This amino acid sequence has been
shown to be the substrate recognition and specifity site
for the enteropeptidase-catalysed activation of trypsinogen (cf. Hesford et al., 1976; Antonowicz, 1979).

METHODOLOGICAL POSSIBILITIES

The methodological possibilities which the detector groups and their peptides offer in protease research
can be summarized as follows: Most substrates, i.e. 2NA,
MNA, AFC, AMC and X peptides are useful for the band development after isoelectric focusing. The special aspects
of the fluorescence detection of the bands are reported

by Sinha et al. (this volume). Homogenate, supernatant
or cell fraction biochemistry can be performed with 2NA,
MNA and AFC substrates, section biochemistry (quantitative histochemistry or cytochemistry) with MNA or X peptides, light microscopic location histochemistry (qualitative histochemistry or cytochemistry) with 2NA, MNA
and X peptides and ultracytochemistry (electron microscopic histochemistry) of proteases with MNA substrates
(cf. Smith and van Frank, 1975; Huseby and Smith, 1980;
Gossrau, 1981; Lojda, 1981). The most valuable substrates
are the MNA peptides; they inform about proteases with
all mentioned methodological procedures and allow,
therefore, the most thorough insight into the biological
meaning of these enzymes.

INVESTIGATED PROTEASES

        Among the proteases which can reliably be studied
at the moment with any of these synthetic peptides only
some are given. They are those which may be employed at
least with two complementary methods, e.g. MNA or AFC peptides in biochemical assays and by isoelectric focusing;
the use of one single procedure often is of rather limited value in protease research due e.g. to specicifity
and localization problems (cf. Gossrau, 1981; Lojda,
1981). The single proteases are aminopeptidase A (an enzyme that is in some locations identical with the angiotensin II degrading angiotensinase A; cf. Hess, 1965;
Kugler, 1982), aminopeptidase B, aminopeptidase M or N,
γ-glutamyltranspeptidase, dipeptidylpeptidase (DPP) I
(cathepsin C), II and IV, enteropeptidase (enterokinase),
trypsin, chymotrypsin, cathepsin B, postproline cleaving
enzymes and proteases that hydrolyse MNA, AMC, AFC and X
peptides for kallikrein, plasmin, thrombin, plasminogen
activator, elastase and cathepsin B. These endopeptidases are only in part identical with the mentioned and
already biochemically classified proteases since they
are hydrolysed by additional proteases; at the moment
these enzymes are identified especially for the kidney
and submandibular gland in our laboratory and are provisionally designated secretory and membrane endopeptidase
I and II (cf. Gossrau, 1981, 1982a; Lojda, 1981).

ISOELECTRIC FOCUSING

        Isoelectric focusing reveals important species, or-
gan (Fig.2), tissue and cell differences of proteases
but is also a useful tool to test the specicifity and
sensitivity of 2NA, MNA, AMC and AFC peptides for each
protease (Sinha and Gossrau, 1982; Smith and Grabske 1982;

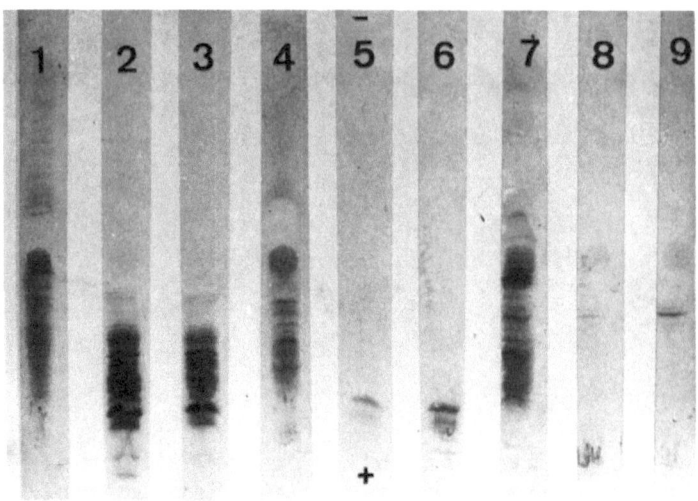

Fig. 2.   Isoelectric focusing and band staining with
          either Coomassie Blue or reaction with Lys-Pro-
          MNA and Fast Blue B for simultaneous coupling.
          Column 1-3 rat kidney, column 1 Coomassie Blue,
          column 2 reaction at pH 7, column 3 see 2 but
          pH 5; column 4-6 rat submandibular gland, column
          4 Coomassie Blue, column 5 reaction at pH 7,
          column 6 see 5 but pH 5; column 7-9 human liver
          biopsy, column 7 Coomassie Blue, column 8 reac-
          tion at pH 7, column 9 see 8 but pH 5

Simultaneous and postcoupling azo-dye techniques (Lojda
et al., 1979) or fluorescence methods serve for the vi-
sualization of the bands. In principle, the methods are
those used for the histochemical protease investigations.
The specicifity of synthetic peptides can be shown e.g.
for lysosomal DPP II and membrane-linked DPP IV in the
jejunum. Gly-Pro-MNA is the most specific peptide for
the investigation of membrane-bound DPP IV which represents cathodic bands. Lysosomal DPP II, which focuses
in the anodal region, hydrolyses the same substrate but
more slowly. The reverse is true for Lys-Ala-MNA because
it is preferentially cleaved by DPP II, i.e. Lys-Ala-MNA
is a relatively specific synthetic peptide for the study
of DPP II. Lys-Pro-MNA and Ala-Pro-MNA enable higher hydrolysis rates than Gly-Pro-MNA or Lys-Ala-MNA in both
the anodal and cathodal region and deliver, therefore,
more bands. However, in contrast to Gly-Pro-MNA or Lys-
Ala-MNA these MNA peptides are hydrolysed by both DPPs

and are in so far less specific substrates.

BZ-Ser-Pro-Phe-Arg-MNA and CBZ-Val-Leu-Arg-MNA for the demonstration of serum and glandular kallikrein are examples for the sensitivity problem of synthetic peptides in proteinase research. After focusing of supernatants from rat submandibular gland the number of bands is the same with both substrates; but the rate of hydrolysis is higher with the BZ-peptide (Fig.3). The highest hydrolysis rates, however, are often seen with D-Val-Leu-Arg-MNA.

Fig. 3.   Isoelectric focusing and band staining with MNA peptides for kallikrein and Fast Blue B for simultaneous coupling

BIOCHEMISTRY

In order to determine the kinetic properties of proteases in homogenates, supernatants or cell fractions MNA and AFC peptides should be prefered in biochemical assays. Methodologically, these assays are one-step reactions whereby the liberation of MNA or AFC can be continuously recorded with a fluorometer. The maximal reaction velocity of the 2NA and AMC substrates often equal those of the MNA and AFC peptides (Zimmerman et al., 1976; Kanaoka et al., 1977; Smith et al., 1980; Gossrau, 1980; Hartmann and Gossrau, 1982). This fact is e.g. demonstrated with the Gly-Pro substrates of DPP IV. The hydrolysis rates of kidney supernatants with Gly-Pro-2NA, Gly-Pro-MNA, Gly-Pro-AMC and Gly-Pro-AFC are rather similar. The same is true with AMC and AFC peptides for the detection of endopeptidases; the AMC peptides for trypsin, plasminogen activator and plasmin are hydrolysed as quickly as the corresponding AFC peptides. However, the fluorescence of 2NA is low compared with that of e.g. MNA and AMC (Fig.4). Moreover, the blank values of the AMC peptides are relatively high since both, the AMC substrates and the liberated AMC show a bluish fluorescence. The advantage of the AFC peptides is the shift of the detector's group fluorescence to yellow green whereas the AFC substrates fluoresce also in the blue region.

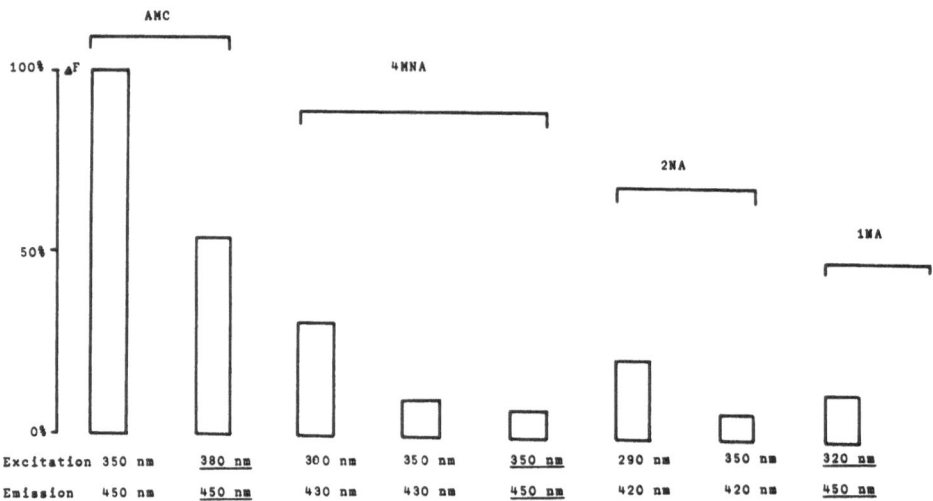

Fig. 4. Relative fluorescence (ΔF) of aequimolar AMC, 4MNA, 2NA and 1NA

SECTION BIOCHEMISTRY

However, homogenate, supernatant and cell fraction biochemistry are of limited value if quantitative informations about protease activities in single cells of heterogenous tissues are desired. They are possible by section biochemistry of proteases; this approach bases on either cytophotometry or microdensitometry, i.e. measurements of absorbance or cytofluorometry, i.e. measurements of fluorescence. In the simultaneous azo-dye reactions for cytophotometric protease determinations (Gossrau and Lojda, 1980; Lojda and Gossrau, 1980; Blenk and Gossrau, 1981; Gutschmidt and Gossrau, 1981; Kugler, 1982; Gutschmidt et al., this volume) the respective protease liberates MNA in a first step; in a second step MNA is simultaneously coupled to Fast Blue B (FBB) and an azo-dye is formed which is then continuously recorded by a microdensitometer (Fig.5). A suitable instrument for continuous or kinetic microdensitometry of membrane, secretory and lysosomal proteases is the flying spot instrument of Vickers; compared with end-point or static microdensitometry this procedure is recommended since it allows the determination of initial reaction rates of proteases in single cells and cell compartments, e.g. secretory granules and lysosomes.

The microdensitometric continuous measurement of

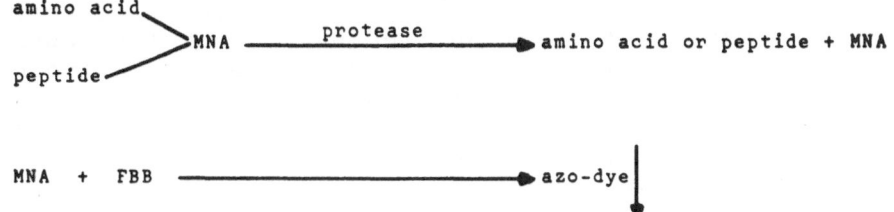

Fig. 5.   Reaction scheme for continuous cytophotometric
           section biochemistry with simultaneous azo-
           coupling

DPP IV in the brush border of intestinal villi illustrates the possibilities of this refined technique.
The pH optimum curve is determined in the brush border
of single enterocytes from the middle part of the villi;
the highest hydrolysis rate occurs at pH 8 (Fig.6); microdensitometry can also be carried out to show the dependency of the hydrolysis rate of DPP IV on different
Gly-Pro-MNA concentrations, The maximal reaction velocity of the enzyme is reached with 3 mM ooncentrations;
the K_m calculated from these curves is in the range of
0.25 mM (Fig.7). The other methodological possibility of

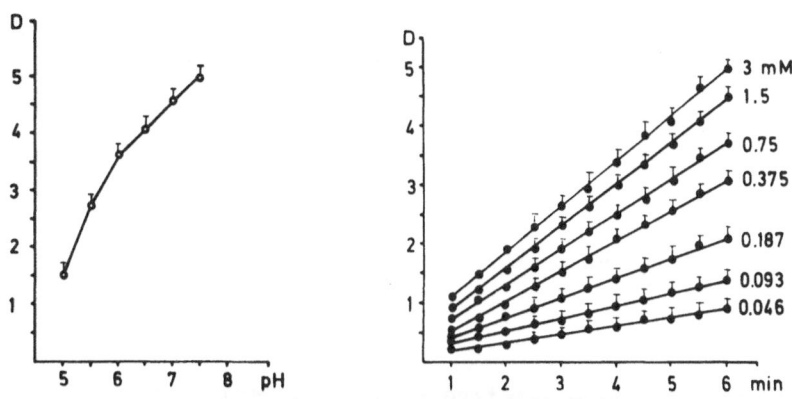

Fig.   6. Dependency of Gly-Pro-MNA hydrolysis of pH.
         Fig. 7. Dependency of hydrolysis rate of Gly-
         Pro-MNA concentration at pH 7.5

section biochemistry is the use of fluorometry. Instead
of a diazonium salt the liberated MNA is simultaneously
coupled to nitrosalicylaldehyde (NSA) forming a fluorescent complex (Fig.8; Dolbeare and Smith, 1977); or X is
liberated; the MNA-NSA complex or X can be recorded by
their fluorescence. The advantage of the fluorescence
procedures is that the protease under investigation is
measured in the absence of the inhibitory Fast Blue B.
The activity of cathepsin B with CBZ-Ala-Arg-Arg-MNA
and NSA in single culture cells in dependency of the reaction time can be measured with this technique; after
a lag phase there is a linear increase of the fluorescent reaction product up to 10 min.

LOCATION HISTOCHEMISTRY

     From section biochemistry with MNA and X peptides
it follows that those substrates are also suitable for
the histochemical location of proteases. Therefore, two
methodological possibilities exist again. The first is
the indirect simultaneous azo-coupling technique; it
corresponds to that for the kinetic cytophotometric
measurements of proteases and is possible, in addition,
with 2NA and X peptides. Furthermore, beside Fast Blue B
other diazonium salts are recommendable like hexazotized
pararosaniline and new fuchsine or Fast Corinth V and
Fast Garnet GBC (cf. Gossrau, 1981; Lojda, 1981). With
this technique the location of enteropeptidase with
Gly-Asp-Asp-Asp-Asp-Lys-2NA as the substrate and Fast
Blue B for coupling in the brush border and its closest
vicinity of the duodenal enterocytes (Fig.9; Lojda and
Gossrau, 1982) or the reaction of DPP I with Gly-Arg-MNA
as the substrate and Fast Corinth V as coupling agent in
liver lysosomes is shown. Secondly, indirect fluores-

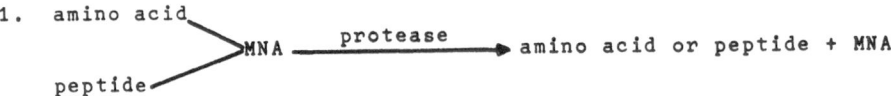

1. amino acid
                     MNA ——— protease ——→ amino acid or peptide + MNA
    peptide

2. MNA + NSA ————————→ fluorescent complex

Fig. 8. Reaction scheme for continuous cytofluorometric
         section biochemistry

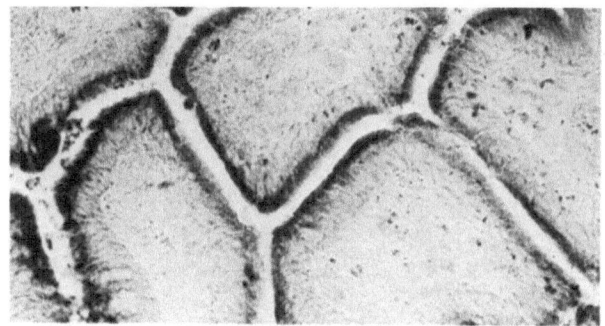

Fig. 9. Location of enteropeptidase

cence histochemistry is useful for protease localization
in cells and tissues. MNA can be coupled to NSA and the
formed MNA-NSA complex is visualized by its fluorescence
(Dolbeare and Smith, 1977; Dolbeare and Vanderlaan, 1979;
Stauber and Ong, 1981). Provided freeze-dried celloidin-
mounted cryostat sections (Lojda et al., 1979) are used
the localization of the reaction product can be nearly
as precise as with azo-dye procedures. Using fluorescence histochemistry DPP IV is localized in the proximal
renal (Fig.10)and jejunal brush border (Fig.11) or cathepsin B with CBZ-Ala-Arg-Arg-MNA or D-Val-Lys-Lys-Arg-
MNA as substrates in the giant lysosomes of the enterocytes of the ilium of suckling rats (Fig.12).

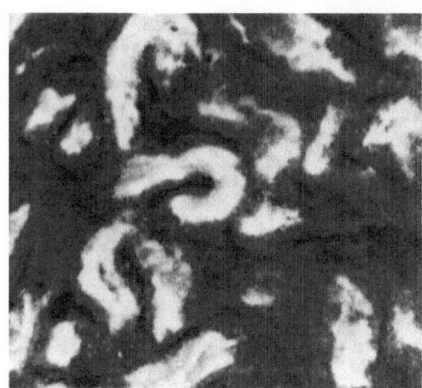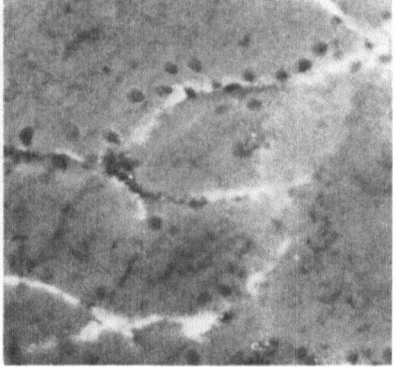

Fig. 10. and 11. Fluorescence microscopic location of
        DPP IV in brush borders

     The future of the fluorescence microscopic locali-
zation of proteases, however, might be their direct location with X peptides. In contrast to MNA the liberated
X is nearly water insoluble and provides, therefore, a
good location especially in freeze-dried celloidin-moun-

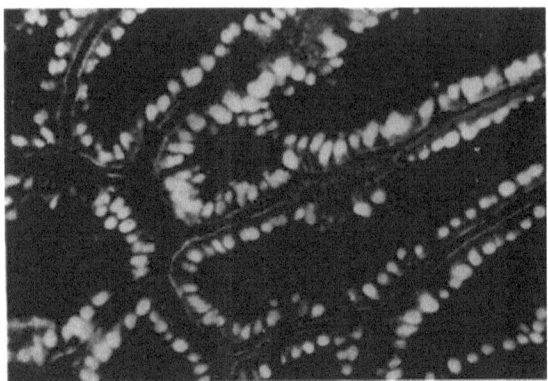

Fig. 12. Fluorescence microscopic location of cathepsin
        B with CBZ-Ala-Arg-Arg-MNA in giant lysosomes

ted but also in fresh frozen cryostat sections after ace-
ton pretreatment. The X procedure allows, e.g. the location of aminopeptidase M by the greenish fluorescence of
the liberated X in the brush border of jejunal enterocytes (Fig.13) and proximal tubule cells.

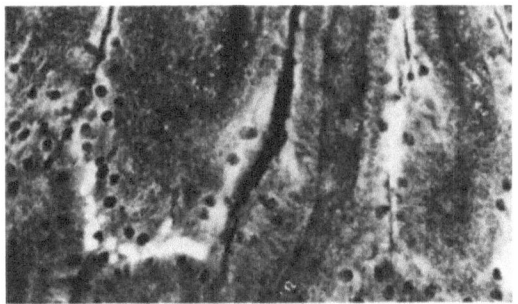

Fig. 13. Fluorescence microscopic location of APM with
        Ala-X in brush border

ULTRACYTOCHEMISTRY

    The draw-back of qualitative light microscopic pro-
tease histochemistry is that it delivers only limited
knowledge about the intra- or subcellular binding sites
of proteases. This knowledge, however, is important to
understand the biological and functional role of these
enzymes. The problem is tried now to be overcome by the
ultracytochemical (electron microscopical) demonstration
of proteases. Ultracytochemistry is possible with simultaneous azo-coupling if hexazotized pararosaniline (HPR)

or new fuchsine (HNF) serve as diazonium salts (Smith
and van Frank, 1975; Gossrau, 1981; Lojda, 1981; Schröder and Gossrau, 1982). This fact bases on the unique
property of their azo-dyes to form an electron dense reaction product by osmification in a third reaction step
after the substrate hydrolysis in the first and the simultaneous coupling of the liberated MNA in the second
step are completed (Fig.14). Using this procedure we are

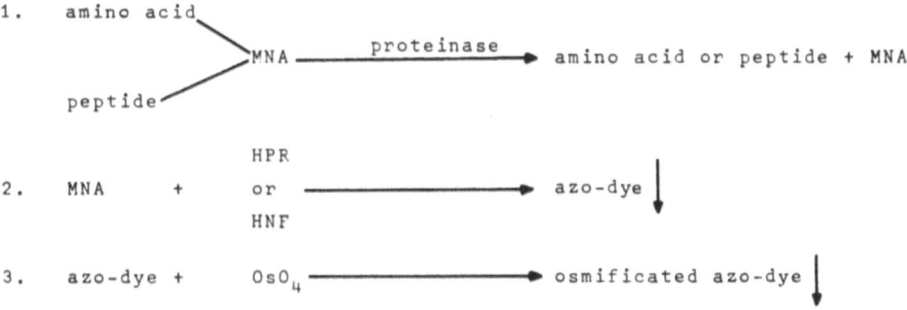

Fig. 14. Reaction scheme for the ultracytochemical
location of proteinases

able to reveal the intracellular binding sites of proteases, e.g. of DPP II in the lysosomes and secretion
granules of the B cells in the endocrine pancreas (Fig.
15,16), of aminopeptidase A in the Golgi apparatus of

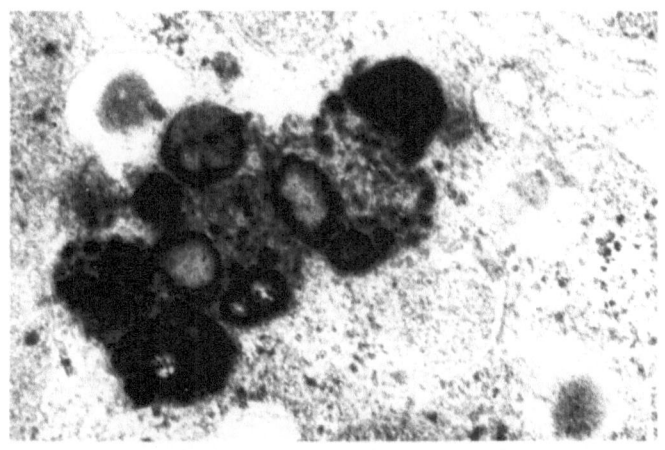

Fig. 15. DPP II in lysosomes (L)

jejunal enterocytes (Fig.17) or of endopeptidase I, which
hydrolyses the elastase substrate Glu-Ala-Ala-Ala-MNA
in the microvilli of the renal proximal tubule.

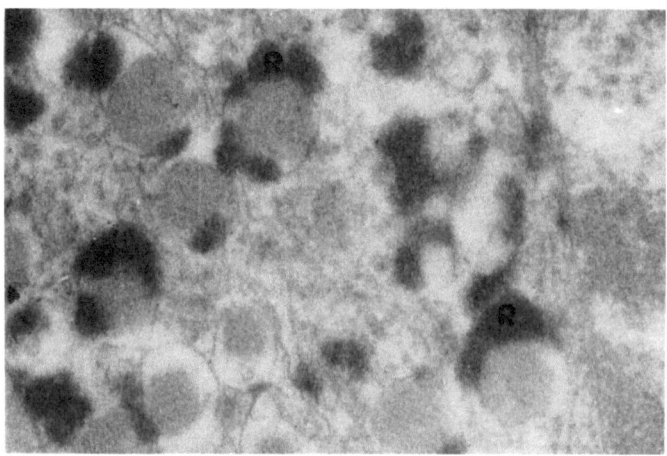

Fig. 16. DPP II in secretion granules, R = reaction pro-
         duct.Figure 15 and 16 are courtsey of Dr. Renate
         Graf (Department of Anatomy, FU Berlin)

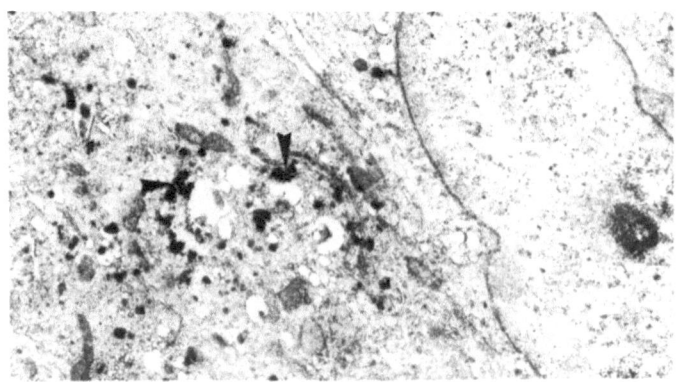

Fig. 17. APA in the Golgi apparatus (arrow heads) of a
         jejunal enterocyte

BIOLOGICAL MEANING

     What can be achieved in the field of protease rea-
search with synthetic peptides from the more applied and
biological standpoint ? This question is briefly discussed (as far it is possible at the moment) using the
rat kidney as an example. The light microscopic and

Table 1. Proteinases (peptidases) and their location in
the kidney of rats

Peptidase	peptidase localization
APA DPP IV	surface membranes of endothelial cells and podocytes of glomerula
APA APM DPP IV GGT endopeptidase I endopeptidase II	microvilli of proximal tubules
DPP I DPP II cathepsin B APA	lysosomes of proximal tubules
DPP I DPP II	lysosomes of distal tubules
DPP IV GGT	surface membranes of descending limbs of Henle
DPP IV GGT	surface membranes of distal cortical tubules
APA	surface membranes of peritubular capillary endothelial cells
DPP IV GGT	surface membranes of vasa recta endothelial cells

ultracytochemical distribution of the binding sites of
proteases obtained with azo-dye and fluorescence methods
are compiled in Table 1 (Gossrau 1982b). If one considers
the role of the kidney in peptide hormone metabolism many of them may be degraded or activated by membrane-
linked and lysosomal proteases in the different segments
of the nephron (Carone and Peterson, 1980; Carone et al.,
1980, b; Emmanouel et al., 1980a, b; Maunsbach et al.,
1980). In addition, proteases bound to the surface membrane of endothelial cells in special parts of the renal
vascular bed, e.g. aminopeptidase A in the capillaries
around the distal straight tubules are involved in degradation processes, too (Lojda and Gossrau, 1980).

However, the question may be asked how relevant are
protease data obtained with synthetic peptides to the
physiological peptides present in the blood and primary
urine of the kidney. To answer this problem inhibitor
experiments are in progress now in our laboratory in

which the proteases are measured and localized by simultaneous incubation with their synthetic and potential physiological substrates or peptides respectively. As already proved by Hess (1965), Nagatsu et al. (1965, 1970) and
Kugler (1982) aminopeptidase A is competitively inhibited
by angiotensin II in the nephron; in addition, it is possible to show competitive inhibition of lysosomal DPP I
and II, membrane-linked DPP IV, membrane-bound endopeptidase I and II and aminopeptidase B by various adrenal,
hypothalamic and gut peptide hormones in the glomerula
and/or tubules. From these findings one may conclude that
those proteases which are primarily localized with synthetic peptides actually hydrolyse peptides normally present in the kidney.

SUMMARY

     2-Naphthylamine, 4-methoxy-2-naphthylamine, 7-amino-
4-methylcoumarin and 7-amino-4-trifluoromethylcoumarin
and X peptides have become a useful tool to study proteases by a battery of complementary methods; they now include biochemistry, section biochemistry, location cytochemistry, ultracytochemistry and isoelectric focusing.
By the use of these procedures we hope that more insight
will be possible in the functional role of proteases in
health and disease in the near future.

REFERENCES

Antonowicz, C., 1979, The role of enteropeptidase in the
        digestion of protein and its development in the
        human fetal small intestine, in:"Development of
        mammalian absorptive processes"(Ciba Foundation
        Symposion 70), Exerpta Medica, Amsterdam.
Blenk, R., and Gossrau, R., 1981, Biochemische und
        quantitativ-histochemische Untersuchungen von
        Proteasen in der Gl. submandibularis von Ratten,
        Acta histochemica Suppl. 24:309.
Carone, F.A., and Peterson, D.R., 1980, Hydrolysis and
        transport of small peptides by the proximal tubule,
        Am. J. Physiol. 238:151.
Carone, F.A., Peterson, D.R., Oparil, S., and Pullman, T.
        N., 1980, Renal transport and catabolism of small
        peptides, in:"Functional ultrastructure of the kid-
        ney,"A.B. Maunsbach, A. Steen Olsen and E.J. Chri-
        stensen, eds., Academic Press, London, New York,
        Toronto, Sydney, San Francisco.
Dolbeare, F.A., and Smith, R.E., 1977, Flow cytometric
        measurement of peptidases with use of 5-nitrosa-
        licylaldehyde and 4-methoxy-ß-naphthylamine deri-

vatives, Clin. Chem. 23:1485.

Dolbeare, F.A., and Vanderlaan, M., 1979, A fluorescent
    assay of proteinases in cultured mammalian cells,
    J. Histochem. Cytochem. 27:1493.

Emmanouel, D.S., Lindheimer, M.D., and Katz, A.J., 1980a,
    Pathogenesis of endocrine abnormalities in uremia,
    Endocr. Rev. 1:28.

Emmanouel, D.S., Lindheimer, M.D., and Katz, A.J., 1980b,
    Role of kidney in hormone metabolism and its impli-
    cation in clinical medicine, Klin. Wochenschr.,
    58:1005.

Gossrau, R., 1980, Conventional techniques for membrane-
    bound enzymes, in:"Trends in enzyme histochemistry
    and cytochemistry" (Ciba Foundation Symposion 73),
    Excerpta Medica, Amsterdam.

Gossrau, R., 1981, Investigation of proteinases in the
    digestive tract using 4-methoxy-2-naphthylamine
    (MNA) substrates, J. Histochem. Cytochem. 29:464.

Gossrau, R., 1982a, Localization and regulation of pro-
    teolytic enzymes in salivary glands, submitted.

Gossrau, R., 1982b, Peptide and protein handling by pro-
    teases in the kidney of laboratory rodents, Acta
    histochem. Suppl., in press.

Gossrau, R., and Lojda, Z., 1980, Study on dipeptidylami-
    nopeptidase II (DAP II), Histochemistry 70:53.

Gutschmidt, S., and Gossrau, R., 1981, A quantitative
    histochemical study of dipeptidylpeptidase (DPP
    IV), Histochemistry 73:285.

Hartmann, K., and Gossrau, R., 1982, Zur Eignung von
    Aminomethylcoumarin- und Naphthylaminsubstraten
    für mikrochemische Peptidasenmessungen, Acta hi-
    stochemica Suppl., in press.

Hesford, F., Hadorn, B., Blaser, K., and Schneider, C.H.,
    1976, A new substrate for enteropeptidase, FEBS
    (Fed. Eur. Biochem. Soc.) Lett. 71:279.

Hess, R., 1965, Arylamidase activity related to angio-
    tensinase, Biochim. Biophys. Acta 99:316.

Huseby, R.M., and Smith, R.E., 1980, Synthetic oligopep-
    tide substrates - their diagnostic application
    in blood coagulation, fibrinolysis, and other pa-
    thological states, Seminars in Thrombosis and
    Hemostasis 6:173.

Kanaoka, Y., Takahashi, T., and Nakayama, H., 1977, A
    new fluorogenic substrate for aminopeptidase,
    Chem. Pharm. Bull. 25:362.

Kugler, P., 1982, Aminopeptidase A is angiotensinase A.
    I. Quantitative histochemical studies in the kid-
    ney glomerulus, Histochemistry 74:229.

Lojda, Z., 1981, Proteinases in pathology. Usefulness
    of histochemical methods, J. Histochem. Cytochem.
    29:481.

Lojda, Z., and Gossrau, R., 1980, Study on aminopeptidase
    A, Histochemistry 67:267.
Lojda, Z., and Gossrau, R., 1982, Histochemical demon-
    stration of enteropeptidase. New method with a
    synthetic substrate and its comparison with the
    trypsinogen procedure, Histochemistry, in press.
Lojda, Z., Gossrau, R., and Schiebler, T.H., 1979, Enzyme
    histochemistry. A laboratory manual, Springer,
    Berlin, Heidelberg, New York.
Lojda, Z., and Kuhlich, J., 1981, The usefulness of ana-
    lytical electrofocusing in a thin-layer polyacryl-
    amide gel (PAG) in the histochemistry of enzymes
    cleaving peptide bonds, Histochemistry 73:311.
Maunsbach, A.B., Steen Olsen, T., and Christensen, E.J.,
    eds., 1980, Functional ultrastructure of the kid-
    ney, Academic Press, London, New York, Toronto,
    Sidney, San Francisco.
Nagatsu, I.T., Gillespie, I.L., George, J.M., Folk, J.E.,
    and Glenner, G.G., 1965, Serum aminopeptidase
    "angiotensinase" and hypertension. - II. Amino
    acid ß-naphthylamide hydrolysis by normal and
    hypertensive serum. Biochem. Pharmacol. 14:853.
Nagatsu, I.T., Yamamoto, T., Glenner, G.G., and Mehl,
    J.W., 1970, Purification of aminopeptidase A in
    human serum and degradation of angiotensin II by
    the purified enzyme, Biochem. Biophys. Acta
    198:255.
Pearse, A.G.E., 1972,"Histochemistry. Theoretical and
    applied,"Churchill Livingstone, Edinburgh.
Schröder, J., and Gossrau, R., 1982, Elektronenmikros-
    kopische Lokalisation von Proteasen mit direkten
    Azofarbstoffmethoden. Acta histochemica Suppl.
    25:147.
Sinha, P., and Gossrau, R., 1982, Histochemical studies
    with MNA substrates using isoelectric focusing,
    J. Histochem. Cytochem. 30:575.
Smith, R.E., Bissell, E.R., Mitchell, A.R., and Pearson,
    K.W., 1980, Direct photometric or fluorometric
    assay of proteinases using substrates containing
    7-amino-4-trifluoromethylcoumarin, Thromb. Res.
    17:393.
Smith, R.E., and van Frank, R.M., 1975, The use of ami-
    no acid derivatives of 4-methoxy-ß-naphthylamine
    for the assay and subcellular localization of
    tissue proteinases, in:"Lysosoms in biology and
    pathology," Vol. 4, J.T. Dingle and R.J. Dean,
    eds., North Holland Publishing Company, Amsterdam.
Smith, R.E., and Grabske, R.J., 1982, Isoenzymes of
    glandular kallikrein revealed by isoelectric

focusing and fluorescence staining with an oli-
gopeptide substrate inpregnated into cellulose,
J. Histochem. Cytochem. 30:575.

Stauber, W.T., and Ong, S.H., 1981, Fluorescence demonstration of cathepsin B in skeletal, cardiac, and
vascular smooth muscle, J. Histochem. Cytochem.
29:866.

Zimmerman, M., Yurewicz, E., and Patel, G., 1976, A new
fluorogenic substrate for chymotrypsin, Anal.
Biochem. 70:258.

KINETIC CHARACTERIZATION OF BRUSH BORDER MEMBRANE PROTEASES IN RELATIONSHIP TO MUCOSAL ARCHITECTURE BY SECTION BIOCHEMISTRY

[1]Siegfried Gutschmidt, [1]Rüdiger Hoper, and [2]Reinhart Gossrau

[1]Department of Internal Medicine and Gastroen-terology, Klinikum Steglitz and [2]Department of Anatomy, Free University
[1]Berlin 45 and [2]Berlin 33, FRG

INTRODUCTION

Proteases are a major constituent of intestinal brush border enzymes. The overall functional response of these enzymes in a given intestinal segment probably depends on changes in their apparent kinetic constants and also on the distribution pattern of their activity in the mucosa. This has been suggested by the results obtained after section biochemistry (quantitative histo-chemistry; Gossrau et al., this volume) analyzing the brush border disaccharidases lactase and α-D-glucosida-se in different experimental models of intestinal adap-tation (Gutschmidt et al., 1982). Using the same tech-nique the apparent kinetic constants of additional brush border enzymes (alkaline phosphatase; aP; Gutschmidt et al., 1980) and dipeptidylpeptidase IV (DPP IV; Gut-schmidt and Gossrau, 1981) have been determined previ-ously at representative sites along the villus (Gut-schmidt et al., 1980).

In the present study, the results of the kinetic characterization of brush border-linked aminopeptidase M (APM) are described. Furthermore, the activity patterns of APM and DPP IV are reported and compared with the ac-cording patterns of disaccharidases and aP under three aspects: The distribution of activity along the proxi-mal to distal intestinal axis, the pattern of Vmax along the vertical villus axis, and the correlation between the Vmax in tissue sections and the villus surface.

MATERIALS AND METHODS

    Samples from proximal, medial and distal rat small
intestine were prepared for the kinetic characterization
of the enzymes by section biochemistry as well as for
three dimensional morphometrical studies (Clarke, 1970;
Gutschmidt et al., 1979).

    In principle, the procedure applied for the detec-
tion of apparent Km- and Vmax-values of APM in the tissue
sections was the same as reported in detail for the characterization of disaccharidases (Gutschmidt et al.,
1979), aP (Gutschmidt et al., 1980) and DPP IV (Gutschmidt and Gossrau, 1981). In brief, the following
steps were carried out: Determination of the absorbance
spectra of the azo-dye in the brush border region, check
of Lambert Beer's law in the tissue section (linear increase of the absorbance as a function of section thickness linear increase of the absorbance ratio's within
a certain wavelength range) and of the absorbance as a
linear function of the incubation time, registration of
the dye-accumumulation in the tissue section as a function of the substrate concentration in the incubation
medium under stopped assay conditions, analysis of this
saturation curve by weighted regression analysis or/and
direct linear plot in order to calculate apparent Km-
and Vmax-values. The distribution pattern of each enzyme activity along the villi was registered by two-
dimensional scanning (Gutschmidt et al., 1980); representative villus sites, which normally show the minimal
(villus base) and the maximal (transition zone between
middle and upper third) section activity, were defined
(Gutschmidt et al., 1980). At these sites the activity
pattern along the proximal to distal intestinal axis was
determined.

    Morphometric measurements were performed in each
intestinal segment and the mean villus surface was calculated (Clarke, 1970; Lorenz-Meyer et al., 1976) after
determining the villus height, width and breadth. The
"apical" section Vmax was thereafter correlated with
the villus surface.

RESULTS

    Using Ala-4-methoxy-2-naphthylamine (MNA; 1 mg dis-
solved in 50 µl dimethylformamide) in 0,1 M phosphate
buffer, pH 7.5 as substrate and 1 mg Fast Blue B (FBB)/
1 ml as coupling agent an azo-dye developed in the brush
border region, which revealed a single-peaked absorbance

spectrum, as shown in Fig. 1. Linear regression lines

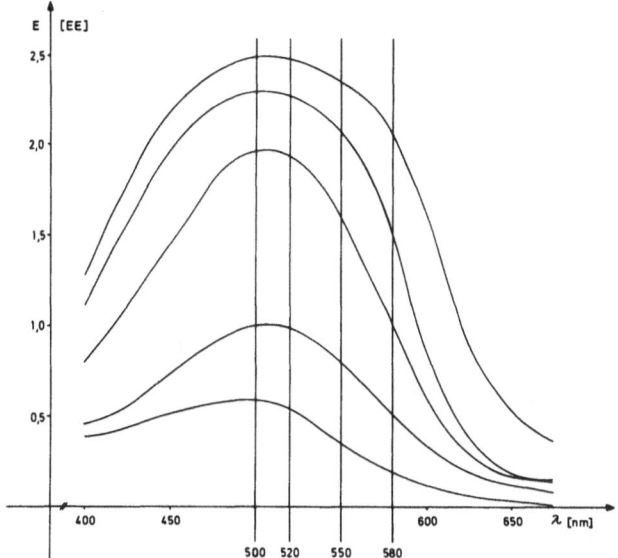

Fig. 1.   Spectrum of the azo-dye developed by APM in
           the brush border region of rat small intestine.
           (EE) = absorbance units, (nm) = wavelength,
           500, 520, 550, 580 = wavelengths used for the
           "wavelengths plot" (Gutschmidt et al., 1981)

were obtained when plotting the absorbance values in the
wavelength range between 500 nm and 580 nm, taken from
spectra at regions of different dye accumulation against
the respective absorbances in the maximum (Gutschmidt
et al., 1981). Thereby a linear increase of the absor-
bance-ratios within this wavelength range was obtained,
thus indicating, that the absorbance is a linear function of the dye concentration. The second stipulation
of the Lambert Beer's law was also fulfilled, since a
linear increase in absorbance could be demonstrated until 10 μm section thickness for both small concentrations of Ala-MNA (0.27 mM) and concentrations in the
range of substrate saturation (0.89-3.76 mM; Fig. 2,
lower panel). In addition, the amount of dye measured
in the sections by its absorbance was a linear function
of the incubation time, at least until 30 sec (Fig. 2,
upper panel). For these reasons steady state conditions
of substrate hydrolysis in sections were assumed when
using a section thickness up to 8 μm and an incubation
time up to 15 sec. The absorbance values were then ta-

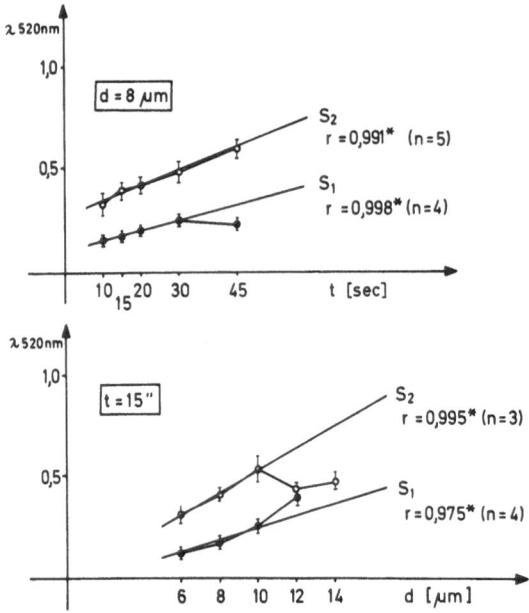

Fig. 2.   Linear increase of dye-absorbance as a func-
           tion of incubation time and section thickness,
           abszissa: t = incubation time, d = section
           thickness, S_1 = 0.27 mM Ala-MNA, S_2 = 0.89 mM
           Ala-MNA, r = regression coefficients ($*$ p$\leqslant$5%),
           ordinate: absorbance units measured at λ520 nm

ken as a measure of the velocity of substrate hydrolysis.
Thus, substrate hydrolysis in sections can be described
as a function of the substrate concentration in the incubation medium as shown in Fig. 3. Thereafter, apparent
Km- (mM) and Vmax- (absorbance units) values were calculated either by weighted regression analysis (Wilkinson,
1961) or by direct linear plot (Eisenthal and Cornish-
Bowden, 1974). The apparent Km-values determined biochemically in tissue sections lay in a similar range as
those calculated from test tube experiments, as is shown
in Table 1 for APM and DPP IV.

       Using two dimensional scanning of the whole dye
deposits along the crypt-villus axis of neutral α-D-
glucosidase, aP and DPP IV peaks of activity were detected between the middle and upper villus third. This became evident when taking the product of the mean absorbance per villus sixth and the respective area covered

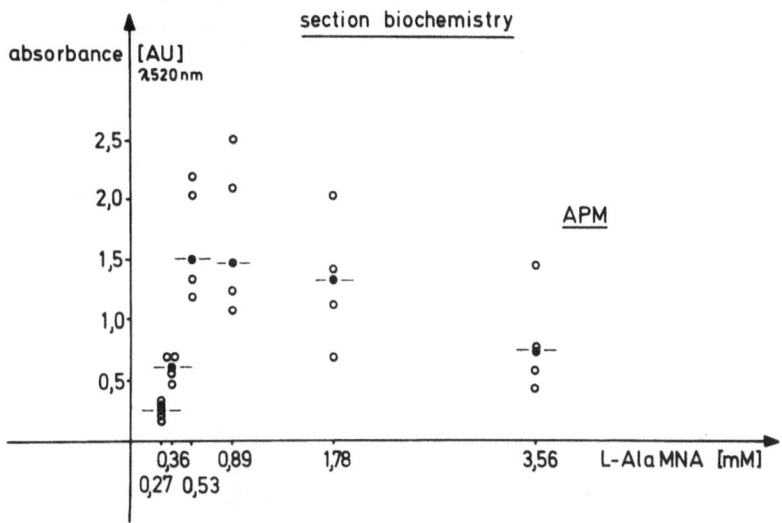

Fig. 3 Ala-MNA hydrolysis in tissue section (measured in absorbance units, AU) as a function of substrate concentration in the incubation medium. Results of five different kinetic assays in the tissue section, "apical" villus site.

Table 1. Comparison of apparent K_m-values of APM and DPP IV determined by section biochemistry and test tube biochemistry from rat small intestine

Enzyme	Method	Substrate	app Km [mM]	References
APM	section biochemistry	L-Ala-MNA	0,67	Hopert and Gutschmidt (1982)
	test tube biochemistry	L-Leu-MNA L-Ala-MNA	0,35 1,32	Hopert and Gutschmidt (1982)
		L-Leu-2NA	0,09	Maze and Gray (1980)
DPPIV	section biochemistry	Gly-Pro-MNA	0,25	Gutschmidt and Gossrau (1981)
	test tube biochemistry	Gly-Pro-4Nitro-anilide	0,24	Svensson et al (1978)
		Gly-Pro-MNA	0,25	Lojda (1979)
		Gly-Pro-2NA	0,25	Gutschmidt and Gossrau (1981)

with dye as a measure for the entire enzyme activity in
that villus segment (Fig. 4). In each case minimal enzyme activity was obtained at the villus base (Fig. 4).
These two regions are called "apical" and "basal", being

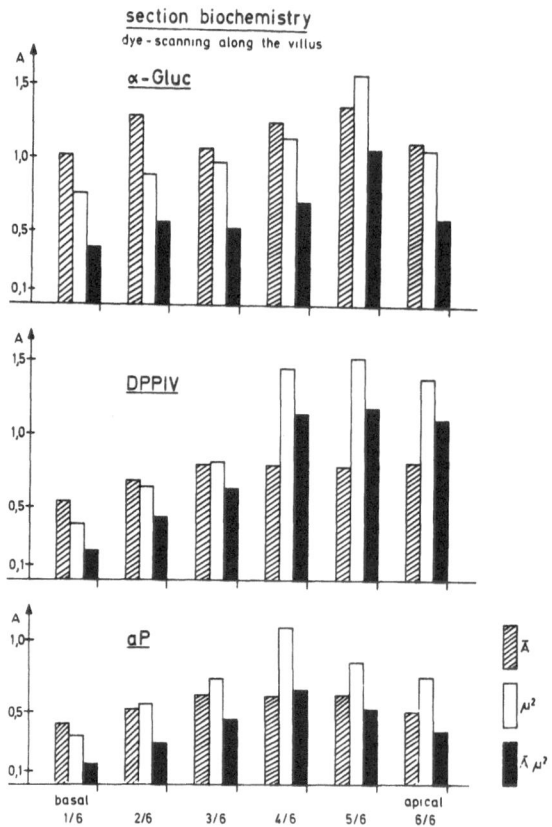

Fig. 4.   Results of two dimensional scanning of the en-
               tire dye-stuff developed under Vmax-conditions
               along the vertical villus axis expressed per
               villus sixth. A = absorbance units registered
               in the respective absorbance maximum, $\bar{A}$ = mean
               absorbance per villus sixth, μ^2 = respective
               area covered with dye, $\bar{A} \cdot \mu^2$ = expression for
               the total amount of dye produced within a vil-
               lus sixth of the brush border region (Gut-
               schmidt et al., 1980), α-Gluc =α-D-glucosida-
               ses

the representative measuring sites for maximum and minimum enzyme activity, respectively. In case of APM, no
gradient could be detected along the intestinal villi,
which means, that this enzyme has already its final level of activity at the villus base.

When measuring enzyme activities at the apical villus site from proximal to distal along the intestinal
axis, a decrease of aP was determined (p ≤5%, KW-test),
whereas both proteases increased towards the terminal
ileum (trend, p ≤10%, Fig. 5). Concerning the correlation with the villus surface, which decreases from pro-

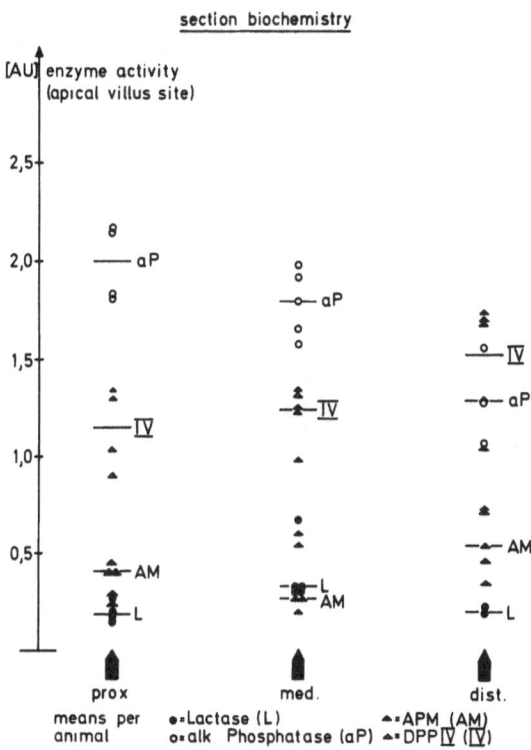

Fig. 5.   Different proximal to distal patterns of the
           section activity at the apical villus site of
           APM and DPP IV, alkaline phosphatase and lac-
           tase along the proximal, medial and distal
           small intestine of the rat, [AU] = absorbance
           units.

ximal to distal, again a striking difference between the
membrane proteases and the other brush border enzymes
is seen: A positive correlation of the Vmax and the villus surface is demonstrated in sections for aP (Fig. 6),
whereas a negative correlation is obtained for the proteases as shown in Fig. 7 for the DPP IV.

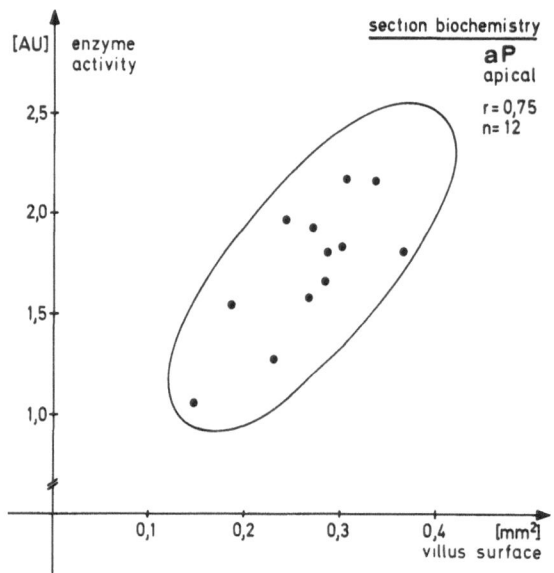

Fig. 6.   Positive correlation between the villus sur-
           face and the activity of alkaline phosphatase
           (aP) at the apical villus region of the proxi-
           mal, medial and distal small intestinal seg-
           ment of tissue sections. r = correlation co-
           efficient. [AU] = absorbance units

CONCLUSIONS

     When applying section biochemistry to the intesti-
nal mucosa of normal rats principle differences were
revealed between membrane proteases on the one hand and
disaccharidases and alkaline phosphatase on the other
concerning their activity pattern. Correlating these
findings with the results of three-dimensional morphometric analysis different regulatory mechanisms are suggested for both groups of enzymes at the level of the
villus as a functional unit. Both aspects of structure
to function interrelationship are of importance for the
interpretation of either physiological or pathological

alterations in the brush border enzyme function.

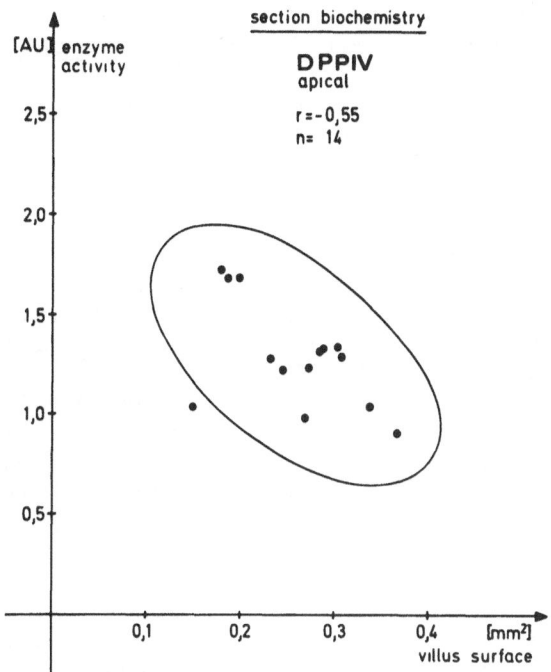

Fig. 7.   Negative correlation between the villus sur-
           face and the activity of DPP IV in tissue sec-
           tions at the apical villus region of the pro-
           ximal, medial and distal small intestinal seg-
           ment. r = correlation coefficient, [AU] =
           absorbance units

REFERENCES

Clarke, R.M., 1970, Mucosal architecture and epithelial
        cell production rate in the small intestine of
        the albino rat, J. Anat. 107:519.
Eisenthal, R., and Cornish-Bowden, A., 1974, The direct
        linear plot, a new graphical procedure for esti-
        mating enzyme kinetic parameters, Biochem, J.
        139:715.
Gutschmidt, S., Emde, C., and Riecken, E.-O., 1980,
        Quantification of α-glucosidases along the villus
        of the small intestine in man, Histochemistry
        67:85.

Gutschmidt, S., and Gossrau, R., 1981, A quantitative
       histochemical study of dipeptidylpeptidase IV
       (DPP IV), Histochemistry 73:285.
Gutschmidt, S., Kaul, W., and Riecken, E.-O., 1979, A
       quantitative histochemical technique for the cha-
       racterisation of α-glucosidase in the brush bor-
       der membrane of rat jejunum, Histochemistry
       63:81.
Gutschmidt, S., Lange, U., and Riecken, E.-O., 1980,
       Kinetic characterization of unspecific alkaline
       phosphatase at different villus sites of rat je-
       junum. A quantitative histochemical study, Histo-
       chemistry, 69:189.
Gutschmidt, S., Lange, U., and Riecken, E.-O., 1981,
       "In situ" measurement of protein contents in the
       brush border region along rat jejunal villi and
       their correlation with four enzyme activities,
       Histochemistry 72:467.
Gutschmidt, S., Menge, H., and Riecken, E.-O., 1982,
       In situ kinetic data of brush border disaccha-
       ridases at different villus sites during intesti-
       nal adaptation, in: Mechanisms of intestinal
       adaptation (R.M. Dowling, J.W.L. Robinson, and
       E.O. Riecken, eds), MTP Press, Lancaster, Boston,
       De Hague.
Lorenz-Meyer, H., Köhn, R,, and Riecken, E.-O., 1976,
       Vergleich verschiedener morphometrischer Metho-
       den zur Erfassung der Schleimhautoberfläche des
       Rattendünndarmes und deren Beziehung zur Funk-
       tion, Histochemistry 49:123.
Wilkinson, G.N., 1961, Statistical estimations in en-
       zyme kinetics. Biochem. J. 80:324.

FLUORESCENCE DETECTION OF PROTEASES WITH AFC, AMC AND MNA PEPTIDES USING ISOELECTRIC FOCUSING

[1]Pranav Sinha, [1]Reinhart Gossrau, [2]Robert E. Smith, and [3]Zdenek Lojda
[1]Department of Anatomy, Free University, [2]Lawrence Livermore Laboratories and [3]Laboratory of Histochemistry, Charles University, [1]Berlin 33, FRG, [2]Livermore, USA and [3]Prague, ČSSR

INTRODUCTION

The use of synthetic peptides in biochemical and histochemical assays has considerably contributed to the progress in protease research and therefore to our knowledge about proteases in healthy and diseased cells and tissues (cf. Huseby and Smith, 1980, Gossrau, 1981; Lojda, 1981; Gossrau et al., this volume).

AZO-DYE METHODS FOR BAND STAINING

Further progress in protease research has been achieved by the combination of isoelectric focusing (IEF) and protease histochemistry with 4-methoxy-2-naphthylamine (MNA) peptides as substrates. In this procedure the bands are stained with simultaneous azo-dye methods (Lojda and Kuhlich, 1981; Sinha and Gossrau, 1982; Smith and Grabske, 1982; Gossrau et al., this volume). Alternatively post-coupling may be used to prevent the inhibitory effect of the diazonium salts. In simultaneous coupling MNA is liberated in a first reaction step from a MNA amino acid or peptide by a respective protease, and in a second step after simultaneous coupling to either Fast Blue B, BB or Garnet GBC an azo-dye occurs. With this technique highly complex band patterns can be revealed, e.g. after focusing of kidney homogenate. When different concentrations of renal homogenates are focused followed by both dipeptidylpeptidase (DPP) staining with the substrate Gly-Pro-MNA and simultaneous

Fast Blue B coupling of MNA and the general protein
staining with Coomassie Blue it becomes evident that
the Gly-Pro-MNA concentration reflects well different
concentrations of homogenate or DPP respectively; furthermore, clear enzyme activity can be observed at sites
where the amount of protein is only low or not detectable by the Coomassie Blue dye stain (Fig.1). Another
example for the potentialities of combined IEF and proteinase histochemistry with azo-dye procedures is the

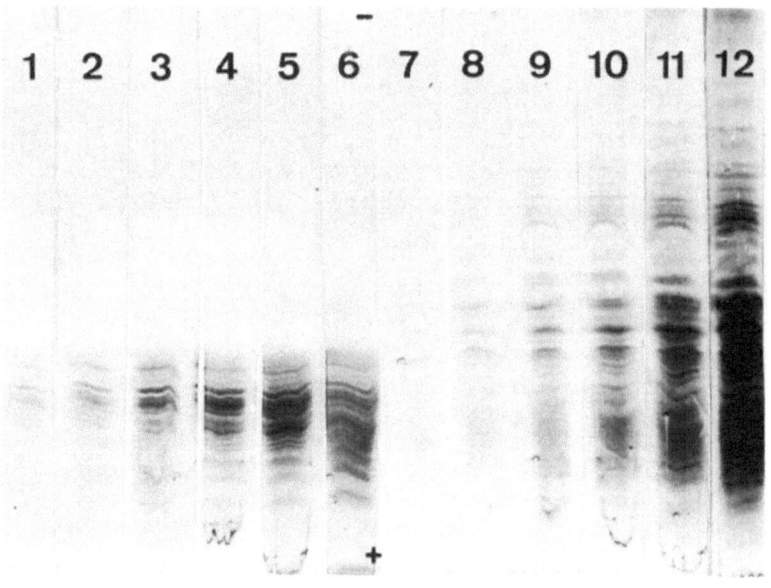

Fig. 1.   Column 1-6 enzyme staining, column 7-12
           Coomassie Blue staining after IEF

band pattern obtained by staining for aminopeptidase M
or APM and APA in renal homogenates. 3 bands exist for
APM and more than 5 for the metalloprotease APA. The
APA inhibitor EDTA and APA activator $CaCl_2$ affect only
the APA bands; this indicates that all these bands are
due to APA and not to APM although at least 3 of them
focus in the same region (Fig.2).

FLUORESCENCE METHODS FOR BAND STAINING

     However, the azo-dye techniques are limited to MNA
peptides, and in simultaneous coupling the diazonium
salts inhibit proteases among which the thiolproteases,
e.g. cathepsin B or DPP I (cathepsin C) are especially
sensitive (cf. Gossrau, 1981). Therefore, methods were
developed which utilize the fluorescence of MNA, 7-amino-

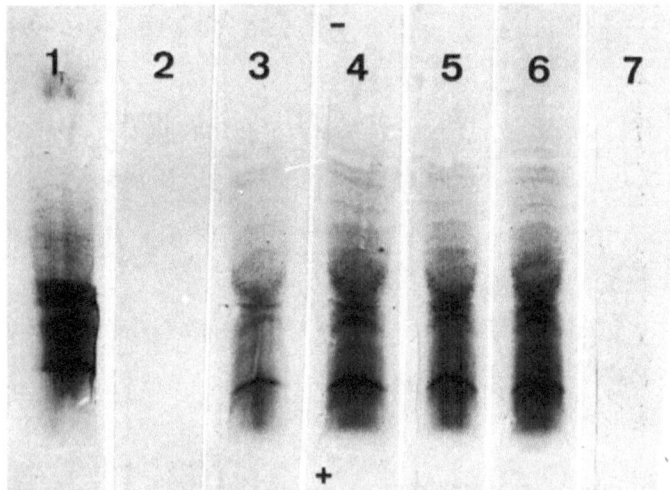

Fig. 2.   Column 1 APM reaction with Ala-MNA, Column 2-7
           APA reaction with α-Glu-MNA, column 2 control
           without substrate, column 3 APA reaction with-
           out $CaCl_2$, column 4-6 APA reaction with in-
           creasing $CaCl_2$ concentrations, column 7 APA
           reaction in the presence of EDTA

4-methylcoumarine (AMC) and 7-amino-4-trifluoromethyl-
coumarine (AFC) without diazonium salts to visualize
different proteases, protease isoenzymes or protease
subunits after IEF; after hydrolysis by a protease the
liberated MNA, AFC or AMC are detected by their fluorescence (Fig.3). In addition, MNA can be simultaneous-

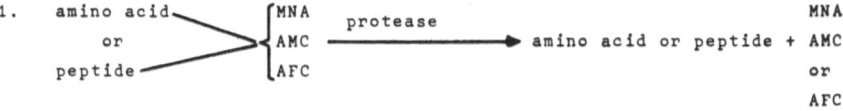

Fig. 3.   Reaction scheme for the fluorescence detection
           of bands after IEF with MNA, AMC, and AFC pep-
           tides

ly coupled to nitrosalicylaldehyde (NSA). The use of AFC
and AMC substrates was in general limited to biochemical
measurements of proteases.

   The proteases and some of their MNA, AFC and AMC
peptides are summarized in Table 1. IEF was carried out
with 10.000 g supernatants from 20% or 5% rat kidney ho-

   Table 1. Investigated proteases and some of their
            substrates

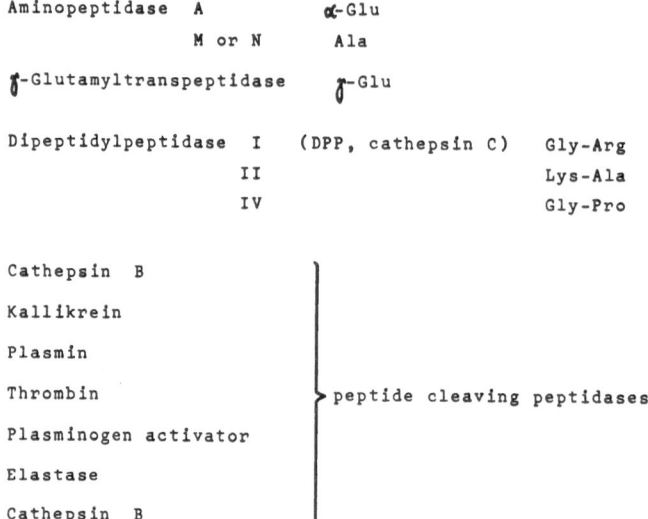

Aminopeptidase A   α-Glu
                M or N   Ala

γ-Glutamyltranspeptidase   γ-Glu

Dipeptidylpeptidase I   (DPP, cathepsin C)   Gly-Arg
                    II   Lys-Ala
                    IV   Gly-Pro

Cathepsin B
Kallikrein
Plasmin
Thrombin   peptide cleaving peptidases
Plasminogen activator
Elastase
Cathepsin B

mogenate. The homogenates were prepared with 5% Triton
X-100 from wet material or freeze-dried cryostat sections (Lojda et al., 1979). After IEF the gels where either incubated directly in the substrate solution or the
overlay technique with agarose gels or cellulose acetate foils impregnated with the substrate was applied.
After incubation the gels and foils were viewed with an
UV lamp and photographed. The best results are obtained
with cellulose acetate foil.

   The fluorescence detection of proteases has the advantage that the enzymes are active in the absence of

the inhibitory diazonium salts which, in addition, may
change their kinetic properties (cf. Gutschmidt and
Gossrau, 1981). Furthermore, many proteinases need activators to develop their maximal activity which may or
actually interfere with the diazonium salts, e.g. cysteine or mercaptoethanol for the activation of thiol
proteinases like cathepsin B. In addition, the pH optimum of the simultaneous azo-coupling reaction is sometimes different of that of the investigated protease,
e.g. of γ-glutamyltranspeptidase. In these cases the
fluorescence staining of the bands is superior and allows informations that are not obtained by simultaneous
azo-coupling. 2 strong fluorescent bands occur e.g. after
incubation with the cathepsin B substrate CBZ-Ala-Arg-
Arg-MNA at pH 7.5. The fluorescence intensity can be enhanced and the number of bands is increased by thiol
groups (Fig.4). Parallel histochemical investigations
already revealed a brush border localization of the reaction product (Gossrau and Gutschmidt, 1982). Using the
elastase substrate Glu-Ala-Ala-Ala-MNA IEF yields negative results but histochemistry a positive brush border
staining (Lojda, 1979). From this it follows that beside
the already known brush border-linked elastase peptides
hydrolysing endopeptidase a second and different thiol-
dependent endopeptidase exists in the renal proximal
microvillous zone. In contrast to the elastase substrate
cleaving endopeptidase which frees MNA indirectly and
needs, therefore, the presence of aminopeptidase M
(Lojda, 1979) the endopeptidase that hydrolyses the cathepsin B substrates liberates MNA directly. - The fluorescence staining with α-Glu-MNA for aminopeptidase A
reveals an additional band (Fig.5) not present with
azo-dye procedures.

The relative disadvantage of the fluorescence staining is that the bands are occasionally less sharp than
with the azo-dye technique (Fig.6). However, if the liberated MNA is coupled to NSA a two-fold effect is observed. Firstly, the fluorescence is more intense and
secondly the single bands can be nearly as sharp as with
the azo-dye staining (Fig.7). - The fluorescence and
staining intensity of the AMC and AFC bands cannot be
improved by NSA due to the uneffective coupling with
this aldehyde. Compared with non-coupled MNA the fluorescence detection with AMC and AFC peptides delivers
identical data. However, ACF fluoresces in the yellow-
green region instead of the blueish AMC fluorescence
and is sometimes more intense than that of AMC.

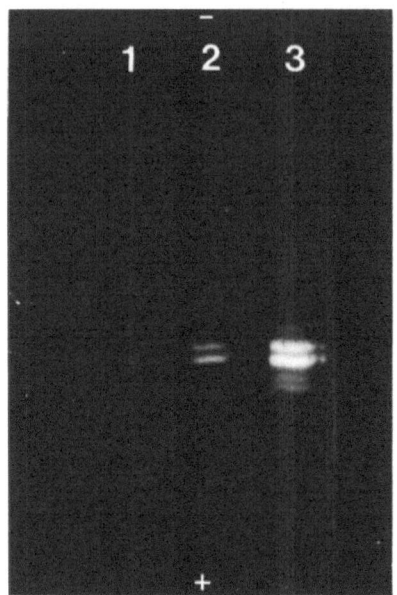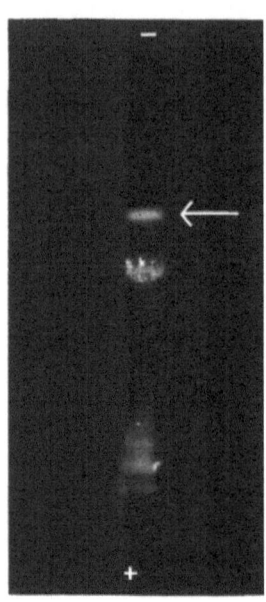

Fig. 4.   Column 1 hydrolysis of CBZ-Ala-Arg-Arg-MNA,
           column 2 as column 1 but with ß-mercaptoetha-
           nol, column 3 as column 1 but with L-Cys. Fig.5.
           APA reaction, arrow = additional band

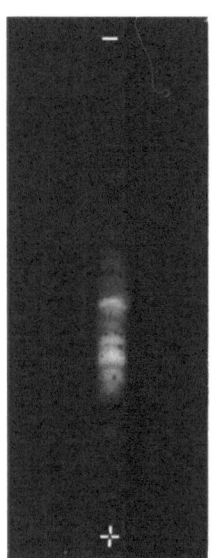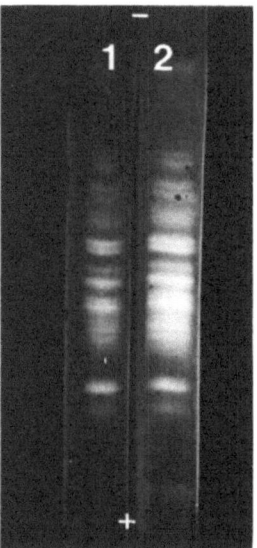

Fig. 6.   Hydrolysis of Gly-Pro-MNA. Fig.7. Hydrolysis of
           Gly-Pro-MNA and simultaneous coupling of MNA
           with NSA; column 1 reaction at pH 5, column 2
           reaction at pH 7

SUMMARY

    The fluorescence detection of proteases is not on-
ly possible with MNA but also with AFC and AMC peptides
which were used up till now only for biochemical protease investigations. The MNA method is more sensitive due
to the coupling capability of MNA with NSA than the AMC
and AFC procedure. In comparison with the simultaneous
azo-dye method the fluorescence technique may be inferior as far as the sharpness of the bands is concerned.
The fluorescence staining is superior, however, if pro-
tenases have to be investigated that are inhibited by
diazonium salts, or where interference of diazonium salts
and activators or inhibitors occurs; finally, it is a
simple procedure to elucidate the value of MNA, AFC and
AMC peptides for one and the same protease. In the field
of applied isoelectric focusing the azo-dye and fluorescence procedure are now used together by us to detect
possible differences in protease isoenzyme patterns in
normal and diseased human biopsies, e.g. from liver,
kidney and small intestine.

REFERENCES

Gossrau, R., 1981, Investigation of proteases in the di-
        gestive tract using 4-methoxy-2-naphthylamine
        (MNA) substrates, J. Histochem. Cytochem. 29:464.
Gossrau, R., and Gutschmidt, S., 1982, Membrane protea-
        ses in the intestine and kidney. Cytochemical
        pattern, kinetic properties and functional sig-
        nificance, Gastroenterol. Clin. Biol. 6:93.
Gutschmidt, S., and Gossrau, R., 1981, A quantitative
        histochemical study of dipeptidylpeptidase IV
        (DPP IV), Histochemistry 73:285.
Huseby, R.M., and Smith, R.E., 1980, Synthetic oligopep-
        tide substrates - their diagnostic application in
        blood coagulation, fibrinolysis, and other patho-
        logical states, Seminars in Thrombosis and Hemo-
        stasis 6:173.
Lojda, Z., 1979, The histochemical demonstration of
        brush border endopeptidase, Histochemistry 64:205.
Lojda, Z., 1981, Proteases in pathology. Usefulness of
        histochemical methods, J. Histochem., Cytochem.
        29:481.
Lojda, Z., Gossrau, R., Schiebler, T.H., 1979, Enzyme
        histochemistry. A laboratory manual. Springer,
        Berlin, Heidelberg, New York.
Lojda, Z., and Kuhlich, J., 1981, The usefulness of
        analytical electrofocusing in a thin layer poly- _

acrylamide gel (PAG) in the histochemistry of enzymes cleaving peptide bonds, <u>Histochemistry</u> 73:311.

Sinha, P., and Gossrau, R., 1982, Histochemical studies with MNA substrates using isoelectric focusing, <u>J. Histochem. Cytochem</u>. 30:575.

Smith, R.E., and Grabske, R.J., 1982, Isoenzymes of glandular kallikrein revealed by isoelectric focusing and fluorescence staining with an oli-gopeptide substrate inpregnated into cellulose, <u>J. Histochem. Cytochem</u>. 30:575.

PATHOPHYSIOLOGY OF THE INTERACTION BETWEEN COMPLEMENT

AND NON-COMPLEMENT PROTEASES

U.E. Nydegger* and S. Suter+

*Blood Transfusion Service of the Swiss Red Cross
 Central Laboratory, Berne
+Department of Medicine, University of Geneva, Switzerland

INTRODUCTION

The complement system is a group of plasma proteins that can
interact with antigen antibody complexes and the surface of microbes
and certain cells for the purpose of protecting the host against
such noxious antigens (1-3). The complement proteins account for
about 5% of the total plasma proteins; activation of the complement
proteins occurs sequentially and is characterized by a conversion
of proteolytic zymogens to active proteinases that catalyze conversion of other zymogens to active enzymes further down the sequence
(4). There are two pathways of activation for complement, the classical pathway that is generally dependent on the formation of specific antibody, and the alternative pathway, that represents a system
for resistance to infection in a non-immune host. One should note
that of the twenty complement components, only five are so far identified as proteinases as shown in the activation scheme on Fig. 1:
The proteinases are $C\bar{1}r$, $C\bar{1}s$, $C\bar{2}a$ in the classical pathway and $\bar{D}$
and $\bar{Bb}$ in the alternative pathway; there is no zymogen form for $\bar{D}$.
All are serine proteases whereby $C\bar{1}r$, $C\bar{1}s$ and $\bar{D}$ possess catalytic
peptide chains of about 25'000 m wt. Trypsin, plasmin, pronase and
chymotrypsin are known to activate $C\bar{1}$ (2); accordingly tryptic and
chymotriptic inhibitors, as well as liquoid and antrypol can be used
to inhibit $C\bar{1}s$ (8), whereby a naturally occuring plasma protein,
the C1 esterase inhibitor inhibits $C\bar{1}r$ and $C\bar{1}s$ but also plasma kallikrein, plasmin, trypsin, chymotrypsin, Hagemann factor (factor
XIIa) and activated thromboplastin antecedent (XIa).

227

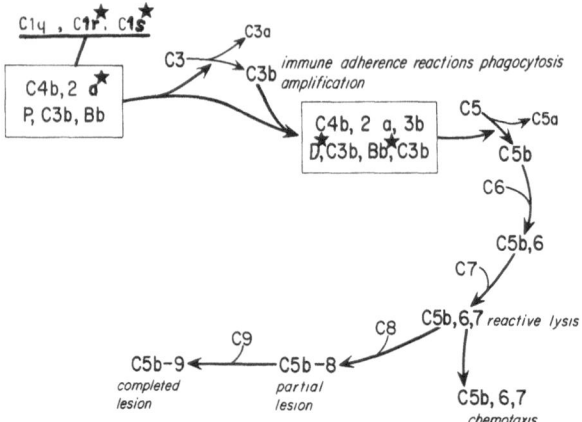

Fig. 1.   The proteins participating in the complement activation
           sequence. The precise mechanisms of classical and alter-
           native pathway activation have been reviewed in detail
           (2, 3). The serine proteases contained in the cascade are
           specially designated with an asterisk. The control pro-
           teins C1 inhibitor, C4 binding protein, H and I are not
           depicted.

THE C3 MOLECULE, TARGET FOR COMPLEMENT AND NON-COMPLEMENT PROTEASES

C3 and its fragments express diverse functions: they serve to
form with Bb the alternative pathway C3b,Bb C3 convertase, and
participate in activation of C5. The C3b fragment is endowed with
capacities to opsonize particles, by serving as a ligand that mediates the adherence of C3b bearing substances to immunological effector cells (immune adherence). C3 is the most abundant complement
component in serum with a concentration of about 1,2 mg/ml. It is
a 9,5 S β-globulin with a m wt of 190'000 and consists of two non-
identical chains α and β linked by disulfide bonds and non-covalent
forces. The m wt of the isolated α- and β-chains are approximately
110'000 and 80'000.

The native C3 molecule is the specific substrate for the clas-

sical, C42, and the alternative C3b,Bb,C3 convertases, with the
proteolytic cleavage of C3 altogether being the most important step
in the activation of the complement system. Cleavage occurs at an
arg-ser peptide bond in position 77 in the α-chain generating a small
fragment with anaphylatoxic activity, C3a, and a major cleavage
fragment C3b (Fig. 2). It should be noted that moderate concentration of non-complement enzymes, such as bovine pancreatic trypsin
used in the original studies (5) or human neutrophil elastase (6)
can also specifically split C3 into C3a and C3b with, as far as
trypsin is concerned, one site of cleavage at the arg-ala bond in
position 69-70 and at the arg-ser bond tentatively positioned at
77-78. When the C3 split products are prepared with elastase (7),
the reaction can either be allowed to proceed for relatively short
time periods, and will then split C3 into C3a and C3b; if the reaction is allowed to proceed for 4 hours or longer, C3b will be converted into C3c and C3d.

Inactivation of C3 occurs through proteolytic fragmentation of
the molecule. The α-chain of C3b is first cleaved by I into two fragments of approximately 60'000 and 40'000 m wt, generating inactive
C3b called iC3b: since the interchain disulfide bonds still hold the
α-chain cleavage fragments together, iC3b has a similar m wt to that
of C3b. Breakdown of iC3b to C3c and C3d is achieved by a variety
of proteases such as trypsin, plasmin and elastase from polymorphonuclear leukocytes (PMNLs). A further cleavage fragment of C3, C3e,
may be cleaved from C3c and exhibits leukocytosis-inducing properties.

Studies describing the detailed chemical structures of human
C5a and C3a and of porcine C3a have delineated the active site of
the C3a molecule. The COOH-terminal arginyl residue is essential to
anaphylatoxin activity and a series of synthetic peptides, based on
the COOH-terminal sequence of human C3a, induce the same variety of
biological responses as does the intact anaphylatoxin. The essential
nature of the COOH-terminal arginyl residue in the human anaphylatoxins provides a simple means of controlling function, since a serum carboxypeptidase of the B type rapidly cleaves this residue rendering the factors inactive as spasmogens. Therefore, in the absence
of an inhibitor to the serum carboxypeptidase, the human anaphylatoxins are rapidly degraded to inactive derivatives referred to as
C3a-des arg77 and C5a-des arg74. Porcine anaphylatoxins are similarly under exopeptidase control except that a small fraction of the
C5a-associated spasmogen activity is carboxypeptidase-resistant, a
behavior of animal C5a anaphylatoxins yet to be explained. A synthetic inhibitor of carboxypeptidase has recently been identified
and characterized to be 2-mercaptomethyl-3-guanidinoethylthiopro-
panoic acid.

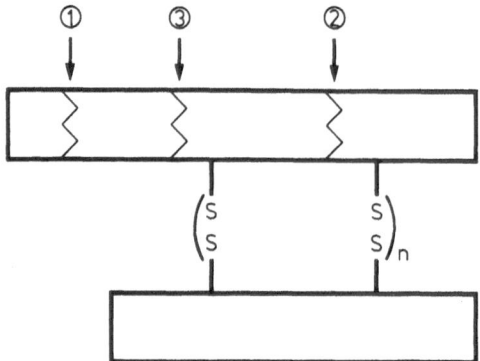

Fig. 2.   Schematic representation of the α- (upper bar) and β-
           (lower bar) chains of the C3 molecule held together by
           disulfide bonds. All major cleavage sites for C3 are lo-
           cated on the α-chain with the following sites of reaction:
           1) for the complement-derived enzymes C42 and P,C3b,Bb,
           to generate C3a and C3b; 2) for I to generate iC3b from
           C3b and 3) for non-complement proteases to generate C3c
           and C3d.

The pathophysiological significance of C3 cleavage by non-complement
proteases

        The deposition of C3b on the surface of microbial organisms is
an efficient and sometimes decisive step in host-defense against
microbes: deposited C3b serves as a ligand between bacteria and C3b
receptor bearing PMNLs and monocytes, in synergy with Fc-receptors
for IgG, thus representing opsonic activity. The deposition of sufficient amounts of C3b on individual microbial strains have been
shown to either depend on prior deposition of antibody on the bacterial surface or to occur directly after activation of the alternative pathway in the case of sialic acid poor strains; very often,
IgG participates in the alternative pathway-mediated opsonization

of pneumococci (9) and a series of other microbes (reviewed in: 10).
In addition encapsulated *S. aureus* bind more C3 when immune antibody is allowed to preopsonize the acidic polysaccharide capsule (11).

While numerous studies have analyzed the biochemical requirements for the microbial surface to allow for efficient deposition
of C3b, we present evidence, that maintenance of effective C3b opsonization before it facilitates phagocytosis through linkeage to
C3bR could be called in question through C3 cleaving activity by
human PMNL-elastase. These data have been published extensively elsewhere (12). The model used was that of PMNL-granular enzyme activity
towards sepharose-bound C3 that was radiolabeled with 125-I. The
finding of a significant 125-I-C3 cleavage was found to be relevant
in a clinical setting through demonstration of similar 125-I-C3
breakdown by human pleural empyema fluids.

MATERIAL AND METHODS

Pleural fluid was drawn by thoracocentesis centrifuged and
stored at -70°C. Granular enzymes from PMNLs were obtained through
homogeneization of the cells and subsequent purification of the
granules by ultracentrifugation and extraction of enzymes at low pH.
C3 was labeled with 125-I using chloramin T as an oxidant. The C3
cleaving activity was tested on 125-I-C3, that was coupled to cya-
nogen-bromide activated sepharose and incubated with the experimental or clinical samples during 9 hours at 37°C. The extent of the
released radioactivity was taken to reflect C3 cleaving activity.
The remainder of techniques used have been published extensively
elsewhere (12).

RESULTS AND DISCUSSION

Pleural empyema supernatants from 10 patients with purulent
effusions and a positive bacteriologic culture for such microbes as
peptococcus, bacteroides, streptococci, *K. pneumoniae* and *S. aureus*
cleaved significantly more 125-I-C3 (18,4%±7,3%, $\bar{X}$ ± 1 SD) than did
19 sterile pleural effusions (2,4%±0,9%, p < 0,001) and sonicates
from bacterial strains commonly occuring in empyema (1,4±0,2%).
Granular enzymes from PMNLs cleaved 78,5% of 125-I-C3. When proteolysis of 125-I-C3 was compared on electrophoretic tracks of sodium
dodeylsulfate polyacrylamide gels that were developed by autoradio-
graphy, the pattern of various sized breakdown products was similar
using either supernatants from infected pleural empyema or PMNL
granular enzymes. The identity of 125-I-C3 breakdown with elastase-
like activity was ascertained using the synthetic elastase sensi-

tive substrate succinyl-Ala-Ala-Pro-7-amino-4-methyl-coumarin
(TGP 928), in which granulocyte elastase selectively cleaves the
7-amino-4-methyl-coumarin part. Three inhibitors of serine proteases,
i.e. phenyl methyl sulfonylfluoride, peptide carbazate inhibitor and
N-acetyl-(Ala)$_4$-chloromethylketone, whereby the latter two are considered to be specific inhibitors of granulocyte elastase (13,14),
efficiently inhibited both, 125-I-C3 cleaving and elastase-like
activities. Human EDTA-serum as source for such inhibitors as α_1 antitrypsin and α_2 macroglobulin almost totally inhibited C3 cleaving
activity by PMNL granular enzymes. Bacterial sonicates had no 125-I-
C3 cleaving activity.

The findings reported here clearly indicate, that products from
inflammatory cells are capable of attacking opsonizing C3,·and evidence was provided recently, that IgG may also become degraded upon
exposure to infected pleural effusions (15). Obviously the C3 used
in this study was deposited on an artificial surface, but Carlo
et al. have recently reported that the cellular intermediate sheep
erythrocyte (E^s)-anti E^s antibody (A), decorated with C1423b, made
using 125-I-C3 also released radiolabel upon incubation with lysosomal lysates from PMNLs (16). In extension to our findings, release
of 125-I-C3 from EAC142-C3b was much better when crude leukocyte
lysosomal lysates or purified leukocyte elastase were allowed to
act on C3 in presence of the alternative pathway control proteins
H and I. By itself the lysates and elastase removed negligible
amounts of radiolabel from the cells. Thus two additional variables
ought to be considered when studying non-complement mediated breakdown of surface deposited C3: the nature of the surface the C3 is
deposited on and the possible helper function of natural complement
inhibitors such as I. Nevertheless ours as the findings of others
strongly suggest that the biological paradox of persistent bacterial
presence and growth in presence of a large number of phagocytic
PMNLs could, at least in part, be due to a C3 cleaving activity,
provided by the phagocytes themselves.

REFERENCES

1. R. R. Porter and K. B. M. Reid, The biochemistry of complement,
    Nature 275:699 (1978).
2. M. Loos, The classical complement pathway: mechanism of activa-
    tion of the first component by antigen-antibody complexes,
    Progr. Allergy 30:135 (1982).
3. M. D. Kazatchkine and U. E. Nydegger, The human alternative com-
    plement pathway: biology and immunopathology of activation
    and regulation, Progr. Allergy 30:193 (1982).

4.   R. R. Porter, The proteolytic enzymes of the complement system,
     in: "Methods in Enzymology, Proteolytic Enzymes," L. Lorand,
     ed., Academic Press, New York (1981).
5.   V. A. Bokisch, H. J. Mueller-Eberhard, and C. C. Cochrane, Iso-
     lation of a fragment (C3a) of the third component of human
     complement containing anaphylatoxin and chemotactic activity
     and description of an anaphylatoxin inactivator of human
     serum, J. exp. Med. 129:1109 (1969).
6.   P. Venge and I. Olsson, Cationic proteins of human granulocytes.
     IV. Effects of the complement system and mediation of chemo-
     tactic activity, J. Immunol. 115:1505 (1975).
7.   B. F. Tack, J. Janatova, M. L. Thomas, R. A. Harrison, and C. H.
     Hammer, The third, fourth and fifth components of human com-
     plement: isolation and biochemical properties, in:"Methods
     in Enzymology, Proteolytic Enzymes," L. Lorand, ed., Academic
     Press, New York (1981).
8.   D. T. Fearon and K. F. Austen, Inhibition of complement-derived
     enzymes, Ann. New York Acad. Sci. 256:441 (1975).
9.   J. A. Winkelstein, H. S. Shin, and W. B. Wood, Jr., Heat labile
     opsonins to pneumococcus. III. Participation of immunoglobulin
     and alternative pathway of C3 activation, J. Immunol. 108:1681
     (1972).
10. M. S. Edwards, A. Nicholson-Weller, C. J. Baker, and D. L. Kasper,
     The role of specific antibody in alternative complement path-
     way-mediated opsonophagocytosis of type III group B strepto-
     coccus, J. exp. Med. 151:1275 (1980).
11. H. A. Verbrugh, P. K. Peterson, B.-Y. T. Nguyen, S. S. Sisson,
     and Y. Kim, Opsonization of encapsulated staphylococcus
     aureus: the role of specific antibody and complement,
     J. Immunol. 129:1681 (1982).
12. S. Suter, U. E. Nydegger, L. Roux and F. A. Waldvogel, Cleavage
     of C3 by neutral proteases from granulocytes in pleural
     empyema, J. infect. Dis. 144:499 (1981).
13. C. P. Dorn, M. Zimmerman, S. S. Yang, E. C. Yurewicz, B. M. Ashe,
     R. Frankshun, H. Jones, Proteinase inhibitors. I. Inhibitors
     of elastase. J. Med. Chem. 20:1464 (1977).
14. J. C. Powers, B. F. Gupton, A. D. Harley, N. Nishino, R. J.
     Whitley, Specificity of porcine pancreatic elastase, human
     leukocyte elastase and cathepsin G: inhibition with peptide
     chloromethyl ketones. Biochim. Biophys. Acta 485:156 (1977).
15. F. A. Waldvogel, P. Vaudaux, P. D. Lew, A. Zwahlen, and U. E.
     Nydegger, Deficient phagocytosis secondary to breakdown of
     opsonic factors in infected exudates, in:"Biochemistry and
     Function of Phagocytes," F. Rossi, P. Patriarca, eds., Plenum
     Publishing Corp., New York (1982).

16. J. R. Carlo, J. K. Spitznagel, E. J. Studer, D. H. Conrad, and
     S. Ruddy, Cleavage of membrane bound C3bi, an intermediate
     of the third component of complement, to C3c and C3d-like
     fragments by crude leukocyte lysosomal lysates and purified
     leukocyte elastase, Immunology 44:381 (1981).

INTERACTIONS BETWEEN THE ALTERNATIVE COMPLEMENT PATHWAY AND

PROTEASES OF THE COAGULATION SYSTEM

M. D. Kazatchkine and M-H. Jouvin

INSERM U28, Hopital Broussais

75014 Paris, France

The Complement system consists of more than twenty components and regulatory plasma proteins involved in both specific (immune) and natural host defense. Upon activation, the classical and the alternative pathways of Complement, both form specific proteolytic complexes named "C3 convertases" that cleave C3 and generate the major cleavage fragment C3b. The classical pathway is often activated after binding of C1 to an IgG or IgM-antigen complex; in contrast, activation of the alternative pathway by activating cell surfaces is dependent on their chemical composition and may occur in the absence of antibody. C3b binds to activating immune complexes or cell surfaces and may interact with the alternative pathway proteins B, D and P to form the amplification C3 convertase that augments C3 cleavage and generation/deposition of C3b. Accumulation of C3b molecules on the target surface changes C3 convertases to C5 convertases and activates the effector sequence C5-C9 generating leukocyte attracting, vasoactive and cytolytic activities of activated Complement. The molecular mechanisms of activation and regulation of the classical and alternative pathway have recently been reviewed in detail[1,2].

Several interactions take place among the classical Complement and coagulation pathways such as the ability of C1 inhibitor to inhibit activated Hageman Factor and plasma kallikrein[3,4]. This paper, however, will only refer to several more recently described interactions between the coagulation system and the human alternative Complement pathway.

THE SERINE ESTERASE Bb

The first step in formation of the amplification C3 convertase and in alternative pathway activation is the binding reaction of Factor B to cell-bound C3b. Complexed B is then cleaved by D to release a small Ba fragment and to uncover the C3 cleaving site on the Bb fragment in the C3b, Bb complex. Although the C3b, B complex exhibits some C3 convertase activity, the proteolytic site for C3 (and C5) in the amplification convertase is only fully expressed on B after its cleavage by D. Proteolytic cleavage of C3 by Bb occurs at an Arg-Ser peptide bond in the alpha chain of the molecule. Whereas Bb exhibits restricted proteolytic specificity for C3/C5 only when it is complexed with C3b, isolated Bb retains limited esterolytic activity on synthetic substrates after it is released from the C3b, Bb complex[5] and was recently shown to cleave plasminogen[6] and prothrombin[7].

Cleavage of purified plasminogen by Bb, whether isolated or complexed with C3b, generates two disulfide-linked polypeptides similar in size to the cleavage fragments of plasminogen by urokinase; incubation of Bb with plasminogen and ^{125}I-fibrin at 37 $^\circ$C results in significant fibrinolysis. Plasminogen activator-activity of Bb is, however,inhibited by dilutions of human serum[6].

In vitro, isolated and complexed Bb also cleaves prothrombin generating similar cleavage fragments to those obtained through the action of Xa. Both plasminogen and prothrombin-cleaving activities of Bb are inhibited by Trasylol.

Amino acid sequence studies of Factor B[8] have shown a 28,000 daltons COOH terminal CNBr peptide from the molecule to exhibit extensive homology in its primary structure with known serine proteases. Structural similarities were highest with plasminogen. A small peptide that was obtained from the 28,000 fragment contains the active site serine sequence Gly-Asp-Ser-Gly-Gly-Pro. Bb differs,however,from other serine proteases in size (60,000 mol. wt.), in that no homology was found between the NH2 terminus of serine proteases and the analogous part of the CNBr cleavage product of Bb and in the requirement for Bb to bind to a cofactor to express proteolytic activity.

THE SERINE ESTERASE D

Factor D, a beta globulin that consists of a single polypeptide chain of 24,000 mol. wt., differs from other plasma proteases in that it is always present in plasma in trace amounts as an active enzyme without known specific inhibitor and unsensitive to plasma protein inhibitors. D converts the C3b, B bimolecular complex into a C3b, Bb convertase by cleaving an Arg-Lys bond in B at a site containing three basic residues Lys-Arg-Lys[9]. Proteinase, trypsin, human plasma kallikrein, plasmin but not thrombin can substitute to some extent for D function[10,11]. However, D has restricted specificity as compared with other serine proteases as it does not hydrolyze synthetic ester substrates except the Factor Xa substrate N-benzoyl-isoleucyl-glutamyl-glycylarginine-p-Nitroanilide HCl[12]. D is inhibitable but relatively insensitive to inactivation by DFP[10].

Amino acid analysis has shown strong homologies between the sequence of the first fifty NH2-terminal amino acids of D and other serine proteases[13]. A 7,000 mol. wt. CNBr peptide from D was found to contain the active site sequence Gly-Asp-Ser-Gly-Gly-Pro of other serine proteases[14].

Human platelets contain D and secrete D when stimulated by ADP, collagen or thrombin[15]. Furthermore, D and DFP-inactivated D was shown to bind to platelets and competitively inhibit binding of thrombin to platelets although the affinity of D for the human platelet thrombin receptor is much less than that of thrombin; thus, D may function as an inhibitor of thrombin-induced platelet aggregation at an early step of thrombus formation[16].

PLASMA KALLIKREIN

In an in vitro system using purified proteins, affinity-purified plasma kallikrein can replace Factor D for cleavage of B within the C3b, B complex although it is ten times less effective than D on molar basis[11]. In contrast with D, kallikrein can also cleave uncomplexed B provided that Magnesium ions are present.

Another interaction of plasma kallikrein with the alternative pathway of Complement is that rabbit, but not human, plasma kallikrein (1 μg/ml) is capable of generating chemotactic activity (C5a) from purified rabbit C5; the reaction is inhibited by antikallikrein antibody and SBTI to the same extent as kinin generation is suppressed[17].

HEPARIN INHIBITS FORMATION OF THE AMPLIFICATION C3 CONVERTASE

Fluid phase heparin inhibits generation of the amplification C3 convertase at final concentrations similar to those required for its antithrombin cofactor activity[18]. The inhibitory site for convertase formation on the heparin molecule is different from the antithrombin binding site since highly active fractions obtained by affinity cycles on insolubilized antithrombin III have similar anticomplementary activity to that of fractions that do not bind antithrombin III. The inhibitory effect of heparin requires substitution of the amino group of the glucosamine and integrity of O-sulfate groups in the molecule[19]. Since O-sulfated N-desulfated-N-acetylated heparin has no anticoagulant activity but similar inhibitory activity on

C3b, Bb formation to that of O-sulfated N-sulfated heparin, the nature of the N-substitution in the heparin molecule distinguishes between the coagulation and the Complement pathways. When bound to a surface, heparin also inhibits alternative pathway activation, although through a different mechanism that involves facilitation of the inhibitory action on C3b of the control proteins H and I[20]. The overall effect of heparin on the Complement system is complex since fluid phase heparin also inhibits C1 activation, binding of C2 to C4, interferes with formation of the C5b67 complex and augments binding of C1q to immune complexes. The pathological relevance of the anticomplementary effect of heparin is more probably to be of significance at the site of inflammation where heparin may act as a mediator rather than in circulating plasma where other proteins than the Complement proteins may compete for binding with heparin.

REFERENCES

1. R. R. Porter, Structure and activation of the early components of Complement, Fed. Proc. 36:2191 (1977).
2. M. D. Kazatchkine, and U. E. Nydegger, The human alternative Complement pathway: biology and immunopathology of activation and regulation, Prog. in Allergy 30:193 (1982).
3. A. D. Schreiber, A. P. Kaplan, and K. F. Austen, Inhibition by C1 inhibitor of Hageman Factor fragment activation of coagulation, fibrinolysis and kinin generation, J. Clin. Invest. 52:1401 (1973).
4. A. P. Kaplan, M. Silverberg, J. T. Dunn and B. Ghebrehivet, Interaction of the clotting, kinin-forming, Complement and fibrinolytic pathways in inflammation, in: "C-Reactive Protein and the Plasma Protein Response to Tissue Injury," I. Kuschner, J. E. Volanakis, H. Gewurz, eds., ANN. N.Y. Acad. Sci., New York (1982).
5. N. R. Cooper and R. J. Ziccardi, The nature and reactions of Complement enzymes, in: "Proteolysis and Physiological Regulation," D. W. Ribbons, K. Bruno, eds., Academic Press, New York (1976).
6. J. S. Sundsmo, and L. M. Wood, Activated Factor B (Bb) of the alternative pathway of Complement activation cleaves and activates plasminogen, J. Immunol. 127:877 (1981).
7. D. S. Fair, J. S. Sundsmo, D. S. Schwartz, T. S. Edgington, and H. J. Müller-Eberhard, Prothrombin activation by Factor B (Bb) of the alternative pathway of Complement, Meeting of the Intern. Soc. Thromb. Haemos., Toronto, p. 301 (1981).
8. J. E. Mole, and M. A. Niemann, Structural evidence that Complement Factor B constitutes a novel class of serine protease, J. Biol. Chem. 255:8472 (1980).
9. P. H. Lesavre, T. H. Hugli, A. F. Esser, and H. J. Müller-Eberhard, The alternative pathway C3/C5 convertase: chemical basis of Factor B activation, J. Immunol. 123:529 (1979).

10. P. H. Lesavre, and H. J. Müller-Eberhard, Mechanisms of action of Factor D of the alternative Complement pathway, J. Exp. Med. 148:1498 (1978).

11. R. G. Discipio, The activation of the alternative pathway C3 convertase by human plasma kallikrein, Immunology 45:587 (1982).

12. A. E. Davis III., C. Zalut, F. S. Rosen, and C. A. Alper, Human Factor D of the alternative Complement pathway. Physicochemical characteristics and N-terminal amino acid sequence, Biochemistry 18:5082 (1979).

13. J. E. Volanakis, A. S. Bhown, J. C. Bennett, and J. E. Mole, Partial amino acid sequence of human Factor D: homology with serine proteases, Proc. Natl. Acad. Sci. USA 77:1116 (1980).

14. A. E. Davis III.,Active site amino acid sequence of human Factor D., Proc. Natl. Acad. Sci. USA 77:4938 (1980).

15. D. M. Kenney, and A. E. Davis III., Association of alternative Complement pathway components with human blood platelets: secretion and localization of Factor D and beta 1 H globulin, Clin. Immunol. Immunopathol. 21:351 (1981).

16. A. E. Davis III., and D. M. Kenney, Properdin Factor D: effects on thrombin-induced platelet aggregation, J. Clin. Invest. 64:721 (1979).

17. R. C. Wiggins, P. C. Giclas, and P. M. Henson, Chemotactic activity generated from the fifth component of Complement by plasma kallikrein of the rabbit, J. Exp. Med. 153:1391 (1981).

18. J. M. Weiler, R. M. Yurt, D. T. Fearon, and K. F. Austen, Modulation of the formation of the amplification convertase of Complement by native and commercial heparin, J. Exp. Med. 147:409 (1978).

19. M. D. Kazatchkine, D. T. Fearon, D. D. Metcalfe, R. D. Rosenberg, and K. F. Austen, Structural determinants of the capacity of heparin to inhibit the formation of the human amplification C3 convertase, J. Clin. Invest. 67:223 (1981).

20. M. D. Kazatchkine, D. T. Fearon, J. E. Silbert, and K. F. Austen, Surface-associated heparin inhibits zymosan-induced activation of the human alternative Complement pathway by augmenting the regulatory action of the control proteins on particle-bound C3b, J. Exp. Med. 150:1202 (1979).

THE CALCIUM-DEPENDENT NEUTRAL PROTEASE OF HUMAN BLOOD PLATELETS:
A COMPARISON OF ITS EFFECTS ON THE RECEPTORS FOR
VON WILLEBRAND FACTOR AND FOR THE Fc-FRAGMENT DERIVED FROM IgG

M.O. Spycher[*], U.E. Nydegger[*] and E.F. Luescher[+]

[*] Central Laboratory of the Swiss Red Cross Blood Transfusion
Service, Berne, Switzerland
[+] Theodor Kocher Institute, University of Berne, Switzerland

SUMMARY

Glycocalicin (Gc) is the large, water soluble fragment, obtained
by cleavage of one of the major membrane glycoproteins, GP Ib, of
human platelets by means of the endogenous, calcium-dependent neutral protease (CNP) obtained from lysed platelets. GP Ib has been
proposed as the receptor for von Willebrand factor (vWF) as well as
for the Fc-receptor of the platelet surface. We have investigated,
whether Gc was involved in a receptor function for aggregated human
IgG, which is a powerful activator of platelets. Neither Gc nor
asialo-Gc inhibited the stimulation of human blood platelets by
bisdiazoniumbenzidine-aggregated human IgG (BDB-IgG). Moreover,
platelets, after treatment with a crude preparation of CNP, which
removes Gc, could be stimulated by BDB-IgG as well as or better
than control platelets, but were unreactive with bovine vWF. We
conclude that the Gc-moiety of GP Ib, which is involved in the bovine vWF binding site, is not the Fc-receptor on platelets. Thus,
the inhibition, by human or rabbit IgG aggregates or monomeric
rabbit IgG, of vWF-induced platelet agglutination, as reported by
other authors, is either due to a steric effect resulting from a
vicinal position of both receptors or involves the residual part
of GP Ib after cleavage of Gc.

INTRODUCTION

Calcium dependent neutral proteases (CNP's) are located in many
cells and tissues. They have been detected first in homogenates of
brain[1] and skeletal muscle[2], later they have been found also in

liver[3], kidney, lung and thymus[4], as well as in the erythrocyte[5,6]
and the platelet[7]. CNP is a non-membrane bound enzyme located in the
cytoplasm of the cells. In the meantime it has been purified to homogeneity from human platelets[8], erythrocytes[9] and from porcine[10,11]
and human[12] skeletal muscle.

For optimal activity the isolated CNP from human platelets,
which accounts for over 95% of the neutral proteolytic activity of
these cells[13], requires reduced sulfhydryl groups, a neutral pH, and
$1 - 2$ mM free Ca^{2+}[7,8]. The enzyme molecule consists of two polypeptide chains of 80'000 and 30'000 daltons molecular weight, respectively[8]. Recently a second rarer form of CNP has been isolated
that is already active with Ca^{2+} in the μM range[14]. In view of the
fact that platelet activation is linked to an increased cytoplasmic
Ca^{2+}-concentration which reaches the μM range[15], this finding has
led to the assumption that CNP plays an important role in the course
of platelet stimulation by activating enzymes such as phospholipases
or kinases[14]. Whether this represents the real physiological function of the CNP remains as yet, this the more as other targets for
the enzyme have also been described. Thus, it has been suggested
that CNP interferes with the platelet cytoskeleton by cleaving
actin-binding protein (ABP), which crosslinks F-actin filaments, into
two high molecular weight fragments[8]. It is interesting that the
disappearance of ABP is not only observed after the disruption of
platelets[8], but also after storage of intact platelets for several
days[16].

Another protein that is altered during storage of platelet
concentrates is glycoprotein Ib (GP Ib), a major GP of the platelet
plasmamembrane[17]. It has been shown that GP Ib is also a substrate
of CNP[18] and its loss from intact cells during storage is probably
due to the action of CNP originating from a few lysed platelets. The
protease cleaves GP Ib specifically, whereby watersoluble Glycocalicin (Gc), which represents almost 90% of the GP Ib-molecule, is released from the membrane[18,29]. This enzymatic attack is not linked
to a stimulation of the platelets as is the case with other proteases such as trypsin, pronase, papain or the physiological activator
thrombin.

Aggregated human IgG is a powerful activator of washed human
platelets acting via an Fc-receptor in the platelet membrane[31]. It
is well established that GP Ib is the receptor for bovine and human
von Willebrand factor (vWF), whereby in the latter case the anti-
biotic ristocetin is required as a cofactor[19,20]. Recently the same
GP has been proposed by Moore et al.,[21] to represent or to be closely associated with the receptor for the Fc-fragment of IgG. This
conclusion was based on the observation that the agglutination or ag-

gregation of human platelets by human vWF in the presence of ristocetin was inhibited by monomeric or aggregated rabbit IgG or its Fc-
fragments, but not by Fab- or pFc-fragments. This inhibition also
occured with heat-aggregated human IgG but not with monomeric human
IgG. On the other hand Pfueller et al.,[22] demonstrated that the binding of bisdiazoniumbenzidine-aggregated human IgG (BDB-IgG) to
lightly fixed, washed platelets which had been treated with proteases (trypsin, pronase, papain) which were shown to cleave GP Ib,
increased rather than decreased. Aggregation of the enzyme-treated
platelets by ristocetin-human von Willebrand factor was however
strongly decreased or abolished. In the experiments described here
the role of GP Ib in platelet Fc-receptor activity has been further
investigated by examining the effect of CNP on the interaction of
BDB-IgG with platelets.

MATERIALS AND METHODS

    Human blood platelets. These were isolated within 20h after
collection from citrated blood collected for the Central Laboratory
of the blood transfusion Service of the Swiss Red Cross in Berne[23].
Three parts of buffy coats were syphoned into one part of 9.6 mM
glucose, 3 mM KCl, 100 mM NaCl, 30 mM sodium citrate, pH 6.5. To
remove most of the remaining leucocytes and erythrocytes, the suspension was centrifuged 6 - 9 min at 150 g and 15°C. The supernatant
containing about $3 \cdot 10^9$ platelets/ml was brought to pH 6.8 and after
1h incubation at 37°C the platelets were washed twice in 0.12 M
NaCl, 0.38% sodium citrate, 0.6% glucose and once in 25 mM TES,
132 mM NaCl, 3 mM KCl, 0.5% glucose, pH 6.8, by centrifugation at
1500 g for 10 min and resuspension. The platelet suspension was diluted to the required concentration with the second wash buffer.

    Treatment of platelets with a crude calcium-dependent neutral
protease extract. The method used was adapted from that of Naim et
al. (submitted for publication). The platelet suspension (16 ml,
$8 \cdot 10^9$ platelets/ml) was sonicated at 4°C with a B-30 Sonifier
(Branson Sonic Power Co., Danbury, CT., USA) for 1 min on setting 7,
50% pulse and then centrifuged at 7'700 g and 4°C. Part of the supernatent (7.5 ml) containing the calcium-dependent neutral protease
(CNP) was added to 30 ml of the platelet suspension ($1 \cdot 10^9$ plate-
lets/ml). The mixture was incubated for 30 min at 37°C after which
the platelets were no longer agglutinable with bovine vWF. The platelets were washed three times as described above and were then resuspended in 0.12 M NaCl, 15 mM KCl, 20 μM $MgCl_2$, 60 μM $CaCl_2$, 26 mM
TES, pH 6.8, to a concentration of $2 \cdot 10^9$ platelets/ml. Control platelets were treated similarly but with 7.5 ml buffer in place of
the supernatent containing the protease.

Measurement of aggregation response and serotonin release of washed platelets. ^{14}C-serotonin was incorporated into washed platelets (10 ml) by incubation in buffer containing 50 μl ^{14}C-serotonin (Amersham International, 8 μCi/ml, 0.14 mM in 70% ethanol) for 1h at room temperature. Aggregation was measured in a Labintec aggregometer. If not otherwise indicated the platelet suspension (100 μl) was mixed with 130 μl of 0.12 M NaCl, 15 mM KCl, 20 μM $MgCl_2$, 60 μM $CaCl_2$, 26 mM TES, pH 6.8, and 50 μl of 0.17 M Tris/HCl, 8 mg/ml BSA, 6 μM imipramine, pH 7.5, in a 1 ml polypropylene tube. To this mixture, which was incubated for 2 min at 37°C in the aggregometer, BDB-IgG or bovine vWF was added in 100 μl 10 mM Tris/HCl, 0.14 M NaCl, pH 8.0 (BDB-IgG), or 100 μl 0.15 M NaCl (bovine vWF). Six min after addition of the stimulant the platelets were sedimented on a Hawksley Microhematocrit centrifuge for 20 secs and 100 μl of the supernatant were removed and the radioactivity determined in the ^{14}C-channel of a liquid scintillation counter.

$$\text{Serotonin release [\%]} = \frac{\text{cpm sample} - \text{cpm blank}}{\text{cpm total} - \text{cpm blank}}$$

Total = 100 μl aliquot, without stimulator, before centrifugation.
Blank = 100 μl aliquot, without stimulator, after centrifugation.
The steepest slope of the aggregation curve was taken as a measure of the "maximal aggregation [%/sec]".

Surface-labelling of washed platelets by the periodate/[^{3}H]-NaBH$_4$ method. The method of Gahmberg and Andersson,[24] as modified by Steiner et al.,[25] was used for surface-labelling of the platelets. Protease treated, washed platelets or control platelets were washed once in 0.13 M NaCl, 10 mM sodium citrate buffer, pH 6.0, and adjusted to a platelet concentration of 6 - 7·10^9/ml by addition of the same buffer. Sodium periodate (0.1 M, 70 μl) was added to 3.4 ml of the platelet suspension which was then left for 10 min at room temperature in the dark. Glycerol (85%, 10 μl) was added, the platelets were washed three times with 20 ml of 0.15 M NaCl, 2 mM EDTA, resuspended in 0.11 M NaCl, 40 mM Tris/HCl, pH 8.0, and treated with 20 μl [^{3}H]-NaBH$_4$ (1 mCi). After 30 min at room temperature the platelets were washed three times with 0.15 M NaCl, 2 mM EDTA and resuspended in the same solution. The specific activity of these labelled platelets was 2 - 3.6·10^6 cpm/mg protein.

Gel electrophoresis. One-dimensional sodium dodecyl sulphate polyacrylamide gel electrophoresis was carried out according to the method of Laemmli,[26] using a 7.5% acrylamide separating gel and a 5% stacking gel of 1.5 mm thickness. After fixing in 20% isopropanol, 10% acetic acid, the gel was prepared for fluorography by the method of Bonner and Laskey,[27]. The gel was then placed in con-

tact with a pre-flashed Kodak X-Omat So-282 film for 1 - 2 days at
-70°C[28]. Densitometry of the fluorogram was carried out in a Kipp
and Zonen densitometer.

Glycocalicin (Gc) prepared according to the method of Cooper
et al.,[19] and asialoglycocalicin (asialo-Gc) prepared according to
Clemetson et al.,[29] were kindly donated by Dr. H.Y. Naim.

Bovine von Willebrand factor (bovine vWF) was prepared by the
method of Santos et al.,[30] and stored in 0.3 M NaCl solution at 4°C.
Before use it was diluted with the same volume of distilled water.

Bisdiazoniumbenzidine-aggregated human IgG (BDB-IgG) was prepared according to Pfueller et al.,[31] .

Tritiated sodium borohydride ($[^3H]$-NaBH$_4$, New England Nuclear)
with a specific activity of 8.8 Ci/mmol was dissolved in 10 mM NaOH
and aliquots of 5 mCi/100 μl were stored at -70°C. The radioactivity was measured in a Packard TriCarb 2450 liquid scintillation
spectrometer using 200 μl of sample with 2 ml of a scintillator
containing Xylene and Triton X-100 (2:1, v/v) and 3 g/l Perma-
blend III (Packard).

Platelet counter : ToA Medical Electronics Co., Ltd, Japan.

RESULTS

In order to study the effect of Gc and asialo-Gc on BDB-IgG
induced aggregation and serotonin release, BDB-IgG and Gc or asialo-
Gc were incubated together at pH 7.4 and 4°C for 30 min, then warmed
to 37°C and added to washed platelets at 37°C. Neither the Gc nor
the purer asialo-Gc at concentrations (mg/ml) of up to about seven
times that of the stimulant inhibited the platelet response to
BDB-IgG (Table 1).

Table 1. Inhibition of BDB-IgG induced platelet aggregation and
serotonin release by glycocalicin and asialoglycocalicin

BDB-IgG [μg/ml]	Glycocalicin [μg/ml]	Asialoclycoca-licin [μg/ml]	Max. Aggrega-tion [%/sec]	Serotonin-release [%]
72	0	–	0.75	66.5
72	542	–	0.76	66.8
31	–	0	0.81	70.8
31	–	238	0.78	68.3

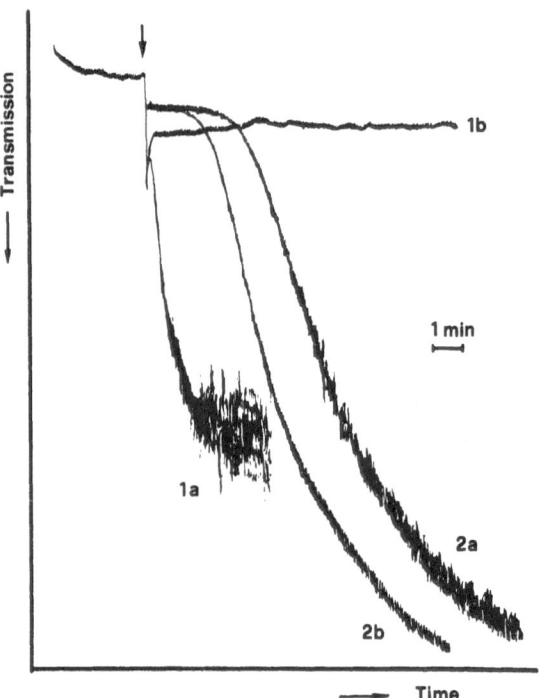

Fig. 1. Aggregation curves of a) normal washed platelets; b) pla-
        telets treated with CNP. Activation was induced by the
        addition of 1) bovine vWF; 2) BDB-IgG. The time at which
        the activators were added is indicated with an arrow.

     The effect of the removal of Gc from the platelet surface by
cleavage with the CNP on BDB-IgG induced platelet aggregation was
studied next. Fig. 1 shows the aggregation curves obtained with
control platelets and with CNP-treated platelets after stimulation
with either BDB-IgG or bovine vWF. The control platelets aggregated
with both agents. A comparison of the effects of BDB-IgG on both
platelet preparations shows that, while the total aggregation and
the rate of aggregation are similar, there was a shorter lag-phase
with the CNP-treated platelets. The protease-treated platelets were
not stimulated by bovine vWF. Fig. 2 shows the densitometric tracing
of the fluorogram of a one-dimensional polyacrylamide gel electro-
pherogram of both types of platelet preparations which had been
surface-labelled using the periodate/$[^3H]$-NaBH$_4$ technique. After
treatment of the platelets with CNP at least 95% of glycoprotein Ib
(GP Ib) are cleaved (there are other minor GP's in this position
which make quantitation difficult). In addition GP V is completely
removed, while the other glycoproteins appear unchanged.

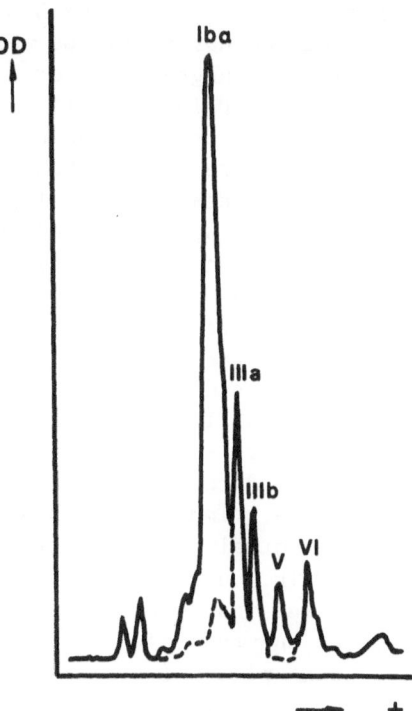

Fig. 2. Densitometric tracing of fluorograms of ^{3}H-labelled platelets solubilized in sodium dodecyl sulphate and reduced
with dithiothreitol; separated by Laemmli-sodium dodecyl
sulphate polyacrylamide gel electrophoresis (7.5% separating gel, 5% stacking gel; approx. 50 μg protein on each
gel).

DISCUSSION

GP Ib is one of the major GP's of the platelet plasmamembrane
and consists of two polypeptide chains α and β. There is general
agreement that it is acting as the receptor for vWF; platelets aggregate upon the addition of human vWF (in the presence of
ristocetin) and of bovine vWF. Fixed platelets are agglutinated by
these two stimulants. Moore et al.,[21] demonstrated that agglutination of formaldehyde-fixed platelets or aggregation of platelets in
plasma induced by ristocetin-human vWF was inhibited by monomeric or
aggregated rabbit IgG as well as by rabbit Fc-fragments or aggregates of human IgG. They concluded that GP Ib was furthermore either
the platelet Fc-receptor or was closely associated with it. This induced us to test, whether Gc, the watersoluble, proteolytic fragment of the α-chain of GP Ib, which represents the major part of

the parent molecule exerts any effect on the BDB-IgG induced aggregation. In our hands neither Gc nor asialo-Gc were capable of inhibiting BDB-IgG induced platelet aggregation or serotonin release.
It should be noted that asialo-GP Ib is still fully active as receptor for vWF; thus Gc and asialo-Gc can be expected to be functionally equivalent. It must be realized, though, that such inhibition experiments offer not necessarily final proof for the identity of a receptor. Thus, Gc has been reported not to interfere
with vWF-induced aggregation[19], unless added in large excess[33];
nonetheless platelets from which Gc has been removed, are no longer
agglutinable with vWF[20]. Platelets from which Gc has been cleaved
however react as well as or better with BDB-IgG. Therefore the conclusion must be drawn that IgG and vWF bind to different sites on
the platelet surface and that at least the Gc-moiety of the GP Ib-
molecule is not the Fc-receptor. Kunicki et al.,[34] reached similar
conclusions as platelets from Bernard-Soulier syndrome patients
which lack GP Ib could not be stimulated with quinine- or quinidine-
dependent antibodies while normal platelets treated with chymotrypsin or trypsin on which GP Ib could no longer be detected, reacted
normally. The possibility remains that GP Ib is not completely
destroyed by the proteases and that part of it, for example the
β-chain or the CNP-resistant piece of the α-chain, remain on the
platelet surface. Indeed, Pfueller et al.,[22] have previously found
that after proteolysis of the platelets surface glycoproteins the
binding of BDB-IgG was rather increased. According to Moore et
al.,[21] this could be ascribed to the improved exposure of the
GP Ib-associated Fc-receptor and it might be that the decrease in
the lag-phase between addition of BDB-IgG and start of aggregation
as observed with CNP-treated platelets is explained in the same way.

REFERENCES

1. G. Guroff, A neutral,calcium activated proteinase from the so-
     luble fraction of rat brain, J. Biol. Chem. 239:149 (1964).
2. R. B. Huston, and E. G. Krebs, Activation of skeletal muscle
     phosphorylase kinase by Ca^{2+}. II. Identification of the ki-
     nase activating factor as a proteolytic enzyme, Biochemistry
     7:2116 (1968).
3. Y. Takai, M. Yamamoto, M. Inoue, A. Kishimoto,and Y. Nishizuka,
     A proenzyme of cyclic nucleotide-independent protein kinase
     and its activation by calcium-dependent neutral protease from
     rat liver, Biochem. Biophys. Res. Commun. 77:542 (1977).
4. L. Waxman, Tissue distribution and comparative properties of a
     calcium-activated protease and its inhibitor, Fed. Proc. 38:
     479 (1979).
5. S. Pontremoli, E. Melloni, F. Salamino, B. Sparatore, M. Michetti,
     U. Benatti, A. Morelli, and A. De Flora, Identification

of proteolytic activities in the cytosolic compartment of mature erythrocytes, Eur. J. Biochem. 110:421 (1980).

6. E. Melloni, F. Salamino, B. Sparatore, M. Michetti, U. Benatti, A. De Flora, and S. Pontremoli, Decay of proteinase and peptidase activities of human and rabbit erythrocytes during cellular aging, Biochim. Biophys. Acta 675:110 (1981).

7. D. R. Phillips, and M. Jakábová, Ca-dependent protease in human platelets. Specific cleavage of platelet polypeptides in the presence of added Ca^{2+}, J. Biol. Chem. 252:5602 (1977).

8. J. A. Truglia, and A. Stracher, Purification and characterization of a calcium dependent sulfydryl protease from human platelets, Biochem. Biophys. Res. Commun. 100:814 (1981).

9. E. Melloni, B. Sparatore, F. Salamino, M. Michetti, and S. Pontremoli, Cytosolic calcium dependent proteinase of human erythrocytes: Formation of an enzyme-natural inhibitor complex induced by Ca^{2+} ions, Biochem. Biophys. Res. Commun. 106: 731 (1982).

10. W. R. Dayton, D. E. Goll, M. G. Zeece, R. M. Robson, and W. J. Reville, A Ca^{2+}-activated protease possibly involved in myofibrillar protein turnover. Purification of porcine muscle, Biochemistry 15:2150 (1976).

11. W. R. Dayton, W. J. Reville, D. E. Goll, and M. H. Stromer, A Ca^{2+}-activated protease possibly involved in myofibrillar protein turnover. Partial characterization of the purified enzyme, Biochemistry 15:2159 (1976).

12. K. Suzuki, S. Ishuira, S. Tsuij, T. Katamoto, H. Sugita, and K. Imahori, Calcium activated neutral protease from human skeletal muscle, FEBS Lett. 104:355 (1979).

13. J. A. Truglia, A. Stracher, and R. C. Lucas, Human platelet cytoskeleton is proteolysed by an endogenous Ca^{2+}-dependent protease, Fed. Proc. 38:469 (1978).

14. M. Sakon, J.-I. Kambayashi, H. Ohno, and G. Kosaki, Two forms of Ca^{2+}-activated neutral protease in platelets. Thromb. Res. 24:207 (1981).

15. J. M. Gerrard, D. A. Peterson, and J. G. White, Calcium mobilization, in:"Platelets in biology and pathology 2," J. L. Gordon, ed., Elsevier/North-Holland Biomedical Press, Amsterdam, New York, Oxford (1981).

16. F. A. Robey, C. M. Freitag, and G. A. Jamieson, Disappearance of actin binding protein from human blood platelets during storage, FEBS Lett. 102:257 (1979).

17. J. N. George, Platelet membrane glycoproteins: Alterations during storage of human platelet concentrates, Thromb. Res. 8:719 (1976).

18. N. O. Solum, I. Hagen, C. Filion-Myklebust, and T. Stabaek, Platelet glycocalicin. Its membrane association and solubilization in aqueous media, Biochim. Biophys. Acta 597:235 (1980).

19. H. A. Cooper, K. J. Clemetson, and E. F. Lüscher, Human platelet membrane receptor for bovine von Willebrand factor (platelet activating factor): An integral membrane glycoprotein, Proc. Natl. Acad. Sci. USA 76:1069 (1979).

20. D. R. Phillips, An evaluation of membrane glycoproteins in platelet adhesion and aggregation, in:"Progress in Hemostasis and Thrombosis 5," T. H. Spaet, ed., Grune and Stratton, New York (1980).

21. A. Moore, G. D. Ross, and R. L. Nachman, Interaction of platelet membrane receptors with von Willebrand factor, ristocetin and the Fc region of immunoglobulin G, J. Clin. Invest. 62: 1053 (1978).

22. S. L. Pfueller, C. S. P. Jenkins, and E. F. Lüscher, A comparative study of the effect of modification of the surface of human platelets on the receptors for aggregated immunoglobulin and for ristocetin von Willebrand factor, Biochim. Biophys. Acta 465:614 (1977).

23. M. Bettex-Galland, and E. F. Lüscher, Studies on the metabolism of human blood platelets with relation to clot retraction, Thromb. Diathes. Haemorrh. 4:178 (1960).

24. C. G. Gahmberg, and L. C. Andersson, Selective radioactive labelling of cell surface sialoglycoproteins by perjodate-tritiated borohydride, J. Biol. Chem. 252:5888 (1977).

25. B. Steiner, K. J. Clemetson, and E. F. Lüscher, Improvement of the perjodate-borohydride surface-labeling method for human blood platelets, Submitted to Thromb. Res.

26. K. Laemmli, Cleavage of structural proteins during the assembly of the head of bacteriophage T4, Nature 227:680 (1970).

27. W. M. Bonner, and R. A. Laskey, A film detection method for tritium-labelled proteins and nucleid acids in polyacrylamide gels, Eur. J. Biochem. 46:83 (1974).

28. R. A. Laskey, and A. D. Mills, Quantitative film detection of ^{3}H and ^{14}C in polyacrylamide gels by fluorography, Eur. J. Biochem. 56:335 (1975).

29. K. J. Clemetson, H. Y. Naim, and E. F. Lüscher, Relationship between glycocalicin and glycoprotein Ib of human blood platelets, Proc. Natl. Acad. Sci. USA 78:2712 (1981).

30. F. Santos, P. R. Johnson Jr., M. Hall, H. R. Clark, and R. H. Wagner, Preparation of bovine platelet aggregating factor (PAF), Thromb. Res. 13:741 (1978).

31. S. L. Pfueller, S. Weber, and E. F. Lüscher, Studies of the mechanism of the human platelet release reaction induced by immunologic stimuli. III. Relationship between the binding of soluble aggregates to the Fc receptor and cell response in the presence and absence of plasma, J. Immunol. 118:514 (1977).

32. J. L. Mc Gregor, K. J. Clemetson, E. James, A. Capitanio, T.
    Greenland, E. F. Lüscher, and M. Dechavanne, Glycoproteins
    of platelet membranes from Glanzmann's Thrombastenia. A com-
    parison with normal using carbohydrate specific or protein
    specific labelling techniques and high-resolution two di-
    mensional gel electrophoresis, Eur. J. Biochem. 116:379 (1981).
33. T. Okumura, and G. A. Jamieson, Platelet glycocalicin: A single
    receptor for platelet aggregation induced by thrombin and
    ristocetin, Thromb. Res. 8:701 (1976).
34. T. J. Kunicki, M. M. Johnson, and R. H. Aster, Absence of the
    platelet receptor for drug-dependent antibodies in the Ber-
    nard Soulier syndrome, J. Clin. Invest. 62:716 (1978).

ALPHA-2-PLASMIN-INHIBITOR INACTIVATION BY HUMAN

GRANULOCYTE ELASTASE

M. Gramse, K. Havemann, and R. Egbring

Department of Hematology, Oncology, Immunology
University of Marburg, FRG

Since the description of the fast reacting α_2-plasmin inhibitor (1-3) many efforts were made to investigate its physiological role as a fibrinolysis inhibitor (4-7). It has been shown that plasmin inactivation by this inhibitor is a two step reaction: The first step, a very rapid binding of plasmin, is followed by its slower inactivation. Whereas the first step does not require the active enzyme, the bound plasmin liberates in the second step a small peptide from α_2-plasmin inhibitor which comparably occurs during other proteinase inhibitor interactions.

Isolated α_2-plasmin inhibitor was found to exist in three forms: a plasmin/plasminogen-binding molecule with a molecular mass of 67-68 kDa, which is described to spontaneously convert into the smaller (65 kDa) non-plasminogen-binding and finally into the inactive inhibitor molecule of 60 kDa (8). A series of proteinases was investigated, but neither of them was found to trigger this conversion (8). There are two reports that granulocyte elastase is able to split α_2-plasmin inhibitor. But, they both do not show a cleavage pattern comparable to the spontaneously occuring conversion (9,10). The presented paper will show that granulocyte elastase causes or at least accelerates the "spontaneous" proteolysis of α_2-plasmin inhibitor which is accompanied by the conversion of the plasminogen-binding into the non-plasminogen-binding form of the inhibitor.

A molar enzyme-inhibitor ratio of 1:60 was chosen as this ratio can be observed during an elastase release for instance in septicemia (11). All incubations were performed at 37°C and elastase activity was stopped after the given time intervals using a granulocyte elastase specific peptide chloromethyl ketone (12).

Figure 1 shows the inactivation of α_2-plasmin inhibitor following
incubation with elastase for 15 min, 60 min, and 24 h. α_2-plasmin
inhibitor activity was measured ba the inhibitory effect against
plasmin and was found to be reduced to 50% after 15 min incubation,
and finally lost after prolonged incubation time. α_2-plasmin inhibitor, incubated in parallel for 24 h with buffer, was yet fully
active. The observed inactivation is not due to competitive inhibition of plasmin - α_2-plasmin inhibitor - complex formation by
elastase. It is otherwise shown that elastase does not form an enzyme-inhibitor complex with α_2-plasmin inhibitor resulting in
elastase inactivation (13).

During α_2-plasmin inhibitor inactivation by elastase a
change in its molecular mass occured as shown in dodecyl sulfate-
polyacrylamide gel electrophoresis in figure 2. Using the indicated marker proteins the estimated molecular mass of the undegraded
plasminogen-binding α_2-plasmin inhibitor was 69 kDa, that of the
non-plasminogen-binding one 66 kDa. A third protein band of 59
kDa , the inactive inhibitor, arose during incubation with elastase for 15 min. After 24 h incubation the plasminogen-binding
form is fully converted into the smaller molecules. α_2-plasmin
inhibitor, incubated for control 24 h with buffer, shows no comparable degradation.

Identical samples were examined in two-dimensional immunoelectrophoresis for their plasminogen-binding properties (fig. 3)
The samples were separated in the first dimension in a plasminogen containing agarose, in the second dimension precipitated with
anti-α_2-plasmin inhibitor serum. Thus, the plasminogen-binding inhibitor is retarded and can be separated from the faster migrating
non-plasminogen-binding inhibitor.

As visible in the figure the inhibitor, already half inactivated by 15 min incubation with elastase, is nearly unchanged in
its plasminogen-binding properties. After 60 min a non-binding
precipitate appears, which is the only form permanent after 24 h
incubation. α_2-plasmin inhibitor, incubated 24 h with buffer,
shows a high percentage of non-plasminogen-binding inhibitor, but
is still fully active.

Surprisingly, if a non-plasminogen-binding α_2-plasmin inhibitor preparation was degraded by elastase, the electrophoretic
mobility of the split products was partially retarded in plasminogen containing agarose (figure 4). That this separation of the
precipitate in two distinct peaks must be due to different plasminogen-binding properties is shown on the left hand site of the
figure, where electrophoresis of identical samples in a plasmino-
gen-free agarose resulted in only one peak.

In summary, α_2-plasmin inhibitor degradation by granulocyte elastase results in α_2-plasmin inhibitor inactivation and in the loss of its plasminogen-binding capacity. These two processes do not depend on each other. It is therefore assumed that elastase cleaves α_2-plasmin inhibitor at at least two sites: First in the inhibitory center, and second at a remote site of the molecule responsible for plasminogen-binding.

Comparable experiments were performed in normal human plasma. Here, however, α_2-plasmin inhibitor inactivation required a thousand-fold excess of elastase.

Plasma was incubated for 15 min with increasing concentrations of elastase (figure 5). Elastase activity was contineously measured during the incubation time and expressed as substrate turnover. The % α_2-plasmin inhibitor activity was measured following specific inhibition of elastase (see also figure 1). The break in the curve of elastase activity at about 800 µg elastase per ml plasma shows the point of α_1-proteinase inhibitor saturation with elastase, α_2-plasmin inhibitor activity in this experiment is reduced only above this elastase concentration. Obviously, α_2-plasmin inhibitor inactivation within 15 min requires free elastase.

Again the inactivation is accompained by a loss of the inhibitor's plasminogen-binding capacity. This is shown in two-dimensional immunoelectrophoresis run in plasminogen containing agarose during the first dimension (figure 6) with four samples from the experiment shown in figure 5: Two plasma samples with the fully active inhibitor which had been incubated with buffer and 700 µg elastase per ml, respectively; one sample with partially (800 µg elastase/ml) and one with fully inactivated α_2-plasmin inhibitor. Whereas plasma, incubated with buffer, separated in equal parts of non-plasminogen and plasminogen-binding α_2-plasmin inhibitor, the fully inactivated plasma was nearly free of the plasminogen-binding inhibitor.

The experiments with normal plasma lead to the conclusion that α_2-plasmin inhibitor is not as highly susceptible to granulocyte elastase degradation as for instance C3 and C5 (14). Thus it is not to expect that α_2-plasmin inhibitor inactivation by elastase will occur in systemic fibrinolytic events. Nevertheless, α_2-plasmin inhibitor may play a role as elastase substrate in local coagulation disorders, where high concentrations of elastase are released during clot formation (15). Plasmin mediated clot lysis may therefore be enhanced by degradation of the fast reacting plasmin inhibitor in this micro-environment.

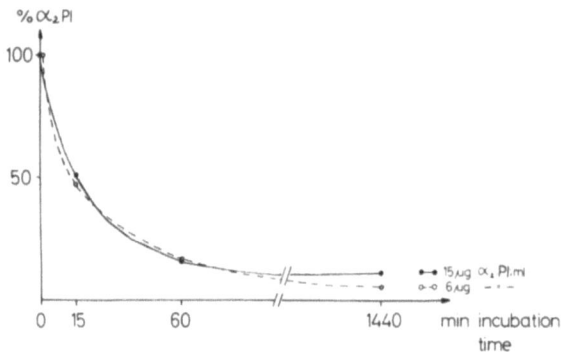

Figure 1:   Time dependence of α_2-plasmin inhibitor inactivation by elastase (1 Mol elastase: 60 Mol α_2-plasmin inhibitor). α_2-plasmin inhibitor was used in two concentrations and tested as inhibitory capacity against plasmin. Standardized human plasma for 100% value and chromogenic substrate for measuring plasmin activity were obtained from Kabi GmbH, Munich.

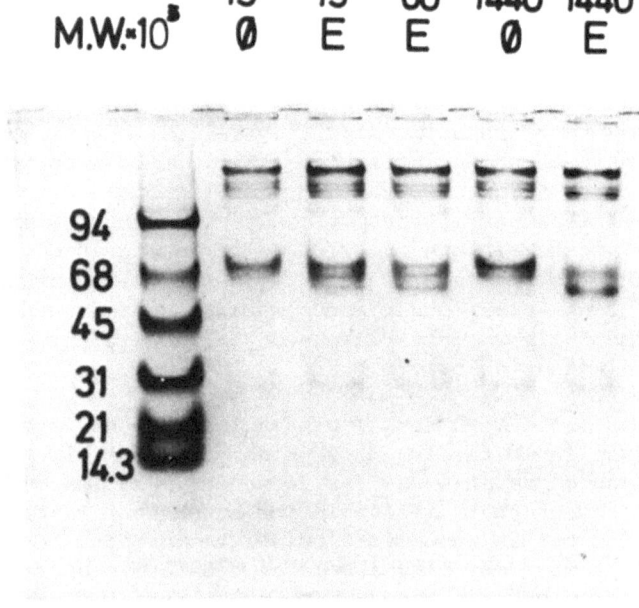

Figure 2:   Dodecyl sulfate-polyacrylamide gel electrophoresis of α_2-plasmin inhibitor previously incubated with buffer (∅) and elastase (E), respectively, for the indicated time intervals. Marker proteins were obtained from BioRad, Munich.

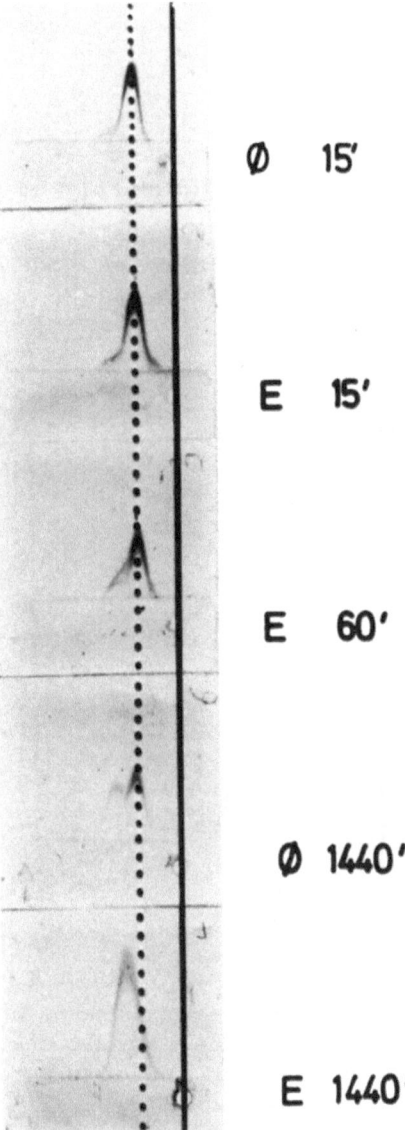

Figure 3: Two-dimensional immunoelectrophoresis of α_2-plasmin inhibitor previously incubated with buffer (∅) and elastase (E), respectively, for 15 to 1440 min. The agarose contained 0.5 CTA U plasminogen/ml during the first dimension and 0.8 % anti-α_2-plasmin inhibitor serum during the second dimension. Plasminogen and the antiserum were obtained from Behringwerke, Marburg.

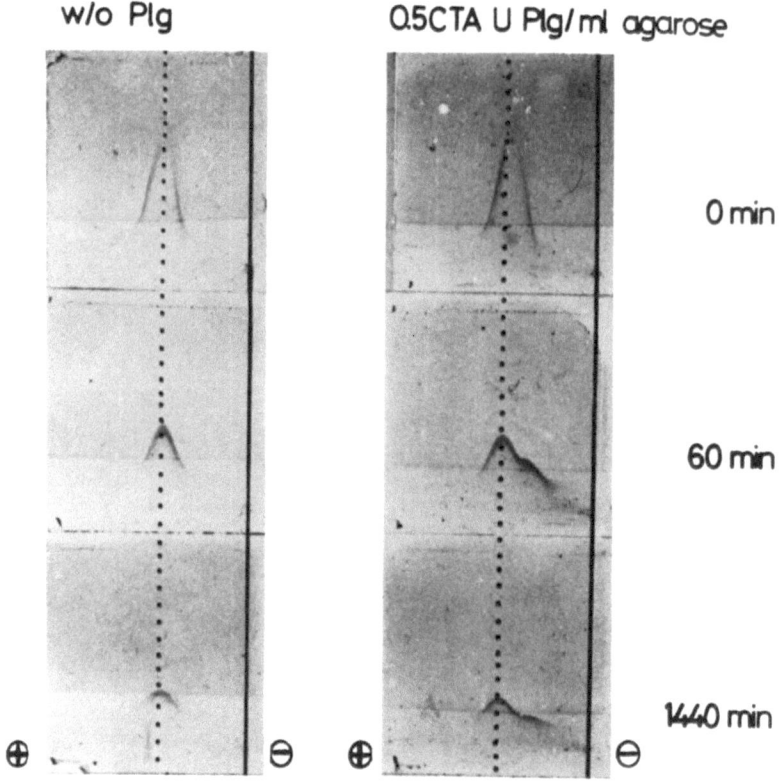

Figure 4: Two-dimensional immunoelectrophoresis of non-plasminogen-binding α₂-plasmin inhibitor. The test conditions are identical to figure 3. The plates on the left hand site were free of plasminogen during the first dimension.

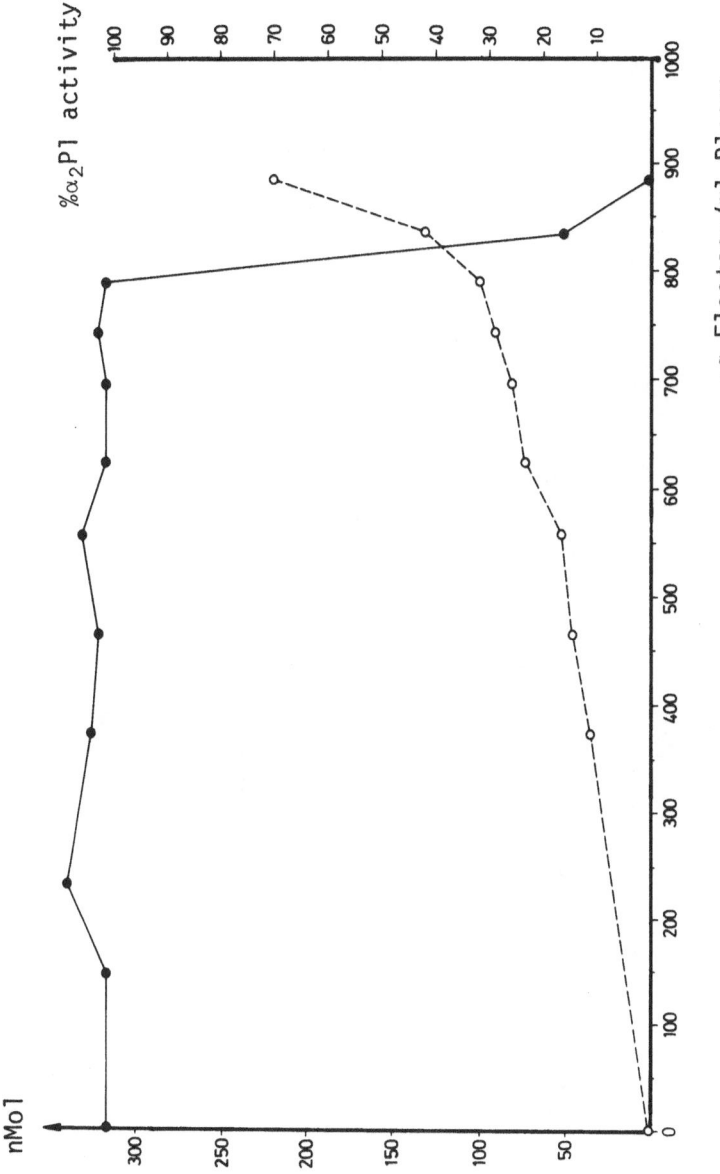

Figure 5: α_2-plasmin inhibitor activity (%) in normal human plasma following 15 min incubation with 0 to 900 µg elastase per ml plasma (●——●). Elastase activity is measured with L-pyroglutamyl-L-prolyl-L-valine-4-nitroanilide and expressed as nM substrate cleaved within 15 min (○--○).

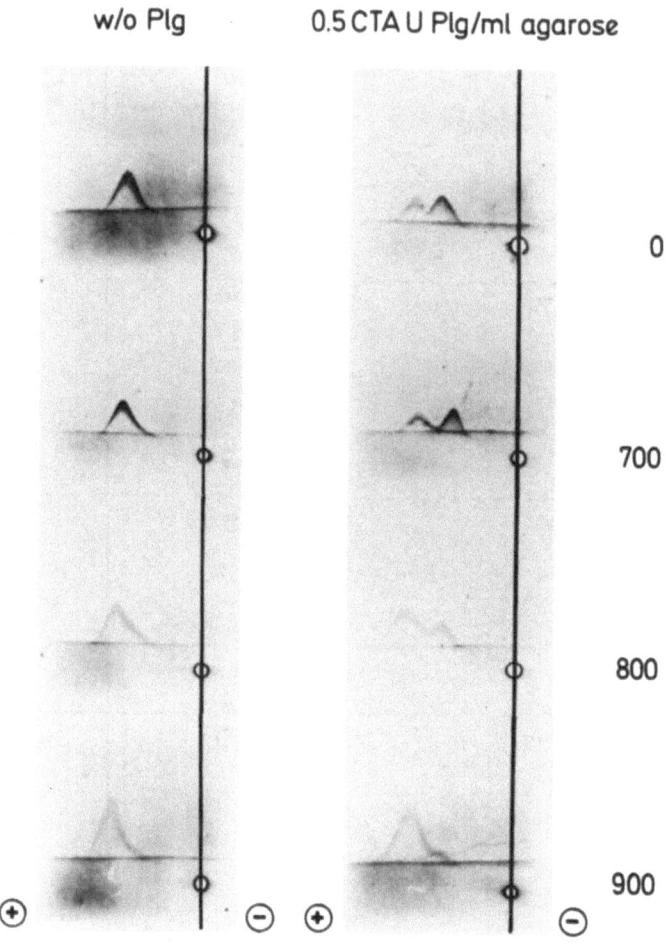

Figure 6: Two-dimensional immunoelectrophoresis of normal human plasma previously incubated for 15 min with 0, 700, 800, and 900 µg elastase per ml plasma. Further test conditions are identical to figure 4.

REFERENCES

1. D. Collen, Identification and some properties of a new fast-
    reacting plasmin inhibitor in human plasma, Eur. J. Biochem.
    69:209 (1976).
2. S. Müllertz, and. I. Clemmensen, The primary inhibitor of plas-
    min in human plasma, Biochem. J. 159:545 (1976).
3. M. Moroi, N. Aoki, Isolation and characterization of α_2-plasmin
    inhibitor from human plasma, J. Biol. Chem. 251:5956 (1976).
4. N. Aoki, M. Moroi, M. Matsudo, and K. Tachiya, The behaviour of
    α_2-plasmin inhibitor in fibrinolytic states, J. Clin. Invest.
    60:361 (1977).
5. Y. Sakata, K. Tateno, T. Tamaki, and N. Aoki, Calcium-dependent
    binding of α_2-plasmin inhibitor to fibrin, Thromb. Res. 16:
    279 (1979).
6. P. C. Harpel, α_2-plasmin inhibitor and α_2-macroglobulin-plasmin
    complexes in plasma. Quantitation by an enzyme-linked differ-
    ential antibody immunosorbent assay, J. Clin. Invest. 68:46
    (1981).
7. H. R. Lijnen, and D. Collen, Interaction of plasminogen inacti-
    vators and inhibitors with plasminogen and fibrin, Seminars
    Thromb. Hemostas. 8:2 (1982).
8. I. Clemmensen, S. Thorsen, S. Müllertz, and L. C. Petersen, Pro-
    perties of three different molecular forms of the α_2-plasmin
    inhibitor, Europ. J. Biochem. 120:105 (1981).
9. H. G. Klingemann, R. Egbring, M. Holst, M. Gramse, and K. Have-
    mann, Digestion of α_2-plasmin inhibitor by neutral proteases
    from human leukocytes, Thromb. Res. 24:479 (1981).
10. M. S. Brower, and P. C. Harpel, Proteolytic cleavage and inacti-
    vation of α_2-plasmin inhibitor and C1 inactivator by human
    polymorphonuclear leukocyte elastase, J. Biol. Chem. 257:
    9849 (1982).
11. R. Egbring, W. Schmidt, G. Fuchs, and K. Havemann, Demonstration
    of granulocytic proteases in plasma of patients with acute
    leukemia and septicemia with coagulation defects, Blood 49:
    219 (1977).
12. J. C. Powers, B. F. Gupton, A. D. Harley, N. Nishino, and R. J.
    Whitley, Specificity of porcine pancreatic elastase, human
    leukocyte elastase and cathepsin G. Inhibition with peptide
    chloromethyl ketones, Biochim. Biophys. Acta 485:156 (1977).
13. K. Ohlsson, and D. Collen, Comparison of the reactions of neu-
    tral granulocyte proteases with the major plasma protease-in-
    hibitors and with antiplasmin, Scand. J. Lab. Clin. Invest.
    37:345 (1977).
14. C. Löffler, Human granulocyte elastase and chymotrypsin: Gene-
    ration of chemotactic activity in presence of serum-antipro-
    teases, in: "Neutral Proteases of Human Polymorphonuclear
    Leukocytes," K. Havemann and A. Janoff, eds., Urban and
    Schwarzenberg, Baltimore/Munich (1978).
15. E. F. Plow, Leukocyte elastase release during blood coagulation,
    J. Clin. Invest. 69:564 (1982).

HEPARIN AND PLASMA PROTEINASE INHIBITORS : INFLUENCE OF

HEPARIN ON THE INHIBITION OF THROMBIN BY α_2 MACROGLOBULIN

P. Lambin*, F. Pochon** and M. Steinbuch***

*Centre National de Transfusion Sanguine -75015 Paris
**U 49 ISERM Institut Curie Biologie -91405 Orsay
***Centre National de Transfusion Sanguine -91400 Orsay

INTRODUCTION

Heparin is a substance extracted from animal tissues, formed by acidic carbohydrate long chains polymers (amino sugar containing sulfated mucopolysaccharides). It is used in therapy for its powerful anticoagulant activity (1). In fact the mucopolysaccharide achieves a dramatical increase of the reaction velocity between proteinases of the clotting system and at least two plasma proteinase inhibitors : antithrombin III (2,3) and heparin cofactor II more recently described (4,5).

α_2 macroglobulin (α_2M) is also a major inhibitor of thrombin (6-8) but whereas several authors (9,10) observed the absence of an accelerating effect of heparin on the inhibitory activity of α_2M, the opposite effect has recently been described by Fischer et al. (11). These authors presume that it is the heparin/thrombin interaction which prevents the enzyme to be inhibited by α_2M. In previous studies we have quantified the interaction between thrombin and heparin and also between thrombin and α_2M (12). We give here new evidences that heparin prevents the thrombin/α_2M interaction.

EXPERIMENTAL PROCEDURES

$\underline{\alpha_2 \text{ macroglobulin}}$: α_2M was prepared by zinc chelate chromatography as described by Kurecki et al.(13) or by several Rivanol precipitations followed by DEAE chromatography as described by Steinbuch et al.(14). Both preparations gave similar results. The protein was homogeneous in polyacrylamide gradient gel electrophoresis.

Thrombin : The supernatant of the cryoconcentrate was used as
starting material. The prothrombin complex was adsorbed onto DEAE-
Sephadex (15) and eluted by 2 M NaCl. After dialysis and concentration to a protein content of 20 mg/ml, prothrombin was activated
by $CaCl_2$ (1/10 of the volume of the solution of 4 M $CaCl_2$). After
activation, the solution was dialysed against 0.025 M phosphate
buffer pH 6.7 and submitted to chromatography on SP-Sephadex as
described by Lundblad et al.(16). The purified enzyme had a specific activity of 1,550 NIH units/mg. Thrombin was labelled with
fluorescein isothiocyanate (FITC) according to a procedure previously
described (17). The dye labelled thrombin preparation (0.7 M FITC
per enzyme molecule) is fully active for BAPNA hydrolysis.
     Thrombin was also labelled with ^{125}I by using the Bolton
Hunter reagent (NEN) as mentioned earlier (18).

Heparin : Heparin (Vitrum) was used either as a commercial preparation or fractionated on Sephadex G 100 according to Jordan et al.
(19). The fraction of Mr 7,000 was used for the stoichiometric
studies. Heparin was labelled by fluorescamin according to Jordan
et al.(20). The fluorescence and polarization data were recorded
by excitation at 390 nm and emission at 500 nm. The polarization
value was computed as previously described (12).

Trypsin and chymotrypsin were puchased from Worthington

Electrophoretic procedures : Protein were analyzed by electrophoresis in linear gradients of polyacrylamide (3 to 20 % and 2 to 10 %).
The gels were prepared as already mentioned (21). Each protein
mixture was treated by 1 % of sodium docecyl sulfate (SDS) in
phosphate buffer 0.02 M pH 7.2 for 3 minutes in a 100° C bath.
Reduction of disulfide bonds was obtained by 1 % of 2-mercaptoetha-
nol. Polyacrylamide gradient gel electrophoreses (PGGE) were
carried out in phosphate buffer (0.01 M, pH 7.2) with 0.1 % SDS
for 16 hr under 40 V and protein stained by Coomassie brillant
blue R 250. In the absence of SDS the same gels were used but the
electrophoreses were performed in Tris-borate-EDTA buffer pH 8.2
(21).

RESULTS

α_2 macroglobulin - thrombin maximum binding ratio

     The α_2M molecule (Mr 720,000) is dissociated by SDS into two
subunits having the same molecular weight (360,000). When thrombin
is added to α_2M, complexes between the enzyme and the inhibitor
are formed. These complexes are not dissociated by SDS (22).
After PGGE in presence of SDS, protein bands of higher molecular
weight that the one of the subunit are observed (Fig. 1).

By adding increasing amounts of thrombin to α_2M from 0.25 to 2
molar ratios we can see that the proportions of these high molecular weight components increase from 0.25 to 1 molar ratios but
remain stable for higher concentration of thrombin.

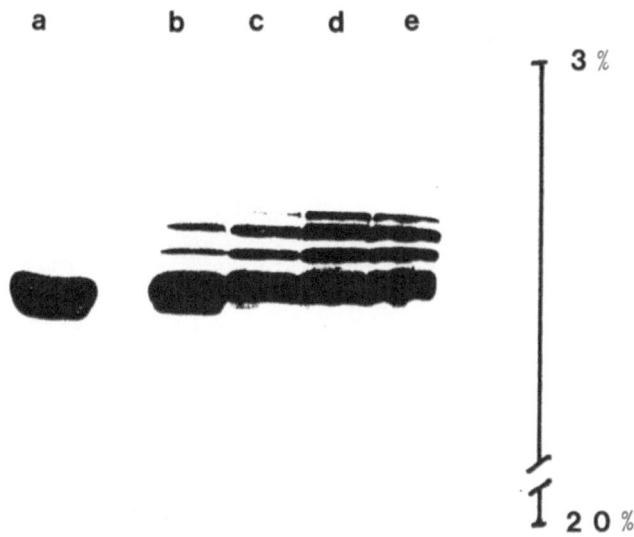

Figure 1 : FORMATION OF HIGH MOLECULAR WEIGHT COMPONENTS FOLLOWING
          THE α_2M/THROMBIN INTERACTION.
          Polyacrylamide gradient gel electrophoresis in SDS of
          a : α_2M subunit (Mr 360,000) b,c,d,e, : α_2M with increas-
          ing proportions of thrombin ; the thrombin/α M molar
          ratios are respectively 0.25 ; 0.5 ; 1 and 2.

     When α_2M is treated by SDS and 2-mercaptoethanol one polypep-
tide chain of 180,000 is observed in PGGE , α_2M being composed by
two equal subunits each consisting of two identical polypeptide
chains bound by disulfide bridges (23). After interaction with
thrombin a 90,000 fragment is obtained resulting from the proteolytic cleavage of α_2M chains in the "bait region" located in the
middle of the chains. The proportion of fragments increases proges-
sively up to a molar ratio α_2M/thrombin of 1/1, whereas further
addition of thrombin does not change the proportion of cleaved
chains. It must be mentioned that thrombin can split a maximum of
50 % of the α_2M chains.
     The covalent linkage of an optical label provides another
mean for determining the stoichiometry of the α_2M-thrombin complex

(17). A three molar excess of FITC-thrombin is mixed with α_2M.
The solution is left for 10 min at room temperature, and the unbound
thrombin is removed by gel filtration on Sephadex G 200. From the
molar concentrations of the dye at 495 nm to that of the protein
at 280 nm, we calculate that α_2M binds 0.80 thrombin molecule.
When ^{125}I thrombin is used instead of FITC thrombin a similar α_2M/
thrombin ratio is obtained. From these results we can conclude
that α_2M binds thrombin in a 1 : 1 molar ratio.

Binding of heparin to thrombin alone or complexed with α_2 M.

When increasing amounts of thrombin are added to fluorescamin
labelled heparin, the fluorescence polarization value (p) of the
labelled heparin increases progressively up to a molar ratio of 2
molecules of heparin (Mr 7,000) per 1 molecule of enzyme (Fig. 2).

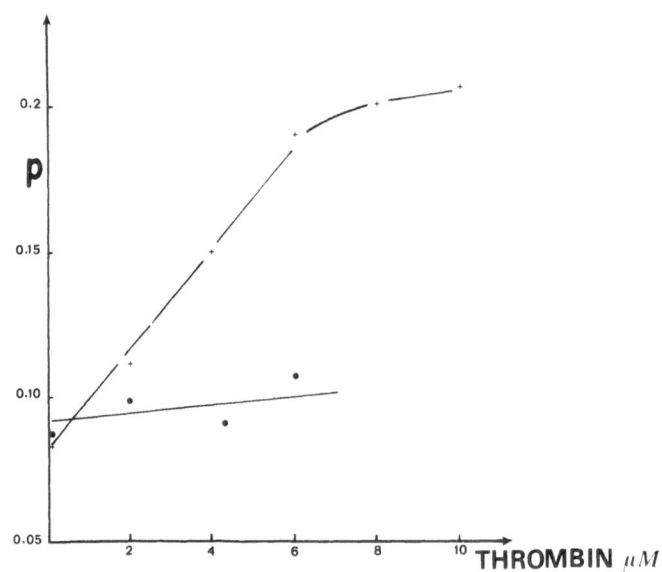

FIGURE 2 : BINDING OF FLUORESCAMIN LABELLED HEPARIN TO THROMBIN
          AND α_2M BOUND THROMBIN.

          To a 2.10^{-5}M fluorescamin labelled heparin solution
          in 50 mM Tris-HCl pH 8.0, thrombin +——+ is added at
          indicated concentrations. The fluorescence polarization
          values (p) are measured after 10 min at 15° C. The α_2M-
          thrombin complex .——. is added to the heparin solution
          and p is determined after 30 min of incubation at 25° C.

On the contrary when the same amounts of heparin are added to thrombin previously incubated with an excess of α_2M, the polarization value remains constant indicating that heparin cannot bind to α_2M complexed thrombin.

Influence of heparin on thrombin/α_2M complex formation.

Heparin can achieve the complete inhibition of α_2M/thrombin interaction. As shown by PGGE in SDS (Fig. 3) the formation of high molecular weight complexes between thrombin and α_2M is completely prevented by previous addition of heparin to thrombin.

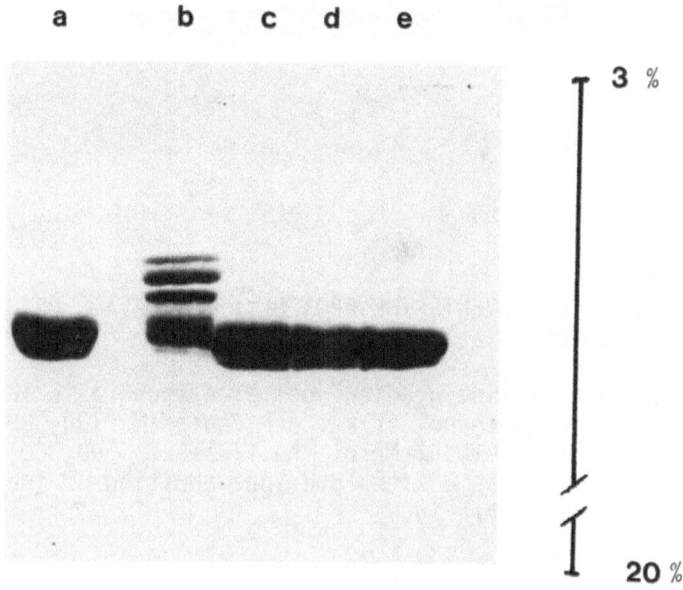

FIGURE 3 : POLYACRYLAMIDE GRADIENT GEL ELECTROPHORESIS IN SDS.

a : α_2M subunit (Mr 360,000); b : high molecular weight complexes appearing after interaction with thrombin; c and d : absence of complexes when thrombin has previously interacted with two different crude preparations of heparin; e : similar result with low molecular weight fraction of heparin.

Moreover after reduction of the disulfide bridges, no cleavage of the polypeptide chains of α_2M can be seen (Fig. 4). Thus the inhibition of the enzyme/inhibitor interaction concerns the first stage i.e. the proteolytic cleavage of α_2M by thrombin.

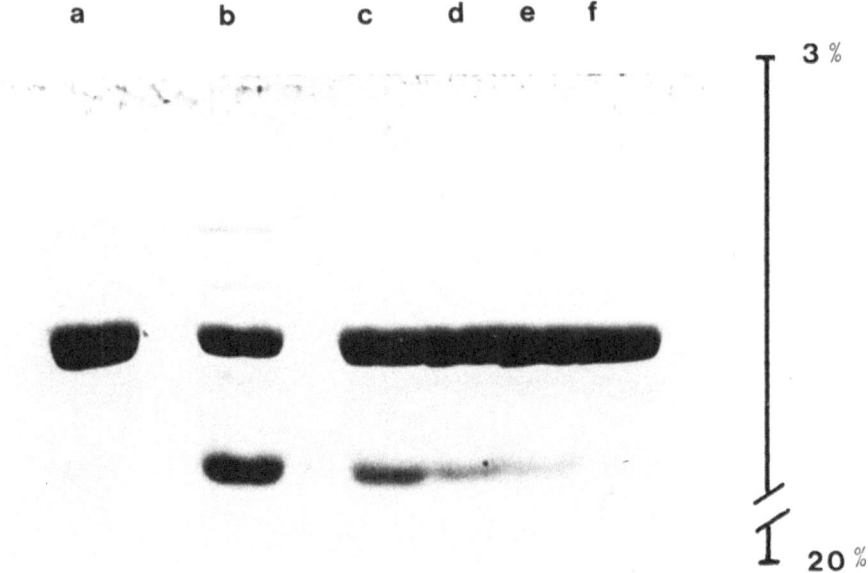

FIGURE 4 : HEPARIN PREVENTS THE PROTEOLYTIC CLEAVAGE OF α_2M CHAINS BY THROMBIN.

Polyacrylamide gradient gel electrophoresis in SDS and 2-mercaptoethanol of a : α_2M chain (Mr 180,000); b : α_2M and thrombin (Mr of the fragment : 90,000); c,d, e, f : the same after previous addition of increasing amount of heparin.

When PGGE is performed in absence of SDS no modification of the mobility of α_2M in the samples containing heparin-thrombin mixtures can be seen whereas thrombin alone induces the appearance of a faster α_2M band (Fig. 5).In addition the incorporation of ^{125}I thrombin to α_2M decreases proportionaly to the amount of added heparin. When a two molar excess of the 7 000 Mr heparin fraction is added to thrombin, a complete inhibition of the α_2M/ thrombin interaction is obtained.

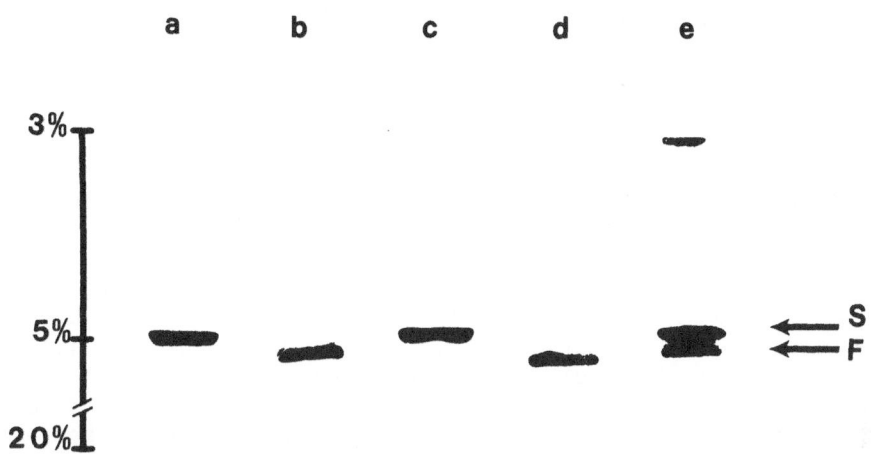

FIGURE 5 : POLYACRYLAMIDE GRADIENT GEL ELECTROPHORESIS IN ABSENCE
OF DENATURING AGENT.

a : native α_2M; b : α_2M and thrombin; c : α_2M with
thrombin/heparin complexes; d : α_2M and trypsin; e :
α_2M region of a serum sample. F : fast form of α_2M.
S : slow form of α_2M.

Similar results are obtained by spectroscopic methods. The
FITC-thrombin fluorescence at 495 nm is 20 % quenched by addition
of a 40 molar excess of heparin. Addition of a two molar excess of
α_2M does not change the fluorescence intensity of the thrombin-
heparin complex whereas addition of α_2M to FITC-thrombin alone
quenches the fluorescence emission by about 30 %.

Influence of heparin on chymotrypsin and trypsin/α_2M complex
formation.
Similar experiments were undertaken with 2 other enzymes :
chymotrypsin and trypsin.
In the case of chymotrypsin no influence of heparin (crude
preparation or low molecular weight fraction) on the α_2M enzyme
interaction could be observed even in the presence of a 10 fold

excess of the amount of heparin giving a clear inhibition of the α_2M/thrombin interaction. Complexes between α_2M and chymotrypsin were observed and the polypeptide chains of the inhibitor were cleaved. On the contrary, a clear inhibition of complex formation between α_2M and trypsin by crude preparations of heparin was observed.

DISCUSSION

Being a highly negatively charged molecule, heparin binds readily to many proteins under physiologic conditions and in particular to thrombin and other serine proteinases of the clotting system. Thus the possibility that heparin achieves a direct inhibition of thrombin has been studied by several authors : Albidgaard (24) has shown that heparin inhibits somewhat the thrombin/fibrinogen interaction and Longas et al (25) has suggested that in the presence of heparin the size of the substrate able to bind to the active site of thrombin may be restricted. Such a phenomenon may be involved as far as the absence of interaction between α_2M and the heparin/ thrombin complex is concerned. In contrast to such a steric hindrance theory of heparin/thrombin complexes versus large substrates, several authors reported that these complexes had a lower activity even for small synthetic substrates than the corresponding thrombin alone (26-28), but Griffith et al.(29,30) showed that the observed inhibition was either negligible or an artifact.

Our observations are thus in accordance with the hypothesis of steric hindrance preventing heparin/thrombin complexes to proteolyse α_2M.

In a similar way Danishefsky and Pixley (31) have shown that heparin interferes in the inhibition of thrombin by α_1 antitrypsin. The inhibitor does not appear to react directly with heparin and the action of heparin is due to the formation of thrombin-heparin intermediate that cannot combine with α_1 antitrypsin. On the contrary the rate constant of inhibition by antithrombin III and by heparin cofactor II is considerably increased in the presence of heparin (4,5, 19).

In conclusion, it appears clearly that heparin prevents inhibition of thrombin by two antiproteinases, α_2M and α_1 antitrypsin, whereas it increases reaction velocity between the same enzyme and both antithrombin III and heparin cofactor II.

REFERENCES

1. L. B. Jacques, Heparin-anionic polyelectrolyte drugs, Pharmacological Reviews 31:99 (1979).

2. U. Abildgaard, Highly purified antithrombin III with heparin cofactor activity prepared by disc electrophoresis, Scand.J. Clin. Lab. Invest. 21:89 (1968).

3. R. D. Rosenberg and P. S. Damus, The purification and mechanism of action of human antithrombin-heparin cofactor, J. Biol. Chem. 248:6490 (1973).

4. D. M. Tollefsen and M. K. Blank, Detection of a new heparin-dependant inhibitor of thrombin in human plasma, J. Clin. Invest. 68:589 (1981).

5. D. M. Tollefsen, D. W. Majerus, and M. K. Blank, Heparin cofactor II, J. Biol. Chem. 257:2162 (1982).

6. M. Steinbuch, C. Blatrix, and F. Josso, $\alpha_2$macroglobulin as progressive antithrombin, Nature 216:500 (1967).

7. A. Takada, T. Koide, and Y. Takada, Interaction of thrombin with antithrombin III and $\alpha_2$macroglobulin in the plasma, Thrombosis Res. 16:59 (1979).

8. A. M. Fischer, J. Tapon-Bretaudière, A. Bros, and F. Josso, Respective roles of antithrombin III and alpha-2-macroglobulin in thrombin inactivation, Thrombos. Haemostas. 45:51 (1981).

9. U. Abildgaard, Purification of two progressive antithrombins of human plasma, Scand. J. Clin. Lab. Invest. 19:190 (1967).

10. M. Steinbuch, C. Blatrix, and F. Josso, Action antiprotease de 1 α_2-macroglobuline II Son rôle d'antithrombine progressive Rev., Franç. Et. Clin. Biol. 13:179 (1968).

11. A. M. Fischer, A. Bros, S. Rafowicz, and F. Josso, Heparin prevents thrombin inhibition by alpha-2-macroglobulin, Thrombosis Res. 18:567 (1980).

12. F. Pochon, P. Lambin, and M. Steinbuch, Heparin and the progressive antithrombin activity of $\alpha_2$macroglobulin, Thrombosis Res. 26:307 (1982).

13. T. Kurecki, L. F. Kress, and M. Laskowski, Purification of human plasma α_2-macroglobulin and α_1-proteinase inhibitor using zinc chelate chromatography, Anal. Biochem. 99:415 (1979).

14. M. Steinbuch, M. Quentin, and L. Pejaudier, Specific technique for the isolation of human α_2-macroglobulin, Nature 205:1227 (1965).

15. J. Heystek, H. G. J. Brummelhuis, and H. W. Krijnen, Contributions to the optimal use of human blood. II. The large scale preparation of prothrombin complex, Vox. Sang. 25:113 (1973).

16. R. L. Lundblad, H. S. Kingdon, and K. G. Mann, Thrombin, in: "Methods in Enzymology," L. Lorand, ed., Academic Press, New York (1976).

17. F. Pochon, V. Favaudon, M. Tourbez-Perrin, and J. Bieth, Localization of the two protease binding sites in human $\alpha_2$macroglobulin, J. Biol. Chem. 256:547 (1981).

18. J. G. Bieth, M. Tourbez-Perrin, and F. Pochon, Inhibition of α_2-macroglobulin-bound trypsin by soybean trypsin inhibitor, J. Biol. Chem. 256:7954 (1981).

19. R. E. Jordan, D. Beeler, and R. D. Rosenberg, Fractionation of low molecular weight heparin species and their interaction with antithrombin, J. Biol. Chem. 254:2902 (1979).

20. R. E. Jordan, G. M. Oosta, W. T. Gardner, and R. D. Rosenberg, The binding of low molecular weight heparin to hemostatic enzymes, J. Biol. Chem. 255:10073 (1980).

21. P. Lambin, Reliability of molecular weight determination of proteins by polyacrylamide gradient gel electrophoresis in the presence of SDS, Anal. Biochem. 85:114 (1978).

22. M. Steinbuch, R. Audran, P. Lambin, and J. M. Fine, Subunit structure of human $\alpha_2$macroglobulin, in: "Protides of the Biological Fluids," M. Peeters, ed., Pergamon Press, Oxford (1976).

23. R. P. Swenson and J. B. Howard, Structural characterization of human α_2-macroglobulin subunits, J. Biol. Chem. 254:4452 (1979).

24. U. Abildgaård, Inhibition of the thrombin-fibrinogen reaction by heparin in the absence of cofactor, Scand. J. Haemat. 5:432 (1968).

25. M. O. Longas, W. S. Ferguson, and T. H. Finlay, Studies on the interaction of heparin with thrombin, antithrombin and other plasma proteins, Arch. Biochem. Biophys. 200:595 (1980).

26. S. M. Strukova, O. A. Semionova, and E. G. Kireeva, The influence of heparin and indol on the catalytic properties of α- and β/γ thrombins, Thrombosis Res. 20:563 (1980).

27. G. F. Smith, The heparin-thrombin complex in the mechanism of thrombin inactivation by heparin, Biochem. Biophys. Res. Comm. 77:111 (1977).

28. I. Danishefsky and R. Pixley, Heparin and protease inhibition. Heparin complexes with thrombin, plasmin and trypsin, Biochem. Biophys. Res. Comm. 91:103 (1979).

29. M. J. Griffith, H. S. Kingdon, and R. L. Lundblad, The interaction of heparin with human-α-thrombin: effect on the hydrolysis of anilide tripeptide substrates, Arch. Biochem. Biophys. 195:378 (1979).

30. M. J. Griffith, H. S. Kingdon, and R. L. Lundblad, Hydrolysis of N-α-benzoyl-L-phenylalanyl-L-valyl-L-arginine-p-nitroanilide by human alpha thrombin in the presence of heparin, Thrombosis Res. 17:83 (1980).

31. I. Danishefsky and R. Pixley, Effect of heparin on the inhibition of thrombin by α_1-proteinase inhibitor, Biochem. Biophys. Res. Comm. 91:862 (1979).

THE INVOLVEMENT OF PLASMATIC AND FIBRINOLYTIC SYSTEMS IN IDIOPATHIC GLOMERULONEPHRITIS (GN)

K. Andrassy, E. Ritz, and R. Waldherr

Departments of Internal Medicine and Pathology, University of Heidelberg, FRG

The concept of a hypothetical role of plasmatic coagulation and fibrinolysis in the genesis of glomerulonephritis (GN) is primarily based on the observation that the course of some, but not all, experimental models of GN is substantially altered by heparin or dicumarol (1, 2, 3). Although the evidence for involvement of plasmatic coagulation and fibrinolytic systems in patients with acute GN (e.g. poststreptococcal GN) is convincing, well documented evidence for their involvement in chronic idiopathic GN is scarce or absent. The controversy is not clarified by the observation that anticoagulation has no convincing effect on the course of several forms of idiopathic GN in controlled trials.

The present study reviews the literature and our own experience in these areas.

1. INVOLVEMENT OF PLASMATIC AND FIBRINOLYTIC SYSTEMS IN EXPERIMENTAL GN

The involvement of coagulation in the genesis of Masugi GN was first suggested by the observation of Silverskjold (1) that in experimental GN of rabbits given nephrotoxic horse serum <u>heparin</u> prevented oliguria, proteinuria and hematuria and mitigated histological lesions. Further evidence was presented by the study of Kleinerman (2) who showed that depot-heparin when given prior to and immediately following the injection of nephrotoxic serum inhibited the production of experimental nephritis in rabbits. No such inhibition was noted when heparin was administered after evidence of

nephritis was present. Vassalli et al. (4) induced intravascular clotting by administration of thrombin or
liquoid and observed swelling and proliferation of endothelial and mesangial cells with accumulation of hyaline or basal membrane-like material leading to glomerular obliteration as in glomerulonephritis. This was
variably accompanied by deposition of fully polymeri-
zed fibrin, as studied by electronmicroscopy. These lesions were prevented by either heparin or dicumarol.
Subsequent studies showed (5) that in a similar fashion
attenuation of proliferative lesions in rabbit Masugi
nephritis (produced by injection of sheep antirabbit
kidney serum) was seen when animals were anticoagulated with heparin or dicumarol. This was later confirmed
by Borero et al. (6). In the study of Humair (7) increased proteinuria and deposition of fibrin(ogen) related antigen (FRA) in glomeruli did not occur in mice
given heparin or urokinase prior to the injection of
liquoid or immunecomplexes and treated with anticoagulants thereafter.

Subsequent studies showed that similar efficacy of
anticoagulation is not observed in all forms of experimental GN. Border (8) was unable to demonstrate that
heparin in the maximal permissible dosage prevented
glomerular deposition of FRA or altered the progression
of experimental chronic immunecomplex glomerulonephritis of the NZB mouse respectively anti-GBM-GN in rabbits. Similar observations were made by Bone (9) and
Briggs (10) who found that heparin neither reduced fibrinogen or fibrin deposition nor improved the histological appearance in nephrotoxic serum nephritis induced in mice by rabbit antiserum. It appears that a
beneficial effect is found in milder and more variable
forms of anti-GBM-GN while little or no effect is found
in severe forms. The lack of efficacy of anticoagulation may be related, amongst others, on the participation of mechanisms of fibrin precipitation in the genesis of GN which are unrelated to classical thrombin
pathway and possibly related to local proteases. A similar lack of efficacy of anticoagulation was also observed in another model of GN, serum sickness nephritis of rabbits (11).

In contrast to the ambiguous results of heparin in
various models of GN, several reports found significant
protection by defibrination with ancrod in nephrotoxic
serum nephritis of rabbits (12-14). Defibrination with
ancrod provided protection, judged by histological and
functional criteria, both when animals were pretreated
or when the drug was administered during the autologous

phase. When further glomerular fibrin deposition was prevented by defibrination, deposited fibrin was rapidly removed indicating that glomerular fibrin clearing mechanisms were retained in this form of crescentic GN. This is also suggested by studies of glomerular fibrinolysis (see below). However, defibrination had no effect on glomerular C3-deposition or the amount of proteinuria.

Fibrinolytic therapy with urokinase decreased fibrinogen or fibrin in glomeruli but did not lessen proteinuria or progressive thickening of the glomerular basement membrane in nephrotoxic serum nephritis induced in mice by rabbit antiserum (10). Fibrinolytic activity of glomeruli of mice with IC-GN was markedly increased by treatment with urokinase (7). Normal glomeruli exhibited little fibrinolytic activity, when tested with the fibrin plate technique (16), although such activity could be demonstrated with somewhat modified techniques (17); in contrast, marked fibrinolytic activity could consistently be demonstrated in glomeruli of kidneys with GN (15).

In isolated rat glomeruli, low fibrinolytic activity was increased by infusion of thrombin and inhibited by simultaneous administration of epsilon-aminocaproic acid (18). This is in agreement with studies of Bernik and Kwaan (19) who were unable to demonstrate fibrinolytic activity in glomeruli of cultured renal tissue in the absence of thrombin pretreatment whereas glomerular fibrinolytic activity markedly increased after administration of thrombin or fibrin. In the study of Giroux (20), increased glomerular fibrinolytic activity measured with the fibrin plate technique or radioassay based on lysis of 125-fibrinogen, was consistantly found in experimental anti-GBM-GN. Such increase was related both to the severity of nephritis (crescent formation) and to the extent of fibrin deposits. It is of note that the persistance of fibrin in glomeruli can therefore not be related to abolished fibrinolytic activity. To the extent that the above studies were performed with plasminogen/fibrinogen plates, lytic activity seems to be "fibrinolysis" and not non-specific protease action, although inhibition studies to document this postulate have not been performed.

2. EVIDENCE FOR INTRAGLOMERULAR DEPOSITION OF FIBRIN IN GLOMERULI OF PATIENTS WITH GN
The first chemically specific demonstration of fibrin in GN came from immunohistological studies of Gitlin (21) and later Vassalli (4). However, with the excep-

tion of extracapillary GN (fig. 1), FRA - as evidence
of local intravascular clotting - is not regularly
found in GN unrelated to systemic diseases.

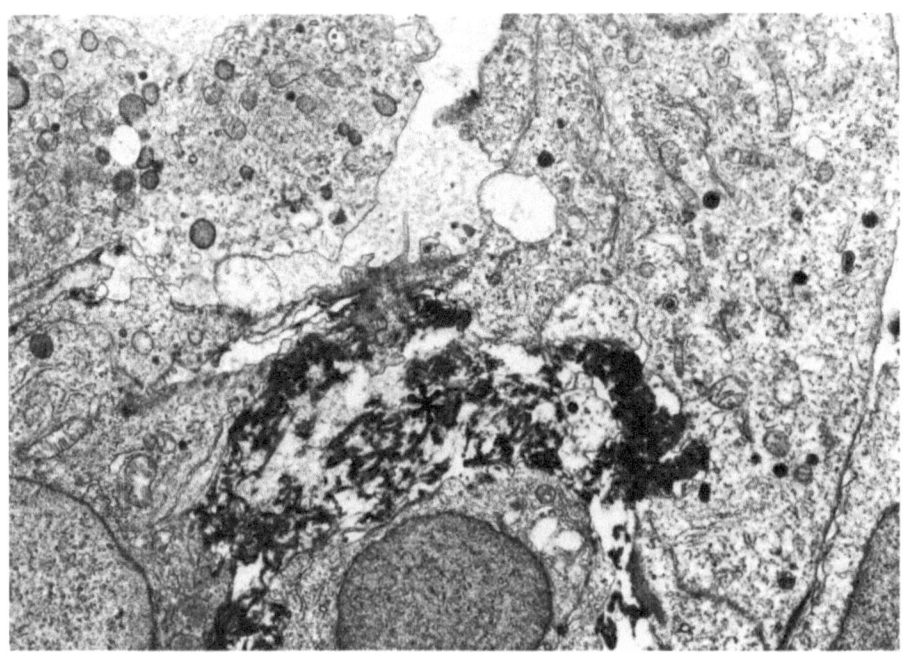

Fig. 1. Fibrin strands (*) between proliferated epithelial cells in a glomerular crescent. Electron
micrograph, x 5.950.

On electronmicroscopy, FRA deposits are usually not
accompanied by deposition of mature fibrin with its
characteristic periodicity but deposition of amorphous
fibrin strands. In contrast, FRA is frequently found
in GN secondary to lupus erythematosus or Schönlein
Henoch's purpura. Finally, FRA is easily demonstrable
in renal vasculature and/or glomeruli in various non-
GN renal diseases, e.g. transplant rejection, hemoly-
tic uremic syndrome, preeclampsia etc.
   Our experience with demonstration of FRA in various
forms of GN is listed in table 1.

A more sophisticated immunohistological approach by
Nagai and Kawai (22) made use of the neoantigen FDP-E
which is unmasked after interaction of fibrin with
plasmin (and possibly other proteases) and does not
cross-react with fibrinogen. This type of FDP-E neoantigen was found not only in most cases of GN (mini-

mal change, poststreptococcal GN, MPGN, Schönlein Henoch, SLE) but also in non-inflammatory glomerular disease (diabetes mellitus, hypertension etc.).

Table 1. Immunhistological demonstration of FRA with
         antifibrin(ogen) antibody (Behring Comp.,
         Marburg) in 139 GN patients

	Pat.with glomerular FRA
Minimal change GN (n=25)	4 (1)
Sclerosing GN (n=6)	4 (1)
Goodpasture's syndrome (n=5)	4 (2)
Wegener's syndrome (n=6)	6 (6)
Membranoproliferative GN (n=11)	4 (1)
Extramembranous GN (n=10)	5 (2)
Focal segmental GN (n=6)	2 (1)
Focal segmental sclerosing GN (n=8)	4 (0)
IgA-GN (n=28)	17 (5)
Mesangioproliferative GN (n=25)	7 (2)
(Interstitial nephritis) (n=9)	O

Marked reaction (+++) in parenthesis.

The relation to the glomerulonephritic process therefore remains uncertain. A clear dissociation of fibrinogen and FDP-E neoantigen was demonstrable in acute GN,
chronic GN and lipoid nephrosis. Another surprising
finding in various forms of GN was the failure to demonstrate f VIII antigen at the site(s) of fibrinogen
antigen, as one would expect for thrombin induced coagulation microclots (23). Such dissociation cautions
against equating glomerular FRA with the presence of
intraglomerular coagulation. Furthermore, in contrast
to reduced platelet half-life, as shown in several
forms of GN (24)
     Wardle et al. (25) failed to find changes of fibri-
nogen half-life in patients with GN. It must be admitted, however, that the sensitivity of the techniques
used may have been too low to detect subtle changes.

3. DEMONSTRATION OF ALTERED PLASMATIC SYSTEM OR FIBRI-
NOLYSIS IN THE CIRCULATION OF PATIENTS WITH GN

     Clearcut evidence of activated coagulation and sub-
sequent fibrinolysis was recently found in children
with poststreptococcal GN (26, 27). High molecular
weight fibrin derivatives, as demonstrated by fibrino-

gen gel chromatography, were detected in the circulation early in the disease. With resumption of diuresis,
variable plasma fibrinogen levels rose uniformly, the
concentration of high molecular weight fibrin derivatives decreased and fibrin degradation products (FDP-X)
were demonstrable in the circulation. This corresponds
to our experience in adult patients with poststreptococcal GN. Fig. 2 shows the clinical course of a patient with poststreptococcal GN. Soluble fibrin mono-
mer complexes (SFMC) could be demonstrated with the beta alanin precipitation/chromatography technique and
high molecular weight FDP with the PAGE technique (28).
Furthermore, antiplasmin-plasmin complexes with a characteristic neoantigen were detected with the Laurell-
technique in all four patients with acute poststreptococcal GN examined. Similar findings were not obtained
in patients with chronic GN.

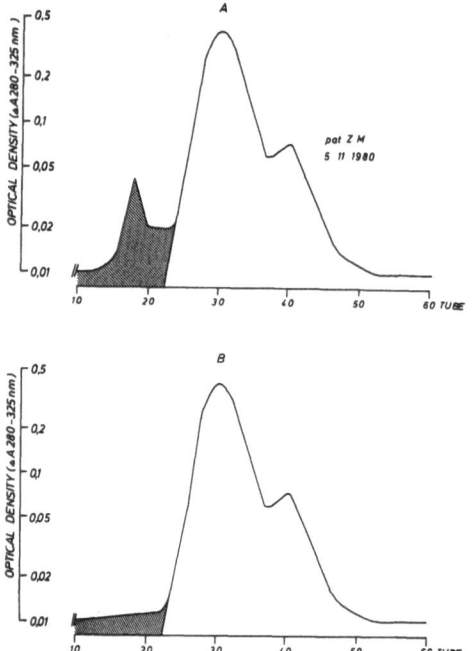

Soluble fibrin mo-
nomer complexes
in a patient with
poststreptococcal
GN
A) acute phase
B) after 3 months.
Note normalisa-
tion of SFMC.

Although previous findings were somewhat contro-
versial (29), the above measurements with more sophisticated techniques tend to confirm previous studies
(30, 31) which described increased systemic fibrinoly-

sis in acute GN, but are at variance with reports making similar observations in chronic GN (32). Tomura et al. (33) measured renal vein FDP in patients with GN. No increase of FDP in renal vein blood was found in non-nephrotic patients with chronic proliferative GN, but high concentrations of FDP, SFMC and fibrinogen were found in chronic proliferative GN with decreased renal function and/or nephrotic syndrome.

It must be admitted that evidence of intravascular coagulation (HMWFC and FDP) in acute GN is not necessarily evidence of thrombin-induced intravascular coagulation. Increased HMWFC and FDP may also result from the action on non-thrombin-proteases, e.g. leucocyte elastase which has been shown to cleave fibrinogen. f VIII and f XIII (34). Our own studies measuring serum elastase concentration with two dimensional electrophoresis suggest that this may indeed occur, at least in some patients. We found elevated elastase antitrypsin complexes in 1 of 4 patients with acute poststreptococcal GN (courtesy Dr. Egbring, Marburg).

4. FIBRINOGEN/FIBRIN SPLIT PRODUCTS IN URINE OF PATIENTS WITH GN

The demonstration of FPD in urine of patients with GN commanded great interest since FDP are known to promote inflammation (30). Furthermore, Naish et al. (31) proposed that fibrin and FDP played an important role in the genesis of proliferative GN and crescent formation. Indeed, with the hemagglutination inhibition test, FDP can be demonstrated in many patients with GN, particularly with proliferative GN (35). Such urinary FDP were variably characterized as high molecular weight and low molecular weight FDP. In some recent studies (34), an effort was made to differentiate between fibrinogen, HMWFC, FDP, fibrin monomer/fibrinogen/FDP complexes and fibrin. It has become clear that the hemagglutination inhibition technique does not differentiate these species. Particularly it does not permit differentiation between circulating FDP filtered into the glomerular filtrate, and degradation products resulting from proteolytic cleavage of fibrinogen filtered into the glomerulus. Fibrinogenuria does indeed occur in GN; this is demonstrated by chromatographic techniques and by precipitation of fibrin in proteinuric urine after addition of thrombin. Proteolysis of urinary fibrinogen can be due either to leucocyte proteases (e.g. in the glomerulus or in the urine) or due to uropeptidases (e.g. uropepsin, urokinase, cathepsin or elastase released from leucocytes). Fibrinogenolysis

and fibrinolysis can be distinguished by the presence
of fibrin D-dimers. The rationale of such distinction
is based on the molecular structure of fibrinogen/fib-
rin and is readily evident from the schema given in
fig. 3.

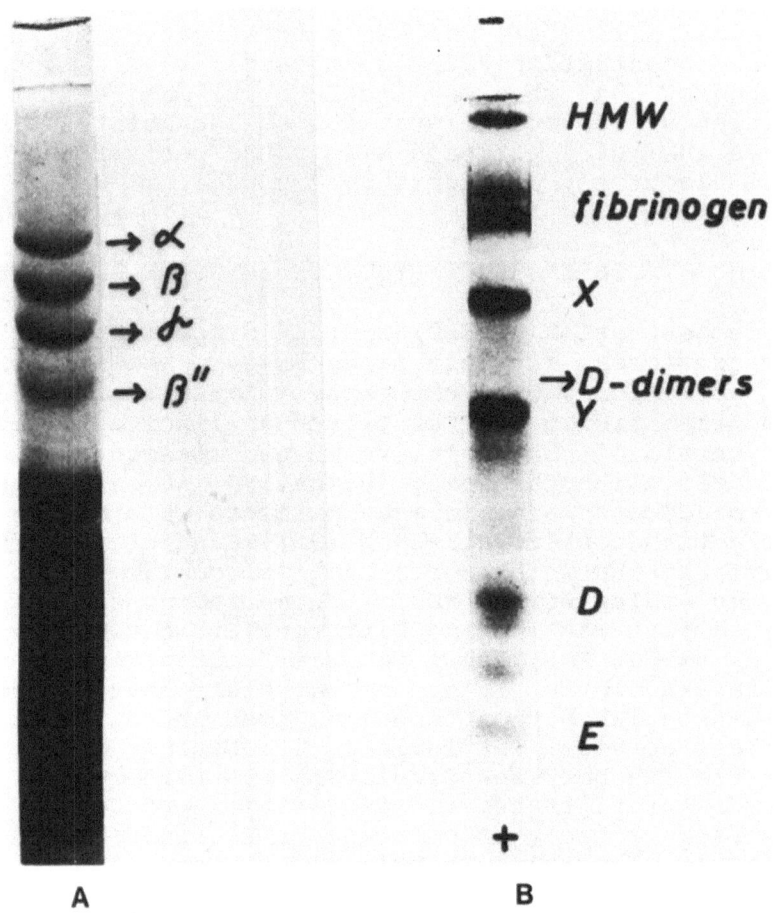

Fig. 3. Disc-gel electrophoresis of so-called "fibrin
split products" of a patient with nephrotic syndrome
(8 g protein/24 h; FDP excretion 380 µg/day). A Chain
analysis of beta-alanin precipitate of urine using SDS
gel analysis (7.5%; pH 7.6 in the presence of mercapto-

ethanol). Note after mercaptoethanol reduction presence
of monomeric alpha, beta- and gamma-polypeptide chains
of fibrinogen (beta" represents proteolytic product of
beta-chain), documenting fibrinogenuria. B SDS-gel analysis of beta-alanin precipitate without mercaptoethanol (5%, pH 7.6). Note presence of fibrinogen and various high molecular weight (X, Y) and low molecular
weight (D, E) FDP. There is no band at the expected position of D dimers (see arrow). D dimers would indicate
fibrinogen-fibrin conversion prior to lysis. No D dimers were demonstrable in the supernatant of the beta-
alanin precipitate step.

As shown in fig. 3, PAGE (or gelchromatography) does
permit distinction between fibrinogen and fibrin degradation products, since the presence of fibrin D-dimers
proves thrombin induced transformation of fibrinogen
to fibrin prior to fibrinolytic degradation.
Studies of Hall et al. (32) and of our own (36) show
that such dimers are demonstrable only in a minority of
patients with urinary "FDP" as measured with the hemag-
glutinine inhibition technique. But the extent to which
this finding reflects intraglomerular fibrinolysis remains undefined and requires further studies. Since
glomerular proteinuria is a reasonable index of the severity of glomerular disease, it is not surprising
that some correlation is found between clinical course,
histological indices of the severity of glomerular damage, immunohistological demonstration of FRA and urinary "FDP" as measured with hemagglutination inhibition. In the absence of the demonstration of D-dimers
the pathogenetic relevance of urinary "FDP" remains in
doubt.
There is no question, however, that in patients with
poststreptococcal GN one can demonstrate FDP-E with
immune precipitation or with PAGE and subsequent identification using anti-E-ab. This complements the demonstration of FDP in the circulation (26).

5. RESULTS OF ANTICOAGULATION IN PATIENTS WITH GN
Following the studies in experimental GN (1-4), a
number of uncontrolled clinical observations suggested
beneficial effects of heparin or warfarine on the clinical course of GN, particularly crescentic GN (37).
An early controlled study (38) failed to show a significant difference between anticoagulated and non-anti-
coagulated patients with rapidly progressive GN and no
convincing effects of anticoagulation was observed in
a later uncontrolled study (39).

One uncontrolled study suggested a beneficial effect
of long-term high dose heparin treatment in patients
with chronic proliferative GN (40). However, this result awaits confirmation by controlled studies.

The Australian working party in GN (41) used a prospective randomized trial and examined the use of warfarine in association with other modalities of treatment in membranous and mesangiocapillary GN. In patients with membranous GN, this regime caused less
proteinuria but did not modify renal function. In MCGN,
a similar reduction of proteinuria was observed, but
morbidity associated with therapy was prohibitive.
In the study of Cattran (42) no beneficial effect
whatsoever could be demonstrated by warfarine treatment
of patients with membranoproliferative GN.

It is evident from the above review that plasmatic
coagulation and fibrinolysis are involved in the genesis of acute GN. The evidence in chronic GN is less
convincing and results of therapeutic trials are frank-
ly disappointing. Further studies will have to determine whether our present techniques are not sensitive
enough to detect the involvement of plasmatic coagulation (and fibrinolysis) in chronic GN, or whether progression of chronic GN is related to mechanisms other
than coagulation.

REFERENCES

1. B. Silfverskjold, Heparin und experimentelle Glomerulonephritis, Skand.Arch.Physiol. 83: 175 (1940).
2. J. Kleinermann, Effects of heparin on experimental
   nephritis in rabbits, Lab. Invest. 3: 495 (1954).
3. B. Halpern, P. Milliez, G. Lagrue and A. Fray, Protective action of heparin in experimental immune
   nephritis, Nature 205: 257 (1965).
4. P. Vassalli, G. Simon and Ch. Rouiller, Electron
   microscopic study of glomerular lesions resulting
   from intravascular fibrin formation, Am.J.Physiol.
   43: 579 (1963).
5. P. Vassalli, R. Mc Cluskey, The pathogenetic role
   of the coagulation process in rabbit masugi nephritis, Am.J.Physiol. 45: 653 (1964).
6. J. Borrero, M. Todd, C. Becker and E. Becker, Masugi nephritis: the renal lesion and the coagulation
   processes, Clin. Nephrol. 1: 86 (1973).
7. L. Humair, H. Kwaan, and E. Potter, The role of fibrinogen in renal disease. II. Effect of anticoagulants and urokinase on experimental lesions in mice,
   J. Lab. Clin. Med. 74: 72 (1969).

8. W. Border, C. Wilson, F. Dixon, Failure of heparin to affect two types of experimental glomerulonephritis in rabbits, Kidney Int. 8: 140 (1975).

9. J. Bone, A. Valdes, F. Germuth, H. Lubowitz, Heparin therapy in anti-basement membrane nephritis, Kidney Int. 8: 72 (1975).

10. J. Briggs, H. Kwaan, E. Potter, The role of fibrinogen in renal disease. III. Fibrinolytic and anticoagulant treatment of nephrotoxic serum nephritis in mice, J. Lab. Clin. Med. 74: 715 (1969).

11. T. Baliah, K. Drummond, The effect of anticoagulation on serum sickness nephritis in rabbits. Proc. Soc. Exp. Biol. Med. 140: 329 (1972).

12. P. Naish, G. Penn, D. Evans and D. Peters, The effect of defibrination on nephrotoxic serum nephritis in rabbits. Clin. Science 42: 643 (1972).

13. N. Thomson, I. Simpson, and D. Peters, A quantitative evaluation of anticoagulants in experimental nephrotoxic nephritis, Clin. Exp. Immunol. 19: 301 (1975).

14. N. Thomson, J. Moran, J. Simpson, and D. Peters, Defibrination with ancrod in nephrotoxic nephritis in rabbits, Kidney Int. 10: 343 (1976).

15. M. Ekberg, M. Pandolfi, Origin of urinary fibrin/ fibrinogen degradation products in glomerulonephritis, Brit. Med. J. II: 17 (1975).

16. K. Andrassy, L. Buchholz, U. Bleyl and E. Ritz, Topography of human urokinase activity in renal tiss. Nephron 16: 213 (1976).

17. J. Bergstein, A. Michael, Cortical fibrinolytic activity in normal and diseased human kidneys, J. Lab. Clin. Med. 79: 701 (1972).

18. T. Watanabe, and K. Tanaka, The role of coagulation and fibrinolysis in the development of rabbit masugi nephritis, Acta Path. Jap. 26: 147 (1976).

19. M. Bernik, Increased plasminogen activator (urokinase) in tissue culture after fibrin deposition, J. Clin. Invest. 52: 823 (1973).

20. L. Giroux, P. Verroust, L. Morel-Maroger, F. Delarue, Glomerular fibrinolytic activity during nephrotoxic nephritis, Lab. Invest. 40: 415 (1979).

21. D. Gitlin, J. Craig, and C. Janeway, Studies on the nature of fibrinoid in collagen diseases. Can. Med. Ass. J. 88: 442 (1957).

22. T. Naigai, and C. Kawai, Fibrinogen and FDP-E-antigen deposits in nephrotic kidneys, Nephron 21: 16 (1978).

23. J. Hoyer, A. Michael, and L. Hoyer, Immunofluorescent localisation of antihemophilic factor antigen

and fibrinogen in human renal disease, J. Clin. Invest. 53: 1375 (1974)·

24. S. George, S. Slichter, G. Quadracci, and G. Striker, A kinetic evaluation of hemostasis in renal disease, New Engl. J. Med. 291: 1111 (1974).

25. E. Wardle, Fibrinogen catabolism studies in patients with renal disease. W. J. Med. XLII: 205 (1973)·

26. N. Alkjaersig, A. Fletcher, M. Lewis, and B. Cole, Pathophysiological response of the blood coagulation system in acute glomerulonephritis, Kidney Int. 10: 315 (1976).

27. Q. Maggiore, B. Jovanovic, G. Baldini, Plasma fibrinolytic hyperactivity in children with acute post-streptococcal glomerulonephritis, Nephron 6: 81 (1969).

28. R. Hafter, H. Graeff, in: Progr. Chem. Fibrin. Thrombolys. Vol. 2, J. Davidson, M. Samama, P. Desnoyers, eds, New York, Raven Press, 137 (1976)·

29. B. Goldschmidt, I. Marosvari, Das fibrinolytische Enzymsystem bei Kindern mit akuter Glomerulonephritis, Klin. Wschr. 46: 421 (1968).

30. A. Clarkson, M. Mac Donald, J. Petrie, and J. Cash, Serum and urinary fibrin/fibrinogen degradation products in glomerulonephritis, Brit. Med. J. II: 447 (1971).

31. P. Naish, D. Evans, and D. Peters, Urinary fibrinogen derivatives excretion and intraglomerular fibrin deposition in glomerulonephritis, Brit. Med. J. I: 544 (1974).

32. C. Hall, J. Blainey, and P. Gaffney, Origin of urinary fibrin-fibrinogen degradation products in renal glomerular disease, Nephron 23: 6 (1979)·

33. S. Tomura, T. Ideura, T. Ida and Y. Osaka, Renal vein fibrin degradation products (FDP's) in glomerulonephritis, Clin. Nephrol. 12: 248 (1979).

34. R. Egbring, W. Schmidt, G. Fuchs, and K. Havemann, Demonstration of granulocytic proteases in plasma of patients with acute leukemia and septicemia with coagulation defects, Blood 49: 219 (1977).

35. U. Hedner, Urinary fibrin/fibrinogen derivatives, Thromb. Diath. Haem. XXXIV: 693 (1975).

36. K. Andrassy, E. Ritz, Th. Mauerhoff, and J. Bommer, What is the evidence for activated coagulation in glomerulonephritis? Am. J. Nephrol. 103: 1982 (in press).

37. P. Kincaid-Smith, Anticoagulants in irreversibel acute renal failure, Lancet 2: 1360 (1968).
38. J. Cameron, A. Leathern, J. Suc, and J. Briggs, Are anticoagulants beneficial in the treatment of rapidly progressive glomerulonephritis? Proc. EDTA 10: 57 (1973).
39. P. Morrin, N. Hinglais, B. Nabarra, and H. Kreis, Rapidly progressive glomerulonephritis, Am. J. Med. 65: 446 (1978).
40. R. Cade, A. De Quesada, D. Shires, D. Levis, The effect of long term high dose heparin treatment on the course of chronic proliferative glomerulonephritis, Nephron 8: 67 (1971).
41. D. Tiller, A. Clarkson, T. Mathew, and N. Thrompson, A prospective randomized trial in the use of cyclophosphamide, dipyridamole and warfarin in membranous and mesangial capillary glomerulonephritis, Proc. 8th Int. Congr. Nephrol., Athen 345, 1981, Karger, Basel.
42. D. Cattran, R. Charron, C. Cardella, J. Rosceo, Controlled trial on mesangiocapillary glomerulonephritis. Proc. 8th Int. Congr. Nephrol., Athen, 287, 1981 , Karger, Basel.

THE EFFECT OF APROTININ ON PLATELET FUNCTION, BLOOD COAGULATION AND BLOOD LACTATE LEVEL IN TOTAL HIP REPLACEMENT - A DOUBLE BLIND CLINICAL TRIAL

S. Haas, R. Ketterl[+], A. Stemberger, P. Wendt,
H.-M. Fritsche[++], H. Kienzle[++], F. Lechner[++],
and G. Blümel
Institute for Experimental Surgery, [+]Surgical
Clinic of the Technical University Munich
[++]Garmisch-Partenkirchen Hospital, FRG

Patients undergoing alloarthroplastic hip replacement are frequently endangered by pulmonary, cardiovascular and thrombotic complications occuring during and
after surgery. This is due to an activation of both the
humoral and cellular blood coagulation by the intravasat-
ion of thromboplastic material. In this context, the reactivity of platelets plays a specific role; e.g. Schlag
et al. have shown by electron microscopical studies of
lung biopsies from shock patients that besides neutrophil
granulocytes, platelets are also trapped in the capillary
and precapillary system[19]. According to Bergentz et al.,
reversible platelet aggregates in the pulmonary circulation are the first morphological substrate of a disturbance
of microcirculation in shock. The resulting pulmonary
failure is not only due to a mechanical obstruction of
the blood vessels in the lung, but rather to the release
reaction of aggregated platelets. These released substances are mainly ADP, serotonin, histamine, catechola-
mines, prostaglandins, hydrolases, and the platelet factors 3 and 4. Thus vasoconstriction, an increased permeability, and procoagulant activity are initiated[1,2]. Even
during an operation, as for example, total hip replacement
and other major bone surgery, an instant stimulation of
the platelet function has been observed in patients[11].
Modig et al. have reported that major bone surgery can
lead to pulmonary complications which are very similar to
those of a "shock lung". An acute pulmonary dysfunction
with an increased vascular resistance in the lung, a con-
striction of the bronchioli and a decrease of PaO_2, have

been seen in patients with total hip replacement[15].

Hyperreactive platelets may be responsible for these pulmonary complications. Therefore, substances affecting certain platelet functions without causing bleeding risks are of great clinical interest. One of these substances seems to be aprotinin (Trasylol[R], Bayer Pharma, Leverkusen, FRG), which has a stabilizing effect on platelets in vitro [7,9]. Referring to this, the effects of a single preoperative prophylactic infusion of Trasylol[R] upon cellular and humoral blood coagulation and blood lactate were studied in patients with total hip replacement under double blind conditions.

MATERIAL AND METHODS

32 patients of either sex, aged between 60 and 75 years and presenting a normal metabolic condition were admitted to our study. The patients underwent an allo-arthroplastic total hip replacement with cemented prostheses and the assignment of the patient was performed by means of a hidden random list. Three times 5000 IU Heparin were given daily as a prophylaxis of thromboembolism, starting from two and three hours before operation. After the induction of anaesthesia, the first blood sample was taken to determine the preliminary values and afterwards a single dose of 20 000 KIU/kg BW Trasylol A or B were infused intravenously within 15 minutes. Further blood samples were taken:
10 minutes after Trasylol A or B infusion, at the moment of the preparation of the femoral shaft, one, two, six and finally 24 hours after Trasylol A or B infusion.
All blood samples were tested for the following parameters:
1) Thromboelastogram; the thromboelastograph from Hellige, Freiburg, FRG, was used.
2) Availability of factor X a following the activation of the endogenous clotting cascade. The test was performed by means of the chromogenic substrate S-2222, Deutsche Kabi Vitrum Munich, FRG.
3) Partial thromboplastin time (PTT); the reagents from Merz and Dade, A.H.S., Munich, FRG, were used.
4) Platelet number; the TOA-platelet counter from Colora Meßtechnik, Lorch, FRG, was used.
5) Ratio of platelet aggregates, according to WU and HOAK [20].
6) Platelet adhesiveness, according to MORRIS[16].
7) Platelet aggregability of platelet rich plasma (PrP) according to BORN[5]. The aggregation was induced by ADP (33 μl of a 5×10^{-5} M solution/300 μl PrP).

8) <u>Spontaneous platelet aggregation</u> according to BREDDIN
et al.[6,12]. 300 µl PrP are rotated at 20 rpm for 10 min,
and the increasing light transmission due to platelet
aggregation is recorded.
9) <u>Blood lactate</u>; the reagents from Boehringer Mannheim,
FRG, were used.

RESULTS

     The reaction time r of the thromboelastogram is the
time needed for the beginning of coagulation after recal-
cification of the citrated blood and this describes the
coagulability of the blood. The coagulation is accelerated
in the placebo- and retarded in the active substance group
(fig. 1). The k-value of the thromboelastogram is the time
needed for the development of a predefined clot strength
and thus describes the velocity of thrombus formation, it
is also accelerated in the placebo- and retarded in the
active substance group.

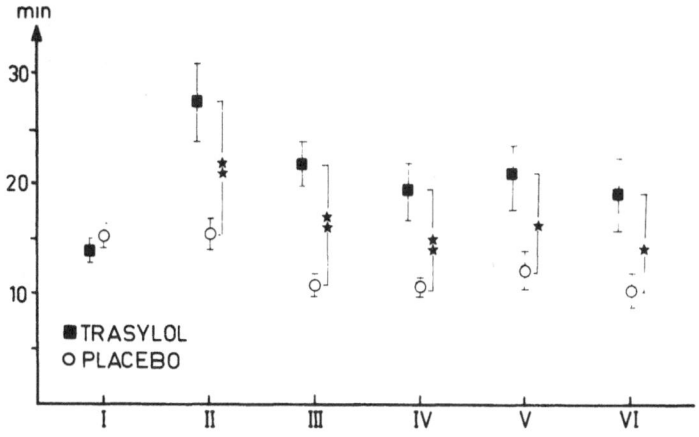

I= Induction of anesthesia   IV=1h following short infusion
II = 10 min following short infusion V = 2h following short infusion
III= Preparation of the femur   VI= 6h following short infusion

Fig. 1: Thromboelastogram reaction time "r" in total hip
replacement; *= $p < 0.05$; **= $p < 0.01$

The maximum amplitude of the thromboelastogram is
mainly influenced by the concentration and quality of
fibrinogen as well as the platelet count and reactivity;
it is significantly elevated in the placebo group and
lowered in the Trasylol-group (fig. 2).

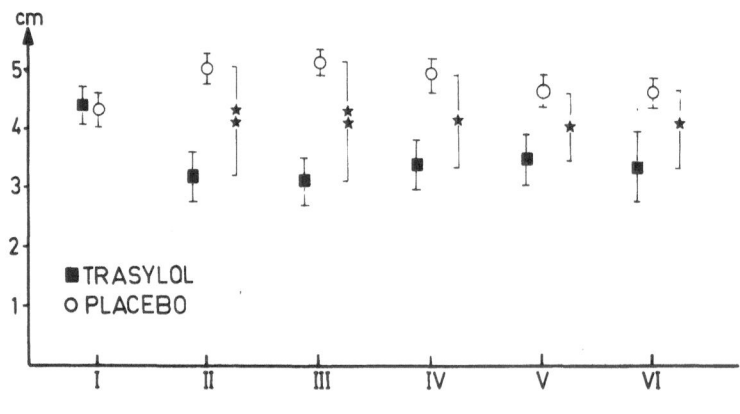

I= Induction of anesthesia   IV=1h following short infusion
II = 10 min following short infusion   V = 2h following short infusion
III= Preparation of the femur   VI= 6h following short infusion

Fig. 2: Thromboelastogram maximal amplitude in total hip
replacement; *= $p < 0.05$; **= $p < 0.01$

    The availability of factor X a following the activat-
ion of the endogenous clotting cascade was determined with
the chromogenic substrate S-2222. It is unchanged in the
placebo- and strongly reduced in the Trasylol group -
probably to the inactivation of plasma kallikrein (fig. 3).
The PTT which describes the activity of the whole endogenous blood clotting cascade following activation of factor XII follows the same pattern. The consumption of platelets during surgery is far more pronounced in the placebo
group (fig. 4). The ratio of platelet aggregates according
to Wu and Hoak[20] indicates the amount of circulating platelet aggregates. The range between 1 and 0 corresponds
to no aggregation at all respectively 100 % of platelets
in form of aggregates. 0.5 would indicate that 50 % of
circulating platelets are aggregated. The normal range
is between 0.8 and 1. During surgery an elevation of the
fraction of circulating platelet aggregates is demonstrable
in the placebo group whereas the opposite was seen in the
Trasylol[R] group (fig. 5).

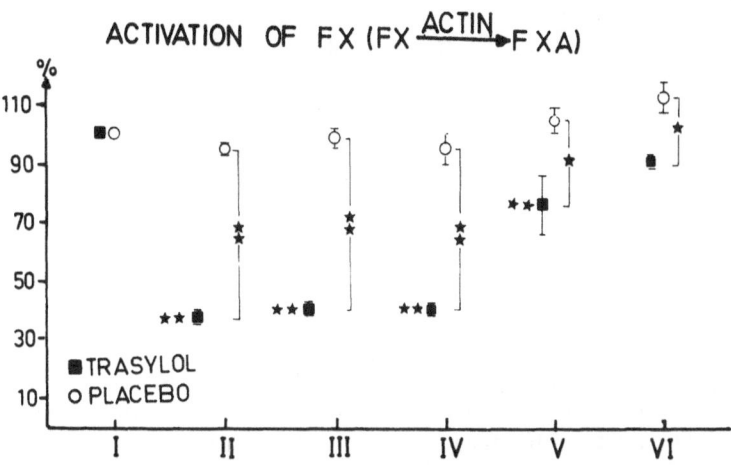

Fig. 3: Availability of endogenously activated F X a in total hip replacement; *= p< 0.05; **= p< 0.01

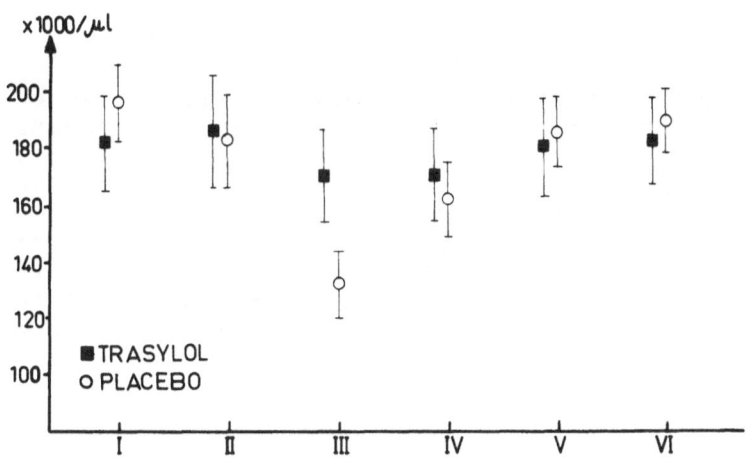

Fig. 4: Platelet number in total hip replacement

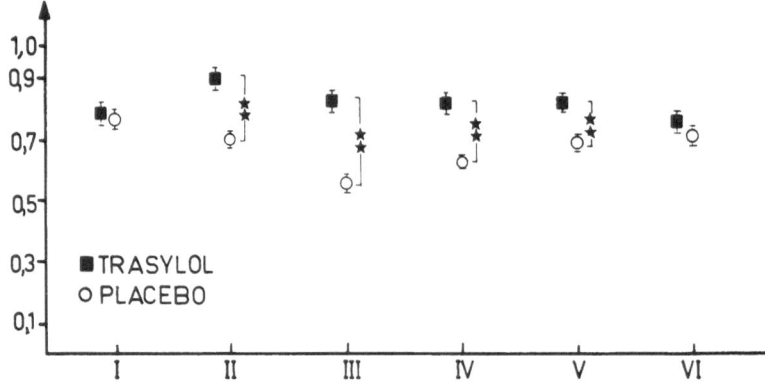

I = Induction of anesthesia   IV = 1 h following short infusion
II = 10 min following short infusion   V = 2 h following short infusion
III = Preparation of the femur   VI = 6 h following short infusion

Fig. 5: Ratio of platelet aggregates in total hip replacement; ** = p < 0.01

        Platelet retention upon contact with glass beads
peaks intraoperatively which indicates an increase in reactivity in the placebo group. In contrast, shortly after
dosage of Trasylol[R], a marked decrease occurs which lasts
throughout surgery - with a tendency towards normalization
after six hours (fig. 6). The maximum amplitude of the
ADP-induced platelet aggregation according to Born corresponds to the amount and the sizes of induced platelet
aggregates. In comparison with the starting values an enhanced reactivity of platelets upon contact with ADP is
demonstrated in the placebo group, whereas the number of
reacting platelets is considerably lower in the active
substance group (fig. 7).

        In addition, the velocity of aggregate formation was
significantly reduced in the Trasylol[R] group which was
shown by a lower angle of the curve in the Trasylol[R]
group (fig. 8).

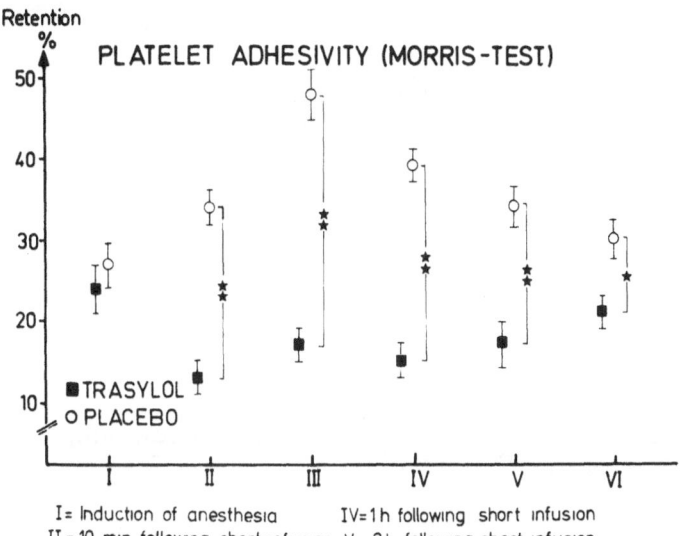

Fig. 6: Platelet adhesivity in total hip replacement;
*= p <0.05; **= p <0.01

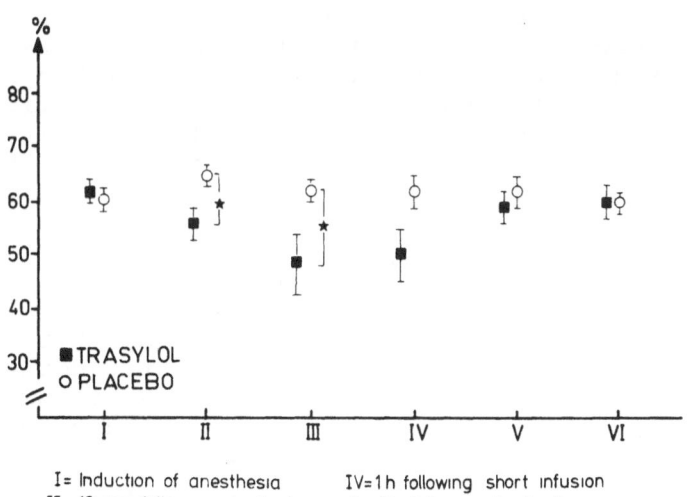

Fig. 7: ADP-induced platelet aggregation maximal amplitude in total hip replacement; *= p <0.05

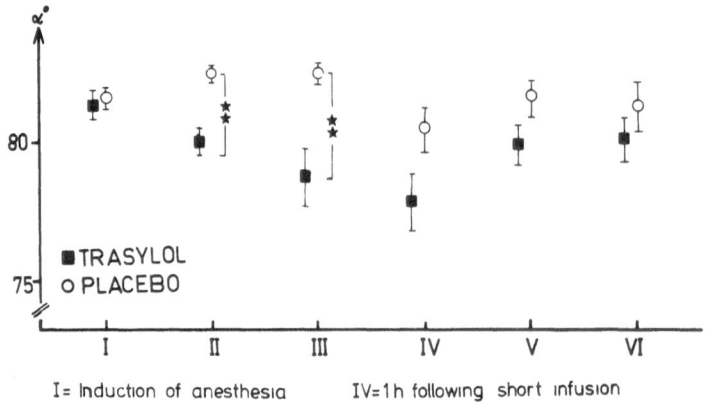

I= Induction of anesthesia   IV=1h following short infusion
II = 10 min following short infusion   V = 2h following short infusion
III= Preparation of the femur   VI= 6 h. following short infusion

Fig. 8: ADP-induced platelet aggregation angle   in total
hip replacement; **= p < 0.01

        Similar effects were obtained with the test of spon-
taneous aggregation. It was significantly enhanced in the
saline group and significantly reduced by aprotinin.

        The blood lactate content - descriptive for the qua-
lity of microcirculation - was followed over 24 h. In the
placebo group the lactate level shows a highly significant
increase to upper normal values with only a slight tendency to normalize even after 24 h. In contrast, the pattern of the Trasylol[R] group looks rather like that of a
non-operated control group.

DISCUSSION

        Summarizing our findings, we can say that the most
critical moment in total hip replacement seems to be the
preparation of the bone marrow in the femur shaft, because
then we found the minimal platelet count and the maximum
of the platelet adhesiveness and aggregability. Aprotinin
in high doses is able to reduce the activation of the
platelet function in these cases. In addition, aprotinin
has an effect on the endogenous system of blood coagulation, as can be seen by the prolongation of the PTT and
the decrease of the endogenously activated factor X a.

This effect may be due to an inhibition of plasma kallikrein, which is known to be a potent activator of the early phase of blood clotting[8]. The above findings concerning platelet function are in agreement with the literature. McMichan et al. have shown that pulmonary insufficiency in patients with multiple trauma is associated with a very distinct and critical fall of platelets, it can be prevented by aprotinin to a certain extent[14]. The important role of the lung as a filter for platelet aggregates after multiple trauma was also shown by Peer and Schwartz[17]. Also, experimental studies in animals have given evidence that aprotinin is able to diminish the trapping of thrombocytes in the lung after polytrauma and burn injury[3,4,13].

Clinically, it is important to state that this effect of aprotinin is reversible. This was found by Harke and Gennrich[10] and Reuter[18]. The underlying principle seems to be an interaction between the substance and membrane bound enzymes. Therefore, aprotinin does not act as other antiaggregating agents do and it can be given without having to fear bleeding risks.

Concluding we can say the significant moderation of the humoral and cellular blood clotting systems and the improvement of the postoperative microcirculation, which is shown by a complete inhibition of the well known postoperative blood lactate increase, suggests the prophylactic application of proteinase inhibitors in order to lower the surgical risks in especially vulnerable patients.

REFERENCES

1. S. E. Bergentz, D. Lewis und U. Ljungquist, Die Lunge im Schock: Thrombozytenanhäufung nach Polytrauma und intravasale Gerinnung, Langenbecks Arch. Klin. Chir. 329:658 (1971).
2. S. E. Bergentz, D. Lewis and U. Ljungquist, Trapping of platelets in the lungs after experimental injury, in: "Microcirculation Approaches to Current Therapeutic Problems," Symp. 6th Eur. Conf. Microcirculation, Aalborg, Karger, Basel (1971).
3. F. W. Blaisdell, R. C. Lim, and R. J. Stallone, The mechanism of pulmonary damage following traumatic shock, Surg. Gynecol. Obstet. 130:15 (1970).
4. F. W. Blaisdell, and R. N. Schlobohm, The respiratory distress syndrome: a review, Surgery 74:251 (1973).

5. G. V. R. Born, and U. J. Cross, The aggregation of platelets, J. Physiol. 168:178 (1963).

6. K. Breddin, H. Grun, H. J. Krzywanek und W. P. Schremmer, Zur Messung der spontanen Thrombozyten-aggregation. Plättchenaggregationstest II. Methodik, Klin. Wschr. 53:81 (1975).

7. S. Haas-Denk, W. Kaunzner und S. v. Sommoggy, Beeinflussung der Thrombozytenfunktion und der Gerinnung von Konservenblut durch Azetylsalizylsäure und Aprotinin, Med. Welt 28:912 (1977).

8. S. Haas, I. Wriedt-Lübbe und G. Blümel, Mechanismus der Fibrinolyseaktivierung und Fibrinolysehemmung, Med. Welt 29:209 (1978).

9. H. Harke, Beeinflussung der Mikroaggregation in lagernden Blutkonserven, Anaesthesist 25:347 (1976).

10. H. Harke und M. Gennrich, Aprotinin-ACD-Blut. Experimentelle Untersuchungen über den Einfluß von Aprotinin auf die plasmatische und thrombozytäre Gerinnung, Anaesthesist 29:266 (1980).

11. R. Ketterl, S. Haas, H.-M. Fritsche, F. Lechner und G. Blümel, Veränderungen der Thrombozytenmorphologie und der Thrombozytenfunktion im Verlauf von Hüft-Totalendoprothesen-Operationen, Med. Welt 31:743 (1980).

12. H. J. Krzywanek, and K. Breddin, Primary shape change of thrombocytes and platelet aggregation after administration of acetylsalicylic acid, Proc. VI Int. Congr. Thromb. Haem., Thromb. Haemost. 38: 251 (1977).

13. D. Loew, K. Breddin, H. Flenker und G. Schnells, Blutplättchen, Gerinnungsfaktoren und morphologische Organveränderungen nach Verbrühungsschock beim Rhesus-Affen, Med. Welt 25:50 (1974).

14. D. S. McMichan, J. Rosengarten, C. McNeur und R. Philipp, Das posttraumatische Lungen-Syndrom. Definition, Diagnose und Therapie, Med. Welt 27: 2331 (1976).

15. J. Modig, S. Olerud, and P. Malmberg, Sudden pulmonary dysfunction in hip replacement surgery, Acta Anaesthesiol. Scand. 17:276 (1973).

16. C. D. W. Morris, Observation on the effect of glass beads on platelet aggregation and its relation to platelet stickiness, Thrombos. Diathes. Haemorrh. 20:345 (1968).

17. R. M. Peer, and S. I. Schwartz, Development and treatment of posttraumatic pulmonary platelet trapping, Ann. Surg. 181:447 (1975).

18. H. D. Reuter, The stabilizing effect of trasylol on platelet membranes, Proc. VII Int. Congr. Thromb. Haem., Thromb. Haemostas. 42:298 (1979).

19. G. Schlag, U. H. Voigt, G. Schnells und A. Glatzl,
     Die Ultrastruktur der menschlichen Lunge im
     Schock. I., _Anaesthesist_ 25:512 (1976).
20. K. K. Wu, and J. C. Hoak, Measurement of platelet
     aggregates in whole blood, _Lancet_ 19:924 (1974).

INTERACTION OF GRANULOCYTE PROTEASES WITH INHIBITORS

IN PULMONARY DISEASES

Kjell Ohlsson, Ulla Fryksmark, Mats Ohlsson
and Hans Tegner
Departments of Surgery, Otorhinolaryngology
and Experimental Research, Malmö General
Hospital, S-214 01 Malmö, Sweden

INTRODUCTION

Sputum granulocytes were implicated as the source
of proteolytic activity found in purulent bronchial
secretions already by Opie.[1] [2] Final proof that a bulk
of these proteases is of granulocytic origin was presented more recently, based on immunochemical studies
utilizing specific antisera against the granulocyte
proteases.[3] The total concentration of these different
granulocyte proteases is extremely high in purulent
bronchial secretions, approaching a level of 0.5-1 g/l.[3] [4]
In such secretions the proteases are present in a free,
active form as well as in complex with the inhibitors.[3]
It seems reasonable to propose that these proteases
are involved in the extensive tissue destructions seen
in bronchiectasis, which is a rather common complication of chronic purulent bronchitis. Furthermore, pulmonary emphysema is currently thought to be due to a
protease-antiprotease imbalance in the lung with uncontrolled digestion of the lung tissue, caused by
granulocyte proteases - particularly elastase.[5] This
hypothesis is supported by the fact that persons with
inherited deficiency of α_1-antitrypsin, the main inhibitor of granulocyte elastase,[6] show an increased
susceptibility to emphysema.[7] Furthermore, recent
results indicate that the emphysema in cigarette smokers might be due in part to local suppression of α_1-
antitrypsin and also antileukoprotease in lung by oxidative agents present in cigarette smoke.[8] [9] In addition, granulocyte homogenates[10] or purified granulo-

299

cyte elastase[11] cause emphysema when administered intratracheally to experimental animals. Macrophage elastase
seems to be a less likely candidate to cause emphysema
- at least as judged from the indirect evidence that
this enzyme is poorly inhibited by α_1-antitrypsin.[12]

These different lines of evidence for the involvement of granulocyte proteases in pulmonary diseases
have focused our interest on the defence mechanism of
the lung against granulocyte proteases.

The purpose of the present paper is to summarize
some of our recent results concerning the interactions
between granulocyte elastase, antileukoprotease and the
major plasma inhibitors of granulocyte proteases, α_1-
antitrypsin and α_2-macroglobulin, as they appear from
in vitro and in vivo studies. In addition, some of our
earlier results are presented as a background - especially those concerning antileukoprotease.

HUMAN NEUTRAL GRANULOCYTE PROTEASES AND THEIR INHIBITORS

The major proteases of human PMN-granulocytes and
their endogenous inhibitors, identified so far, are
compiled in Table 1. The daily normal turnover of neutral protease, elastase and cathepsin G has been estimated to about 1 g[13].   This normal turnover of granulocyte proteases engages to some extent the plasma protease inhibitors, as judged from the demonstration of
granulocyte elastase in serum by radioimmunoassay.[14]
We have also found neutral protease[15] in normal plasma
with the aid of a newly developed enzyme immunoassay.
Both enzymes are identified in complex with α_1-anti-
trypsin, which indicates that they are both released
in an active form from the granulocytes. The normal
extracellular turnover of granulocyte proteases is
greatly enhanced in inflammatory processes, as judged
from increased plasma levels and the tremendous local

Table 1.   Major endogenous inhibitors of granulocytic neutral
proteases

NAME	MOL.WT.	INHIBITION OF			
		COLLAGENASE	ELASTASE	NEUTRAL PROTEASE ("COLLAGENASE")	CATHEPSIN G (CHYMOTRYPSIN-LIKE CATIONIC PROTEIN)
		METALLO	SERINE	SERINE	SERINE
α_1-ANTITRYPSIN	55,000	-	++	+	(+)
ANTICHYMOTRYPSIN	69,000	-	-	-	+
α_2-MACROGLOBULIN	725,000	++	+	++	++
β_1-ANTICOLLAGENASE	40,000	+	-	-	-
ANTILEUKOPROTEASE	10,500	-	++	-	+

accumulation of granulocyte protease.[3][4] α_2-Macroglobu-
lin-bound proteases are not identified by radioimmunoassay because of steric hindrance for the antibody to
reach the α_2-macroglobulin-bound protease.[14] The levels
of α_2-macroglobulin protease complexes in plasma are,
however, judged to be rather low because of the short
half-life of these complexes in the circulation.

   The circulating inhibitor system is complimented
by potent inhibitors against granulocyte elastase and
cathepsin G in human mucous secretions. These inhibitors are of low molecular weight ($\sim 11,000$) and have
been purified from seminal plasma,[16] cervical mucus[17]
and bronchial secretion.[18] Based on identical protein
characteristics, antienzymatic properties and immunological reactions, we concluded that the same inhibitor was present in all these mucous secretions and
suggested the term 'antileukoprotease' for this inhibitor, which is not related to the inter-α-trypsin inhibitor.[19] We have later, using our antibody prepared
against antileukoprotease isolated from bronchial secretions, identified the inhibitor in bronchial and
maxillary sinus mucosa and secretions,[20] cervical mucosa and secretions,[21] middle ear mucosa and secretion[22] and also in the parotid and mixed saliva
(Table 2). In contrast with an earlier report,[23] we
also could identify it in serous cells of the parotid
(Fig. 1) and submandibular glands using a procedure ·
based on the PAP-technique. But in agreement with a
preliminary finding by the same group,[23] we could demonstrate the presence of antileukoprotease in secretory cells in the epithelium of the bronchioles(Fig. 2).
This certainly indicates that the inhibitor also plays
a role in the distal respiratory tract. In addition,
we[24] and another group[25] have found in contradiction with

Table 2.   Antileukoprotease concentration in
           mixed and parotid saliva from healthy
           individuals (N = 30)

	Antileukoprotease mg/l Mean ± SEM
Mixed saliva	0.90 ± 0.06
Parotid saliva	3.80 ± 0.10

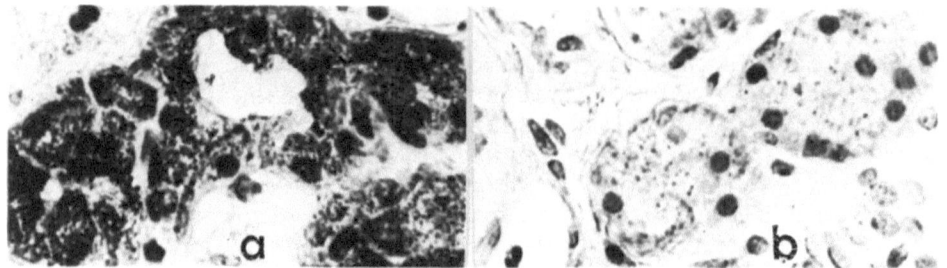

Fig. 1 a and b.   a) Peroxidase immuno staining for antileukoprotease in the serous cells of
the parotid gland. b) Section of parotid gland incubated with rabbit anti-
human antileukoprotease absorbed with
antileukoprotease. PAP-technique.
Antiserum dilution 1:1000.

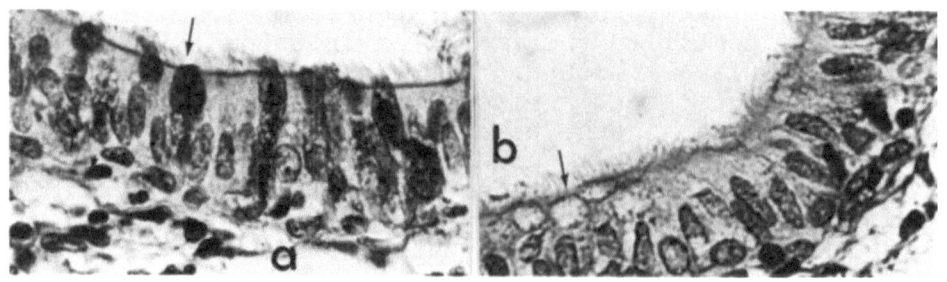

Fig. 2 a and b.   a) Bronchiolar epithelium. Positive
peroxidase immuno staining for antileukoprotease of the non-ciliated secretory cells (→). PAP-technique.
Antiserum dilution 1:1000. b) Control.
Secretory cells (→).

earlier reports,[26] independent evidence indicating that
the acid stable urinary trypsin inhibitor (mol.wt ∿
40,000 as judged from SDS gel electrophoresis) is a relatively potent inhibitor of granulocyte elastase. We
could also demonstrate that the uropepsine-induced
cleavage product of this inhibitor (mol.wt. around 20,000)
retained the inhibiting capacity against granulocyte
elastase. The urinary trypsin inhibitor shows immunologic cross reactivity with the inter-α-trypsin inhibitor.
We have some evidence indicating that this inhibitor
may also be present in bronchial secretions and other
body fluids. Taken together, this newly discovered pro-

perty of the urinary trypsin inhibitor may perhaps help
explain some conflicting reports concerning acid stable
inhibitors in bronchial secretions[26] - especially as a
further cleavage of the urinary trypsin inhibitor into
active fragments of the molecular size of about 10,000-
14,000 seems possible.[27]

Antileukoprotease is normally present in serum as
a trace protein (125 μg/l) in a free form. Serum antileukoprotease appears immunologically identical to the
standard used in the radioimmunoassay as judged from
the parallel dilution curves. The immunoreactivity of
antileukoprotease complexes with granulocyte elastase
is, however, clearly altered and diminished.[28]

SERUM AND SPUTUM ANTILEUKOPROTEASE IN PATIENTS WITH
CHRONIC BRONCHITIS

Thirty patients with chronic cough and sputum production were studied. All had obstructive airway disease with FEV/FVC ratio of 40.5 ± 12 (mean value $\pm$ SEM).
All had pyrexia and purulent sputum with positive bacterial culture. Serum and sputum samples were taken
within 5 days after admittance to hospital. The serum
level of antileukoprotease showed about 100 % increase
in these patients (Table 3). The immunoreactive antileukoprotease in serum was, however, present as a free
and active inhibitor as revealed by analysis of gel
filtration fractions from serum samples with and without the addition of granulocyte elastase in excess of
the inhibitory capacity of the plasma specimen. Also the
serum level of granulocyte elastase was increased two-
to three-fold. All elastase was identified in complex
with α_1-antitrypsin. The value for antileukoprotease
in sputum, although high, is underestimated, as gel

Table 3.   Antileukoprotease concentration in serum
           and sputum from patients with chronic
           bronchitis (N = 30)

	Antileukoprotease mg/l Mean ± SEM
Sputum	31.6 ± 4
Serum	0.24 ± 0.02

filtration of the sputum samples showed totally complexed inhibitor, although in most cases free and
active elastase was demonstrated.

SERUM LEVELS OF ANTILEUKOPROTEASE AND GRANULOCYTE
ELASTASE IN BRONCHOPNEUMONIA

    Serum samples from 15 patients with bronchopneumonia
were collected every two days during the hospital stay.
In addition, control samples were obtained 1 month
after recovery at the earliest. All patients showed
pyrexia with temperatures above 38.5 °C and X-ray findings typical of bronchopneumonia. They all had respiratory symptoms with shortness of breath and/or thoracic
pain. Antileukoprotease (Fig. 3) as well as granulocyte elastase showed high serum levels in the acute
phase, in some cases exceeding 10 times the normal mean
value. The return to normal serum levels was parallel
for both proteins and followed the clinical recovery.
Gel filtration analysis of serum samples from the
acute phase (Fig. 4) as well as from the recovery phase
demonstrated the antileukoprotease in a free, active
form, while all immunoreactive elastase in serum was
bound by α_1-antitrypsin.
    The presence of elastase-antileukoprotease com-
plexes in large amounts in local airway secretions,
simultaneously with their total absence in serum in
acute chest infections, appeared at first surprising.
Especially as very high levels of free antileukoprotease

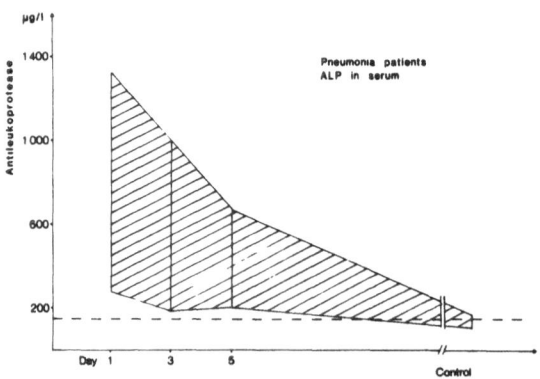

Fig. 3.   Concentration of antileukoprotease in serum
           samples from 15 patients with pneumonia
           during their hospital stay. Control samples
           were obtained after recovery.

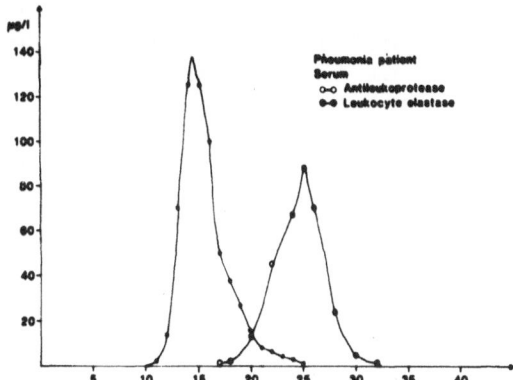

Fig. 4.   Distribution of immunoreactive antileukoprotease (o - o) and leukocyte elastase (● - ●)
in the fractions obtained on gel filtration on
Sephadex G-75 of the serum sample from a
pneumonia patient in the acute phase. The
immunoreactive elastase was α_1-antitrypsin-
bound as it was precipitated also by antibodies against this inhibitor. Antileukoprotease eluted in a volume corresponding to
the free inhibitor and showed inhibiting capacity against leukocyte elastase.

were present in serum during the acute phase. A complete
local metabolism of the complexes seemed a less likely
explanation. On the other hand, the antileukoprotease
complexes were supposed to be very stable, arguing
against dissociation ($K_i = \sim 10^{-9}$).[19]

STABILITY OF THE ELASTASE ANTILEUKOPROTEASE COMPLEX

     Equimolar elastase-antileukoprotease complexes
were produced by gel filtration on Sephadex G-75 of
mixtures of elastase and antileukoprotease with antileukoprotease in molar excess. Elastase and antileukoprotease were ^{125}I-labeled alternatively during the
experiments. Gel chromatography after 24 hours incubation of the elastase-antileukoprotease complex at room
temperature revealed no dissociation of the complex
at all. Elastase antileukoprotease complexes were also
mixed with plasma at room temperature. The elastase
content of the complex added corresponded to about 1 %
of the total inhibitory capacity of the serum sample.
The reaction mixtures were submitted to gel filtration
on Sephadex G-200 columns immediately and after 24 hours

incubation at room temperature. Also without special incubation the major part (about 80 %) of the complex was
cleaved. About 90 % of the released elastase was found
in complex with α_1-antitrypsin, as judged from the gel
filtration and the immunoprecipitation experiments.
The remaining 10 % was bound by α_2-macroglobulin. The
major part of antileukoprotease was obtained in a free
form, but a few per cent was bound by elastase-α_2-macro-
globulin complex (Fig. 5).

We have earlier obtained very similar results with
the pancreatic secretory trypsin inhibitor (PSTI),
another acid stable low molecular weight inhibitor.

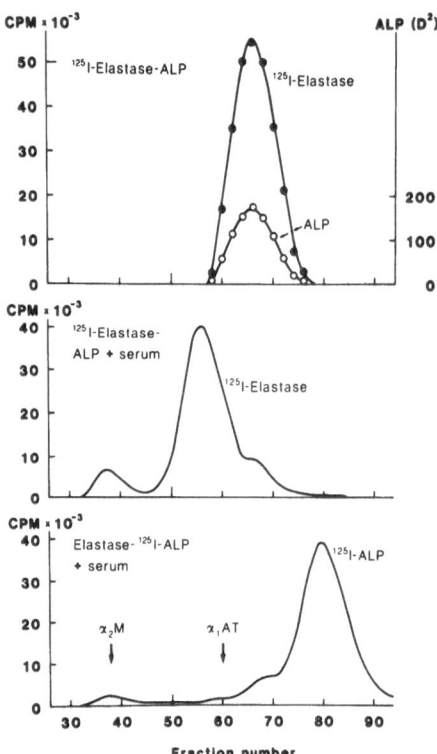

Fig. 5.   Distribution of radioactivity and immuno-
           reactive antileukoprotease (o - o) among the
           fractions obtained by gel filtration on
           Sephadex G-75 of equimolar ^{125}I-elastase-
           antileukoprotease complexes and reaction mix-
           tures of trace amounts of elastase-antileuko-
           protease complexes and human serum. Elastase
           and antileukoprotease were ^{125}I-labeled alter-
           natively.

The PSTI-trypsin complexes, like the antileukoprotease
elastase complexes, may be cleaved by acidification.
More interestingly after the addition of PSTI-trypsin
complexes to serum, the PSTI of the complex was immediately recovered in a free form. About 75 % of the
trypsin from the complexes was bound by α_2-macroglobu-
lin and the rest by α_1-antitrypsin. Also similarly to
the elastase-antileukoprotease experiment, a small part
of the PSTI was recovered bound by trypsin-α_2-macroglo-
bulin complexes.[29]

     Detailed kinetic data concerning the inhibition of
leukocyte elastase by antileukoprotease was recently
obtained in a collaboratory study with Bieth.[30] It was
then shown that α_1-antitrypsin alone is able to dissociate the elastase antileukoprotease complex in a pure
system with the rate constant of $1.3 \times 10^{-4} \times S^{-1}$. The
K_i of the elastase antiprotease complex was, however,
found to be even lower than estimated before, $1.2 \times
10^{-11}$ as compared to 10^{-9}.[30] The rate constant $(1.1 \times
10^{-7} \times S^{-1})$ for the association between granulocyte
elastase and antileukoprotease was, however, 6 times
lower than the one reported for the association of
granulocyte elastase with α_1-antitrypsin.

INTERACTION OF ANTILEUKOPROTEASE WITH ELASTASE-α_2-
MACROGLOBULIN COMPLEXES

     The regular demonstration of antileukoprotease
binding to elastase α_2-macroglobulin complexes in the
preceding experiments stimulated a direct titration of
elastase-α_2-macroglobulin complexes with antileukoprotease. 1:1 and 2:1 elastase-α_2-macroglobulin complexes
were produced. The capacity of antileukoprotease to
inhibit the Suc(Ala)$_3$-pNA cleaving activity of the
complexes was studied using varying incubation times
for antileukoprotease with the complexes. The inhibition
of the α_2-macroglobulin-bound elastase by antileukoprotease was found to be a rather slow process. However,
the 2:1 molar complex is much more resistent to inhibition than the 1:1 complex (Fig. 6). Similar results
have earlier been reported for the Aprotinin[31] and the
soybean trypsin inhibitor[32] interactions with the tryp-
sin-α_2-macroglobulin complex. This inhibition of the
elastase-α_2-macroglobulin complex by antileukoprotease
may, however, play a minor biologic role because, firstly, the inhibition is a rather slow process, and, secondly,
the α_2-macroglobulin concentration in bronchial mucous
secretions is very low.
     In conclusion, the results presented indicate that

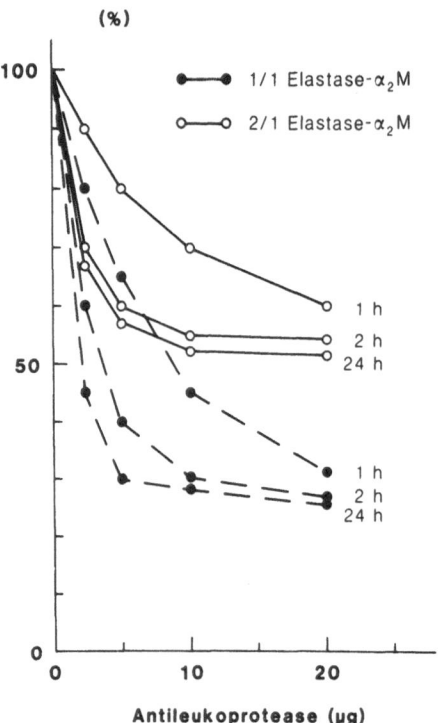

Fig. 6.   Time dependency of the inhibition of α_2-macro-
           globulin-bound granulocyte elastase by anti-
           leukoprotease. The inhibition is much more
           pronounced if α_2-macroglobulin is incompletely
           saturated with elastase than if it is comple-
           tely saturated.

antileukoprotease, the dominating inhibitor of the respiratory tract mucosa, functions as a local inhibitor
directed primarily against granulocyte elastase. Elastase antileukoprotease complexes are rapidly cleaved
in the interstitial tissues and blood. Elastase is
transferred to plasma protease inhibitors, mainly α_1-
antitrypsin. Antileukoprotease in blood is only demonstrated in a free form, although a small part could
be bound by elastase-α_2-macroglobulin complexes as
judged from *in vitro* experiments.

ABSTRACT

     The elastase-antielastase hypothesis of lung tis-
sue destruction has focused our interest on the two

main inhibitors of granulocyte elastase in the lung,
α_1-antitrypsin dominating blood, interstitial tissue
and alveolar fluid lining and antileukoprotease dominating the respiratory tract mucosa. Antileukoprotease
as well as elastase and α_1-antitrypsin show increased
serum levels during bronchitis and bronchopneumonia,
α_1-antitrypsin because it is an acute phase reactant,
elastase and antileukoprotease because of influx from
the inflamed tissues. Elastase is identified in the
bronchial expectorates, mainly in complex with antileukoprotease, but often also in a free, active form.
The granulocyte elastase in serum from these patients
is, however, only found in complex with α_1-antitrypsin.
The increased amounts of antileukoprotease in serum
are always in a free and largely active form. The explanation for the absence of elastase-antileukoprotease
complexes in serum is offered by some of our recent
results. The elastase-antileukoprotease complexes are
rapidly dissociated when mixed with serum *in vitro*,
although the equilibrium dissociation constant K_i of
the complex is 1.2×10^{-11} M. Furthermore, in a pure
in vitro system, α_1-antitrypsin is able to dissociate
a leukocyte elastase-antileukoprotease complex with
the rate constant of $1.3 \times 10^{-4} \times S^{-1}$. A small part of
the antileukoprotease released from the elastase-antileukoprotease complex on mixture with serum is recovered
bound by elastase-α_2-macroglobulin complexes. Antileukoprotease inhibits the enzymatic activity of elastase-
α_2-macroglobulin complex relatively slowly. 1:1 elas-
tase-α_2-macroglobulin complexes are, however, inhibited more readily than 2:1 saturated complexes.

REFERENCES

1. E. L. Opie, Enzymes and anti-enzymes of in-
    flammatory exudates, J Exp Med 7:316 (1905).
2. E. L. Opie, Intracellular digestion. The en-
    zymes and anti-enzymes concerned, Physiol
    Rev 2:552 (1922).
3. K. Ohlsson and H. Tegner, Granulocyte colla-
    genase, elastase and plasma protease inhibi-
    tors in purulent sputum, Europ J Clin Invest
    5:221 (1975).
4. D. Y. Twumasi and I. E. Liener, Proteases
    from purulent sputum. Purification and proper-
    ties of the elastase and chymotrypsin-like
    enzymes, J Biol Chem 252:1917 (1977).

5. C.-B. Laurell, Is emphysema in α_1-antitrypsin deficiency a result of autodigestion? Editorial, Scand J Clin Lab Invest 28:1 (1971).

6. K. Ohlsson and I. Olsson, Neutral proteases of human granulocytes. III. Interaction between human granulocyte elastase and plasma protease inhibitors, Scand J Clin Lab Invest 34:349 (1974).

7. C.-B. Laurell and S. Eriksson, The electrophoretic α_1-globulin pattern of serum in α_1-antitrypsin deficiency. Scand J Clin Lab Invest 15:132 (1963).

8. A. Janoff and H. Karp, Possible mechanisms of emphysema in smokers. Cigarette smoke condensate suppresses protease inhibition in vitro, Amer Rev Resp Dis 116:65 (1977).

9. K. Ohlsson, U. Fryksmark, and H. Tegner, The effect of cigarette smoke condensate on α_1-antitrypsin, antileukoprotease and granulocyte elastase, Europ J Clin Invest 10:373 (1980).

10. B. Mass, T. Ikeda, D. R. Meranze, G. Weinbaum, and P. Kimbel, Induction of experimental emphysema: cellular and species specificity, Amer Rev Resp Dis 106:384 (1972).

11. R. M. Senior, H. Tegner, C. Kuhn, K. Ohlsson, C. Barry, and J. A. Pierce, The induction of pulmonary emphysema with human leukocyte elastase, Amer Rev Resp Dis 116:469 (1977).

12. L. M. Hinman, C. A. Stevens, R. A. Matthay, and J. B. L. Gee, Elastase and lysosome activities in human alveolar macrophages, Amer Rev Resp Dis 121:263 (1980).

13. K. Ohlsson, Granulocyte proteases: their release and inhibition in the body, in: "Regulatory Proteolytic enzymes and their Inhibitors," S. Magnusson et al., eds., Pergamon Press, Oxford and New York (1978).

14. K. Ohlsson and A.-S. Olsson, Immunoreactive granulocyte elastase in human serum, Hoppe-Seyler Z Physiol Chem 359:1531(1978).

15. K. Ohlsson, Polymorphonuclear leucocyte collagenase, in: "Collagenase in Normal and Pathological Connective Tissues," D. E. Woolley and J. M. Evanson, eds., John Wiley & Sons Ltd. (1980).

16. H. Schiessler, M. Arnhold, and H. Fritz, Characterization of two proteinase inhibitors from human seminal plasma and spermatozoa, in: "Bayer-Symposium V Proteinase Inhibitors," H. Fritz, H. Tschesche, L. J. Greene, and E. Truscheit, eds., Springer-Verlag, Berlin (1974).

17. O. Wallner and H. Fritz, Characterization of
     an acid stable proteinase inhibitor in human
     cervical mucus, Hoppe-Seyler Z Physiol Chem
     355:709 (1974).
18. K. Ohlsson, H. Tegner, and U. Åkesson, Isola-
     tion and partial characterization of a low
     molecular weight acid stable protease inhibi-
     tor from human bronchial secretion, Hoppe-
     Seyler Z Physiol Chem 358:583 (1977).
19. H. Schiessler, K. Hochstrasser, and K. Ohls-
     son, Acid-stable inhibitors of granulocyte
     neutral proteases in human mucus secretions:
     Biochemistry and possible biological func-
     tion, in: "Neutral Proteases of Human Poly-
     morphonuclear Leukocytes," K. Havemann and
     A. Janoff, eds., Urban and Schwarzenberg,
     Inc., Baltimore-Munich (1978).
20. U. Fryksmark, K. Ohlsson, Å. Polling, and
     H. Tegner, Distribution of antileukoprotease
     in upper respiratory mucosa, Ann Otol Rhinol
     Laryngol 91:268 (1982).
21. B. Casslén, M. Rosengren, and K. Ohlsson,
     Localization and quantitation of a low
     molecular weight proteinase inhibitor, anti-
     leukoprotease, in the human uterus. Hoppe-
     Seyler Z Physiol Chem 362;953 (1981).
22. B. Carlsson, On granulocyte proteases and pro-
     tease inhibitors in otitis media, Thesis,
     Stockholm (1982).
23. J. A. Kramps. C. Franken, C. J. L. Meyer, and
     J. H. Dijkman, Localization of a low mole-
     cular weight protease inhibitor in serous
     secretory cells of the respiratory tract,
     J Histochem Cytochem 29:712 (1981).
24. B.-M. Jönsson, C. Löffler, and K. Ohlsson,
     Human granulocyte elastase is inhibited
     by the urinary trypsin inhibitor, Hoppe-
     Seyler Z Physiol Chem 363:1167 (1982).
25. B. J. Bromke and F. Kueppers, The major uri-
     nary protease inhibitor; simplified purifi-
     cation and characterization, Biochem Med
     27:56 (1982).
26. K. Hochstrasser, H. Feuth, and K. Hochgesand,
     Proteinase inhibitors of the respiratory
     tract: Studies on the structural relation-
     ship between acid-stable inhibitors present
     in the respiratory tract, plasma and urine,
     in: "Bayer-Symposium V Proteinase Inhibitors,"
     H. Fritz, H. Tschesche, L. J. Greene, and
     E. Truscheit, eds., Springer-Verlag, Berlin,
     Heidelberg, New York (1974).

27. K. Hochstrasser and E. Wachter, Kunitz-type
     proteinase inhibitors derived by limited pro-
     teolysis of the inter-α-trypsin inhibitor, I,
     Hoppe-Seyler Z Physiol Chem 360:1285 (1979).
28. U. Fryksmark, K. Ohlsson, M. Rosengren, and
     H. Tegner, A radioimmunoassay for measurement
     and characterization of human antileukopro-
     tease in serum, Hopp-Seyler Z Physiol Chem
     362:1273 (1981).
29. A. Eddeland and K. Ohlsson, Studies on the
     pancreatic secretory trypsin inhibitor in
     plasma and its complex with trypsin in vivo
     and in vitro, Scand J Clin Lab Invest 38:507
     (1978).
30. F. Gauthier, U. Fryksmark, K. Ohlsson, and
     J. Bieth, Kinetics of the inhibition of leuko-
     cyte elastase by the bronchial inhibitor, Bio-
     chem Biophys Acta 700:178 (1982).
31. P. O. Ganrot, Inhibition of the trypsin α_2-
     macroglobulin complex by the protease inhibi-
     tor from bovine lung, Arkiv för Kemi 26:583
     (1967).
32. J. G. Bieth, M. Tourbez-Perrin, and F. Fochon,
     Inhibition of α_2-macroglobulin-bound trypsin
     by soybean trypsin inhibitor, J Biol Chem
     256:7954 (1981).

ACKNOWLEDGEMENTS

     This investigation was supported by grants from the
Swedish Medical Research Council (project no. B83-17X-
03910-11B), the Medical Faculty, University of Lund,
Malmö General Hospital Foundation for Medical Research,
the Foundation of Greta and Johan Kock, the Swedish
Association against Heart and Chest Diseases, the Swedish Association against Rheumatism, the Swedish Tobacco Company, the Foundation of Alfred Österlund, and the
Foundation of King Gustav Vth.

LEUKOPROTEINASES AND PULMONARY EMPHYSEMA: CATHEPSIN G AND OTHER CHYMOTRYPSIN-LIKE PROTEINASES ENHANCE THE ELASTOLYTIC ACTIVITY OF ELASTASE ON LUNG ELASTIN

Christian Boudier, Philippe Laurent and Joseph G. Bieth

INSERM Unité 237 - Faculté de Pharmacie - Université
Louis Pasteur - 67048 Strasbourg (France)

INTRODUCTION

Leukocyte elastase has caused more and more interest in recent years because of its involvement in pulmonary emphysema[1]. Polymorphonuclear leukocytes contain high amounts of another neutral proteinase, cathepsin G whose pathological role has not yet been elucidated. This enzyme might participate in the degradation of lung tissue components other than elastin[2]. We wish to report in this paper that, despite of its very low elastolytic activity[3,4], cathepsin G plays an important role in elastolysis since it strongly stimulates the elastase-catalyzed solubilization of lung elastin. We also demonstrate that this stimulating property is related to the chymotrypsin-like character of cathepsin G since other chymotrypsin-like enzymes such as bovine α-chymotrypsin or human pancreatic elastase are also able to enhance the activity of leukocyte elastase.

STIMULATION OF THE ELASTOLYTIC ACTIVITY OF ELASTASE BY CATHEPSIN G

Figure 1 shows that leukocyte cathepsin G markedly stimulates the elastolytic activity of leukocyte elastase. The stimulation is more pronounced with human lung elastin than with bovine ligamentum nuchae elastin.

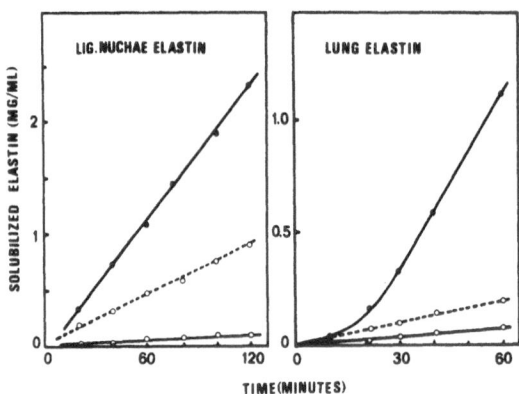

Fig. 1 Kinetics of solubilization of bovine ligamentum nuchae elastin (Worthington) and human lung elastin (prepared by hot
alkali treatment[5]) by 1 μM human leukocyte elastase[6] (o——o),
2 μM human leukocyte cathepsin G (o----o) and the mixture
of the two enzymes (●——●). Assay conditions : elastin =
3 mg/ml, 50 mM Tris-HCl pH 8.0, 37°C. Solubilized peptides
were assayed by absorbance measurements at 280 nm
after acidification of aliquots of reaction mixtures[7]. The
concentration of solubilized elastin (ordinates) was
calculated from the absorbance of completely digested
elastin ($E_{1cm}^{1\%}$ = 13.3 and 6.0 for lung elastin and ligamentum
elastin respectively). Other technical details may be found
in ref. 8.

        The observed enhancement of elastolysis is not due to an arti-
factual solubilization of non-elastin peptides by cathepsin G since
the aminoacid composition of the partial elastin digest is the same
whether pure elastase or elastase + cathepsin G is used for enzymatic digestion.

        When the two proteinases are added sequentially to elastin, the
enhancement of elastolysis is much less pronounced. Cathepsin G and
elastase therefore must act simultaneously on elastin in order to
achieve maximum stimulation. The enhancing effect occurs also at a
physiological ionic strength (50 mM Tris + 100 mM NaCl). Cathepsin
G is unable to activate the elastase-catalyzed hydrolysis of N-suc-
cinyl-(L-alanine)$_3$-p-nitroanilide, a synthetic elastase substrate[9].
On the other hand, cathepsin G has almost no stimulating effect on
the solubilization of elastin by porcine pancreatic elastase.

STIMULATING EFFECT OF OTHER CHYMOTRYPSIN-LIKE PROTEINASES

Since cathepsin G is a chymotrypsin-like enzyme[3], we also investigated the potentiating effect ob bovine α-chymotrypsin and human pancreatic elastase 2 (a gift from Dr. Michel Rabaud, Pessac, France), a proteinase with chymotrypsin-like specificity[10]. Table 1 shows that these two enzymes are also able to stimulate the elastase-catalyzed solubilization of lung elastin. Their stimulating efficiency is however lower than that of cathepsin G.

Table 1 Stimulation of the elastolytic activity of leukocyte elastase on lung elastin by chymotryspin-like proteinases.

Proteinases used to stimulate leukocyte elastase activity[a]	Stimulation factor[b]
Human leukocyte cathepsin G	4.9
Human pancreatic elastase 2	3.7
Bovine pancreatic α-chymotrypsin	2.4

[a] the final concentration of each enzyme was 1 µM (including leukocyte elastase)

[b] elastolysis was measured as described in the legend to fig. 1. The stimulation factor is given by $(v(e,c)-v(c))/v(e)$, where $v(e)$ is the elastolysis rate in the presence of leukocyte elastase alone, $v(c)$ is the rate in the presence of the chymotrypsin-like proteinase alone and $v(e,c)$ is the rate in the presence of leukocyte + chymotrypsin-like enzyme. The rates were calculated from the slopes of the linear parts of the kinetic curves (see fig. 1).

In addition, all three chymotrypsin-like proteinases exhibit a low but distinct elastolytic activity on lung elastin. Bovine α-chymotrypsin, for instance, is three times more active on lung elastin than cathepsin G. However, the former enzyme does not solubilize bovine ligamentum nuchae elastin whereas the latter does (see fig. 1).

DISCUSSION

Cathepsin G does not enhance the catalytic activity of elastase as demonstrated with the synthetic elastase substrate. Therefore, its stimulation of elastolysis must be ascribed to its ability to cleave certain peptide bonds of elastin. Cathepsin G is a relatively non-specific protease. It cleaves peptide bonds adjacent to Tyr, Phe, Leu, Ile and Met residues[3,12,11]. Aminoacid analysis of our lung elastin preparation[8] shows that these residues represent 2.1 %, 2.2 %, 5.9 %, 2.1 % and 0.1 % respectively of the total aminoacides. Therefore, elastin contains a large number of bonds susceptible to cathepsin G attack. By contrast, leukocyte elastase has a very narrow specificity. It cleaves preferentially Val-X bonds and to a lesser extent Ala-X links which represent 12.8 % and 25.8 % of total bonds[13]. In some parts of fibrous elastin the hydrolysis of such bonds may be sufficient to bring peptides into solution. In other parts this may not be case and solubilization may require the additional action of a proteinase with broader specificity like cathepsin G or other chymotrypsin-like enzymes.

Porcine pancreatic elastase cleaves not only Ala-X and Val-X bonds but also Gly-X links which represent 33.1 % of total bonds[14] This broader specificity may explain first why this enzyme has a much higher elastolytic activity than leukocyte elastase[4] and second why its activity is not significantly enhanced by cathepsin G.

Many workers have reported that leukocyte elastase is poorly efficient in solubilizing elastin compared to pancreatic elastase[3,4]. In addition, animal studies have confirmed that the amount of leukocyte elastase needed to induce emphysema is much higher than that of pancreatic elastase[1]. These data would cast some doubt on the role of leukocyte elastase in the pathogenesis of pulmonary emphysema. Polymorphonuclear leukocytes contain about equimolar concentrations of elastase and cathepsin G. According to our data, the elastinolytic activity of such a mixture is 5 times higher than that of elastase alone (table 1). Hence, effective degradation of lung elastin may occur in vivo after liberation of leukocyte neutral proteinases. Our findings therefore assign a clear-cut role to cathepsin G in the development of pulmonary emphysema.

We have shown that enhancement of elastolysis may also be brought about by other proteinases (table 1). It is likely that other proteinases with broad substrate specificity will also do so. In vivo degradation of proteins may therefore result from the concerted action of several proteinases. This complicates the search for therapeutic agents (e.g. synthetic proteinase inhibitors) designed to prevent proteolysis in vivo.

REFERENCES

1. R.M. Senior, H. Tegner, C. Kuhn, K. Ohlsson, B.C. Starcher and
   J.A. Pierce, The induction of pulmonary emphysema with human
   leukocyte elastase, Amer. Rev. Resp. Dis. 116:469 (1977).
2. P.M. Starkey, A.J. Barrett and M.C. Burleigh, The degradation
   of articular collagen by neutrophil proteinases, Biochim. Biophys.
   Acta 483 : 386 (1977).
3. W. Schmidt and K. Havemann, Isolation of elastase-like and chymotrypsin-like neutral proteases from human granulocytes, Hoppe-
   Seyler's Z. Physiol. Chem. 355 : 1077 (1974).
4. C.R. Reilly and J. Travis, The degradation of human lung elastin
   by neutrophil proteinases, Biochim. Biophys. Acta 621 : 147
   (1980).
5. A.T. Lansing, M.A. Rosenthal and E.V. Dempsey, The structure and
   chemical characterization of elastic fibers as revealed by elastase and by electron microscopy, Anat. Rec. 114 : 555 (1952).
6. R.R. Martodam, R.J. Baugh, T.Y. Twumasi and I.E. Liener, A
   rapid procedure for the large scale purification of elastase and
   cathepsin G from human sputum, Prep. Biochem. 9 : 15 (1979).
7. W. Ardelt, S. Ksienzy and N. Nidzwiecka, Spectrophotometric method for the determination of pancreatopeptidase E activity,
   Analyt. Biochem. 34 : 180 (1970).
8. C. Boudier, C. Holle and J.G. Bieth, Stimulation of the elastolytic activity of leukocyte elastase by leukocyte cathepsin G,
   J. Biol. Chem. 256 : 10256 (1981).
9. J.G. Bieth, B. Spiess and C.G. Wermuth, The synthesis and analytical use of a highly sensitive and convenient substrate of elastase, Biochem. Med. 11 : 350 (1974).
10. C. Largman, J.W. Brodrick and M.C. Geokas, Purification and
    characterization of two human pancreatic elastases, Biochemistry
    15 : 2491 (1976).
11. A.M.J. Blow and A.J. Barrett, Action of human cathepsin G on the
    oxidized B chain of insulin, Biochem. J. 161 : 17 (1977).
12. K. Nakajima, J.C. Powers, B.M. Ashe and M. Zimmerman, Mapping
    the extended substrate binding site of cathepsin G and human
    leukocyte elastase. Studies with peptide substrates related
    to the α_1-protease inhibitor reactive site, J. Biol. Chem. 254 :
    4027 (1979).
13. A.M.J. Blow, Action of human lysosomal elastase on the oxidized
    B chain of insulin, Biochem. J. 161 : 13 (1977).
14. A. Sampath Narayanan and R.A. Anwar, The specificity of purified
    porcine pancreatic elastase, Biochem. J. 114 : 11 (1969).

ADULT RESPIRATORY DISTRESS SYNDROME (ARDS):

EXPERIMENTAL MODELS WITH ELASTASE AND THROMBIN INFUSION IN PIGS

H. Burchardi[1], T. Stokke[1], I. Hensel[1], H. Köstering[2],
G. Rahlf[3], G. Schlag[4], H. Heine[5], and W. H. Hörl[6]

[1]Dept. of Anesthesiology , [2]Int. Med., [3]Pathology, Univ.
Göttingen, FRG; [4]Ludwig-Boltzmann Ist.f.Exp.Traum.Wien,
Austria; [5]Dept. of Anatomy, [6]Int.Med.Univ.Würzburg,FRG

The "adult respiratory distress syndrome" (ARDS) is a dramatic
event in intensive care medicine. In spite of a considerable increase in knowledge during the last decade, mortality is still over
50 %.

CHARACTERISTICS OF ARDS

ARDS is a clinical syndrome. It represents an acute, progressive respiratory failure with hypoxemia (due to increased intrapulmonary shunt), increased dead space ventilation and decreased lung
compliance.

Progression necessitates intensive care including oxygenation
and artificial ventilation. Finally, complete deterioration of pulmonary gas exchange results in fatal hypoxemia.

ARDS is apparently not a primary disease, but a secondary respiratory complication following severe pulmonary and (even more frequently) non-pulmonary diseases. It appears to be an almost uniform
reaction of the lung to various affectious f.e.: shock, trauma,
sepsis, pancreatitis, intoxications (as bromcarbamide, paraquat).

Some typical early morpho-pathological findings (beginning
within hours) are: pulmonary leukostasis, lesion of the pulmonary
capillary endothelium, pulmonary interstitial edema.

If ARDS persists for a longer period of time (one week and
longer), hyaline membranes and a severe, progressive interstitial
fibrosis develop. Lungs, then, become heavy and stiff with a dark-
ly red spotted surface.

PATHOGENETIC HYPOTHESES

Among a great number of hypotheses two pathogenetic mechanisms have been discussed intensively to play a role in early ARDS development:

1,16 • Pulmonary microembolisation of platelet aggregations and fibrin - the disseminated intravascular clotting (DIC).
• Selective pulmonary sequestration of polymorphonuclear neutrophils (PMNs) - the pulmonary leukostasis.

Aggregations of platelets and fibrin emboli and/or aggregates of PMNs may damage lung tissue either directly (f.e. by occlusion) or indirectly by releasing mediators. Today it seems that the pulmonary leukostasis play the primary role in early ARDS, perhaps triggered by complement activation (alternative pathway)[3,6]:

MALIK and coworkers recently demonstrated in animal experiments that even the effects of the disseminated intravascular clotting depended on the presence of the pulmonary leukostasis[15]. Aggregated polymorphonuclear neutrophils may release mediators in the lungs: f.e. free toxic radicals[19] or lysosomal proteases, like f.e. elastase and others[17].

According to this, LEE and coworkers[13] found significantly high levels of elastolytic activity in the bronchial lavage fluid from ARDS patients.

We therefore studied in an experimental model the following two hypothetical pathomechanisms: the effect of induced DIC and the effect of elastase on lung function in minipigs.

METHODS

Experimental procedure

23 minipigs were studied in anaesthesia (1 mg/kg·h azaperon, 2.5 mg/kg·h metomidate) and myorelaxation (hexacarbacholin-bromid) during artificial ventilation (Engström-Respirator).

3 groups were examined:
• a control group (series C, n = 3) without treatment
• a thrombin group (series T, n = 11) with a continuous intravenous infusion of thrombin (TOPOSTASIN(R)) (75-150 U/kg·h) in order to induce DIC
• an elastase group (series E, n = 9) with a continuous intravenous infusion of elastase (porcine pancreatic elastase, Serva, Ordering N≛ 20930) (330 U/kg·h). One unit elastase digests at 25°C and pH 8.5 1 μmol N-acetyl-(L-alanin)$_3$-methylester in one minute.

Intravascular lines were inserted into the aorta, pulmonary artery, and v. femoralis and jugularis for hemodynamic measurements and blood sampling.

Measurements were repeated in short intervals when a stable hemodynamic state was achieved: Cardiovascular parameters were measured by conventional catheter-technique: a systemic arterial pressure (P_{art}) and pulmonary artery pressure (P_{AP}); cardiac output ($\dot{Q}_{Fick}$) (by Fick's principle).

Total thoracic compliance (C) and respiratory resistance (R) were measured continuously using pneumotachography (Dr. Fenyves & Gut, described in detail in 2). Pulmonary capillary perfusion ($\dot{Q}_{N_2O}$) and lung volume (FRC) were determined by a new N_2O rebreathing technique (described elsewhere[22,23]).

Arterial and mixed venous blood gas analyses were performed by routine laboratory methods (Radiometer, ABL 2; for O_2 - content: Lexington, Lex-O_2-Con). Inspiratory and expiratory gas mixtures were seperated exactly (described in 20) and gas concentrations (O_2, CO_2, N_2, Ar, N_2O) were measured continuously by mass spectrometry (Perkin-Elmer, MGA 1100A). Venous admixture ($\dot{Q}_S/\dot{Q}_T$) was calculated at F_IO_2 = 0.4; physiological dead space (V_D/V_T) was calculated with the Bohr-Enghoff formula after sampling of mixed exspired gas in a special mixbox[5].

Control measurements were performed after hemodynamic parameters had stabilized and animals were then treated with thrombin or elastase, respectively, for 3 - 5 hours and then sacrified. Plasma samples were taken for determination of proteolytic activity and for examination of coagulation parameters[11,12] in short intervals (5 - 30 minutes) during the whole experiment.

Tissue samples (lung, heart, liver, kidney, muscle) were obtained immediately for determinations of the proteolytic activity before sacrifying the animals, frozen by liquid nitrogen and stored at -70ºC before analysis.

Proteolytic activity in plasma and tissue samples was measured, as previously described, using as substrates phosphorylase kinase (from rabbit skeletal muscle) or azocasein[8,9,10]. The organs were homogenized (3 times 30 sec) with 10 vol. buffer containing 50 mM Tris/HCl, 1 M KCl, pH 8.5. The homogenate was decanted through glass wool, stirred for 12 hours at 4ºC and centrifuged for 30 minutes at 15,000 rpm (4ºC). The supernatant was dialyzed against 50 mM Tris/HCl, 4 mM EDTA, pH 7.0[4]. Polyacrylamide gel electrophoresis in the presence of sodium dodecylsulfate was carried out according to Weber and Osborn[25] Protein was determined by the Lowry method[14]

RESULTS AND DISCUSSION

Control group

Control animals (n = 3) were prepared and treated identically to those of the other groups but did not receive thrombin or elastase infusion. Hemodynamic parameters did not change significantly in the control group suggesting that influences of the basic experimental procedure (anaesthesia, artificial ventilation etc.) remained stable during the experiments.

In contrast, both thrombin as well as elastase infusion caused remarkable hemodynamic and morphological changes. Since time course of these effects varied somewhat between the individual animals, the results will be presented by demonstrating single, but typical experiments instead of mean values of the total group.

Thrombin group

A remarkable phenomenon was the immediate influence of thrombin infusion on the pulmonary circulation: Few minutes after onset of the thrombin infusion pulmonary artery pressure increased markedly and remained elevated (Fig. 1 and 2).

This might, of course, have been a direct effect of the occlusion of the pulmonary vascular bed. It is, however, more likely that early after thrombin infusion the rise of pulmonary artery pressure is due to a release of vasoactive substances. Later on, the increase in pulmonary vascular resistance was probably caused by a combination of both, humoral effects as well as mechanical occlusion by clotting elements. The pulmonary artery pressure remained elevated in spite of a progressive decrease in cardiac output ($\dot{Q}_T$ (thermo-dilution)) and pulmonary capillary perfusion ($\dot{Q}_{N_2O}$) (see Fig. 2) as disseminated intravascular clotting went on. Note that $\dot{Q}_{N_2O}$, measured by the N_2O-rebreathing method, was only that part of pulmonary blood flow which perfuses capillaries being in contact with ventilated alveoli. Intrapulmonary shunt perfusion in "shunting" capillaries without contact to ventilated alveoli did not enter into the parameter $\dot{Q}_{N_2O}$.

Oxygenation deteriorated progressively during thrombin infusion. This is illustrated by the significant increase in alveolo-arterial oxygen pressure difference ($AaDO_2$) despite inspiratory O_2 concentration was kept constant at 40 % ($F_IO_2 = 0.4$). This gas exchange defect was caused by an increase in venous admixture as well as by a decrease in pulmonary diffusing capacity (measured by a CO-rebreathing-method, described elsewhere[21,24]).

Respiratory mechanics also deteriorated very soon: The distensibility of the lung, the lung compliance (C) decreased progressively during the induced DIC (Fig. 2). This may have been due to a slight interstitial edema which could be demonstrated morphologically; but it may also have been due to an increase in pulmonary blood volume (congestion) and stiffness of the congested capillaries.

Consequently, the lung volume (functional residual capacity = FRC) decreased likewise (Fig. 2) (which also may have caused an increase in intrapulmonary blood shunting and a progressively disturbed oxygenation). Although both C and FRC decreased the specific compliance (C/FRC) was reduced, too, indicating that the reduction of compliance was not only due to a decrease in lung volume.

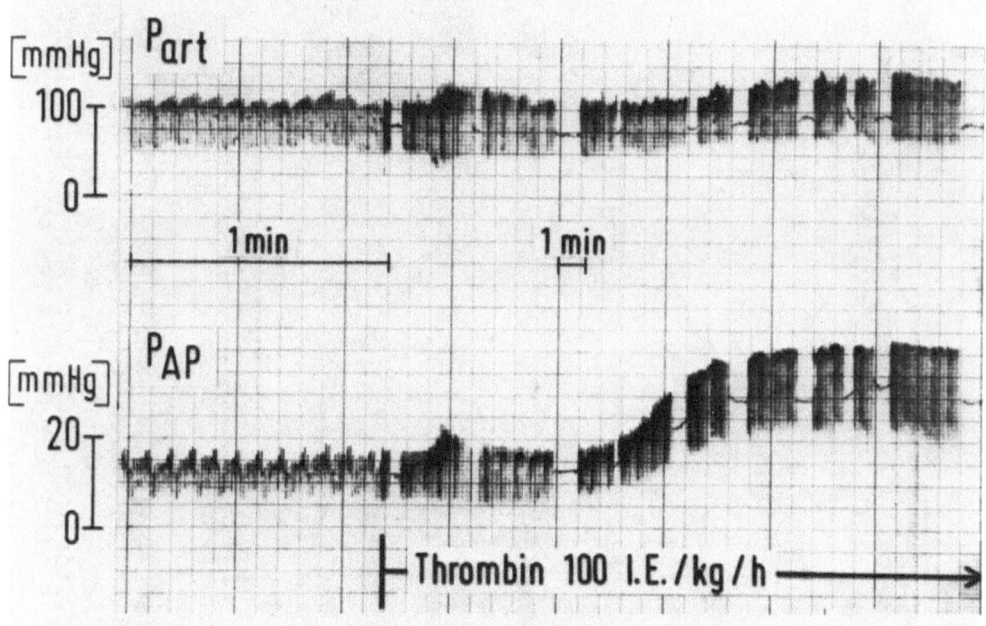

Fig. 1. Original registration of systemic artery pressure (Part) and pulmonary artery pressure (P$_{AP}$) at the beginning of a thrombin infusion in animal experiment (mini-pig). Few minutes after onset of the thrombin infusion, pulmonary artery pressure increases markedly.

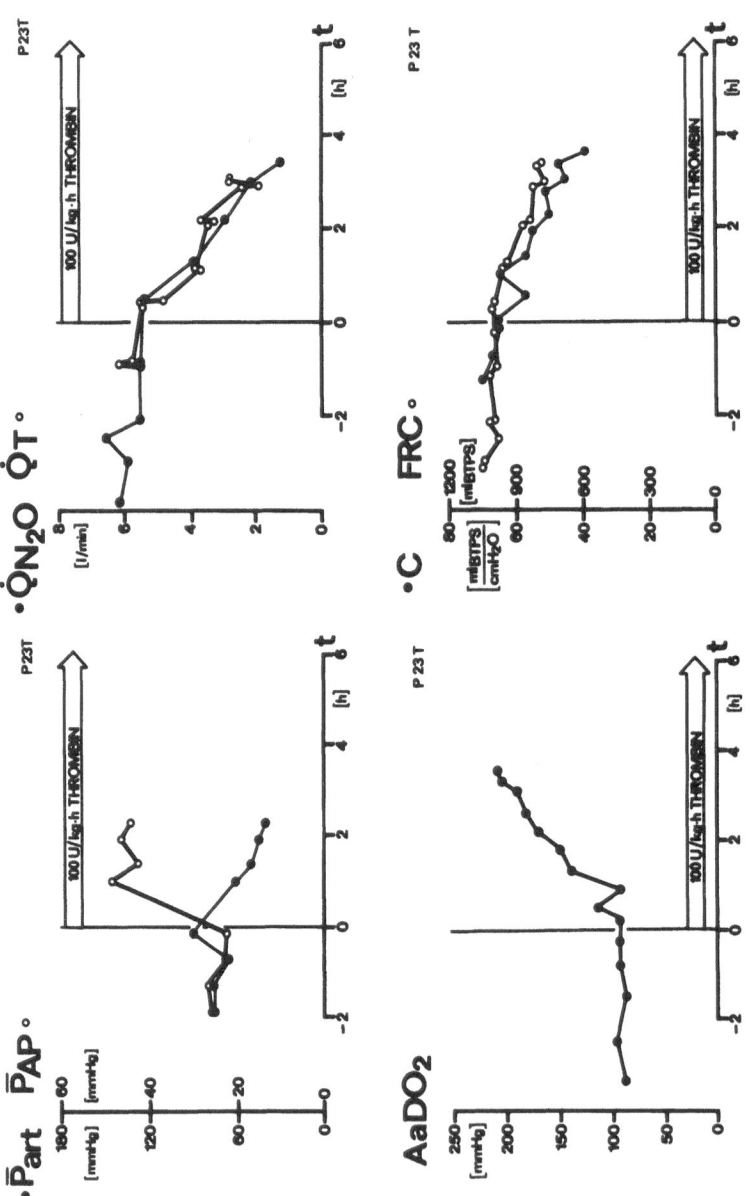

Fig. 2. Time course of mean systemic pressure ($\bar{P}_{art}$), mean pulmonary artery pressure ($\bar{P}_{AP}$), pulmonary capillary perfusion ($\dot{Q}_{N_2O}$), cardiac output ($\dot{Q}_T$), alveolo-arterial oxygen partial pressure difference (AaDO$_2$), total compliance (C) and functional residual capacity (FRC) before and during thrombin infusion in a single animal experiment (mini-pig). Onset of the thrombin infusion at t = 0. Time in hours. For details see text.

On the other hand, respiratory resistance was not affected by thrombin induced DIC in our experiments[7]. Rådegran[18] found an increase in airway pressure during thrombin infusion in dogs. He interpreted this as a result of an increased airway resistance, but corresponding to our findings it may be fully explained by a decrease in pulmonary compliance.

Impaired pulmonary capillary perfusion, disturbed gas exchange function and decreased distensibility of the lung was adequately explained by the histological picture of the thrombin lung (Fig. 3): the interstitial space was broadened and congested; the capillaries were filled with leukocytes.

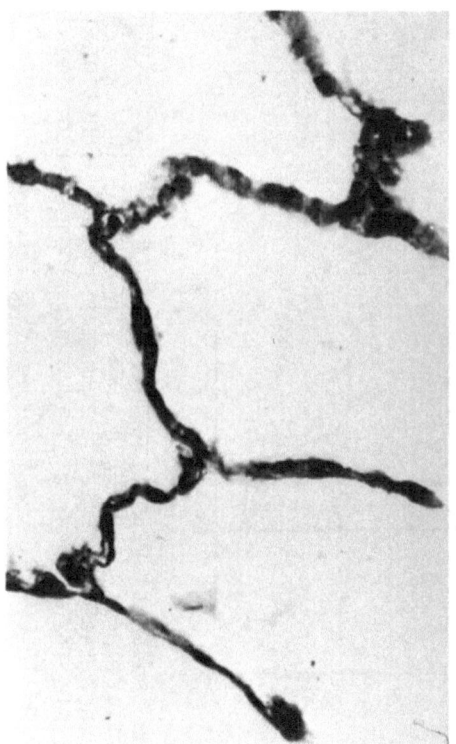 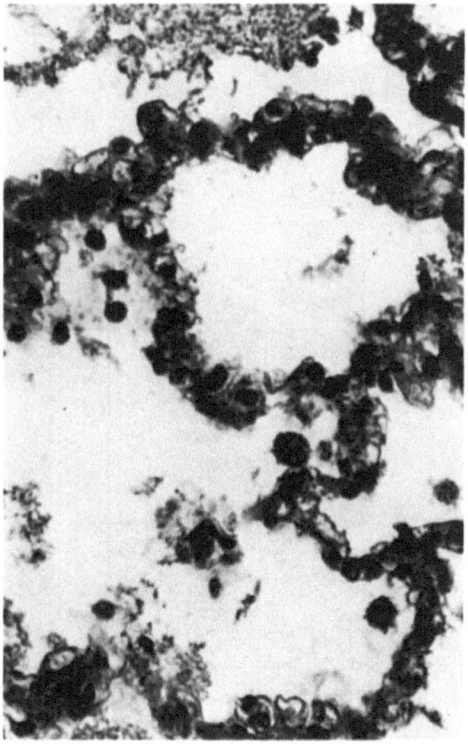

Fig. 3. Histology of mini-pig lung. Left part: normal (control group). Right part: interstitial edema and capillary congestion after 3 hours of continuous IVinfusion of 100 IU/kg · h thrombin.

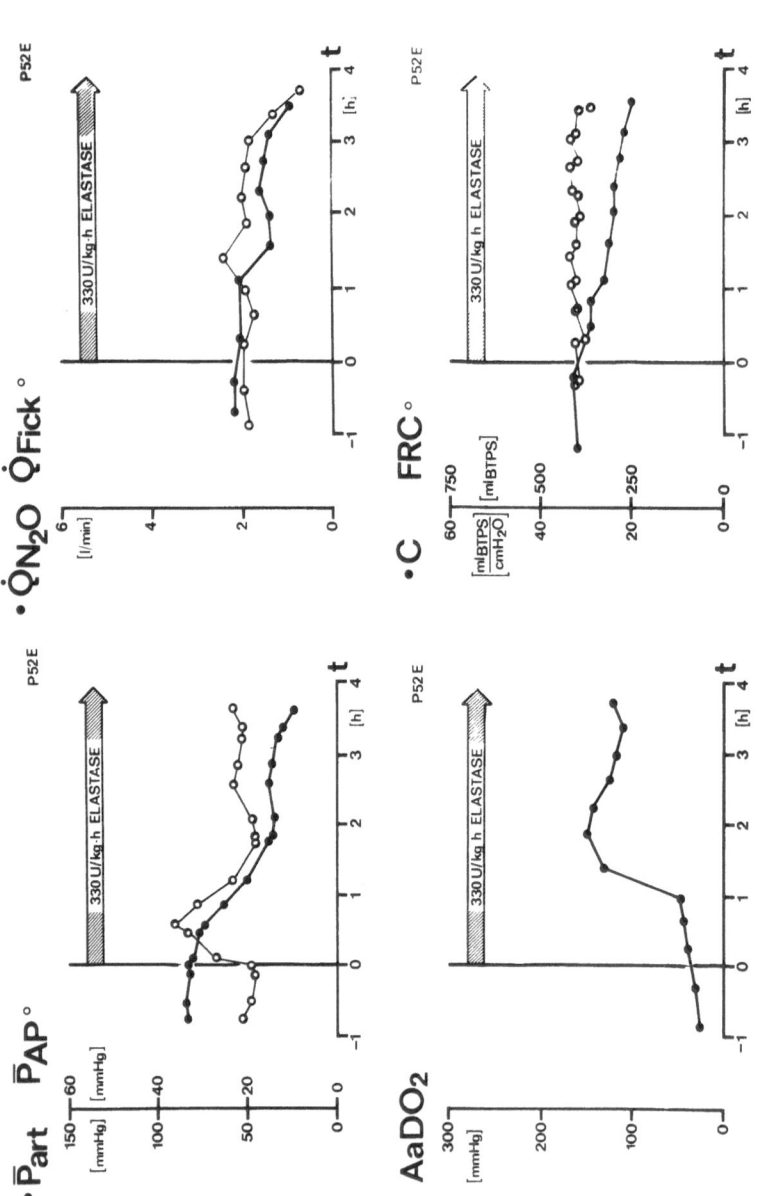

Fig. 4. Time course of mean systemic pressure ($\bar{P}_{art}$), mean pulmonary artery pressure ($\bar{P}_{AP}$), pulmonary capillary perfusion ($\dot{Q}_{N_2O}$), cardiac output ($\dot{Q}_{Fick}$), alveolo-arterial oxygen partial pressure difference ($AaDO_2$), total compliance (C) and functional residual capacity (FRC) before and during elastase infusion in a single animal experiment (mini-pig). Onset of the elastase infusion at t = 0. Time in hours. For details see text.

Elastase group

In the elastase series, we got almost the same functional results:

Pulmonary circulation: Elastase infusion caused an immediate rise of pulmonary artery pressure (P_{AP}) (Fig. 4) similar to thrombin representing an increase in pulmonary vasculary resistance. But in this case, the increase was only temporary. It fell later on when systemic blood pressure progressively decreased. Thus, the increase of pulmonary vascular resistance probably was a direct effect of elastase or an indirectly triggered phenomenon affected by activation of other mediators.

Pulmonary capillary perfusion ($\dot{Q}_{N20}$) (Fig. 4) deteriorated significantly less following elastase infusion than following thrombin induced intravascular clotting. This may be explained by the fact that in induced DIC intrapulmonary microemboli could indeed produce a certain intravascular obstruction which did not occur after elastase infusion. But, inspite of that, oxygenation deteriorated after elastase infusion as well as after thrombin; the $AaDO_2$ increased likewise (Fig. 4).

Lung compliance (C) decreased after elastase as well as after thrombin demonstrating the progressive stiffness of the lung. But, nevertheless, lung volume (FRC) remained unchanged during constant artificial ventilation. So, here the reduction of compliance was not due to a decrease of lung volume but mainly caused by a reduced distensibility of the lung itself (as the specific compliance (C/FRC) fell).

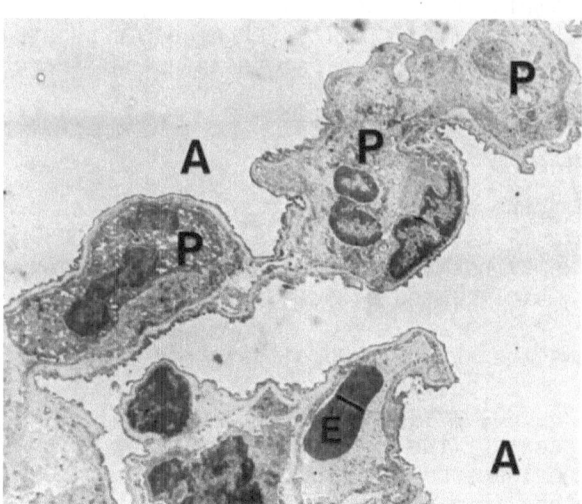

Fig. 5. Electron microscopic lung picture after 4 hours of continuous IV infusion of elastase in minipig: pulmonary leukostasis.
P: Degranulated polymorphonuclear neutrophils
E: Red blood cell   A: Alveolar space

Coagulation parameters

During the first 30 minutes of thrombin infusion, the coagulation parameters demonstrated the signs of hypercoagulability, as seen in the time course of the thrombin generation test, the decrease in the r- and the k-time of the thrombelastogram and the increase in partial thromboplastin time (PTT) and thrombin time (TT). The number of platelets and leukocytes in peripheral blood decreased significantly. Plasma concentration of fibrinogen decreased markedly whereas the fibrin monomere complexes had their highest concentrations at about 15 minutes after onset of thrombin infusion.

Later on, the coagulation parameters showed the typical signs of disseminated intravascular coagulation with consumption of coagulation factors; PTT and TT increased significantly and the plasma concentrations of all coagulation factors decreased markedly.

In the elastase treated animals, too, the coagulation parameters showed the typical signs of hypercoagulability, but not as pronounced as in the thrombin group. The concentrations of platelets and leukocytes in peripheral blood decreased markedly. In addition, endogenous heparin, antithrombin III, alpha-2-antiplasmin, and alpha-2-macroglobulin significantly decreased. At the end of the experiment, the concentration of alpha-2-macroglobulin was about one third of the initial value. This indicates, that the inhibition capacity of elastase was not totally exhausted by the intravenously infused elastase. The concentration of plasma fibrinogen decreased. In contrast to the thrombin group, the fibrin monomere complexes increased only slightly. The plasmin activity (chromogenic substrates) rose immediately after the onset of elastase infusion. There was an increase in fibrinolytic activity in both groups.

Protease investigations

Fig. 6 shows the effect of elastase infusion on proteolytic activity of the plasma using azocasein as a substrate. Proteolytic activity, however, was not detectible in plasma fractions of control animals and thrombin treated pigs.

Fig. 7 shows the typical subunit structure of phosphorylase kinase with the three polypeptide chains alpha (Mol.wt. 135,000), beta (Mol.wt. 120,000) and gamma (Mol.wt. 42,000) before incubation (0h). Crude extracts of the lungs of elastase or thrombin treated animals destroyed the subunit structure of the enzyme within 12 hours of incubation (Fig. 7) whereas no proteolytic digestion was determined incubating the enzyme with crude extracts of the lungs of control animals. Furthermore, crude extracts of the liver, heart, skeletal muscle and the kidney of elastase as well as thrombin treated pigs did not affected the subunit structure of phosphory-

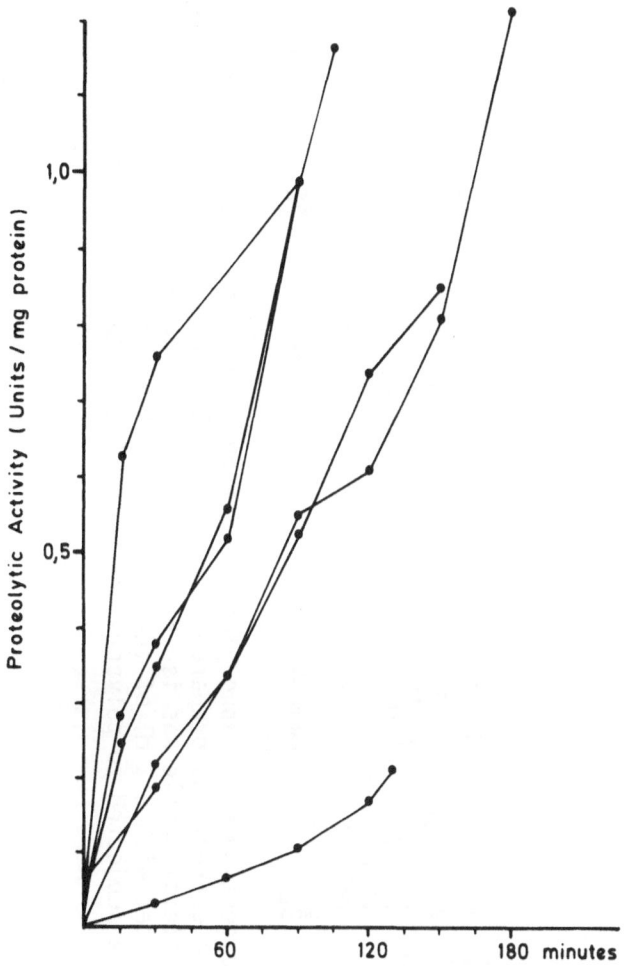

Fig. 6. Effect of elastase infusion on proteolytic activity of the plasma. Samples of six minipigs were taken at the times indicated in the figure. Activities were determined using azocasein as a substrate[4].

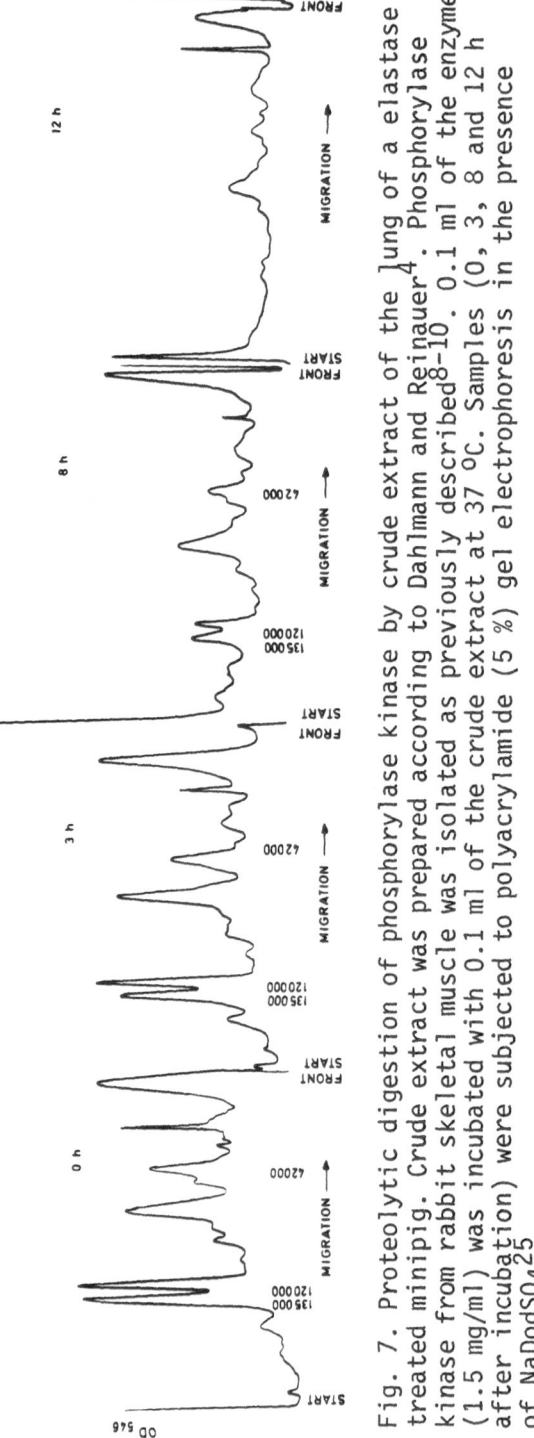

Fig. 7. Proteolytic digestion of phosphorylase kinase by crude extract of the lung of a elastase treated minipig. Crude extract was prepared according to Dahlmann and Reinauer[4]. Phosphorylase kinase from rabbit skeletal muscle was isolated as previously described[8-10]. 0.1 ml of the enzyme (1.5 mg/ml) was incubated with 0.1 ml of the crude extract at 37 ^{0}C. Samples (0, 3, 8 and 12 h after incubation) were subjected to polyacrylamide (5 %) gel electrophoresis in the presence of NaDodSO$_4$[25].

lase kinase during 12 hours of incubation at 37 $^{\circ}$C indicating the key role of lung tissue in both experimental models.

CONCLUSIONS

Both thrombin as well as elastase produce in minipig:

- increase in pulmonary vascular resistance
- pulmonary leucostasis
- progressive respiratory failure
- hypercoagulability with disseminated intravascular coagulation
- high proteolytic activity selectively in the lung.

So, deduced with caution, in human ARDS it may be possible that lysosomal proteases released from degranulated PMNs in the lung may per se

- initiate further pulmonary PMN-sequestration by neutrophil activation and
- directly or indirectly damage lung tissue and produce respiratory failure

and thus play an important role in the pathology of this disease.

REFERENCES

1. F. W. Blaisdell, and R. M. Schlobohm, The respiratory distress syndrome. A review, Surgery 74:251 (1973).
2. H. Burchardi, and I. Hensel, Measurement of respiratory resistance and ventilation distribution in artificially ventilated patients, in: "Cardiac Lung," G. Giutini, P. Panuccio, eds., Piccin Medical Book, Padova (1979).
3. Ph. R. Craddock, D. E. Hammerschmidt, Ch. F. Moldow, O. Yamada, and H. S. Jacob, Granulocyte aggregation as a manifestation of membrane interactions with complement: possible role in leukocyte margination, microvascular occlusion, and endothelial damage, Seminars in Hematology 16:140 (1979).
4. B. Dahlmann, and H. Reinauer, Purification and some properties of an alkaline proteinase from rat skeletal muscle, Biochem. J. 171:803 (1978).
5. N. J. H. Davies, and D. M. Denison, The measurement of metabolic gas exchange and minute volume by mass spectrometry alone, Resp. Physiol. 36:261 (1979).
6. D. E. Hammerschmidt, L. J. Weaver, L. D. Hudson, Ph. R. Craddock, and H. S. Jacob, Association of complement activation and elevated plasma-C5a with adult respiratory distress syndrome, Lancet I:947 (1980).

7. I. Hensel, H. Burchardi, T. Stokke, P. Hallecker, J. Jörck, E.
    Turner, D. Weber und K.-H. Wencker, Thrombininduzierte intra-
    vasale Gerinnung am Zwergschwein als Modell zur Schocklunge
    - Veränderungen der Atemmechanik und der ventilatorischen
    Verteilung, in: "Anaesthesiologie und Intensivmedizin," K.H.
    Weis,G.Cunitz, eds., Springer, Berlin (1980).
8. W. H. Hörl, and A. Heidland, Enhanced proteolytic activity -
    cause of protein catabolism in acute renal failure, Am. J.
    Clin. Nutr. 33:1423 (1980).
9. W. H. Hörl, J. Stepinski, C. Gantert, M. Hörl, and A. Heidland,
    Evidence for the participation of proteases on protein cata-
    bolism during hypercatabolic renal failure, Klin. Wochenschr.
    59:751 (1981).
10. W. H. Hörl, C. Gantert, I. O. Auer, and A. Heidland, In vitro
    inhibition of protein catabolism by alpha$_2$-macroglobulin in
    plasma from a patient with posttraumatic acute renal failure,
    Am. J. Nephrol. 2:32 (1982).
11. U. Kasten, U. Artmann, T. Kaethner, H. Burchardi, and H.
    Köstering,Influence of elastase on blood coagulation factors:
    Studies in vitro, in rats and in Göttinger miniature pigs,
    Thrombos. Haemostas. 46:352 (1981).
12. H. Köstering, R. Rahlf, R. Rüfer, B. Schmidt, I. Marschelke,
    G. Hasenbein und H. Burchardi, Tierexperimentelle Unter-
    suchungen zur Schocklunge: Induzierte disseminierte intra-
    vasale Gerinnung, in: "Anaesthesiologie und Intensivmedizin,"
    Bd. 125, J. B. Brückner, ed., Springer, Berlin (1980).
13. C. T. Lee, A. M. Fein, M. Lippmann, H. Holtzmann, P. Kimbel,
    and G. Weinbaum, Elastolytic activity in pulmonary lavage
    fluid from patients with adult respiratory distress syndrome,
    New Engl. J. Med. 304:192 (1981).
14. O. H. Lowry, N. J. Rosebrough, A. L. Farr, and R. J. Randall,
    Protein measurement with the Folin phenol reagent, J. Biol.
    Chem. 193:265 (1951).
15. A. B. Malik, M. V. Tahamont, and A. Johnson, Effects of fibrin
    and formed elements on pulmonary hemodynamics and fluid ex-
    change, Conference on mechanisms of lung microvascular in-
    jury, New York (1981).
16. C. Mittermayer, W. Vogel, H. Burchardi, H. Birzle, K. Wiemers
    und W. Sandritter, Pulmonale Mikrothrombosierung als Ursache
    der respiratorischen Insuffizienz bei Verbrauchskoagulopathie
    (Schocklunge), Dtsch. med. Wschr. 95:1999 (1970).
17. K. Ohlsson, Granulocyte proteases, their release and inhibition
    in the body, in: "Regulatory Proteolytic Enzymes and their
    Inhibitors," S. Magnusson, ed., Pergamon Press, Oxford, New
    York (1978).
18. K. Radegran, Circulatory and respiratory effects of induced
    platelet aggregation. An experimental study in dogs, Acta
    chir. scand. Suppl. 420 (1971).

19. Th. Sacks, Ch. F. Moldow, Ph. R. Craddock, T. K. Bowers, and H. S. Jacob, Oxygen radicals mediate endothelial cell damage by complement-stimulated granulocytes, J. Clin. Invest. 61: 1161 (1978).
20. T. Stokke und H. Burchardi, Einfache, exakte Trennung von In- und Exspirationsgas während maschineller Beatmung, Anaesthesist 31:293 (1982).
21. T. Stokke, H. Burchardi, I. Hensel, P. Hallecker, E. Turner und D. Weber, Thrombininduzierte intravasale Gerinnung am Zwerg- schwein als Modell zur Schocklunge. - Veränderungen der pulmonalen Kapillarperfusion und der Diffusionskapazität, in: "Anaesthesiologie und Intensivmedizin," K.H. Weis, G. Cunitz, eds., Springer, Berlin (1980).
22. T. Stokke, H. Burchardi, I. Hensel, W. Ohrdorf und H. Boch- Fiola, Nicht-invasive Bestimmung der pulmonalen Kapillar- perfusion unter Beatmung, Anaesthesist 31:510 (1982).
23. T. Stokke, I. Hensel und H. Burchardi, Eine einfache Methode für die Bestimmung der funktionellen Residualkapazität während Beatmung, Anaesthesist 30:124 (1981).
24. T. Stokke, W. Röhrborn, I. Hensel, O. Hilfiker, U. Braun und H. Burchardi, Bestimmung der pulmonalen Diffusionskapazität während Beatmung, Anaesthesist 30:602 (1981).
25. K. Weber, and M. Osborn, The reliability of molecular weight determinations by dodecylsulfate polyacrylamide gel electro- phoresis, J. Biol. Chem. 244:4406 (1969).

INTERACTIONS OF GRANULOCYTE PROTEASES WITH INHIBITORS

IN RHEUMATOID ARTHRITIS

Lars Ekerot and Kjell Ohlsson

Departments of Hand Surgery, Clinical Chemistry
and Experimental Research, Malmö General
Hospital, S-214 01 Malmö, Sweden

INTRODUCTION

Cumulative evidence indicates that articular destruction in rheumatoid arthritis (RA) might be mediated by proteolytic enzymes[1,2,3]. The hypothesis of enzymatic degradation is emphasized by the presence in
the arthritic joint of a multitude of proteolytic enzymes with potential ability to attack the macromolecules of cartilage matrix. These enzymes can originate
from nearly all cell types found in the inflamed joint
as proliferating synovial lining cells,[4] polymorphonuclear leucocytes (PMNs)[5,6] invading the joint space
and macrophage-like cells[7,8] infiltrating the synovial
membrane. Released extracellularly, several of these
enzymes have their optimum activity at the neutral pH
prevailing within the arthritic joint and the concerted
action of these neutral proteases implies a considerable destructive potential. One reason, however, for
doubting the role of these enzymes in articular deterioration is the control mechanism exerted by the protease inhibitors present in the synovial fluid
(SF)[9,10,11].

PROTEASE - INHIBITOR INTERACTIONS

Rheumatoid SF showed significantly increased concentrations of most plasma protease inhibitors when compared with control SF[12], however, the concentration of
the low molecular weight inhibitor, antileukoprotease[13], was remarkably enough significantly lower
in RA (Table 1).

Table 1. Plasma/serum and SF concentrations of
          protease inhibitors. The level of anti-
          leukoprotease is given in µg/l, the
          levels of the remaining inhibitors in
          per cent of the concentration of a
          standard. Significant values of the nor-
          mally distributed test statistic are
          given (n.s. = non significant)

Inhibitor		RA (n=29)		Controls (n=23)		Mann-Whitney test
		range	mean	range	mean	
Antileukoprotease	serum	60-400	166	61-200	112	+3.39
	SF	10- 54	39	32- 88	52	-3.12
α_1-Antitrypsin	plasma	82-228	153	44-152	86	+4.88
(α_1-AT)	SF	20-127	82	18- 84	42	+5.19
α_1-Antichymotryp-	plasma	80-335	194	50-120	84	+5.84
sin (α_1-ACHY)	SF	20-260	94	6- 42	24	+5.84
α_2-Macroglobulin	plasma	44-125	86	44-188	93	n.s.
(α_2-M)	SF	13- 50	28	4-35	17	+3.96

   α_1-AT together with α_2-M account for more than
90% of the protease inhibiting capacity of plasma[14].
According to table 1, they retain their quantitative
dominating role also in rheumatoid SF, but the normal
plasma molar ratio of α_1-AT to α_2-M (10:1) was increased
to approximately 25:1 (Table 2).

Table 2. Plasma and SF molar concentrations
          (µmol/l) of α_1-AT and α_2-M in RA (n=29)
          and controls (n=23) (mean ± SEM)

Inhibitor		RA		Controls	
α_1-AT	plasma	36.0	± 2.0	20.3	± 1.3
	SF	19.3	± 1.0	10.0	± 0.8
α_2-M	plasma	2.15	± 0.11	2.30	± 0.21
	SF	0.700	± 0.004	0.416	± 0.005

The inhibitory capacity of SF depends not only on its molar concentration of inhibitors but also on their biologic state, i.e. the extent of consumption in complexes and the inhibitory reactivity of the non-complexed inhibitors. Studies of the electrophoretic homogeneity of α_1-AT and α_2-M in a series of SFs from seropositive rheumatoid arthritics thus showed a striking discrepancy in the degree of complexation of these inhibitors: a low degree of complexation of α_1-AT contrasted with an almost complete saturation of α_2-M ($\sim$ 90%) [14,15] (Fig. 1).

This difference could not be explained by selective inactivation of SF α_1-AT [16,17], as approximately 90% of the free α_1-AT retained its reactivity on addition of increasing amounts of human granulocyte elastase [15]. This electrophoretically demonstrable capacity of α_1-AT to bind elastase was equivalent to an enzymatic inhibitory effect shown by inhibition of the elastolytic activity on addition of increasing amounts of rheumatoid SF to a constant amount of granulocyte elastase (Fig. 2).

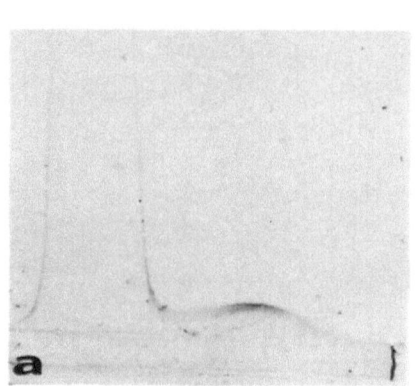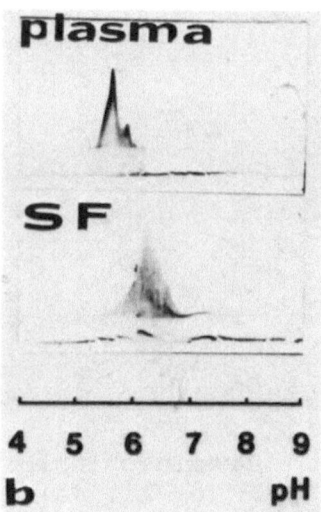

Fig. 1.   Crossed immunoelectrophoresis of rheumatoid SF (a) with antiserum against human α_1-AT and (b) following isoelectric focusing in polyacrylamide gel with antiserum against human α_2-M. (a) The bulk of α_1-AT is still in free form, while the altered isoelectric point in (b) indicates extensive complexation of α_2-M.

The electrophoretically high degree of complexa-
tion of α_2-M was corroborated by titration of α_2-M in a
series of SFs with human cathodal trypsin and BAPNA as
substrate showing some 90% of the inhibitor to be non-
reactive and therefore probably complexed. This high
degree of saturation of SF α_2-M was interpreted to be
well compatible with its polyvalent ability inhibiting
all types of endoproteases in contrast to α_1-AT, which
is a specific serine protease inhibitor[18] regulating
in particular the activity of granulocyte elastase [19].
Electrophoretically, however, it was only possible to
prove complexation of the PMN neutral protease earlier
demonstrated as a constituent of the azurophil granules[6]
with α_1-AT[15], but the use of a radioimmunoassay[20] enabled also the demonstration of granulocyte elastase
in α_1-AT-containing fractions on gel filtration of

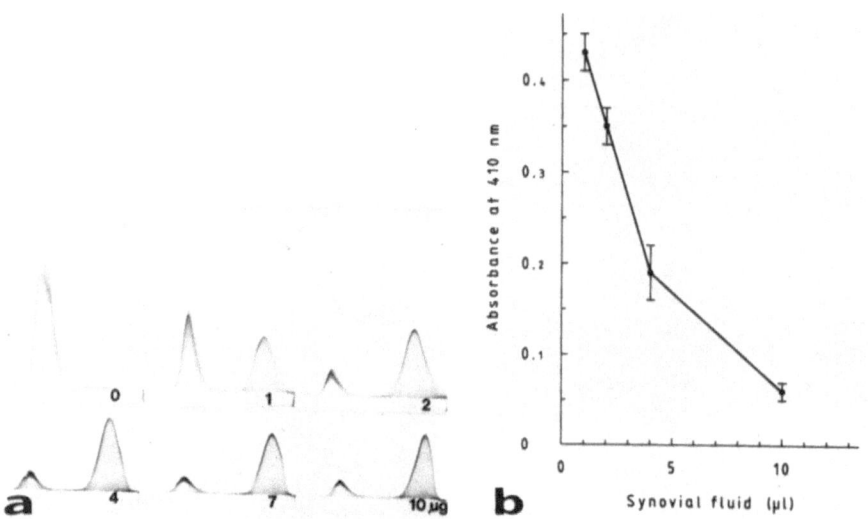

Fig. 2.   Retained inhibitory capacity of rheumatoid
           SF α_1-AT. (a) Crossed immunoelectrophoresis
           with antiserum against α_1-AT of reaction mix-
           tures of rheumatoid SF and increasing amounts
           of granulocyte elastase. The remaining peak
           of free α_1-AT on addition of elastase in
           excess represents the non-reactive inhibitor
           ($\sim$10% out of the free α_1-AT). (b) Residual
           activity of 5 mg granulocyte elastase on
           Suc-(Ala)$_3$-pNA plotted against increasing
           amounts of SF.

rheumatoid SF [21]. The immunoreactive elastase was also
precipitated by antibodies against α_1-AT and therefore
interpreted to be complexed with this inhibitor. This
quantitative discrepancy in demonstrability seemed surprising in view of the capacity of α_1-AT to bind elastase and the release of equal amounts of this enzyme
and the PMN neutral protease from PMNs on phagocytosis[22] . Several tentative explanations could be given
to this such as a rapid articular elimination of elas-
tase-α_1-AT complexes or a binding of the enzyme in an
immunochemically non-detectable form. These suggestions
were supported by the finding of a rapid dissociation
of elastase-α_1-AT complexes in rheumatoid joints (preliminary data). Thus, ^{125}I-elastase-α_1-AT complexes
were injected intra-articularly 1-2 hours prior to synovectomy and their enzyme constituent was partly recovered
in the α_2-M-containing fractions on gel filtration of
SF (Fig. 3).
    Since the radioactivity was precipitated with α_2-M
antibodies, a transfer of labeled elastase from α_1-AT
and complexation with α_2-M seemed probable. This mechanism of temporary α_1-AT inhibition would partly explain
the low degree of complexation of SF α_1-AT lacking electrophoretically demonstrable elastase-α_1-AT complexes
and might indicate the nature of the free, non-reactive
SF α_1-AT as representing released, inactivated inhibitor
molecules. A transfer of elastase would put additional

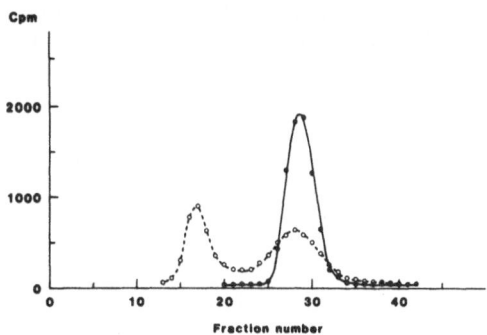

Fig. 3.   Dissociation in vivo of ^{125}I-elastase-α_1-AT
           complexes with transfer of labeled elastase to
           α_2-M. Elution patterns on gel filtration of the
           elastase-α_1-AT preparation injected (——) and
           the rheumatoid SF aspirated (---) (CPM). The
           first peak corresponds to the α_2-M-containing
           fractions, the second to the α_1-AT containing
           fractions.

stress on α_2-M at the same time as the α_2-M-bound elastase escapes immunochemical detection by blockage of
its antigenic sites.

The α_2-M-bound proteases could not be identified by
the immunochemical methods used.

The biologic function of antileukoprotease and
α_1-ACHY in rheumatoid SF is not fully recognized. Antileukoprotease is a strong inhibitor of granulocyte
elastase [12], but the absence of elastase-antileukoprotease complexes in SF was compatibel with the preserved
ability of SF α_1-AT to inhibit elastase, as elastase
is rapidly taken over from antileukoprotease by α_1-AT[23].
However, because of its low molecular weight and basic
isoelectric point, it may be assumed that antileukoprotease fulfills a physiologic function bound to the
cartilage via an ion exchange effect. α_1-ACHY is a very
fast acute phase reactant and it might serve an important biologic function in the early leukocytic phase
of inflammation as a potent inhibitor of cathepsin G[24].

ELIMINATION OF COMPLEXES

It may, however, be assumed that the intrasynovial
protease-inhibitor interactions are regulated not only
by enzymes and inhibitors but also by the resulting
complexes. This implies that the inhibitory capacity
would be determined by a balance between the flux of
free inhibitor and its elimination in complexed form,
i.e. a slow articular elimination rate would mean an
additional risk of saturation. Assuming α_2-M to play a
central role in the protease-inhibitor interactions,
it therefore seemed important to study the intra-articular fate of α_2-M complexes. Thus, intra-articularly
injected radioactive complexes were found to be eliminated with a half-life shorter than 2 hours in experimental arthritis [25].The resorption was achieved via
hematogenous and lymphatic drainage and was accelerated by joint movements, which seemed to force complexes
directly through the synovial membrane into the blood.
Retention of radioactivity in macrophage-like cells in
the regional lymph nodes indicated lymphatic clearance
of complexes in a manner similar to that of clearance
of circulating complexes by Kupffer cells in the
liver [26]. In addition, a local articular uptake of
complexes was demonstrated by the retention of radioactivity in the same type of synovial lining cells as
those containing immunohistochemically demonstrable α_2-M
(Fig. 4).

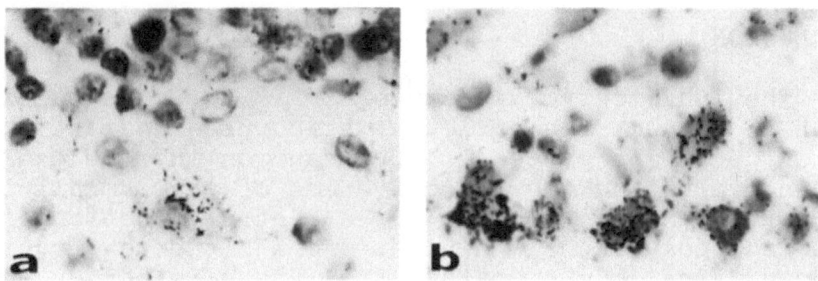

Fig. 4.   Microautoradiographs demonstrating intracel-
           lular retention of radioactive solutes in
           sections from (a) regional lymph node and
           (b) synovial membrane.

        This local phagocytic mechanism of α_2-M complex
elimination seemed to be valid also in RA. Thus, intracellular α_2-M could immunohistochemically be demonstrated in synovial membrane sections from rheumatoid arthritics [27] . This finding did not warrant any conclusion
about the origin of the immunoreactive deposits, since
the α_2-M-containing macrophage-like cells are capable
also to synthesize the inhibitor. However, following
pre-operative injection of ^{125}I-elastase-α_2-M complexes
into rheumatoid joints, the phagocytic ingestion of
complexes was corroborated by microautoradiographs
demonstrating retention of radioactivity in macrophage-
like synovial cells identical in distribution throughout the membrane with those containing immunoreactive
α_2-M [28] (Fig. 5).

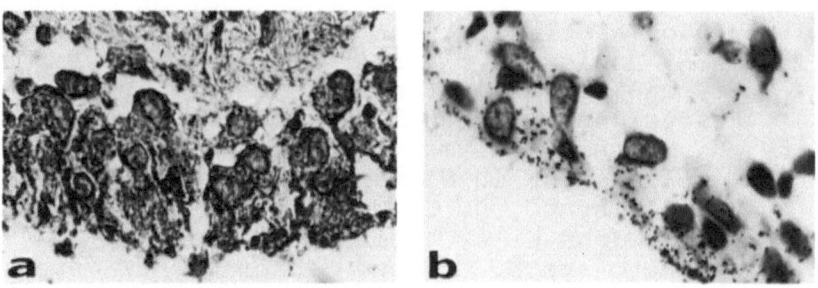

Fig. 5.   Sections of rheumatoid synovial membrane (a)
           demonstrating immunoreactive α_2-M as indica-
           ted by the dark staining and (b) showing
           labeled cells in the surface layer.

CONCLUSION

    The inhibitory capacity of SF α_1-AT seems suffi-
cient to protect the articular structures against the
action of serine proteases, but caution must be exercised in evaluating the effectiveness of the overall inhibitory mechanisms comprising a central function of
α_2-M as a polyvalent inhibitor and a hitherto largely
unknown "final common pathway" in the elimination of
proteases. The high degree of SF α_2-M saturation despite
its rapid elimination in complexed form is indicative
of a highly active process with a pronounced release of
endoproteases and an impending breakdown of their control mechanisms. This, however, applies only to free SF
and does not take into account the possibility of an
uneven articular distribution of PMNs and inhibitors
and the influence of the substrate on the enzyme-inhibitor interactions. Thus, the conditions may by quite different in the micromilieu close to the substrate, where
concentrations of proteases exceeding also that of the
local inhibitory capacity may be reached.

REFERENCES

    1. J. T. Dingle, Articular damage in arthritis
        and its control, Ann Int Med 88:821 (1978).
    2. A. J. Barrett, The possible role of neutrophil
        proteinases in damage to articular cartilage,
        Agents Actions 8:11 (1978)
    3. W. Mohr, G. Köhler, and D. Wessinghage, Poly-
        morphonuclear granulocytes in rheumatoid tis-
        sue destruction, Rheumatol Int 1:21 (1981).
    4. S. M. Krane, Collagenase production of human
        synovial tissues, Ann NY Acad Sci 256:289 (1975).
    5. N.M. Hadler, J. K. Spitznagel, and R. J. Quinet,
        Lysosomal enzymes in inflammatory synovial
        effusions, J Immunol 123:572 (1979).
    6. K. Ohlsson, Polymorphonuclear collagenase, in:
        "Collagenase in Normal and Pathological Connec-
        tive tissue", D. E. Woolley and J. M. Evanson,
        eds., John Wiley & Sons Ltd (1980).
    7. Z. Werb and S. Gordon, Secretion of a specific
        collagenase by stimulated macrophages, J Exp
        Med 142:346 (1975).
    8. Z. Werb and S. Gordon, Elastase secretion by
        stimulated macrophages, J Exp Med 142:361 (1975).
    9. W. Pruzanski, M. L. Russel, D. A. Gordon, and
        M. A. Ogryzlo, Serum and synovial fluid pro-
        teins in rheumatoid arthritis and degenerative
        disease, Am J Med Sci 265:483 (1973).

10. G. Shtacher, R. Maayan, and G. Feinstein, Proteinase inhibitors in rheumatoid arthritis, Biochim Biophys Acta 303:138 (1973).

11. D. Brackertz, J. Hagmann, and P. Kueppers, Proteinase inhibitors in rheumatoid arhtritis, Ann Rheum Dis 34:225 (1975).

12. L. Ekerot, K. G. Sjöblom, K. Ohlsson, and F. A. Wollheim, Proteinase inhibitors in rheumatoid synovial fluid. A quantitative analysis, In manuscript.

13. U. Fryksmark, K. Ohlsson, M. Rosengren, and H. Tegner, A radioimmunoassay for measurement and characterization of human antileukoprotease in serum, Hoppe Seylers Z Physiol Chem 362:1273 (1981).

14. K. Ohlsson, α_1-Antitrypsin and α_2-macroglobulin interactions with human neutrophil collagenase and elastase, Ann NY Acad Sci 256:409 (1975).

15. L. Ekerot and K. Ohlsson, Protease inhibitors in rheumatoid synovial fluid. Analysis of electrophoretic homogeneity and inhibitory capacity, Rheumatol Int 2:21 (1982).

16. N. R. Matheson, P. S. Wong, and J. Travis, Enzymatic inactivation of human alpha$_1$-proteinase inhibitor, Biochem Biophys Res Comm 88:402 (1979).

17. P. S. Wong and J. Travis, Isolation and properties of oxidized alpha$_1$-proteinase inhibitor from human rheumatoid synovial fluid, Biochem Biophys Res Comm 96:1449 (1980).

18. C. B. Laurell and J. O. Jeppsson, Protease inhibitors in plasma, in: "The Plasma Proteins", F. W. Putnam, ed.. Academic Press, New York, San Francisco, London (1975).

19. K. Ohlsson and I. Olsson, Neutral proteases of human granulocytes. III. Interaction between human granulocyte elastase and plasma protease inhibitors, Scand J Clin Lab Invest 34:349 (1974).

20. K. Ohlsson and A. S. Olsson, Immunoreactive elastase in human serum, Hoppe Seylers Z Physiol Chem 359:1531 (1978).

21. L. Ekerot and K. Ohlsson, Immunoreactive granulocyte elastase in rheumatoid synovial fluid and membrane, Scand J Plast Reconstr Surg, In press.

22. K. Ohlsson and I. Olsson, The extracellular release of granulocyte collagenase and elastase during phagocytosis and inflammatory processes, Scand J Haematol 19:145 (1977).

23. F. Gauthier, U. Fryksmark, K. Ohlsson, and
     J. G. Bieth, Kinetics of the inhibition of
     leukocyte elastase by the bronchial inhibi-
     tor, Biochim Biophys Acta 700:178 (1982).
24. K. Ohlsson and U. Åkesson, α_1-Antichymotrypsin
     interaction with cationic proteins from granu-
     locytes, Clin Chim Acta 73:285 (1976).
25. L. Ekerot and K. Ohlsson, The elimination of
     α_2-macroglobulin complexes from the arthritic
     joint. An experimental study in dogs, Scand
     J Plast Reconstr Surg, In press.
26. K. Ohlsson, Elimination of ^{125}I-trypsin-α-
     macroglobulin complexes from blood by reti-
     culoendothelial cells in dog, Acta Physiol
     Scand 81:269 (1971).
27. L. Ekerot and K. Ohlsson, Immunoreactive
     α_2-macroglobulin in rheumatoid synovial mem-
     brane, Scand J. Plast Reconstr Surg, In press.
28. L. Ekerot, K. Ohlsson, and F. Rank, The eli-
     mination of α_2-macroglobulin complexes from
     the arthritic joint. A clinical study in
     rheumatoid arthritis. Scand J Plast Reconstr
     Surg, In press.

ACKNOWLEDGEMENTS
     This investigation was supported by grants from
the Swedish Medical Research Council (project no.
B83-17X-03910-11B), the Medical Faculty, University of
Lund, the Swedish Association against Rheumatism, the
Foundation of King Gustav Vth, the Foundation of Greta
and Johan Kock, the Foundation of Malmö General Hospi-
tal for Medical Research, the Foundation of Alfred
Österlund, and the Swedish Society of Medical Research.

QUANTITATION OF HUMAN LEUKOCYTE ELASTASE, CATHEPSIN G, α-2-MACRO-
GLOBULIN AND α-1-PROTEINASE INHIBITOR IN OSTEOARTHROSIS AND
RHEUMATOID ARTHRITIS SYNOVIAL FLUIDS

G. D. Virca[1], R. K. Mallya[2], M. B. Pepys[3]
and H. P. Schnebli[1]

[1]Research Department, Pharma Division, CIBA-GEIGY,
CH-4002 Basel, [2]Department of Rheumatology, Kings
College Hospital, London, SE 5, [3]Immunological
Medicine Unit, Department of Medicine, Royal Post-
graduate Medical School, London, W12, OH5

INTRODUCTION

In both rheumatoid arthritis and osteoarthrosis the site of
disease related damage is the articular cartilage. There is ample
evidence supporting the involvement of lysosomal proteinases in
articular cartilage destruction, as seen in inflammatory joint
diseases (1-5). Barrett has made the rather startling estimate
that an inflamed joint must deal with approximately 8 mg of PMN-
released enzymes per day (5). This figure was derived from the
following assumptions: (a) neutrophils survive only 3-4 hours in
the joint (6), (b) there are approximately 4 μg elastase (and
probably an equivalent amount of cathepsin G) present per 10^6 cells
(7), and (c) synovial fluids from knee joints have shown upward to
and exceeding 10^8 cells (5).

Neutrophil enzymes have been shown to be present, and active
in rheumatoid synovial fluid (1,3,8,9). Some investigators have
also reported increased levels in cartilage of osteoarthrotic
joints and synovial membranes (10-14). Thus the presence and
degradative abilities of leukocyte proteinases make them a logical
choice to explain the destruction observed in the joints of
osteoarthrosis and rheumatoid arthritis patients.

Although there is an abundance of proteolytic enzymes present
in rheumatoid joints, there are also generous supplies of plasma
proteinase inhibitors (5,15,16), in particular α-1-proteinase inhibitor (α_1PI) and α-2-macroglobulin (α_2M). Here, we sought to
determine levels of elastase and cathepsin G activity in synovial
fluids from patients suffering from rheumatoid arthritis or osteo-

345

arthrosis, and to compare these activities with normal plasma controls. In conjunction with this we also determined the proteinase inhibitory capacity of these synovial fluids. From this, and direct measurements, the amounts of active $\alpha_1 PI$ and $\alpha_2 M$ were assessed.

PATIENTS

Seven patients were diagnosed as having typical O.A. both by clinical and radiological criteria. Age range of patients: 52-73 years; duration of the disease: 3-7 years. All O.A. patients had erythrocyte sedimentation rates of less than twenty mm/first hour. One patient was on Brufen, the others were receiving no treatment. All synovial fluids we removed for therapy.

Eight patients were diagnosed as having classical R.A. (American Rheumatology Association Classification), with active, erosive, seropositive disease and erythrocyte sedimentation rates of greater than 60 mm/first hour in all cases. Age range of patients: 61-66 years; duration of the disease: 2-12 years. All R.A. patients were being treated with non-steroidal anti-inflammatory drugs, one was also receiving gold treatment, and another penicillamine; again, all fluids were removed for therapy.

METHODS AND MATERIALS

Synovial fluid samples were drawn into EDTA coated syringes, centrifuged immediately to remove cells and stored frozen until assayed. Plasma samples were treated the same way, except only a portion of the plasma was frozen, the remainder was assayed immediately.

Leukocyte elastase and cathepsin G

Leukocyte elastase activity in synovial fluid and plasma samples was determined in the presence of a metal chelator using MeO-suc-Ala-Ala-Pro-Val-pNA as substrate. Twenty to 40 µl of synovial fluid was diluted to 160 µl with 0.9% NaCl; 40 µl o-phenanthroline (100mM) was added and the mixture preincubated at 37° for 5 min. The reaction was started by the addition of 200 µl of substrate solution (2mM MeO-suc-Ala-Ala-Pro-Val-pNA in 0.2 M Tris-HCl, pH 7.5). After an incubation period of two hours at 37° the reaction was stopped by adding 1.0 ml trichloroacetic acid (7.5%). After removal of precipitated material by centrifugation, the supernatant was poured off and the released p-nitroaniline converted to an azo dye by the sequential addition of 25 µl of $NaNO_2$ (0.18 g/10 ml), 25 µl ammonium sulfamate (0.9 g/10 ml) and 25 µl N-1-naphtyl ethylenediamin-HCl (0.18 g/10 ml). The resulting color was measured at 544 nm, and results expressed in nmoles p-nitroaniline released per ml synovial fluid (or plasma) per hour.

The determination of cathepsin G activity in synovial fluid and plasma was performed in a similar manner, except that o-phenanthroline was omitted, and Suc-Ala-Ala-Pro-Phe-pNA (Bachem Fein-chemikalien AG 4416-Bubendorf, Switzerland) in 0.1 M Tris-HCl, pH 7.4, containing 1.0 M NaCl was used as substrate (final concentration in the assay was 0.75 mM). Again, results are expressed in nmoles p-nitroaniline released/ml/hr.

α_1-proteinase inhibitor

Quantitation of total levels of α_1PI in synovial fluids and plasmas was performed via radial immunodiffusion using M-partigen plates (Behring Werk, Marburg, BRD). The amount of active α_1PI present in synovial fluids and plasma was determined by titration with trypsin. This approach was a modification of the method developed by Bieth et al. (17) for the assessment of functional α_1PI in human plasma. Varying amounts of synovial fluid (or plasma) were incubated with a constant amount of trypsin and the residual enzymatic activity was then measured and plotted as shown in Figure 1. From the linear portion of the plot, the volume of plasma or synovial fluid that contains the amount of α_1PI equivalent to the amount of trypsin used in the assay, can be extrapolated. For details and theoretical backround of this method see Bieth et al. (17).

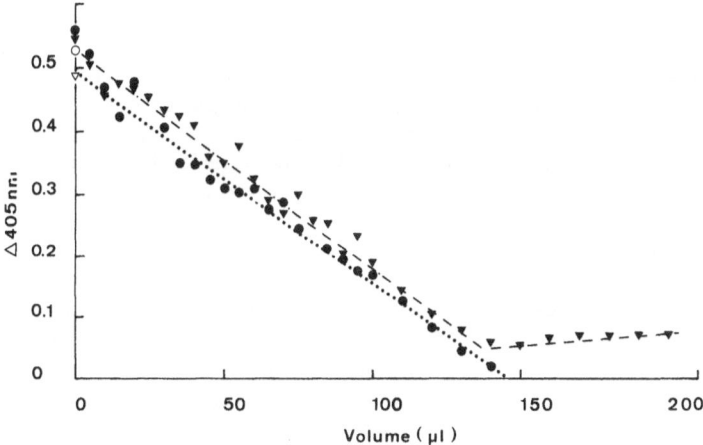

Figure 1

Trypsin titration of human plasma. A constant amount of trypsin (2.4 µg) was incubated with increasing amounts of diluted plasma (1:50 with 0.9% NaCl). Residual trypsin activity was assayed as described in methods, and is given here as O.D.405. Circles show titration performed with plasma pretreated with methylamine to eliminate α_2M activity, and triangles show titration with untreated plasma.

Figure 1 also shows the same titration following pretreatment of plasma with methylamine (0.2 M final concentration, at
25°C for one hour) to inactivate α_2M (18,19). It can be seen
that the contribution of α_2M in this assay is small, and, for
practical purposes, can be ignored.

In the experiment, fifty microliters of trypsin (3.1 μg active
enzyme) were incubated with 5-50 μl of synovial fluid that had been
diluted ten-fold with 0.9% NaCl in a total reaction volume of 200 μl
for 30 minutes at 37°. One milliliter of prewarmed (37°) substrate
solution was added (final concentrations: 50 mM Tris-HCl, pH 8.0
containing 50 mM $CaCl_2$ and 1 mM α-N-benzoyl-DL-arginine-p-nitro-
anilide HCl (BZ-ARG-pNA; Sigma Chem. Co., St.Louis, Mo. 63178)),
and the assay mixture incubated another 10 minutes. The reaction
was stopped by addition of 0.1 ml of glacial acetic acid, and released p-nitroaniline monitored at 405 nm. Plasma titrations were
conducted in a similar manner, using 2.4 μg trypsin and 1:50 diluted plasma.

All synovial fluid and plasma titrations utilized active site
titrated bovine trypsin (Worthington Corp., Freehold, N.J.).
Trypsin titration was performed by the method of Chase and Shaw
(20), using p-nitrophenyl- p^1-guanidinobenzoate HCl (Fluka, AG
Chem. Fabrik, CH-9470 Buchs) as titrant. The enzyme preparation
used in this study was 66% active.

$\alpha_2$macroglobulin

The total level of α_2M in synovial fluid and plasma was
assessed by radial immunodiffusion using M-partigen plates (Behring
Werke, Marburg, BRD). Assessment of active α_2M utilized a modification of the method of Ganrot (21), the α_2M colorimetric procedure developed by Boehringer Mannheim (Mannheim, BRD).

RESULTS AND DISCUSSION

α_1PI and α_2M were detected in all synovial fluids. The total
(immune reactive) amounts of α_1PI in the synovial fluids from O.A.
and R.A. patients were significantly different (Table 1). The α_1PI
concentrations in synovial fluid from O.A. patients ranged from
0.67-1.45 mg/ml with a mean value of 1.13 mg/ml, slightly below
normal plasma concentrations of this inhibitor. R.A. joint fluid
concentrations of this inhibitor ranged from 1.51-3.86 mg/ml
(mean=2.49 mg/ml). This greater than two fold increase in R.A.
synovial fluid levels of α_1PI relative to O.A. synovial fluid
concentrations is reflective of the acute phase response of α_1PI
in the chronic inflammation associated with this disease. In each
instance (R.A. and O.A.) synovial fluid α_1PI concentrations were
at least 50% of their respective plasma levels, thereby indicating
unrestricted passage of α_1PI from plasma to the inflamed joint.

Table 1

Alpha-1-proteinase inhibitor activity in R.A. (N=8) and O.A. (N=7) synovial fluids and normal plasma controls (N=10). Values are presented as range, mean and standard deviation.

	R.A. SYNOVIAL FLUIDS	O.A. SYNOVIAL FLUIDS	NORMAL PLASMA
IMMUNE REACTIVE α_1PI (MG/ML)	1.51 - 3.86 (2.49 ± 0.85)	0.67 - 1.45 (1.13 ± 0.25)	1.79 - 2.0 (1.90 ± 0.08)
TRYPSIN REACTIVE α_1PI (MG/ML)	1.38 - 2.47 (1.93 ± 0.42)	0.88 - 1.68 (1.21 ± 0.36)	1.68 - 1.81 (1.74 ± 0.04)
(TRYPSIN/IMMUNE) REACTIVE α_1PI (x 100)	59 - 100% (80.2 ± 12.0%)	82 - 100% (94.6 ± 7.1%)	87 - 100% (92.0 ± 4.0%)

In contrast, synovial fluid α_2M levels from O.A. and R.A. patients were less varied (Table 2). Although, again concentrations of this inhibitor were higher in the R.A. synovial fluids: O.A. synovial fluid range, 0.32-1.02 mg/ml, mean 0.64 mg/ml, R.A. synovial fluid range 0.70-1.42 mg/ml with a mean value of 1.06 mg/ml. These values are approximately one fourth to one third normal plasma levels. Since α_2M is not an acute phase protein in man there can not be a corallary to the above arguement to explain higher concentrations of this inhibitor in R.A. synovial fluids relative to the O.A. joint fluids. However, one plausible explanation could be that increased vascular permeability during inflammation more readily permits the very large (725,000 dalton) α_2M molecules to enter the synovial fluid in R.A. patients.

Table 2

Alpha-2-macroglobulin activity in R.A. (N=8) and O.A. (N=7) synovial fluids and normal plasma controls (N=10). Values are presented as range, mean and standard deviation.

	R.A. SYNOVIAL FLUIDS	O.A. SYNOVIAL FLUIDS	NORMAL PLASMA
IMMUNE REACTIVE α_2M (MG/ML)	0.70 - 1.42 (1.06 ± 0.22)	0.32 - 1.02 (0.64 ± 0.25)	1.63 - 2.76 (2.25 ± 0.37)
TRYPSIN REACTIVE α_2M (MG/ML)	0.03 - 0.65 (0.27 ± 0.20)	0.15 - 1.00 (0.46 ± 0.28)	1.55 - 2.42 (2.01 ± 0.30)
(TRYPSIN/IMMUNE) REACTIVE α_2M (x 100)	3 - 48% (24.5 ± 14.8%)	46 - 98% (65.7 ± 17.1%)	78.6 - 100% (89.5 ± 6.3%)

The proportion of inactive inhibitors was much greater in the synovial fluids from R.A. patients than in those from O.A. patients. The amount of α_1PI exhibiting trypsin inhibitory activity (when compared to total, immune reactive α_1PI) revealed that all syno-

vial fluid samples examined contained between 59-100% active α_1PI
(Table 1). O.A. synovial fluid levels ranged from 82-100% trypsin
reactive α_1PI (mean=94.6%), these values coincide nicely with
plasma controls and indicate the presence of negligible amounts
of α_1PI-enzyme complex (or otherwise inactivated α_1PI) in synovial
fluids of patients suffering from O.A.

The synovial fluids from R.A. patients on the other hand revealed trypsin-inhibiting α_1PI ranging from 59-100% with a mean
value of 80.2%; indicating that on average approximately 15% of
R.A. synovial fluid α_1PI was unable to inhibit trypsin (when compared to the inhibitory capacity of plasma controls). We assume
that this 15% unreactive α_1PI is due to α_1PI-enzyme complexes
and/or oxidized α_1PI, as had been previously reported (22).

Assays to determine the amount of active α_2M present in the
synovial fluid samples likewise revealed O.A. synovial fluids to
contain higher proportions of active inhibitor than R.A. samples.
In each instance values were derived by comparing trypsin-reactive
α_2M with total, immunereactive levels of the inhibitor (Table 2).
Active α_2M in O.A. synovial fluids ranged from 46-100% (mean=65.7%),
whereas R.A. synovial fluids revealed 3-48% active α_2M (mean=24.5%).
Again, as was the case with α_1PI, a much greater percentage of the
α_2M present in R.A. synovial fluid was inactive compared to O.A.
joint fluid and normal plasma α_2M. Although α_2M can be inactivated
by means other than formation of α_2M-enzyme complex, e.g. inactivation by small molecular weight nucleophiles (25,26), we believe
these results reflect a greater percentage of α_2M-proteinase complexes to be present in R.A. relative to O.A. synovial fluids.
The presence of α_2M-elastase complexes was confirmed by monitoring
elastase activity in the presence and absence of 150 µg/ml active
α_1PI. The added α_1PI did not inhibit the elastase activity, thus
confirming that this activity is α_2M-bound.

Reactive site specific substrates enabled us to detect both
leukocyte elastase and cathepsin G in all synovial fluid samples,
control plasmas from healthy donors and in plasmas from O.A. and
R.A. patients. The results are shown in Table 3. R.A., O.A. and
normal plasma elastase and cathepsin G activities were similar,
with the patient plasmas having slightly higher activities; this
is consistent with previous studies (23). Activitites of both
elastase and cathepsin G were approximately 5-10 fold higher
in the O.A. synovial fluids than in the plasma samples, about 100
times greater in R.A. synovial fluids than in plasma controls.

Our study has confirmed that R.A. and O.A. synovial fluids
possess adequate levels of active α_2M and α_1PI; nevertheless,
proteolytic destruction of articular cartilage in these disease
states is apparent. A possible explanation is that these inhibitors are unable to gain ready access to the released lysosomal

Table 3

Leukocyte elastase and cathepsin G activities in R.A. and O.A. synovial fluids and plasmas, and normal plasma controls. All activities are expressed as nmoles/ml/hr.

	ELASTASE	CATHEPSIN G
NORMAL PLASMA (N=10)	0.41 - 1.45 (0.9 ± 0.3)	1.83 - 4.64 (3.0 ± 0.84)
R.A. PLASMA (N=8)	0.32 - 7.97 (1.26 ± 1.22)	1.32 - 7.39 (4.36 ± 4.29)
O.A. PLASMA (N=7)	0.48 - 3.44 (1.23 ± 1.12)	1.34 - 7.06 (2.63 ± 2.19
R.A. SYNOVIAL FLUIDS (N=8)	65.7 - 254.7 (109.4 ± 62.7)	91.6 - 863.9 (297.7 ± 231.1)
O.A. SYNOVIAL FLUIDS (N=7)	3.4 - 24.8 (11.6 ± 7.2)	2.1 - 32.9 (17.7 ± 11.6)

enzymes. Presumably the pannus is capable of creating a demarcation between the inhibitor-rich synovial fluid and the bulk of pannus contents. Thus lysosomal enzymes released on the synovial fluid surface of pannus may be effectively inhibited, whereas those proteinases at the interface of pannus with cartilage are for the most part inaccessible to the synovial fluid inhibitors. This inaccessibility could help explain the presence of both active and enzyme-comlexed inhibitor in the synovial fluid at the same time as articular cartilage destruction is ongoing.

The presence of high concentrations of proteinase-inhibitor complexes in R.A. synovial fluids is not surprising, because significant amounts of PMN proteinases are released within the rheumatoid joint. Therefore the observed levels of complexes may merely reflect a steady state situation in the inflammed joint. Furthermore, high levels of enzyme-inhibitor complexes could possibly reflect an aberration in the clearance of these complexes from the rheumatoid joint. Half-times of $\alpha_2 M$ and $\alpha_1 PI$ complexes in plasma are very short (12 and 60 minutes respectively (24)), indicating rapid rates of removal from the system. Thus extended life time of complexes in R.A. (and to a lesser degree in O.A.) synovial fluids could contribute to chronic inflammation.

REFERENCES

1. A. Janoff, G. Feinstein, C. J. Malemud, and J. M. Elias, Degra-
    dation of cartilage proteoglycan by human leukocyte granule
    neutral proteases - a model of joint injury - I. J. Clin.
    Invest. 57:615 (1976).
2. P. M. Starkey, A. J. Barrett, and M. C. Burleigh, The degra-
    dation or articular collagen by neutrophil proteinases, Bio-
    chem. Biophys. Acta 483:386 (1977).
3. H. Keiser, R. A. Greewald, G. Feinstein, and A. Janoff, Degra-
    dation neutral proteases - a model of joint injury - II. J.
    Clin. Invest. 57:625 (1976).
4. D. S. Howell, A. I. Sapolsky, J. C. Pita, and J. F. Woessner,
    The pathogenesis of osteoarthritis, Semin. Arthritis. Rheum.
    5:365 (1976).
5. A. J. Barrett, The possible role of neutrophil proteinases in
    damage to articular cartilage, Agents and Actions 8:11 (1978).
6. S. Ruddy, Synovial fluid: Mirror of the inflammatory lesion in
    rheumatoid arthritis, in: "Rheumatoid Arthritis," E. D. Har-
    ris, ed., Medcom Press, New York (1974).
7. K. Ohlsson,and M. Delshammar, Interactions between granulocyte
    elastase and collagenase and the plasma proteinase inhibitors
    in vitro and in vivo, in: "Dynamics of connective Tissue
    Research," P. M. C. Burleigh and A. R. Poole, eds., North-
    Holland Publishing Co., Amsterdam (1975).
8. G. Weissmann, Lysosomal mechanisms of tissue injury in arthritis,
    N. Engl. J. Med. 286:141 (1972).
9. C. J. Malemud, and A. Janoff, Human polymorphonuclear leukocyte
    and cathepsin G mediate the degradation of lapine articular
    cartilage proteoglycan, Ann. N.Y. Acad. Sci. 256:254 (1975).
10. M. G. Ehrlich, H. J. Mankin, and B. V. Treadwell, Acid hydrolase
    activity in osteoarthritic and normal human cartilage, J.
    Bone Joint Surg. 55A:1068 (1973).
11. A. J. Anderson, Lysosomal enzyme activity in rate with adjuvant -
    induced arthritis, Ann. Rheum. Dis. 29:307 (1970).
12. P. Barland, A. B. Novikoff, and. D. Hammerman, Lysosomes in the
    synovial membrane in rheumatoid arthritis: a mechanism for
    cartilage erosion, Trans. Ass. Am. Phy. 77:239 (1964).
13. M. I. Barnhart, C. Quintana, and H. L. Lenon, Proteases in in-
    flammation, Ann. N.Y. Acad. Sci. 146:527 (1968).
14. G. P. Kerby, and S. M. Taylor, Enzymatic activity in human synov-
    ial fluid from rheumatoid and non-rheumatoid patients, Proc.
    Soc. Exp. Biol. Med. 126:865 (1967).
15. M. Verwart, K. Fehr, A. Baici, G. Sommermeyer, M. Knöpfel, M.
    Cancer, P. Salgam, and A. Böni, Degradation in vivo of ar-
    ticular cartilage in rheumatoid arthritis by leukocyte elas-
    tase from polymorphonuclear leukocytes, Rheumatol. Int. 1:
    121 (1981).

16. K. Kleesiek, S. Neumann, and H. Greiling, Determination of the elastase activity and proteinase inhibitors in the synovial fluid, Fresenius Z. Anal. Chem. 311:434 (1982).

17. J. Bieth, P. Metais, and J. Warter, Activation, inhibition and protection of tryptic activity and α-chymotryptic activity by normal human serum, Clin. Chim. Acta. 20:69 (1968).

18. M. Steinbuch, L. Pejaudier, M. Quentin, and V. Martin, Molecular alteration of α_2-macroglobulin by aliphatic amines, Biochim. Biophys. Acta. 154:228 (1968).

19. G. S. Salvesen, and A. J. Barrett, Covalent binding of proteinases with α_2-macroglobulin, Biochem. J. 187:695 (1979).

20. T. Chase, and E. Shaw, p-Nitrophenyl-p-guandinobenzoate HCL: a new active site titrant for trypsin, Biochem. Biophys. Res. Comm. 29:508 (1967).

21. P. O. Ganrot, Determination of α_2-macroglobulin as trypsin-protein esterase, Clin. Chim. Acta 14:493 (1966).

22. P. S. Wong, and J. Travis, Isolation and properties of oxidized alpha-1-proteinase inhibitor from human rheumatoid synovial fluid, Biochem. Biophys. Res. Comm. 96:1449 (1980).

23. H. P. Schnebli, P. Christen, M. Jochum, R. K. Mallya and M. B. Pepys, Plasma levels of inhibitor bound leukocytic elastase in rheumatoid arthritis patients, in: "Proteases: Potential Role in Health and Disease," W. H. Hörl and A. Heidland, eds., Plenum Publ. Corp., New York, in press (1983).

24. K. Ohlsson, and C.-B. Laurell, The disappearance of enzyme-inhibitor complexes from circulation of man, Clin. Sci. Mol. Med. 51:87 (1976).

PLASMA LEVELS OF INHIBITOR-BOUND LEUKOCYTIC ELASTASE IN

RHEUMATOID ARTHRITIS PATIENTS

H. P. Schnebli[1], P. Christen[1], M. Jochum[2],
R. K. Mallya[3] and M. B. Pepys[4]

[1] Research Department, Pharma Division, Ciba-Geigy,
CH-4002 Basel, [2] Abt. Klinische Chemie, Chirurgische
Klinik, Universität München, D-8000 Munich. [3] Dept.
Rheumatology, Kings College Hospital, London, SE 5. [4]
Immunological Medicine Unit, Dept. Medicine, Royal
Postgraduate Medical School, London, W12, OH5

INTRODUCTION

It is generally agreed that: (a) massive numbers of PMN leukocytes accumulate at sites of chronic inflammation (1), and in
inflammatory synovial fluids (2), (b) these PMN leukocytes release elastase (3) as well as a number of other enzymes and inflammatory mediators (4) during phagocytosis or other mechanisms (5),
(c) substantial quantities of elastase have been identified in the
synovium (6,7) or, complexed to naturally occuring inhibitors
in synovial fluids of RA patients (8-12) or penetrated into the
articular cartilage (13,14).

Without entering the debate as to whether or not leukocytic
elastase is responsible for tissue destruction in RA and other
inflammatory diseases (for a review see Barrett, ref. 15), we
have taken the findings listed above to suggest that leukocytic
elastase released at sites of inflammation may lead to elevated
plasma levels of this enzyme.

Indeed, elevated plasma levels of leukocyte elastase - α_1PI
complexes have previously been observed (16) and subsequently
followed up in more detail (17) in patients with leukemia or
septicaemia. Increased levels of elastase - α_2M complexes were
also found in septicaemic patients (18).

In the present study we measured the plasma levels of elastase - α_1PI complexes by an immunological technique (ELISA)
(Method A) and of α_2M bound elastase with a synthetic chromogenic substrate (Method B) in RA patients (n=87) and healthy
controls (n=24).

PATIENTS AND METHODS

Twenty seven male and 60 female patients suffering from rheumatoid arthritis according to the criteria of the American Rheumatism Association were studied. Their disease activity was assessed
by a new multivariate analysis comprising the following criteria:
morning stiffness, pain scale, grip strength, articular index,
haemoglobin concentration and erythrocyte sedimentation rate (19).
On the basis of their global score in the analysis the patients
were assigned to one of four grades of disease activity (MDAG 1-4).

Blood samples were drawn into EDTA-coated tubes and centrifuged
immediately in order to remove cells; plasma was then stored at
-20° until assayed.

Levels of immunreactive elastase in plasma were determined
with a newly developed enzyme-linked immunoassay (ELISA) (12,20).
This method (Method A) measures exclusively the elastase complexed to α_1PI. Plasma samples were incubated in plastic tubes
coated with antibodies against elastase. After washing with
buffer, the surface fixed elastase-α_1PI complexes were reacted
with anti-α_1PI antibodies labelled with alkaline phosphatase.
Under the conditions used, the enzymatic activity of the alkaline
phosphatase (assayed with p-nitrophenyl phosphate) was proportional to the concentration of elasatase-α_1PI complex in the
sample. (For details of this assay, see ref. 20).

Leukocyte elastase activity in plasma was measured by a
modification (18) of the procedure described earlier (21). In
this assay (Method B) the release of p-nitroaniline from Methoxy-
Succinyl-L-Ala-L-Ala-L-Pro-L-Val-p-Nitroanilide (MeO-Suc-Ala-Ala-
Pro-Val-pNA), a highly specific chromogenic substrate for leukocytic elastase (22), was determined photometrically; o-phenanthroline was included to inhibit an unrelated metal dependent elastase-like enzyme (18, 23). Briefly, 0.2 ml of plasma was diluted
with 0.2 ml of buffer containing the substrate and the metal
chelator; final concentrations: 0.1 M Tris-HCl, pH 7.5, 1 mM
MeO-Suc-Ala-Ala-Pro-Val-pNA and 10 mM o-phenanthroline. The
mixture was incubated at 37° for 1 to 6 hours and the reaction
was stopped by the addition of 1 ml trichloroacetic acid (7.5 %).
Precipitated protein was removed by centrifugation and the
released nitroaniline was transformed to an azo-dye by the
sequential addition of 25 µl each of sodium nitrite (1.8 %),

ammonium sulphamate (9 %) and N-1-naphtyl ethylene-diamine di-HCl
(1.8 %) to the supernatant. The resulting color was measured at
550 nm. The corresponding enzyme activity is expressed in nmoles/
ml/h of substrate hydrolyzed.

RESULTS AND DISCUSSION

     The present study was undertaken to determine whether the re-
lease of leukocyte elastase at sites of active inflammation is reflected in the plasma levels of this enzyme, and whether this may
be diagnostically useful in RA. Two independent methods for the
determination of HLE plasma levels were employed.

     The design of the enzyme-linked immunoassay (20) ensures that
only HLE bound to α_1-proteinase inhibitor was measured (Method A).
In our hands, the HLE levels in control plasma were compatible
with the previously published values of 84 $\pm$ 25 ng/ml (20) and
60 ng/ml (median value) (24) using the same assay system.

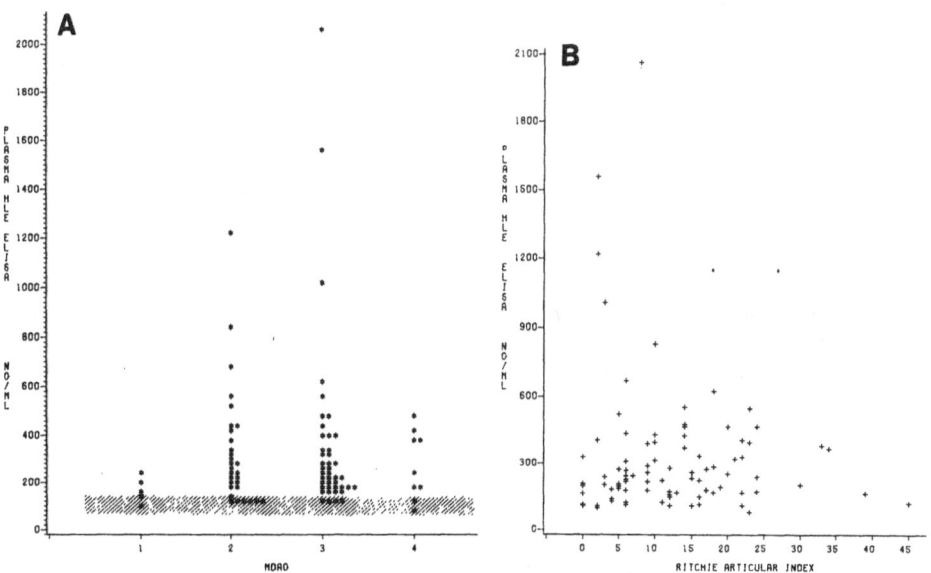

Fig. 1   HLE plasma levels measured immunologically as α_1PI
           complex in 87 RA patients: comparison with MDAG (A),
           a weighted index of disease activity, or with the
           Ritchie articular index (B).

     The majority of the plasma samples from RA patients had
clearly elevated HLE-α_1PI levels (Fig. 1A); 65 of 87 samples
revealed values more than 2 standard deviations above the normal
mean value. However, as can be seen in the same figure, the HLE
concentration in plasma does not correlate with the "mean disease
activity grade (MDAG)" (19), a weighted index based on 6 disease

parameters (see patients and methods) or with the Ritchie articular index (25) alone (Fig. 1 B). Interestingly, in the same
patient collective, C-reactive protein (CRP) levels correlated
well, both with the MDAG and the Ritchie index (26).

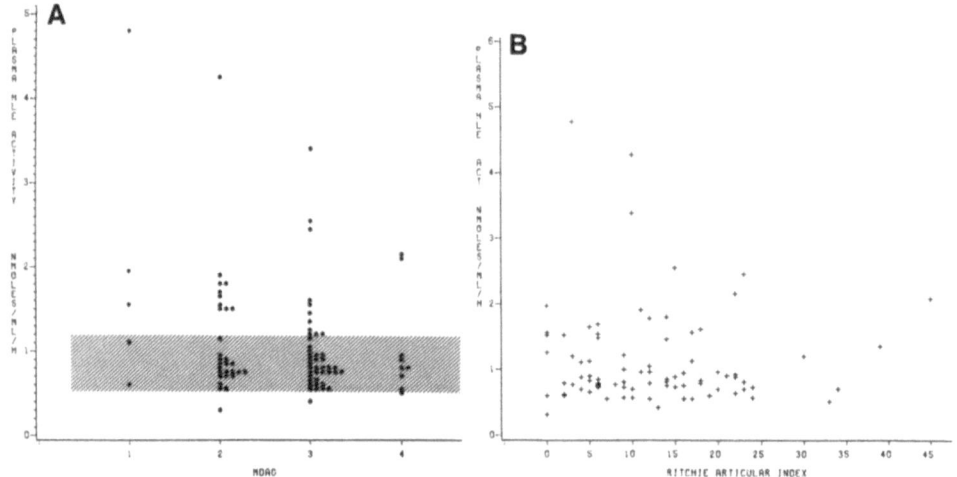

Fig. 2   HLE plasma levels measured as enzymic activity of α_2M
          complexes in 87 RA patients: comparison with MDAG (A)
          or with the Ritchie articular index (B).

      The measurement of elastase activity in plasma (Method B) is
based on the fact that α_2M bound proteinases retain hydrolytic
activity towards low M_r substrates, though they are unable to
attack most large (protein) substrates. As shown previously with
plasma from septicaemic patients (18), this elastase activity can
be separated from the bulk of plasma proteins on Sephacryl S300.
The apparent molecular weight (approx. 800 000 Da), the substrate
specificity (MeO-Suc-Ala-Ala-Pro-Val-pNA) and the inhibition
pattern (insensitive to EDTA and o-phenanthroline, sensitive to
PMSF and the elastase specific inhibitor Eglin C) identifies this
activity as leukocyte elastase-α_2M complex. Compared with plasma
from 24 healthy controls, HLE activity was elevated in a number
of plasma samples from RA patients (20 out of 87 samples were
elevated more than 2 standard deviations above the normal mean)
(Fig. 2A).

      However, again no correlation could be found between plasma
HLE activity levels and disease activity (Fig. 2). Furthermore
the HLE-α_1PI levels (Method A) and the HLE-α_2M levels (Method B)
did not correlate (R<0.1, Fig. 3).

      It must be stressed that the levels of elastase activity in
plasma are very low, requiring a sensitive assay and prolonged
reaction times (see methods). Normal plasma (n=24) hydrolyze
approximately 0.9 ± 0.3 nmoles MeO-Suc-Ala-Ala-Pro-Val-pNA per

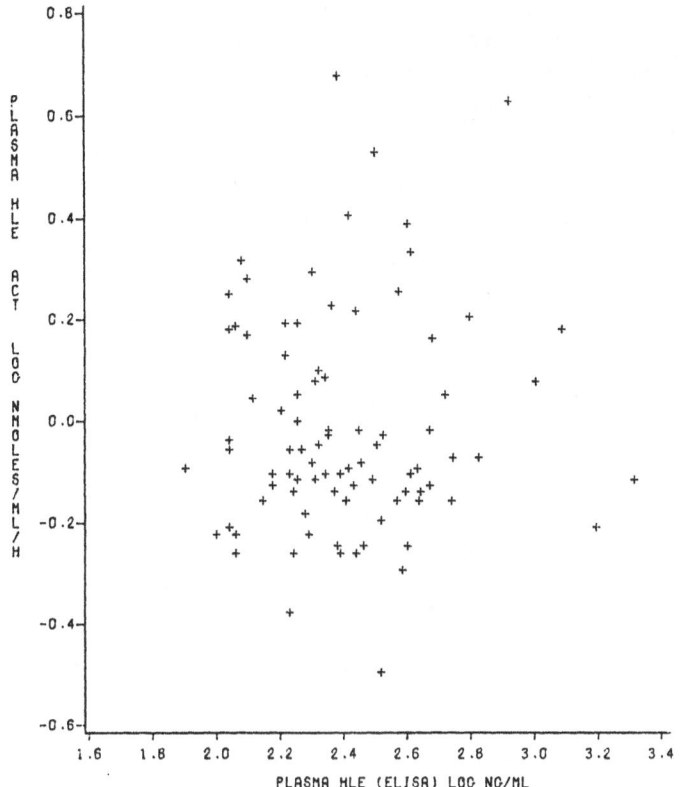

Fig. 3:   HLE-α_1PI levels and HLE-α_2M activities in plasma of
           87 RA patients (note logarithmic transformation).

ml per hour. Although an extrapolation from this value to the
amount of enzyme actually present as α_2M complex is not permissible, it can be estimated to be about 1-2 ng elastase/ml
plasma. This is much less than the 84 ng elastase/ml bound to α_1PI
in normal plasma (see above). However, these values are quite
compatible, when the known rates of complex formation and elimination are taken into account: the association rate constant of
elastase/α_1PI is 1.5 to 2 fold higher than that of elastase/α_2M
(27); together with the much higher plasma concentration of α_1PI
(approx. 10 times in molar terms) this accounts for a 15-20 fold
higher rate of formation of the elastase/α_1PI complex compared
with the formation of the elastase/α_2M complex in plasma. Indeed,
it was found that 90 % of a small amount of radiolabelled elastase
added to human serum was bound to α_1PI (28). The half-time of elimination of elastase-α_2M complexes in man is approx. 12 min, that
of elastase -α_1PI is approx. 60 min (29).

Using these numbers for a simple computer simulation experiment we
calculated an influx of 47 ng/ml/h of elastase into plasma (of healthy
controls) creating steady state levels of elastase bound to α_1PI and
to α_2M of 90 ng/ml and 0.9 ng/ml, respectively, - surprisingly close
to our estimated values of 84 and 1 to 2 ng/ml.

These calculations, although to be interpreted with caution,
point to another important consideration. The flux of 47 ng of
elastase per ml plasma per hour would indicate a total turnover of
HLE of 2 to 3 mg per day, which is far below the estimated value of
300-400 mg elastase assumed to be released as a consequence of the
normal life span of PMN leukocytes in healthy persons (28). It must
be concluded, therefore, that only a small part of the elastase present in the neutrophil granules ever reaches the circulation. This
might be one reason for the lack of correlation of plasma HLE levels
and disease activity as observed here. Other reasons may be inflammation related alteration of the clearance rate (reticulo endothelial
system) and altered rates of synthesis and consumption of proteinase
inhibitors (α_1PI is an acute phase reactant).

It is concluded that, although HLE is substantially elevated in
the plasma of many RA patients, the plasma concentration of this enzyme by itself cannot serve as an indicator of disease activity. It
is possible, however, that in longitudinal studies (within the same
patients) changes in plasma HLE levels may be indicative of disease
progression; indeed, changes in plasma levels of HLE have previously
been shown to be indicative of disease progression in septicaemia(17).

REFERENCES

1. W. Mohr, and D. Wessinghage, The relationship between polymorpho-
    nuclear granulocytes and cartilage destruction in rheumatoid
    arthritis, Z. Rheumatol. 37:81 (1978).
2. S. Ruddy, Synovial fluid: Mirror of the inflammatory lesion in
    rheumatoid arthritis, in: "Rheumatoid arthritis," E. D. Harris,
    ed., Medcom Press, New York (1974).
3. L. J. Ignarro, Regulation of lysosomal enzyme secretion: Role
    in inflammation, Agents and Actions 4:241 (1974).
4. P. M. Henson, Mechanisms of mediator release from inflammatory
    cells, in: "Mediators of Inflammation," G. Weissmann, ed.,
    Plenum Press, New York (1974).
5. S. Honig, S. Hoffstein, and G. Weissmann, Leukocyte lysosomes
    and inflammation: The example of arthritis, Pathobiol.
    Annu.   8:315 (1978).
6. J. Saklatvala, A. J. Barrett, Identification of proteinases in
    rheumatoid synovium. Detection of leukocyte elastase, cathep-
    sin G and another serine proteinase, Biochim.Biophys. Acta
    615:167 (1980).

7. H. Menninger, R. Putzier, W. Mohr, D. Wessinghage, and K. Till-
    mann, Granulocyte elastase at the site of cartilage erosion
    by rheumatoid synovial tissue, Z. Rheumatol. 39:145 (1980).
8. F. S. Steven, A. Torre-Blanco, and J. A. A. Hunter, A neutral
    protease in rheumatoid synovial fluid capable of attacking
    the telopeptide regions of polymeric collagen fibrils, Bio-
    chim.Biophys. Acta 405:188 (1975).
9. G. D. Virca, H. P. Schnebli, R. K. Mallya, and M. B. Pepys,
    Quantitation of leukocyte elastase, α_1-proteinase inhibitor
    and α_2-macroglobulin in synovial fluid, in preparation.
10. G. P. Kerby, and S. M. Taylor, Enzymatic activity in human
    synovial fluid from rheumatoid and non-rheumatoid patients,
    Proc. Soc. Exp. Biol. Med. 126:865 (1976).
11. N. M. Hadler, J. K. Spitznagel, and R. J. Quinet, Lysosomal en-
    zymes in inflammatory effusions, J. Immunol. 123:572 (1979).
12. K. Kleesiek, S. Neumann, and H. Greiling, Determination of the
    elastase α_1-proteinase inhibitor complex, elastase activity
    and proteinase inhibitors in the synovial fluid, Fresenius
    Z. Anal. Chem. 311:434 (1982).
13. J. D. Sandy, A. Sriratana, H. L. G. Brown, and D. A. Lowther,
    Evidence for polymorphonuclear-leucocyte-derived proteinases
    in arthritic cartilage, Biochem. J. 193:193 (1981).
14. M. Velvart, K. Fehr, A. Baici, G. Sommermeyer, M. Knöpfel, M.
    Cancer, P. Salgam, and A. Böni, Degradation in vivo of arti-
    cular cartilage in rheumatoid arthritis by leucocyte elastase
    from polymorphonuclear leucocytes, Rheumatol. Int. 1:121
    (1981).
15. A. J. Barrett, The possible role of neutrophil proteinases in
    damage to articular cartilage, Agents and Actions 8:11 (1978).
16. R. Egbring, W. Schmidt, G. Fuchs, and K. Havemann, Demonstration
    of granulocytic proteases in plasma of patients with acute
    leukemia and septicaemia with coagulation defects, Blood 49:
    219 (1977).
17. M. Jochum, K. H. Duswald, E. Hiller and H. Fritz, Plasma levels
    of human granulocytic elastase-α_1-proteinase inhibitor com-
    plex (E-α_1PI) in patients with septicemia and acute leukemia,
    in: "Progress in Clinical Enzymology II," D. M. Goldberg and
    M. Werner, eds., Masson Publ., N.Y., in press (1982).
18. L. P. Nelles, M. Jochum, K. H. Duswald and H. P. Schnebli, In-
    creased levels of two distinct elastase like hydrolases in
    plasma during septicaemia, in: "Progress in Clinical Enzymo-
    logy II," D. M. Goldberg and M. Werner, eds., Masson Publ.,
    N.Y., in press (1982).
19. R. K. Mallya, and B. E. W. Mace, The assessment of disease ac-
    tivity in rheumatoid arthritis using a multivariate analysis,
    Rheumatol. and Rehabil. 20:14 (1981).
20. S. Neumann, N. Hennrich, G. Gunzer, and H. Lang, Enzyme linked
    immuno-assay for human granulocyte elastase in complex with
    α_1-proteinase inhibitor in plasma, J. Clin. Chem. Biochem.
    19:232 (1981).

21. H. P. Schnebli, Adjuvant-induced inflammatory disease in the
     rat: Plasma levels of peptide hydrolases and protease inhi-
     bitors reflect disease activity, Agents and Actions 9:497
     (1979).
22. K. Nakajima, J. C. Powers, B. M. Ashe, and M. Zimmermann, Mapp-
     ing the extended substrate binding site of cathepsin G and
     human leucocyte elastase, J. Biol. Chem. 254:4027 (1978).
23. J. Saklatvala, Hydrolysis of the elastase substrate succinyl-
     trialanine nitroanilide by a metal-dependent enzyme in rheu-
     matoid synovial fluid, J. Clin. Invest. 19:794 (1977).
24. D. Neumeier, A. Fateh-Moghadam und G. Menzel, Humane Granulo-
     zyten-Elastase. I. Zur Methodik einer enzymimmunologischen
     Bestimmung des Elastase-α_1-Proteinaseinhibitor-Komplexes,
     Fresenius Z. Anal. Chem. 311:389 (1982).
25. D. M. Ritchie, J. A. Boyle, J. M. McInnes, M. K. Jasani, T. G.
     Dalcos, P. Grieveson, and W. W. Buchanan, Clinical studies
     with an articular index for the assessment of joint tender-
     ness in patients with rheumatoid arthritis, Q. J. Med. 147:
     393 (1968).
26. R. K. Mallya, F. C. deBeer, H. Berry, E. D. B. Hamilton, B. E.
     Mace, and M. B. Pepys, Correlation of clinical parameters of
     disease activity in rheumatoid arthritis with serum concen-
     tration of C-reactive protein and erythrocyte sedimentation
     rate, J. Rheumatol., in press.
27. G. D. Virca, and J. Travis, Kinetics of association of human
     leukocyte elastase and cathepsin G with human α_2-macroglobu-
     lin, in preparation.
28. K. Ohlsson, and A. S. Olsson, Immunreactive granulocyte elastase
     in human serum, Z. Physiol. Chem. 359:1531 (1978).
29. K. Ohlsson, and C. B. Laurell, The disappearance of enzyme-in-
     hibitor complexes from circulation of man, Clin. Sci. Molec.
     Med. 51:87 (1976).

ACKNOWLEDGEMENT

    We are grateful to Dr. S. Neumann and Dr. H. Lang (Merck, Darm-
stadt) for providing us with the materials for the ELISA assays.

a_2M-PASEBIC ASSAY : A SOLID PHASE IMMUNOSORBENT ASSAY TO CHARACTERIZE ALPA$_2$-MACROGLOBULIN - PROTEINASE COMPLEXES AND THE PROTEINASE BINDING-CAPACITY OF ALPHA$_2$-MACROGLOBULIN

Wolfgang Borth

Institute of Immunology of the University
Borschkegasse 8 A
A-1090 Vienna, Austria

INTRODUCTION

Alpha$_2$-macroglobulin (a_2M) is a major plasma antiproteinase, which displays unique properties : (i) it is a scavenger of virtually all proteinases (E.C.3.4.21-24) (Barrett,1973); (ii) for some of the proteinases it is the major inhibitor (e.g.:metallo-proteinases) (Borth et al.,1981); (iii) a_2M does not bind to the active center of the enzyme , and small peptides (M_r <10,000) have access to the catalytic site of the proteinase (Ganrot, 1966) ; (iv) a_2M.Proteinase (a_2M.P) complexes are rapidly cleared by macrophages (Debanne et al., 1975; Flory and Vischer,1981) and fibroblasts (Mosher and Vaheri, 1980 ; Dickson et al., 1981) via highly discriminative receptors; (v) macrophages were reported to be activated by a_2M.P to secrete fibrinolytic proteinases (Vischer and Berger, 1980).
Complexed a_2M occurs in synovial effusions of different arthritides and other altered biological fluids (Abe et al.,1973; Shtacher et al.,1973; Harpel,1981; Heckl et al.,1982). Amount and nature of inactive a_2M, however, remained obscure as yet. The aim of this study is to establish an assay, which allows concentration of a_2M and a_2M.P from biological fluids (e.g.: synovial effusions) in order to quantify and characterize a_2M.P by means of chromogenic substrates.

MATERIALS AND METHODS

Polystyrene cuvettes (Makro-Küvette 2000) and the photometer 4010 were from Boehringer Mannheim GmbH. S-2288, a chromogenic substrate for a broad spectrum of serine proteinases, was from Kabi. Trypsin (from bovine pancreas,type III) was from Sigma. Chemicals were of analytical grade.

Isolation of plasma a_2M for standardization of the assay: a_2M
was isolated from fresh male plasma in the presence of soybean trypsin inhibitor (type II-S,Sigma). The method is detailed elsewhere
(Borth and Susani, this volume).

Coating of polystyrene cuvettes with antibody: 1 ml of phosphate
buffered saline (PBS) containing 30 µg of the immunoglobulin fraction
of rabbit anti-human a_2M antiserum (Dako) was pipetted into the bottom of each cuvette and incubated over night in the presence of NaN_3
at 4°C. Subsequently the solution was discarded and the cuvettes washed 3 times with PBS. The cuvettes were coated with 0.5% (W/V PBS)
chicken egg albumin (OVA,2x cristalized;Fluka A.G.) in order to pre-
clude any unexpected nonspecific binding of a_2M.

Immunoimmobilization of a_2M by the solid phase antibody: Serial
dilutions of plasma a_2M (0.1-100 µg plasma a_2M containing traces of
a_2M. ^{125}I-Trypsin / ml PBS) were added to the cuvettes and left for
2 hours at 25°C. The solution was withdrawn and the cuvettes washed
3 x with o.25% (W/V PBS) OVA . The cuvettes were counted in a gamma-
counter.

Determination of the trypsin binding capacity (=proteinase binding capacity) of solid phase immunoimmobilized a_2M: A constant
amount of trypsin (3 µg/ 1ml PBS) was added to increasing amounts
of solid phase bound a_2M in order to record the incorporation of
the enzyme as expressed by the hydrolytic activity of the final immunoimmobilized a_2M.T. In other experiments increasing amounts of
trypsin (0.1-50 µg T/ 1ml PBS) were added to constant amounts of
immunoimmobilized plasma a_2M in order to follow up the effect of increasing enzyme activity on the stability of the solid phase assay.

After addition of the trypsin ,the system was allowed to react
for 10 min at 25°C. The reaction was stopped by addition of a 4 fold
molar excess of soybean trypsin inhibitor (SBTI) and left again for
10 min. The reaction solution was removed and the cuvettes washed
3 times.

1.5 ml reaction buffer (0.1 M Tris.HCl, 0.15 M NaCl, 0.05 M
$CaCl_2$, pH 8.2) were placed into the cuvettes and the chromogenic
substrate added (150 µg S-2288 in 50 µl deionized water / cuvette).[*]
The hydrolytic reaction was performed at 25°C for 1-2 hours in a
rocking water bath. The reaction was stopped by pouring the reaction
mixture into the cuvette of the photometer, which was equiped with
automated suction and E read at 405 nm. Blanks without trypsin but
otherwise identical were incubated simultaneously.

Concentration of a_2M was determined by $A_{280,1cm}^{1\%}$=8.93 and by
rocket immunoelectrophoresis.

RESULTS AND DISCUSSION

For standardization of the assay we have designed experiments
in order to determine the amount of solid phase bound plasma a_2M
in dependence of the amount added to the system (Fig.1.) .

---
[*]Complete hydrolysis of 150 µg S-2288 gives $\Delta E_{405,1cm,1.5ml}$=1.1

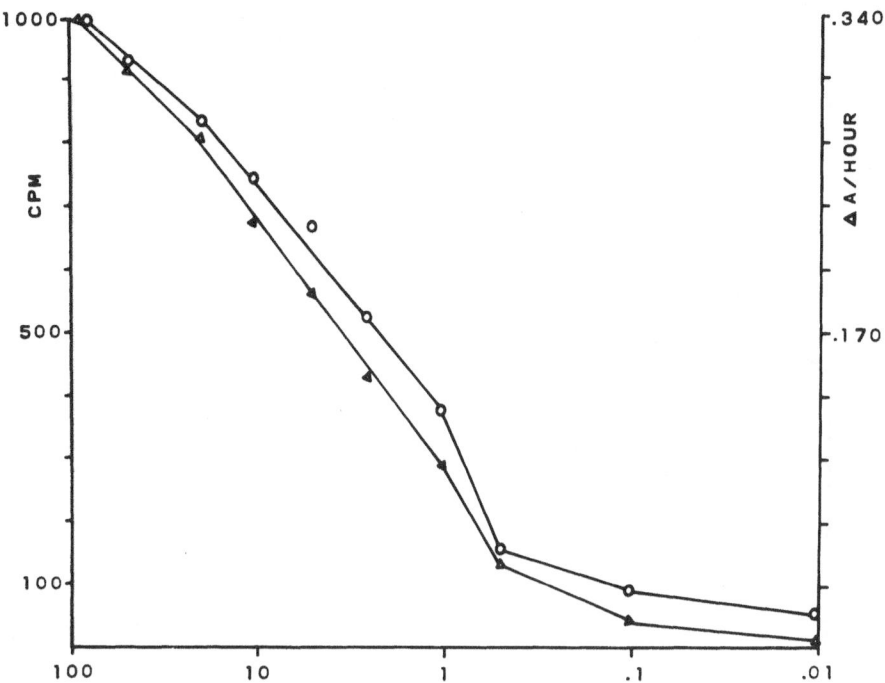

Fig. 1.   The solid phase anti human a₂M antibody was incubated with
          serial dilutions of purified plasma a₂M (abscissa ; 0.1-
          100 μg) containing traces of a₂M. ^{125}I-T (left ordinate).
          Ater three washes 2.5 μg trypsin were added to complex
          the immunoimmobilized a₂M. After repeated washes the
          radioactivity bound to the insoluble antibody was mea-
          sured (Δ-Δ). Simultaneously, the hydrolytic activity
          (right ordinate) was determined using S-2288 (o-o). The
          amount of bound a₂M.^{125}I-T was found to be paralleled
          with the hydrolytic activity of the immunoimmobilized
          a₂M.T complex. The linearity of the assay is within 1 to
          15 μg of added a₂M. 50% saturation of the solid phase
          antibody by a₂M is estimated as the amount of a₂M neces-
          sary , which results in a 50% binding of the tracer as
          well as 50% of the hydrolytic activity relative to the
          maximum observed. This corresponds to the amount of
          2.5 μg a₂M. Repetitive assays of the immunoimmobilized
          a₂M.T complex revealed an equivalent loss of radioactivity
          and hydrolytic activity of approximately 15%,but no change
          of specific activity was observed (after 3 assays). This
          was, however, observed previously and described more in
          detail (Harpel and Hayes,1980).

In addition to the first experiment it was necessary to follow up
the influence of varying amounts of enzyme on the solid phase assay
(Fig.2.).

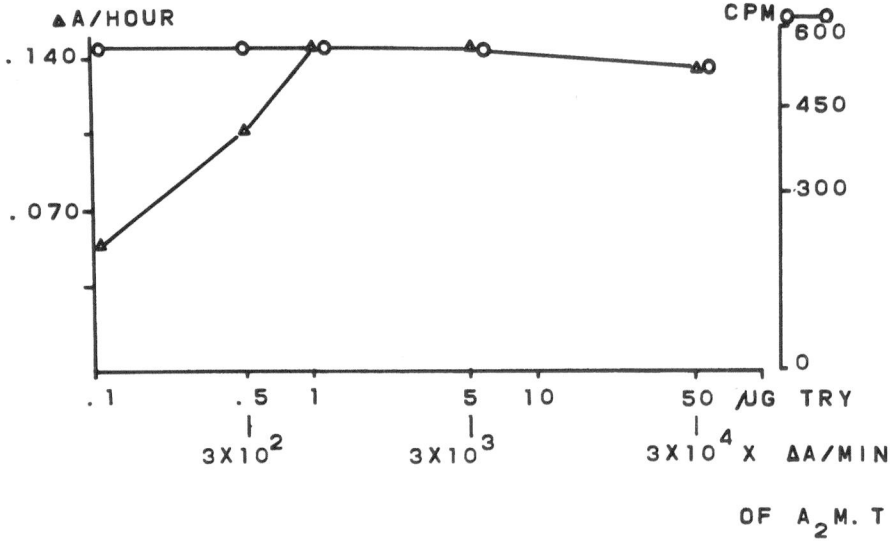

Fig. 2.   Increasing trypsin concentrations (0.1-50 µg trypsin cor-
            responding to $3x10^1$-$3x10^4$ times the hydrolytic activity
            of the final immunoimmobilized a_2M.T) were added to
            constant amounts of solid phase bound a_2M (2.5 µg) con-
            taining traces of a_2M.^{125}I-T (o-o). Following incubation
            for 1o min. at room temperature the reaction was stopped
            with SBTI. The reaction mixture was left again for 10'
            in a rocking water bath. The supernatant was withdrawn
            and the cuvettes washed 3 times. Finally the hydrolytic
            activity was determined and the solid phase bound a_2M.
            ^{125}I-T counted in a gamma-counter(o-o). As illustrated,
            addition of 1-3 µg of trypsin were sufficient to fully
            complex a_2M as expressed in the maximal obtainable hydro-
            lytic activity. Less than 1 µg did not fully complex
            a_2M after 10' of incubation. On the other hand , a
            small equivalent loss of radioactivity and hydrolytic
            activity (approx. 10%) was recorded after addition of
            excess proteinase (50 µg trypsin).

APPLICATION OF THE a_2M PASEBIC ASSAY TO SYNOVIAL FLUIDS

    Synovial fluids (n=14) were treated with hyaluronidase in the
presence of SBTI. a_2M concentration was estimated by electroimmunoassay using plasma a_2M ($A^{1\%}_{280,1cm}$ =8.93) as a standard. The aliquot
of synovial fluid a_2M used was with due regard to the linearity of
the assay.

    The native hydrolytic activity of the immobilized a_2M from
synovial fluids was estimated using a chromogenic substrate with
broad sensitivity for serine proteinases (P_{serine}) (S-2288). This
activity was regarded as $a_2M.P_{serine}$ complex present in the synovial
effusion. The activity (ΔE of native $a_2M.P_{serine}$) was expressed
as percentage of the maximal ΔE (=ΔE of total $a_2M.P_{serine}$)obtained
by the reference value of the standard curve. Subsequently, trypsin
was added and the total $a_2M.P_{serine}$ activity determined. The ΔE of
total $a_2M.P_{serine}$ -ΔE of native $a_2M.P_{serine}$ was considered as the
actual proteinase binding capacity of a_2M (a_2M PASEBIC). There was,
however, a remarkable discrepancy between the ΔE of total $a_2M.P_{serine}$
of synovial fluids and the maximal ΔE as expressed by the respective
reference values of the standard curve (Table 1.). This inactive a_2M
was tentatively termed $a_2M.P_{non-serine}$. For this,several explanations
are justified: i, complexation of a_2M with other proteinases
than P_{serine}; ii, inactive a_2M not related to interaction with a
proteinase; iii, considerable lower activity of the trapped P_{serine}
than that of trypsin towards the substrate.

CONCLUSIONS

    The a_2M PASEBIC assay allows concentration and characterization
of a_2M and its complexes. Minutes amounts as low as ng of trapped
proteinase can be detected by prolonging incubation times of the
complex with the substrate. The major limiting factors of sensitivity are homogeneity of the reference plasma a_2M (no complexation
should be detectable after isolation) and stability of the chromogenic substrate (the substrate should be sensitive, but should not
display autolysis).

    Circunstantial evidence indicates that availability of a_2M is
of crucial importance in regulation of proteolytic enzymes under
pathophysilogical conditions: the role of a_2M gains so much more
importance since a_1-antiproteinase,antithrombin III,$\overline{C1}$ inactivator
and a_2-plasmin inhibitor can be enzymatically inactivated under
'in vitro' conditions by myeloperoxidase (Matheson et al.,1981) and
human leukocyte elastase (Banda et al.,1980;Brower and Harpel,1982;
Gramse et al.,this volume). Alpha$_1$-antiproteinase deficiency apparently does not influence incidence and course of chronic inflammatory
disease (Sjöblom and Wollheim,1977). By contrast, there have been no
reports of individuals lacking a_2M suggesting the fundamental role
in proteolytic regulation by a_2M.

Table 1. a_2M PASEBIC in synovial fluids from different arthritides

patient[b]	a_2M mg/dl plasma/SF	native[a] a_2M.P$_{serine}$	a_2M[a] PASEBIC
1 recurrent effu- sions,?,38a,m	186/24	4.2	95.9
2 RA,44a,f	211/60	3.6	87.6
3 RA,50a,m	140/24	6.4	80.3
4 Gonococcal arth. 82a,m	220/15	0.6	99.5
5 RA,41a,f	174/58	0.7	99.3
6 RA,56a,f	170/60	5.0	62.2
7 traumatic,55a,m	n.d/90	0.9	99.1
8 RA,67a,f	276/107	1.6	79.8
9 RA,69a,f	210/85	4.2	69.1
1o RA,60a,f	175/75	4.9	89.1
11 RA,63a,f,	n.d/115	0.7	98.9
12 j RA,11a,f	n.d/150	3.1	84.9
13 RA,73a,f	236/40	0.5	97.8
14 villonodular synovitis,36a,m	333/129	5.4	81.7

[a] expressed as percentage of total a_2M.
[b] Arthrocentesis was performed before synovectomy in all patients.

From these and other observations it is concluded that (i) a_2M complexes to P$_{serine}$ but also complexes to a high degree with other subtypes of proteinases,(ii) the occurence of a_2M.P is not disease specific,(iii) the relative amount of a_2M.P seems to increase with activity of inflammation,(iv)the lower the amount of active a_2M the lower the antiinflammatory potential of the environment as evidenced by the distinct increase of C3d,and fibronectin split products in SF. It is considered that a_2M.P may affect target cells (Borth and Susani,this volume) in synovial membrane. In view of the present results and discussion it should be apparent that the a_2M PASEBIC assay may be a valuable aid in examination of synovial fluid and evaluation of joint disease.

REFERENCES

Abe,S.,Nagai,Y.,1973,Evidence for the presence of a complex of col-
      lagenase with a_2-macroglobulin in human rheumatoid synovial

fluid:A possible regulatory mechanism of collagenase activity in vivo,J.Biochem. 73:897.

Banda,M.J.,Clark,E.J.,Werb,Z.,1980,Limited proteolysis by macrophage elastase inactivates human a_1-antiprotease,J.Exp.Med.152:1563.

Barret,A.J.,Starkey,P.M.,1973,The interaction of a_2-macroglobulin with proteinases,Biochem.J.133:709.

Brower,M.S.,Harpel,P.C.,1982,Proteolytic cleavage and inactivation of a_2-plasmin inhibitor and C1 inactivator by human polymorphonuclear leukocyte elastase,J.Biol.Chem.257:9849.

Borth,W.,Menzel.E.J.,Salzer,M.,Steffen,C.,1981,Human serum inhibitors of collagenase as revealed by preparative isoelectric focusing,Clin.Chim.Acta,117:219.

Debanne,M.T.,Bell,R.,Dolovich,J.,1975,Uptake of a_2-macroglobulin proteinase complex by macrophages,Biochim.Biophys.Acta.411:295

Dickson,R.B.,Willingham,M.C.,Pastan,I.,1981,Binding and internalization of ^{125}I-a_2-macroglobulin by cultured fibroblasts, J.Biol.Chem.256:3454.

Flory,E.,Vischer,T.L.,1981,1981,a_2-macroglobulin as an inclusion in synovial fluid monocytes,Rheumatol. Int.1:61.

Ganrot,P.O.,1966,Determination of a_2-macroglobulin as trypsin-protein esterase,Clin.Chim.Acta.14:493.

Harpel,P.C.,Hayes,M.B.,1980,Immunoimmobilization of a_2-macroglobulin-ß trypsin complexes: a novel approach for the biochemical characterization of modulator-proteinase interaction, Analyt.Biochem.108:166.

Harpel,P.C.,1981,a_2-plasmin inhibitor and a_2-macroglobulin-plasmin complexes in plasma: Quantitation by an enzyme-linked differential antibody immunosorbent assay,J.Clin.Invest.68:46.

Heckl,F.,Borth,W.,Menzel,J.,Schwägerl,W.,1982,Characterization of the catalytic potential of a_2-macroglobulin.proteinase complexes and proteinase binding capacity of a_2M from synovial effusions:correlation with serological and clinical findings, VIII th meeting of the Federation of European connective tissue Societies,Copenhagen,abst.134.

Matheson,N.R.,Wong,P.S.,Shuyler,M.,Travis,J.,1981,Interaction of human a_1-proteaseinhibitor with neutrophil myeloperoxidase, Biochemistry.20:331.

Mosher,D.,Vaheri,A.,1980,Binding and degradation of a_2-macroglobulin by cultured fibroblasts,Biochim.Biophys.Acta 612:113.

Shtacher,G.,Maayan,R.,Feinstein,G.,1973,Proteinase inhibitors in human synovial fluid,Biochim.Biophys.Acta.303:138.

Sjöblom,K.G.,Wollheim,F.A.,1977,Alpha$_1$-antitrypsin phenotypes and rheumatic disease,Lancet ii:41.

Vischer,T.L.,Berger,D.,1980,Activation of macrophages to produce neutral proteases by endocytosis of a_2-macroglobulin-trypsin complexes,J.Reticuloendothel.Soc.28:427.

ROLE OF ALPHA$_2$-MACROGLOBULIN: PROTEINASE COMPLEXES IN PATHOGENESIS OF INFLAMMATION - 'F' a$_2$M BUT NOT 'S' a$_2$M INDUCES SYNOVITIS IN RABBITS AFTER REPEATED INTRA-ARTICULAR ADMINISTRATION*

Wolfgang Borth and Martin Susani

Institute of Immunology of the University and Department
of Pathology of the Poliklinik, A-1090 Vienna , Austria

INTRODUCTION

The interaction of proteinases with alpha$_2$-macroglobulin (a$_2$M) involves a conformational change, which results in the steric entrap-ment of the enzyme. In addition to the functional importance sugges-ted by the binding, internalization, and subsequent degradation of a$_2$M.Proteinase (a$_2$M.P) complexes by macrophages and fibroblasts , a$_2$M has a number of other interesting and unusual characteristics (Borth and references therein, this volume).Early studies by Blatrix et al.(1974) and Ohlsson (1974) revealed rapid clearence of a$_2$M.P from circulation by the rethiculo-endothelial system in man and dog. This was not the case for other antiproteinase proteinase complexes. Debanne et al. (1975) corroborated this concept by 'in vitro' stu-dies: uptake of a$_2$M by macrophages was specific for a$_2$M.P and was not valid for a$_1$PI.P complexes. The receptor recognition site of a$_2$M is therefore a structure, either revealed during or assembled by the formation of complexes with proteinases or by reaction with primary amines. For the time being, there is only one report that macro-phages respond actively to receptor mediated endocytosis of a$_2$M.P (Vischer and Berger,1980). Alpha$_2$-M.P occur in synovial effusions from different arthritides indicating that a$_2$M controls proteolytic activity in the extravascular space (Shtacher et al.,1973; Ekerot and Ohlsson; Virca et al.; Borth ; this volume). Cumulative evidence indicates that a$_2$M.P are taken up by free and tissue fixed macro-phages in synovial fluid and tissue (Flory and Vischer,1981;Ekerot

---

* 'S' a$_2$M : the electrophoretically slow form of a$_2$M corresponds to the fully active plasma a$_2$M; 'F' a$_2$M : the electrophoretically fast form of a$_2$M, which is more compact when complexed with proteinases or when reacted with primary amines (e.g.:methylamine; Barrett et al.,1979)

371

and Ohlsson, this volume; Borth et al.,1983). Furthermore, fibroblasts were recorded to take up a_2M.P 'in vivo' (Borth and Viehber-
ger,1983). We guess that endocytosis of a_2M.P by these cells is not
only to remove and degrade a_2M.P but also a way to influence cellular
performance , particularly in inflammation.

In pilot studies we observed that serum derived but not plasma
derived a_2M induced synovitis in rabbits after repeated administration to the knee joint without prior immunization. This finding
prompted us to study in detail the intra-articular role and fate of
a_2M.P in view of its clinical importance.

MATERIALS AND METHODS

Animals: Female rendombred rabbits weighing between 3.0 and 3.5
kg were used.

Reagents: Trypsin was from Miles (pig pancreas; commercial ac-
tivity:4,500 U/mg). Collagenase was isolated from rheumatoid arthri-
tis synovial cultures (Borth et al.,1981). Methylamine was from
Sigma. Ovalbumin, 5x cristalized, from Fluka A.G. Epoxy-activated
Sepharose 6B for Zn^{2+} chelate affinity chromatography was from
Pharmacia.

Isolation of serum- and plasma-a_2M: A 5-12.5% serum or plasma
PEG precipitate was used as starting material for Zn^{2+} chelate
affinity chromatography (Porath et al.,1975). Subsequently,the a_2M
fraction (approx. 90-95% purity) was freed from contaminating plasma
proteins by means of chromatofocusing. The column (1x14 cm) was
developed in a pH gradient from pH 6.0 to 4.5 according to the
manufacturer's instructions (Pharmacia). a_2M was collected between
pH 5.38-4.8 (4°C). This material showed homogeneous purity as determined by SDS-PAGE electrophoresis at pH 9.58 (Barrett et al., 1979).
The serum derived a_2M, however, was recorded to be 12% complexed,
apparently generated during coagulation of plasma.

Preparation of plasma a_2M.Proteinase complexes: Purified plasma
a_2M was affinity immobilized on the Zn^{2+} column and the active enzymes loaded onto the column. a_2M was reacted for 10 min at 25°C
and excess enzyme eluted. This procedure converted 's' a_2M into 'f'
a_2M to 96% as revealed by SDS-PAGE pH 9.58.

Methylamine reaction of plasma a_2M: This was done according to
Howard (1981). The reaction fully converted 's'a_2M into 'f'a_2M .

Animal experiments: All protein solutions in their final form
were sterile filtered and stored at -70°C.
150 µg 'f' a_2M in o.5ml buffer were administered on day 0,3 and 6
to the experimental (=right) joint cavity. The contralateral (=con-
trol) knee joint received native plasma a_2M (150 µg/0.5 ml) or a
4-fold molar excess of active proteinase adjusted to a overall
protein concentration of 150 µg protein/0.5ml with ovalbumin. An
other 'control' was run with 150 µg serum derived a_2M, which contained partially complexed a_2M. A group of animals received a fourth
injection on day 7 with 250 µg rhodamine-(TRITC) labelled 'f' a_2M
into the experimental joint , in order to view the fate of the agent

by means of fluorescence microscopy.This group of animals was sacrified 4 h after last challenge. Animals were sacrified one, 7 and 14
days after last challenge. Histology and estimation of inflammation
was performed as described elsewhere (Steffen et al., 1980).

RESULTS

Incidence and specificity of 'f' a_2M induced synovitis: An acute
to chronic inflammatory synovial membrane developed in all experi-
metal joints . Serum-derived a_2M and active proteinases evoked a
moderate,acute to subacute synovitis, which was not observed to turn
into a chronic form. In contrast, plasma derived-a_2M did not cause
inflammation. A summary of the intra-articular effect of the experimental agents is given in table 1.

Table 1 . Effect of 's' and 'f' a_2M after
intra-articular administration

inflammation[a]	agent[b]
severe	a_2M.Trypsin a_2M.Collagenase a_2M.Methylamine
moderate	serum derived a_2M trypsin collagenase
none	plasma-derived a_2M ovalbumin buffer

[a] Severe inflammtion is considered here as an acute
inflammtion , which turned into chronic synovitis
in some experimental joints.
[b] All protein solutions were adjusted to 150 µg protein / 0.5 ml buffer after sterile filtration.

Pathology: The synovial fluid cell count rose rapidly in the
experimental joint s after last challenge, but fell to low levels
after 1 week. A significantly larger number of white cells were in
the experimental joint.

The moderate to gross changes observed in the experimental
synovial membrane were typified by edema , concentration of hetero-
phils , lining cell enlargement and hyperplasia and increased vas-

cularization of the capsule as indicated by an enlargement of post-
capillary venules and by an increased number of capillaries (Fig.1)

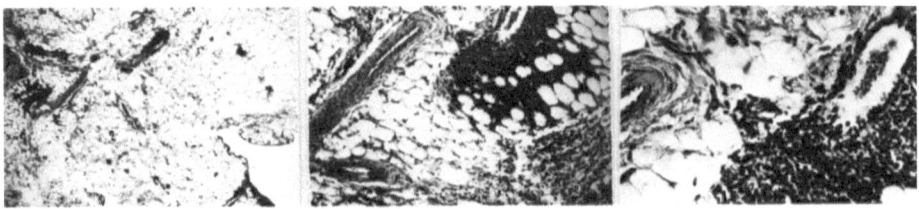

Fig.1   Rabbit synovial membrane (RSM) 1 day after last challenge
         of an experimental knee joint: acute inflammation of the
         subsynovial tissue. Severe leukodiapedesis is observed
         around postcapillary venules, wheras the adjacent arterial
         vessel is almost without evidence of alteration. It is note-
         worthy that leukocytes are oriented to the perimeniscal
         synovium rather than to the synovial cavity.

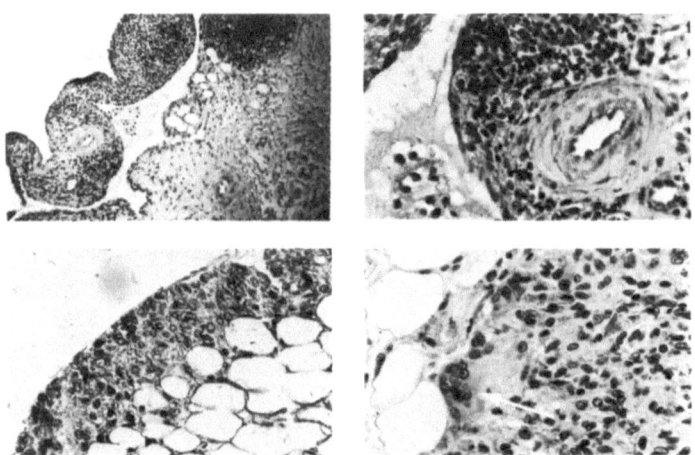

Fig.2   RSM 14 days after last challenge of an experimental joint:
         chronicity of inflammation is evident. Focal accumulation
         of mononuclear cells just under the mesothel is recorded
         (fig. at the top). Areolar tissue is extensively replaced
         by fibroblasts and collageneous stroma. Lining cell hyper-
         plasia up to 6 layers thick (fig. bottom left). Giant cell
         formation is observed in some specimen (fig. bottom right,
         arrow).

In later stages of inflammation the infiltrate became more diffuse with predominant appearance of mononuclear cells, which showed tendency to larger focal accumulations just under the mesothel (Fig.2). Fibroplasia, sometimes with early onset, was seen in some of the experimental joints. In some specimen giant cell formation was recorded (Fig.2).

Fluorescence microscopy revealed the TRITC-labelled 'f' a_2M as inclusion of macrophage-like cells in synovial and subsynovial tissue The image of the granular staining resembles that found in human arthritides (own observation)(Fig. 3).

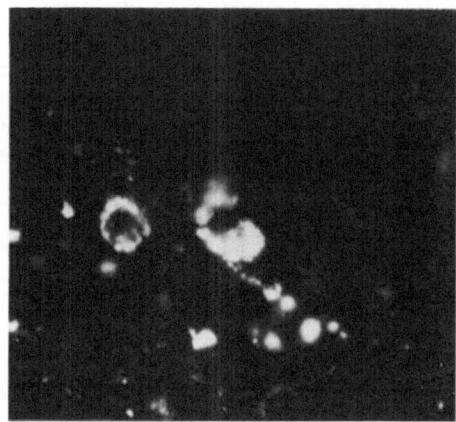

Fig.3   Perinuclear localization of TRITC-labelled 'f' a_2M in some
         macrophage-like cells in the proliferating lining cell-
         layer of rabbit synovial membrane. Thus, a limited number of
         macrophage-like cells are involved in the specific uptake
         of 'f' a_2M.

DISCUSSION

     The present results demonstrate that $a_2M.P$ complexes and me-
thylamine treated a_2M, respectively, when injected intraarticular-
ly, evoke rheumatoid-like lesions in rabbit synovial membrane. The
experimental data suggest that the conformational alteration is
capable to challenge free and tissue fixed macrophages. The bound
enzyme is not essential for this challenge. This fully supports
earlier 'in vitro' findings: molecular constraints necessary for
receptor recognition of 'f' a_2M are independent of identity and
retained activity of the bound proteinase (Imber et al.,1981).

     Specific receptors for $a_2M.P$ complexes are also ascribed to
fibroblasts. Recent studies in this laboratory by means of immuno-
electronmicroscopy revealed a_2M as inclusion of fibroblast-like
cells in RA synovial tissue (Borth et al.,1983;Borth and Viehber-

ger, 1983). Since fibroblasts provide connective tissue with many
molecular constituents, qualitative and quantitative changes in
synthesis and secretion in the presence of a_2M.P should be envi-
sioned. The early onset of synovitis with subsequent fibroplasia
in some experimental joints may indicate direct effect of 'f' a_2M
on fibroblasts.

Evidence has been gathered in support of the view that a_2M
represents the major inhibitor and elimination form of proteinases:
a_2M is the major inhibitor of proteinases in complex biological
fluids even in the presence of 'specific' plasma proteinase inhibitors (Harpel,1981;Shapira et al.,1982;Ekerot and Ohlsson,this
volume). While the biological halflife of a_2M.P is short in ircula-
tion, the amount of complexes to be eliminated by the local (e.g.:
synovial membrane) or systemic RES seems to be limited. Inflammatory
states with augmented liberation and activation of proteinases may
generate consequently huge amounts of a_2M.P (Borth;Ekerot and
Ohlsson; this volume). This may cause an 'overload' of the RES with
subsequent deleterious secondary reactions. Thus, paradoxically, a
general scavenger of proteinases may perpatuate inflammation, when
its complexes rise to pathological levels. Activation of macrophages
by a_2M.P seems to parallel other 'in vitro' stimuli such as immuno-
complexes and activated complement components (Cardella et al.,
1974;Schorlemmer et al.,1976).

In view of the foregoing discussion it should be apparent that
determination of a_2M.P may be a valuable aid, all the more, as a_2M
binds rapidly to active proteinases thereby reflecting pathological
states with increased liberation of proteinases early. A convenient
method for the quantitative analysis of a_2M.P and the proteinase
binding capacity of a_2M in biological fluids has been introduced
(Borth,this volume).

REFERENCES

Barret,A.J.,Brown,M.A.,Sayers,C.A.,1979,The electrophoretically
     'slow' and 'fast' forms of the a_2-macroglobulin molecule,
     Biochem.J.181:401.
Blatrix,C.,Israel,J.,Audran,R.,Drouet,J.,Amouch,P.,Steinbuch,M.,1974,
     Plasma clearence of human antiproteinase/proteinase complexes,
     in:"Proteinase Inhibitors",H.Fritz,H.Tschesche,L.J.Greene and
     E.Truscheit,ed.,Springer,Berlin.
Borth,W.,Menzel,E.J.,Salzer,M.,Steffen,C.,1981,Human serum inhibi-
     tors of collagenase as revealed by preparative isoelectric
     focusing,Clin.Chim.Acta.117:219.
Borth,W.,Viehberger,G.,1983,On the role of alpha$_2$-macroglobulin in
     collagen degradation,Coll.Rel.Research., in the press.
Borth,W.,Susani,M.,Viehberger,G., Immunolocalization of a_2M in

synovial membrane by means of light-and electronmicroscopy,
submitted.

Cardella,C.J.,Davies,P.,Allison,A.C.,1974,Immunocomplexes induce
selective release of lysosomal hydrolases from macrophages,
Nature.247:46.

Debanne,M.T.,Bell,R.,Dolovich,J.,1975,Uptake of a_2-macroglobulin.
proteinase complex by macrophages,Biochim.Biophys.Acta.411:295

Flory,E.,Vischer,T.L.,1981,a_2-macroglobulin as an inclusion in
synovial fluid monocytes,Rheumatol.Int.1:61.

Harpel,P.C.,1981,a_2-plasmin inhibitor and a_2-macroglobulin-plasmin
complexes in plasma:Quantitation by an enzyme-linked differential immunosorbent assay.J.Clin.Invest.68:46.

Howard,J.B.,1981,Reactive site in human a_2-macroglobulin:circum-
stantial evidence for a thiolester.Proc.Natl.Acad.Sci.USA.
78:2235.

Imber,M.J.,Pizzo,S.V.,1981,Clearence and binding of two'fast'forms
of human a_2-macroglobulin.J.Biol.Chem.256:8134

Ohlsson,K.,1974,Interaction between endogeneous proteases and plasma
protease inhibitors in vitro and in vivo,in:"Proteinase Inhibitors",H.Fritz,H.Tschesche,L.J.Greene and E.Truscheit,ed.,
Springer,Berlin.

Porath,J.,Carlson,J.,Olsson,G.,Belfrage,G.,1975,Metal affinity
chromatography,a new approach to protein fractionation,
Nature.258:598.

Schapira,M.,Scott,F.,Colman,R.W.,1982,Contribution of protease inhibitors to the inactivation of kallikrein in plasma,J.Clin
Invest.69:462

Schorlemmer,H.V.,Bitter-Sauermann,D.,Allison,A.C.,1977,Complement
activation by the alternative pathway and macrophage enzyme
secretion in the pathogenesis of chronic inflammation,
Immunology.32:929.

Shtacher,G.,Maayan,R.,Feinstein,G.,1973,Proteinase inhibitors in
human synovial fluid,Biochim.Biophys.Acta.303:138.

Steffen,C.,Menzel,E.J.,Zeitelhofer,J.,Smolen,J.,Lanzer,G.,1980,
Experimental arthritis induced by granulocyte collagenase,
Scand.J.Rheumatol.9:179.

Vischer,T.L.,Berger,D.,1980,Activation of macrophages to produce
neutral proteases by endocytosis of alpha₂-macroglobulin-
trypsin complexes,J.Reticuloendothel.Soc.28:427.

ENZYME-LINKED IMMUNOASSAY FOR HUMAN GRANULOCYTE ELASTASE

IN COMPLEX WITH α_1-PROTEINASE INHIBITOR

Siegfried Neumann, Norbert Hennrich,
Gerhard Gunzer, and Hermann Lang

Biochemical Research Institute
E. Merck
D-6100 Darmstadt, FRG

INTRODUCTION

Human polymorphonuclear neutrophilic leukocytes contain large
amounts of neutral proteinases which are released when the granulocytes are exposed to a phagocytic stimulus. In inflammatory
diseases local imbalance between these proteinases and available
proteinase inhibitors may cause tissue injury as well as degradation of plasma proteins (1,2). Granulocyte elastase
(E.C.3.4.21.11) is of special pathological interest because of
its high concentration and its broad specificity for a variety of
connective tissue components (i.e. elastin, collagen, proteoglycans)
and plasma proteins (i.e. IgG, C_3, C_5, various clotting factors).
Though it seems obvious that a quantitative determination of
elastase in inflamed tissue and in the circulating blood may give
information on disease activity, up to now there have been published only few reports on quantitative assays for the detection and
quantitation of elastase in biological fluids (3-6).
The main antagonizing agent in plasma for elastase is α_1-protein-
ase inhibitor (7). Upon the interaction of elastase with the
inhibitor a complex with a molar ratio of 1 to 1 and a molecular
weight of approximately 80,000 Daltons is formed (7,8).
We developed a solid-phase enzyme-linked immunoassay for the determination of the complex of elastase with α_1-proteinase inhibitor
(9,10). By using this assay remarkedly elevated levels of the
complex were demonstrated in plasma samples of septicemic patients
(11) and in synovial fluids of patients with rheumatoid arthritis
(12,13).
In this paper we want to report on a modified version of our assay
and on the performance characteristics of the test.

MATERIALS AND METHODS

Reagents and materials used in this study were essentially those as described in our previous report (10).

In short, elastase was isolated from granule extracts of normal human granulocytes by a three-step procedure involving copper-chelate$^{(R)}$ chromatography and ion-exchange chromatography on SP-Sephadex$^{(R)}$ (from Deutsche Pharmacia, Freiburg, FRG), and CM-Cellulose. The preparation was free of contaminant granular or serum proteins as shown by polyacrylamide gel electrophoresis at pH 4.3, and by immunoelectrophoresis with antisera against granular extract and human serum. Human α_1-proteinase inhibitor was isolated from outdated blood bank plasma. The purification procedure consisted of ion-exchange chromatography on DEAE-Sepharose$^{(R)}$ CL-6B, affinity chromatography on Concanavalin A-Sepharose$_{(R)}$ and Blue-Sepharose$^{(R)}$ CL-6B, and gelfiltration of Sephacryl$^{(R)}$ S-200(all media from Deutsche Pharmacia, Freiburg, FRG). The protein preparation was homogeneous as judged by poly-acrylamide gel electrophoresis at pH 8.6, and by immunoelectro-phoresis. Antisera against elastase were raised in sheep and were shown to be antigen-specific by immunoelectrophoresis. The immunoglobulin G fraction was isolated with the caprylic acid procedure (14).

Plastic supports were coated with antibodies against elastase by incubation of polystyrene tubes 10.5 mm x 40 mm (No. 655061, Greiner, Nürtingen) with 50 μg immunoglobulin per ml overnight at 4 $^{\circ}$C.

Antibodies against α_1-proteinase inhibitor were purified from rabbit antiserum by immunadsorption on agarose-fixed α_1-proteinase inhibitor. The antibody molecules were coupled to calf intestinal alkaline phosphatase by a one-step glutaraldehyde procedure (15). The enzyme-antibody conjugate was isolated by gel-filtration.

Complex Formation

The elastase/α_1-proteinase inhibitor complex was produced by mixing elastase with an excess of the inhibitor. 0.5 mg elas-tase and 2.8 mg α_1-proteinase inhibitor in 2 ml 30 mM Tris (hydroxymethyl)aminomethane/HCl buffer pH 8.0 with 85 mM NaCl were incubated for 30 min at 37 $^{\circ}$C. Under these conditions elastase was totally inactivated by complex formation as controlled by enzyme assay using methoxysuccinyl-L-Ala-L-Ala-L-Pro-L-Val-p-nitro-anilid (13), immunoelectrophoresis and polyacrylamide gel electro-phoresis at pH 4.3.

Standard Solutions and Control Plasma

Standard solutions with defined levels of complexed elastase were prepared by diluting the reaction mixture with phosphate-buffered saline containing 10 g/l bovine serum albumin.

Control plasma with a known level of the complex was produced as follows: Citrated plasma from unimmunized sheep was inactivated at 56 $^{\circ}$C for 45 min, and mixed with a defined amount of elastase-inhibitor complex.

The Immunoassay

The principle of the assay is given in Figure 1.

Test Procedure

The performance of the assay in microculture plates was reported previously (9,10). Here we describe the performance in antibody-coated tubes.

Step 1   Dilute plasma 1:25 or 1:50 in phosphate-buffered saline containing 10 g/l bovine serum albumin and 20 mM EDTA
Step 2   Incubate 5oo µl of diluted sample or standard per tube 1 h at room temperature

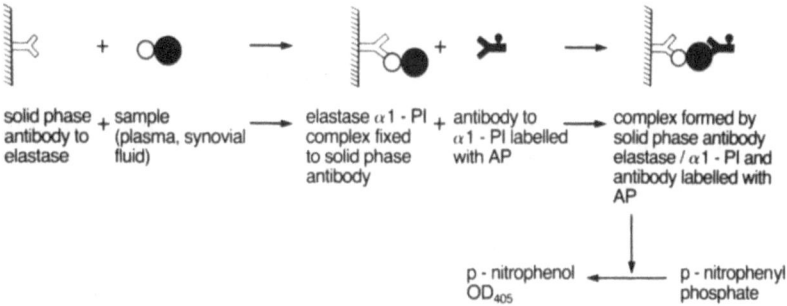

solid phase + sample   →   elastase α1 - PI + antibody to   →   complex formed by
antibody to   (plasma, synovial   complex fixed α1 - PI labelled   solid phase antibody
elastase   fluid)   to solid phase   with AP   elastase / α1 - PI and
    antibody   antibody labelled with
    AP

p - nitrophenol ←   p - nitrophenyl
OD_{405} phosphate

Figure 1 :   Principle of the Enzyme Immunoassay

Step 3   Wash with distilled water containing 0.5 g/1 polyoxyethylen-
         sorbitanmonolaurat
Step 4   Incubate with 5oo µl of alkaline phosphatase labelled
         rabbit-IgG against α_1-proteinase inhibitor 1 h at room
         temperature
Step 5   Wash with destilled water containing 0.5 g/1 of polyoxyethyl-
         ensorbitanmonolaurat
Step 6   Incubate with 500 µl of 10 mMoles/1 p-nitrophenyl phosphate
         in 1 M diethanolamine/HCl-buffer pH 9.8 containing 0.5
         mMoles $MgCl_2$/1 90 min at 25 $^\circ$C
Step 7   Stop enzyme reaction by adding 500 µl 2 M NaOH
Step 8   Read absorbance at 405 nm

Sample Collection

     Citrated or EDTA plasma was used. Plasma was separated from
blood within 30 min after sample collection to prevent interference by leakage of leukocytes (16). Tests were performed either
on fresh plasma or on samples stored frozen at -20 $^\circ$C or -70 $^\circ$C.

RESULTS

Standard Curve

     The absorbance values of the blank and the standard
solutions were plotted versus concentration on a linear diagram
(Figure 2). The intercept at 0 ng/ml and the slope were calculated
by linear regression. Results from unknown samples were referred
to the calibration curve.
     The working range of the assay was 0.5 to 5.0 ng of complexed
elastase in the assay tube. This is equal to a range of 25 to 250
µg/1 or 50 to 500 µg/1 in plasma when testing plasma samples of
20 or 10 µl in 500 µl final volume, respectively.

Detection Limit

     The lower limit of sensitivity as defined by the mean ab-
sorbance of the blanks plus three times standard deviation was
0.25ng per test tube.

Precision

     Precision was controlled with different plasma pools (Table 1).
Within-run variability (coefficient of variation) was 4 to 8 %
and between-run imprecision was in the range of 3 to 8%.

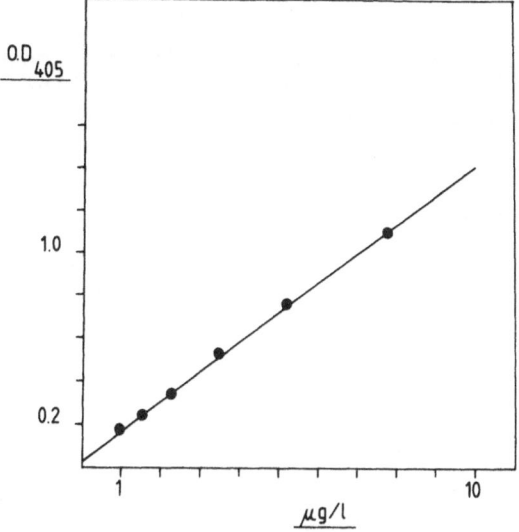

Figure 2 :   Calibration Curve

Absorbance is plotted as a function of
Concentration (given as µg of elastase
in complex with α_1-proteinase inhibitor
per 1). By linear regression analysis
a function of y = 0.037+0.138x (r= .999)
was calculated.

Table 1 : Precision of the Assay

Sample	Sample Volume	Series	Exp.per Series	C.V. (%)
Within-series Imprecision				
Pooled Normal Plasma 1 (52 µg/1)	2o µl	1	15	7.7
Pooled Normal Plasma 2 (206 µg/1)	2o µl	1	15	3.9
Pathological Plasma (389 µg/1)	1o µl	1	15	4.0
Between-series Imprecision				
Pooled Normal Plasma 1 (47 µg/1)	2o µl	15	2	8.2
Pooled Normal Plasma 2 (191 µg/1)	2o µl	15	2	7.4
Pathological Plasma (339 µg/1)	1o µl	15	2	2.9

Stability

Citrated plasma could be stored at least 6 month at -20 $^{\circ}$C or 14 days at +4 $^{\circ}$C without appreciable loss of the antigenic reactivities of the elastase/ α_1-proteinase inhibitor complex.

Linearity and Parallelity

Citrated plasmas were pooled to give plasma specimen with levels of 60 or 500 µg of complexed elastase/l. Linear results were obtained with samples from 10 to 80 µl plasma from the first pool and with samples from 1 to 8 µl from the second pool. The plasma dilution curves had slopes very similar to that of the calibration curve (Figure 3).

Recovery

Aliquots of the elastase/ α_1-proteinase inhibitor complex as produced in vitro were added to three different citrated human plasma samples. The concentration of the complex in the native and the spiked plasma samples were assayed. The recovery in the three plasma specimen was near 100 percent.

Normal Plasma Range

The plasma concentration of the elastase/ α_1-proteinase inhibitor complex was measured in citrated plasma from 43 randomly selected, apparently healthy laboratory workers. The group was composed of 40 males and 3 females, age ranged from 47 to 65 years. The concentration of complexed elastase ranged from 50 to 181 µg/l, mean was 97.5 µg/l with a standard deviation of 25.8 µg/l.

DISCUSSION

In the last years and again at this symposium ample evidence has been presented for a major role of the neutral proteinases from granulocytes in the pathogenesis of inflammatory diseases like lung emphysema, septicemia, inflammatory joint disease and others. Granulocyte elastase deserves special interest. Previous reports showed the presence of granulocyte elastase in plasma of patients with septicemia or leukemia (3), in serum of patients with myeloid leukemia (17), in peritoneal fluids from patients with peritonitis (18), in sputum from bronchitic patients (19) and in

synovial exudates from some patients with rheumatoid arthritis (20). Granulocyte elastase was also demonstrated in inflamed joint tissue by immunohistochemical techniques (21,22).
     In spite of the growing information on granulocyte elastase as a main pathogenetical factor in inflammation there have been only a few attempts with respect to development of quantitative laboratory assays for the determination of elastase concentration in biological fluids. These methods were based on electroimmuno-diffusion (3), radioimmunoassay (4,6) or on enzyme immunoassay for canine granulocyte elastase (5). None of these assays for elastase have yet found broad access to clinical studies.

     It was previously reported that in plasma granulocyte elastase is complexed with plasma proteinase inhibitors, mainly with α_1-proteinase inhibitor (7,23). Elastase in complex with this inhibitor is devoid of enzymatic activity. A small fraction of the enzyme will be bound to α_2-macroglobulin (7) and can be detected by enzymatic activity on small substrates (12).

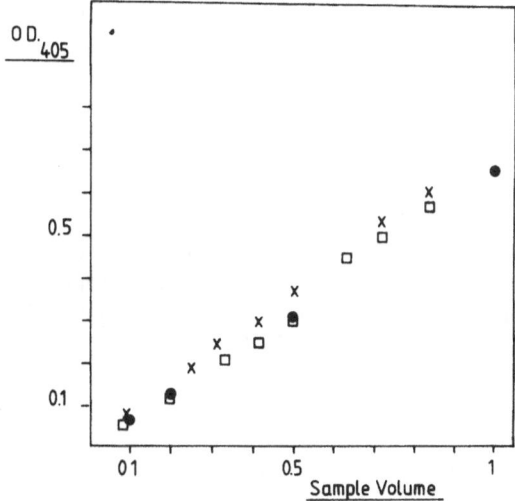

Figure 3 : Linearity and Parallelity of Plasma in the Assay

          Varying quantities of standard solution or plasma
          were measured in the assay, and absorbance is
          plotted as a function of the sample volume.
          Definition of sample volume:
          • Calibrator (10 µg/1), 1 = 500 µl
          ☐ Pooled human plasma (60 µg/1), 1 = 100 µl plasma
          ✕ Pooled human plasma (500 µl/1), 1 = 10 µl plasma
          Final volume in the test tube was brought to 500 µl
          by addition of diluent, if necessary.

Because of the small percentage of this fraction and the low
halftime of the complex with α_2-macroglobulin, determination of
elastase activity in plasma does not appear to be a sensitive
parameter for the level of the enzyme in the circulation.

We have therefore chosen an alternative approach, i.e. the
immunochemical quantitation of the complex of elastase with
α_1-proteinase inhibitor. The assay makes use of the dual antigenic specificities of the complex of an enzyme with its protein-
eaceous inhibitor. The procedure follows the performance route of
a solid phase enzyme-linked immunoassay (sandwich-type). The
basic principle and the performance in microculture plates were
described in a previous paper (10).
The test version as detailed in this report is a modified
assay in test tubes, which can be performed with pipettors,
diluters and photometers of a routine clinical chemistry laboratory.
As in the microculture plate version sensitivity compares with that
of radioimmunoassays. Precision and accuracy are within reasonable
limits. Linearity of plasma samples is seen within a broad range
of plasma concentration within the test tube. Recovery of the
enzyme inhibitor complex added to plasma is near the expected
concentration.

The working range of the assay allows the quantitation of
plasma levels in normal subjects, which are below the detection
limit of the electroimmunodiffusion technique as used by others (3).
A range of 50 to 181 µg of elastase in complex with α_1-proteinase
inhibitor per litre of plasma was found in a group of 43 normal
persons, with a mean of some 98 µg/litre. In comparison, a serum
level of 135 ± 59 µg/litre was found by radioimmunoassay in a group
of 66 normal individuals (4) and a plasma level of 25 ± 6 µg/litre
was determined by radioimmunoassay in 7 individuals (16).
Additional work is needed to define the reference range in normal
plasma more precisely.

What kind of information can this assay give in clinical
situations ?
First, it can be shown that the increase of plasma levels of
the elastase/α_1-proteinase inhibitor complex is a sensitive signal
for septicemia in surgical patients, and plasma levels are meaning-
ful for follow-up of patients and control of patient management
(11, 24).
Second, plasma levels of the complex are also elevated in acute
renal failure (25, 26). This is discussed as evidence for a causal
role of elastase in protein hypercatabolism in the disease. If this
hypothesis could be substantiated concepts for therapy should be
reconsidered (25, 26).
Third, enormous concentrations of the elastase inhibitor
complex were demonstrated in synovial exudates from patients with

inflammatory joint diseases as opposed to low levels in specimen from patients with degenerative joint afflictions (12). In plasma from patients at an acute stage of rheumatoid arthritis a significant rise of the concentration can be observed, whereas practically normal values were seen in non-inflammatory joint disease (13). The clinical significance of measuring plasma levels of elastase inhibitor levels in rheumatoid arthritis has been discussed by others (27). However, to evaluate the clinical meaning of elastase levels in plasma in rheumatoid arthritis conclusively, further studies with a close follow-up of the patients and definition of antiphlogistic treatment at the time of blood sampling are needed.

In conclusion, we report on a solid-phase enzyme-linked immunoassay for elastase/α_1-proteinase inhibitor as a practicable means for the clinical evaluation of leukocyte elastase as a marker of inflammatory disease activity. Increased plasma levels are consistently observed in acute inflammation like septicemia or acute renal failure. It is anticipated that these observations lead to consequences in therapy. The clinical meaning of increased plasma levels in rheumatoid arthritis will be discussed elsewhere (13).

ABSTRACT

Granulocyte elastase in complex with its main inhibitor in plasma, i.e. α_1-proteinase inhibitor, was quantitatively determined by incubating the sample with solid-phase fixed antibodies against elastase first and reacting then with alkaline phosphatase-labelled antibodies against α_1-proteinase inhibitor. In normal plasma a level of 97.5 ± 25.8 µg elastase/l (mean ± s.d., n = 43) was found, whereas moderately to markedly increased plasma concentrations were demonstrated in a variety of patients with inflammatory diseases like septicemia or rheumatoid arthritis.

ACKNOWLEDGEMENTS

The skillfull technical assistance of Ursula Götzmann-Fuhren and Norbert Avemarie is gratefully appreciated. The investigation was supported by the Bundesministerium für Forschung und Technologie, FRG (project No. 01 ZR 089-ZK/NT/MT 295).

REFERENCES

1. K. Havemann and A. Janoff, eds., "Neutral Proteases of Human Polymorphonuclear Leukocytes", Urban and Schwarzenberg, Baltimore and Munich (1978).
2. G. Weissmann, ed., "The Cell Biology of Inflammation", Elsevier - North Holland, Amsterdam, New York, and Oxford (1980).

3. R. Egbring, W. Schmidt, G. Fuchs, and K. Havemann, Demonstration of granulocytic proteases in plasma of patients with acute leukemia and septicemia with coagulation defects, Blood 49:219 (1977).

4. K. Ohlsson and A.-S. Olsson, Immunoreactive granulocyte elastase in human serum, Hoppe-Seyler's Z. Physiol. Chem. 359:1531 (1978).

5. U. Kuchich, W.R. Abrams, and H.L. James, Solid-phase immunoassay of dog neutrophil elastase, Anal. Biochem. 109:403 (1980).

6. E.F. Plow and J. Plescia, Non-plasmin mediated fibrinolysis, in: "Progress in Fibrinolysis V", J.F. Davidson, I.M. Nilsson, and B. Astedt, eds., Churchill Livingstone, Edinburgh, London, Melbourne, and New York (1981), pp. 70 - 74.

7. K. Ohlsson and I. Olsson, Neutral proteases of human granulocytes. III. Interaction between human granulocyte elastase and plasma protease inhibitors, Scand. J. clin. Lab. Invest. 34:349 (1974).

8. R.J. Baugh and J. Travis, Human leukocyte granule elastase: rapid isolation and characterization, Biochemistry 15:836 (1976).

9. S. Neumann, N. Hennrich, G. Gunzer, and H. Lang, Enzyme-linked immunoassay for elastase from leukocytes in human plasma, J. Clin. Chem. Clin. Biochem. 19:232 (1981).

1o. S. Neumann, N. Hennrich, G. Gunzer, and H. Lang, Enzyme-linked immunoassay for complexes of human granulocyte elastase with α_1-proteinase inhibitor in plasma, in: "Progress in Clinical Enzymology II ", D.M. Goldberg and M. Werner, eds., Masson Publ., New York (1983), in press.

11. M. Jochum, K.-H. Duswald, E. Hiller, and H. Fritz, Plasma levels of human granulocytic elastase - α_1-proteinase inhibitor complex (E-α_1PI) in patients with septicemia and acute leukemia, in: "Selected Topics in Clinical Enzymology", D.M. Goldberg and M. Werner, eds. W.de Gruyter, Berlin-New York (1983); in press.

12. K. Kleesiek, S. Neumann, and H. Greiling, Determination of the elastase α_1-proteinase inhibitor complex, elastase activity and proteinase inhibitors in the synovial fluid, Fresenius Z. Anal. Chem. 311:434 (1982)

13. K. Kleesiek, D. Brackertz, S. Neumann, and H. Greiling Elastase from granulocytes in chronic joint diseases: Determination of elastase/α_1-proteinase inhibitor complex in plasma and synovial fluid. In preparation.

14. M. Steinbuch and R. Audran, The isolation of IgG from mammalian sera with the aid of caprylic acid, Arch. Biochim. Biophys. 134:279 (1969).

15. E. Engvall, K. Jonsson, and P. Perlmann, Enzyme-linked immunosorbent assay. II. Quantitative assay of protein antigen, immunoglobulin G, by means of enzyme-labelled antigen and antigen-coated tubes, Biochim. Biophys. Acta 251:427 (1971).

16. E.F. Plow, Leukocyte elastase release during blood coagulation. A potential mechanism for activation of the alternative fibrinolytic pathway, J. Clin. Invest. 69:564 (1982)

17. L. Olsson, T. Olofsson, K. Ohlsson, and A. Gustavsson, Serum and plasma myeloperoxidase, elastase and lacto-ferrin content in acute myeloid leukemia, Scand. J. Haematol. 22:397 (1979).

18. K. Ohlsson, Collagenase and elastase released during peritonitis are complexed by plasma protease inhibitors, Surgery 79:652 (1976)

19. R.A. Stockley and D. Burnett, Alpha$_1$-antitrypsin and leukocyte elastase in infected and noninfected sputum, Amer. Rev. Resp. Dis. 120:1081 (1979).

2o. H. Menninger, R. Putzier und F. Hartmann, Die Bedeutung von Proteasen für die rheumatische Gelenkdestruktion, Verh. Dtsch. Ges. Rheumatol. 6:229 (1980).

21. K. Fehr, A. Baici, M. Velvart, M. Knöpfel, M. Rauber, G. Sommermeyer, P. Salgram, G. Cohen und A. Böni, Die chronische Polyarthritis: Rolle der polymorph-kernigen Leukozyten bei der Zerstörung von Pannus-freiem Gelenkknorpel, Bull. Schweiz. Akad. Med. Wiss. 35.317 (1979).

22. H. Menninger, R. Putzier, W. Mohr, D. Wessinghage, and K. Tillmann, Granulocyte elastase at the site of cartilage erosion by the rheumatoid synovial tissue, Z. Rheumatol. 39:145 (1980).

23. K. Ohlsson and C.-B. Laurell, The disappearance of enzyme-inhibitor complexes from the circulation of man, Clin. Sci. Mol. Med. 51:87 (1976).

24. M. Jochum, K.-H. Duswald, S. Neumann, J. Witte, and H. Fritz, Proteinases and their inhibitors in septicemia - Basic concepts and clinical implications - This issue.

25. W.H. Hörl, R.M. Schäfer, K. Scheidhauer, M. Jochum, and A. Heidland, Proteolytic activity in patients with hypercatabolic renal failure. This issue.

26. A. Heidland, W.H. Hörl, N. Heller, H. Heine, S. Neumann, and E. Heidbreder, Release of granulocyte proteases in patients with acute and chronic renal failure, This issue.

27. H.P. Schnebli, P. Christen, M. Jochum, R.K. Mallya, and M.B. Pepys, Plasma levels of inhibitor bound leukocytic elastase in rheumatoid arthritis patients, This issue.

PROTEINASES AND THEIR INHIBITORS IN SEPTICEMIA
- BASIC CONCEPTS AND CLINICAL IMPLICATIONS

M. Jochum[2], K.-H. Duswald[1], S. Neumann[3],
J. Witte[1], H. Fritz[2]

[1]Surgical Clinic, [2]Department of Clinical
Chemistry & Biochemistry, University of
Munich, [3]Department of Biochemical Research
E. Merck, Darmstadt, FRG

INTRODUCTION

The Role of Lysosomal Enzymes in Inflammation

During the inflammatory response various systemic
or local tissue cells are activated thereby releasing
internal, mostly lysosomal enzymes. They trigger the activation of the clotting, fibrinolysis and complement
cascades, the disrupture of cell membranes and tissue
structures, and the release of toxic peptides (Fig. 1).

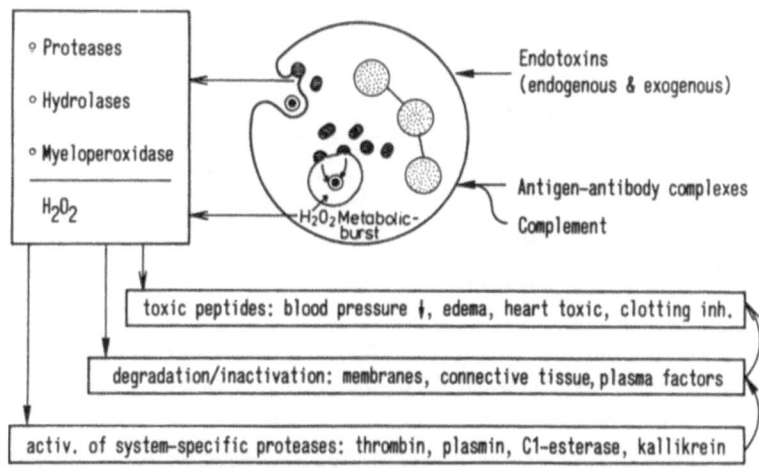

Fig. 1: Liberation and effects of lysosomal factors. For details see text.

Phagocytes, especially the granulocytes and monocytes or macrophages, but also fibroblasts, endothelial
cells and mast cells are known to be very rich in such
internal or lysosomal enzymes.

So far, only the properties and pathobiochemical
effects of enzymes of the azurophilic and specific lysosomes of polymorphonuclear granulocytes (neutrophils)
have been investigated in more detail. Such enzymes, for
example the neutral elastase and cathepsin G as well as
the acidic cathepsins are preformed and stored in the
lysosomes in fully active form[1,2]. In this way, they can
respond immediately to perform their biological function,
namely the degradation of extracellular and intracellular material after phagocytosis.

Outside the phagocytes, proteolytic action of the
lysosomal enzymes is normally prevented or balanced by
proteinase inhibitors present in plasma, interstitial
fluid and body secretions (Fig. 2). These inhibitors

Abbr.	Inhibitor	M.W. x 1 000	Mean conc.	
			mg/100 ml	μmol/l
α_2M	α_2-Macroglobulin	725	260	3.6
α_1PI	α_1-Protease inhibitor	50	260	52
α_1AC	α_1-Antichymotrypsin	70	45	6.4
β_1CI	β_1-Collagenase inh.	40	1.5	0.4
ITI	Inter-α-trypsin inh.	160	45	2.8
AT III	Antithrombin III	65	26	4.0
α_2PI	α_2-Plasmin inhibitor	70	6	0.9
C1 INA	C1-Inactivator	100	24	2.4

Fig. 2: Plasma inhibitors directed against lysosomal proteinases of various body cells
or plasma-derived proteinases.

comprise together with inhibitors of clotting, fibrinolysis and complement proteinases more than 10 % of the
plasma proteins. Non-lysosomal proteinases like, for
example, thrombokinases and plasminogen activators are

faced only with a very low inhibitory potential in body fluids. They are, therefore, privileged candidates for the activation of clotting and fibrinolysis if released into the circulation after increased production due to an inflammatory stimulus.

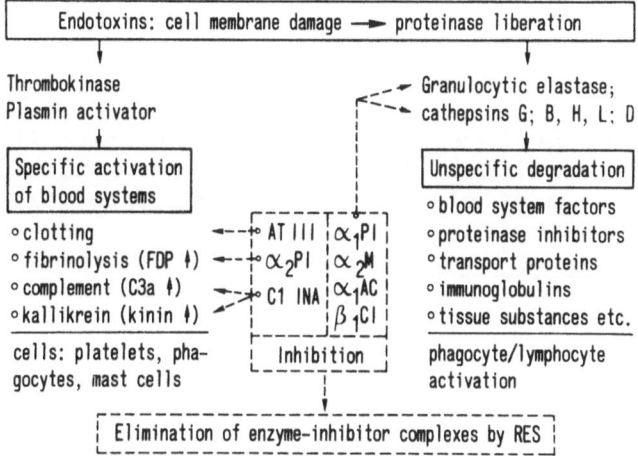

Fig. 3: Consumption of plasma factors during inflammation. For details see text.
Abbreviations: antithrombin III (AT III), α_2-plasmin inhibitor (α_2PI), C1 inactivator (C1 INA), α_1-proteinase inhibitor (α_1PI, formerly α_1-antitrypsin), α_2-macroglobulin (α_2M), α_1-antichymotrypsin (α_1AC), β_1-collagenase inhibitor (β_1CI), fibrin (-ogen) degradation products (FDP), complement factor (C 3a), reticuloendothelial system (RES).

Lysosomal enzymes liberated during severe inflammation like septicemia or septic shock can enhance, together with thrombokinases and plasminogen activators, the inflammatory response via two major routes characterized by either substrate-specific or substrate-unspecific proteolysis (Fig. 3).

The system-specific proteinases, thrombokinases and plasminogen activators, trigger the activation of the clotting, fibrinolysis and complement cascades (summarized as 'blood systems') by substrate-specific proteolysis of proenzymes and cofactors. The activated enzymes are subsequently inhibited by their natural inhibitors; the enzyme-inhibitor complexes thus formed are rapidly eliminated from the circulation by the reticuloendothelial system (RES). Hence, in a series of steps based on highly specific interactions not only the proteinases respectively their zymogens are consumed but also their natural antagonists, the system-specific inhibitors. Until recently, the given sequence of reactions was assumed to be exclusively responsible for the development of disseminated intravascular coagulation (DIC).

Results obtained very recently indicate that an additional reaction path may contribute considerably to the consumption of plasma factors during severe inflammations. This implies inactivation of plasma factors by substrate-unspecific proteolysis due to liberated lysosomal proteinases. Egbring and coworkers observed in animals a significant decrease in several clotting factors after infusion of endotoxin or human neutrophil elastase[3]. Similar results were obtained by Ohlsson and coworkers[4,5] in canine endotoxemia. Moreover, in patients suffering from sepsis or septic shock a striking consumption of blood system factors, immunoglobulins and proteinase inhibitors has been found by the teams of Egbring[6], Aasen[7], Gallimore[8] and Witte[9].

The potency of lysosomal enzymes for the degradation of native plasma protein inhibitors was also demonstrated in vitro by the inactivation of antithrombin III due to catalytic amounts of neutrophil elastase[10]. α_1-Protease inhibitor (formerly α_1-antitrypsin) is inactivated in a similar way by the macrophage elastase[11] and, in addition, by oxidation of its reactive-site methionine residue via the lysosomal myeloperoxidase/hydrogen peroxide system[12].

RESULTS

Neutrophil Elastase and Plasma Factors in Sepsis

Assay of liberated elastase. Due to the presence of an excess of the endogenous inhibitors α_1-protease inhibitor (α_1PI) and α_2-macroglobulin (α_2M), direct measurement of the neutrophil-derived proteinase activities in plasma or other body fluids is not feasible. However, increased levels of the elastase-α_1PI complex would be already a clear indication for elastase liberation. Quantitative estimation of the plasma levels of the elastase-α_1-protease inhibitor complex was carried out with a highly sensitive enzyme-linked immunoassay[13]. Briefly, the E-α_1PI complex of the plasma sample was bound to surface-fixed antibodies directed against neutrophil elastase. After washing, a second alkaline phosphatase-labelled antibody directed against α_1PI was fixed to the complex. Under suitable conditions, the activity of fixed alkaline phosphatase towards p-nitrophenylphosphate is proportional to the concentration of the E-α_1PI complex in the sample.

In a first approach, we were interested to see whether a relationship exists between the plasma levels of E-α_1PI and the severity of postoperative infections. To achieve this purpose, besides other factors the levels of factor XIII (Faktor XIII-Schnelltest, Behringwerke AG Marburg) and antithrombin III (S-2238, Deutsche Kabi Munich) were continuously monitored because both clotting factors are known to be easily degraded by neutrophil elastase in vitro.

Patients. In the clinical trial, patients subjected to major abdominal surgery were included if the operation time exceeded 120 min. Diagnosis of septicemia in the postoperative course was confirmed by prospectively established septic criteria. In the prospective study more than 120 patients were included. Thirty of them fulfilled the defined septic criteria during the postoperative course. Of these patients, fourteen survived the infection (group B) whereas sixteen died as a direct result of septicemia (group B). Eleven patients being without infection after abdominal surgery served as controls (group A) (Fig. 4).

○ Defined infection site & pos. bact. culture

○ Body temperature > 38.5°C

○ Leukocytosis   with > 15 000 cells/mm^3 or
   Leukocytopenia with < 5 000 cells/mm^3

○ Platelets < 100 000/mm^3 or drop > 30 %

○ (Positive blood culture)

Fig. 4: Prospectively defined sepsis criteria. Diagnosis of septicemia in the postoperative course was confirmed in patients fulfilling all these conditions.

Elastase-α_1PI levels. With the enzyme-linked immunoassay elastase levels between 60 and 110 ng/ml were found in 153 healthy individuals. In patients without preoperative infection (group A and B), the operative trauma was followed by an increase of the E-α_1PI level up to 3-fold of the normal value. Patients suffering from preoperative infections (6 out of 16 in group C) showed already clearly elevated preoperative E-α_1PI levels. Immediately after surgery a slight decrease was observed, probably due to elimination of the infection focus. Before onset of sepsis, the E-α_1PI concentrations of group B and C showed a moderate elevation but no significant changes compared to the postoperative levels. However, at the beginning of septicemia a highly significant increase of the E-α_1PI levels could be detected: up to 6-fold in group B and up to 10-fold in group C. Peak levels were found above 2 500 ng/ml in both groups. The E-α_1PI levels of septic patients who recovered showed a clear tendency towards normal values. In patients with persisting septicemia, high levels of E-α_1PI were measured until death. (Fig. 5).

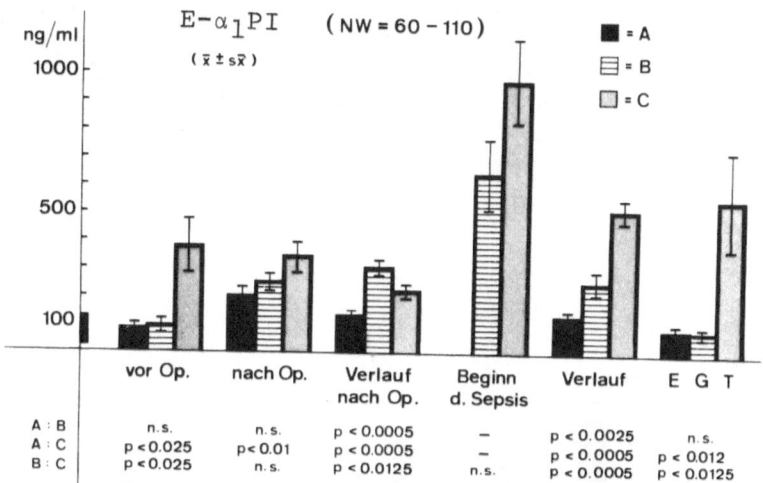

Fig. 5: Plasma levels of elastase-α_1-proteinase inhibitor complex (E-α_1PI) in patients subjected to major abdominal surgery.

A = patients (n=11) being without postoperative infection

B = patients (n=14) surviving postoperative septicemia

C = patients (n=16) dying as a result of septicemia

The E-α_1PI levels are given as mean values ($\pm$ SEM) for the day before operation, the day after operation as well as for the postoperative phase before sepsis, at onset of sepsis and during septicemia. Last determinations were done on day of discharge (D) for group A, on day of recovery (R) for group B, and before death (D or †) for group C.

nr = normal range.

Antithrombin III activity. In non-infected patients the activity of AT III, the most important inhibitor of the clotting system, was in the normal range during the whole observation period. In infected patients, however, the AT III activity was found already below the clinically critical concentration of about 75 % of the standard mean value before onset of septicemia. This low value normalized in all patients overcoming the infection,

whereas a further significant decrease (up to 45 % of
the norm) was found in group C patients with lethal outcome. Probably, the extremely low AT III activity in the
latter patients, having permanently elevated E-α_1PI levels, may be due to a significant degree to degradation
by lysosomal enzymes and especially by elastase.

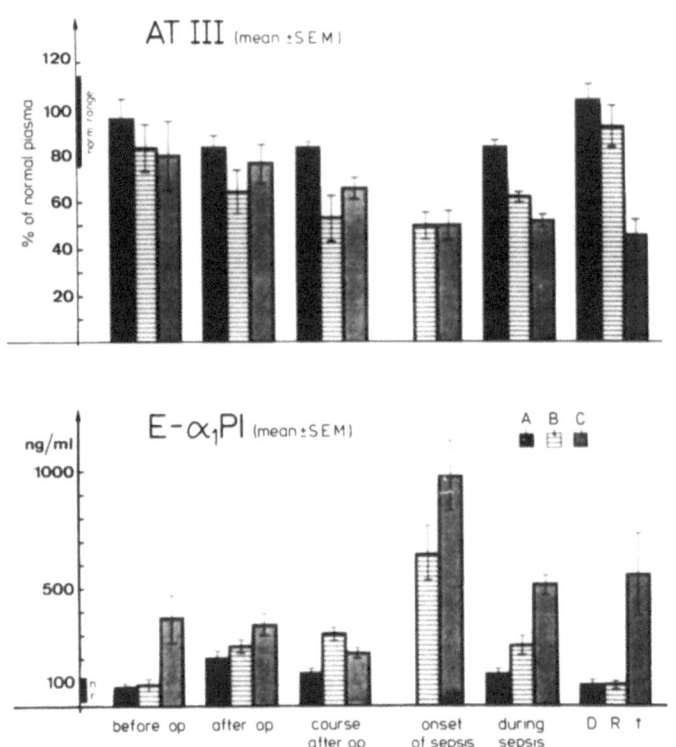

Fig. 6: Plasma levels of the inhibitory activity of antithrombin III (AT III) compared to
the amount of the elastase-α_1-proteinase inhibitor complex (E-α_1PI) in patients subjected
to major abdominal surgery.
For details, see legend to Fig. 5

Factor XIII activity and A & S subunit levels. Similar results were obtained for F XIII, the fibrin stabilizing coagulation factor. In plasma of patients who
did not survive septicemia, the F XIII activity decreased

up to 28 % of the standard mean value. As measured by
immunoelectrophoresis, these patients also had very low
concentrations of both subunit A, comprising the active
enzyme, and subunit S, representing the carrier protein
(data not shown). In contrast, group A patients with uncomplicated postoperative course showed normal or only
slightly decreased concentrations of subunit S, although
subunit A and fibrin stabilizing activities were often
significantly reduced.

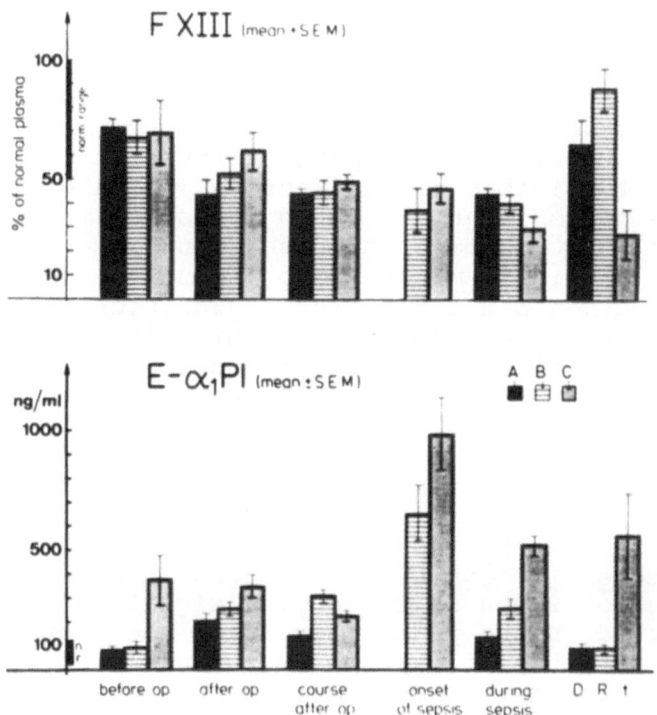

Fig. 7: Plasma levels of the fibrin stabilizing
activity of factor XIII (F XIII) compared to
the amount of the elastase-α_1-proteinase inhibitor complex (E-α_1PI) in patients subjected to
major abdominal surgery.
For details, see legend to Fig. 5

As demonstrated formerly by Egbring and coworkers[6] and
Ikematsu and coworkers[14], reduction of both subunits of

F XIII cannot be due to activation of the clotting cascade alone. During clotting, that means by the action of
thrombin only subunit A is consumed simultaneously with
the F XIII activity but not subunit S. Elastase, however,
is able to degrade both subunits to a similar degree.
These data and the results presented in our clinical
trail suggest that in the patients suffering from septicemia, unspecific proteolytic degradation by granulocytic elastase and/or other lysosomal proteinases is involved to a significant degree in the depletion of F XIII.

     Conclusions from the Clinical Studies. The results
of the clinical studies show that in inflammatory diseases a correlation exists between the release of a lysosomal enzyme marker, the neutrophil elastase, and the
clinical situation of the patient respectively the consumption of selected plasma factors. We take this as a
clear indication that liberated lysosomal factors and
especially neutrophil proteinases contribute significantly to the inflammatory response of the organism by
substrate-u n s p e c i f i c degradation of plasma and
other factors. Early application of suitable and potent
exogenous proteinase inhibitors should prevent or at
least diminish, therefore, such destructive proteolytic
processes.

Inhibitor therapy in experimental endotoxemia

     To confirm this assumption, we established an endo-
toxemia model in dogs by intravenous infusion of E. coli
endotoxin for 2 hours. Thereby, a significant decrease
was observed in the plasma levels of the clotting factors antithrombin III, prothrombin and factor XIII, of
the fibrinolysis factors plasminogen and α_2-antiplasmin,
and of the complement factor C3. The levels were followed
up over an experimental period of fourteen hours and
their alterations checked for statistical significance[15]
(Fig. 8).

     Simultaneous intravenous administration of a rela-
tively specific inhibitor of neutrophil elastase and
cathepsin G, the Bowman-Birk inhibitor from soybeans,
clearly reduced the endotoxin-induced decline of the
tested plasma factors. The 7 000 dalton inhibitory miniprotein was effective in dosages ranging from 3 - 8 mg
(i.e. 10 - 25 trypsin inhibiting units) per kilogram
body weight. A reasonable assumption would be, therefore, that the endogenous inhibitor was able to prevent
or reduce the neutrophil proteinase-induced consumption
reactions very effectively.

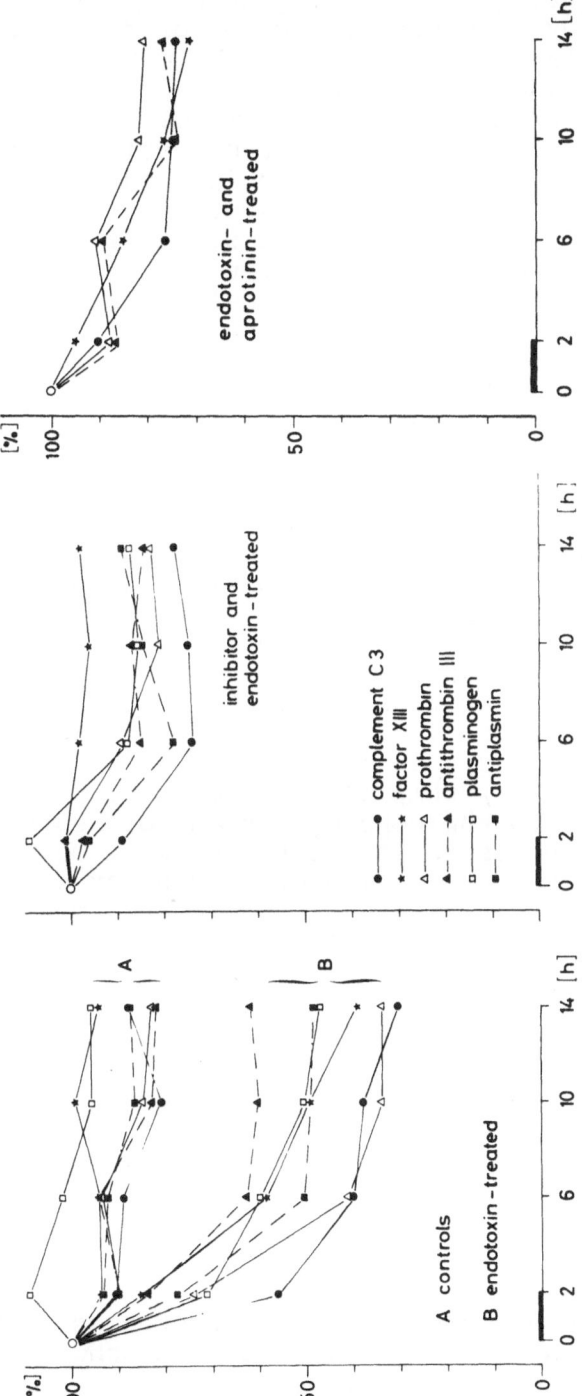

Fig. 8: Changes in the plasma levels of various plasma factors during the acute phase of experimental endotoxemia in dogs (n=6 for each group) without or with inhibitor treatment. Applied inhibitor dosages: 10 - 25 trypsin inhibiting units (Bowman Birk inhibitor) or 80 000 kallikrein inhibitor units (aprotinin) per kg body weight over the observation period of 14 hours. Data are given as mean values in percentage of the individual starting values. Thick line at the abscissa = endotoxin infusion period. For further details see text and ref. 15.

Surprisingly, intravenous administration of the broad-spectrum proteinase inhibitor aprotinin (Trasylol) reduced the endotoxin-induced reduction of the plasma factors to a similar degree (Fig. 8). We must emphasize, however, that the administered aprotinin dose of 80 000 kallikrein inhibitor units (i.e. 9.6 mg of the basic 6 500 dalton miniprotein from bovine organs) per kilogram body weight over the observation period of 14 hours would correspond to 5.6 Mill. kallikrein inhibiting units in the case of a 70 kg patient or to 9.6 Mill KIU/24 h. Although we do not know at present the lowest dosage of aprotinin effective in our animal model, we speculate that a relatively high dose is necessary to inhibit effectively endotoxin-induced activation of the intrinsic clotting cascade and thus also systemic fibrinolysis as well as complement reactions.

GENERAL CONCLUSIONS

In severe inflammatory processes, multiple trauma or shock various cells like neutrophils, macrophages, endothelial cells, and mast cells are stimulated or desintegrated. In this way a high potential of lysosomal enzymes is released of which the proteinases are of special pathogenetic effectiveness. Recent studies in our laboratory and by others indicate strongly that substrate-unspecific proteolysis by lysosomal proteinases and especially by the neutrophil elastase contributes to a significant degree to the consumption and/or degradation of extracellular substances in such diseases. On the other hand, early administration of convenient exogenous inhibitors directed against the lysosomal enzymes and/or plasmin should have a positive therapeutic effect also in humans.

ACKNOWLEDGEMENT

This work was supported by the Deutsche Forschungsgemeinschaft, Sonderforschungsbereich 51. The excellent technical assistance of Mrs. U. Hof and Mrs. C. Seidl is greatly appreciated.

REFERENCES

1. K. Havemann and A. Janoff: Neutral Proteases of Human Polymorphonuclear Leukocytes. Urban and Schwarzenberg Verlag, Baltimore-Munich (1978).

2. S.J. Klebanoff and R.A. Clark: The Neutrophil Function and Clinical Disorders. Elsevier/North-Holland Biomedical Press, Amsterdam, The Netherlands (1978).

3. R. Egbring, M. Gramse, N. Heimburger, K. Havemann: In vivo effects of human granulocyte neutral proteinases on blood coagulation in green monkeys (cercopithecus aetiops). 6th Int. Congr. Thromb. Hemostr. Philadelphia 1977, Abstr. Diath. Haemorrh. 38:222 (1977).

4. A.O. Aasen, K. Ohlsson, M. Larsbraaten, E. Amundsen: Changes in plasminogen levels, plasmin activity and activity of antiplasmin during endotoxin shock in dogs. Eur. surg. Res. 10:63-72 (1978).

5. A.O. Aasen, K. Ohlsson: Release of granulocyte elastase in lethal canine endotoxin shock. Hoppe Seyler's Z. Physiol. Chem. 359:683-690 (1978).

6. R. Egbring, W. Schmidt, G. Fuchs, K. Havemann: Demonstration of granulocytic proteases in plasma of patients with acute leukemia and septicemia with coagulation defects. Blood 49:219-231 (1977).

7. A.O. Aasen, N. Smith-Erichsen, M.J. Gallimore, E. Amundsen: Studies on components of the plasma kallikrein-kinin system in plasma samples from normal individuals and patients with septic shock. Advances in Shock Res. 4:1-10 (1980).

8. M.J. Gallimore, A.O. Aasen, K. Lyngaas, M. Larsbraaten, N. Smith-Erichsen, E. Amundsen: Studies on the coagulation and fibrinolytic systems in septicemia. Microvasc. Res. 18:292 (1979).

9. J. Witte, M. Jochum, R. Scherer, W. Schramm, K. Hochstrasser, H. Fritz: Disturbances of selected Plasma Proteins in Hyperdynamic Septic Shock. Intensive Care Med. 8:215-222 (1982).

10. M. Jochum, S. Lander, N. Heimburger, H. Fritz: Effect of human granulocytic elastase on isolated human antithrombin III. Hoppe Seyler's Z. Physiol. Chem. 362: 103-112 (1981).

11. M.J. Banda, E.J. Clark, Z. Werb: Limited proteolysis by macrophage elastase inactivated human α_1-proteinase inhibitor. J. Exp. Med. 152:1563-1570 (1980).

12. K. Beatty, J. Beith, J. Travis: Kinetics of association of serine proteinases with native and oxidized

α_1-proteinase inhibitor and α_1-antichymotrypsin. J. Biol. Chem. 255:3931-3934 (1980).

13. S. Neumann, N. Hennrich, G. Gunzer, H. Lang: Enzyme-linked immunoassay for human granulocyte elastase α_1-proteinase inhibitor complex. In: D.M. Goldberg and M. Werner (eds.) Progress in Clinical Enzymology -II, 1982 (in press) and this issue.

14. S. Ikematsu, R.P. McDonagh, H.M. Reisner, C. Skrzynia, J. McDonagh: Immunochemical studies of human factor XIII: Radioimmunoassay for the carrier subunit of the zymogen. J. Lab. Clin. Med. 97:662-671 (1981).

15. M. Jochum, J. Witte, H. Schiessler, H.K. Selbmann, G. Ruckdeschl, H. Fritz: Clotting and other plasma factors in experimental endotoxemia: Inhibition of degradation by exogenous proteinase inhibitors. Eur. surg. Res. 13:152-168 (1981).

PROTEOLYTIC ACTIVITY IN PATIENTS WITH

HYPERCATABOLIC RENAL FAILURE

Walter H. Hörl[1], Roland M. Schäfer[1], Klemens Scheidhauer[1],
Marianne Jochum[2], and August Heidland[1]

[1]Department of Medicine, University of Würzburg,
[2]Surgical Clinic, Department of Clinical Chemistry
and Biochemistry, University of Munich, FRG

INTRODUCTION

Despite several advances in dialysis and medical therapy, the
mortality rate for patients with acute renal failure (ARF) remains
distressingly high. When ARF is associated with major surgery or
trauma, the mortality rate is about 50 to 70 %. Such patients are
often hypercatabolic as a result of sepsis, hemorrhage, or open-
draining wounds. They may be wasted or malnourished from underlying
illnesses. Losses of glucose, amino acids and proteins during hemodialysis or peritoneal dialysis contribute to wasting.

Total parenteral nutrition with amino acids and hypertonic
glucose has been reported to: stabilize or reduce serum levels of
urea nitrogen, potassium and phosphorous improve wound healing; enhance survival from ARF and possibly increase the rate of recovery
of renal function[1]. However, nitrogen balance remains negative in
all patients[2] suggesting that decreased protein synthesis is not
the sole cause of loss of body nitrogen. Several studies have shown
that proteinases participate in protein catabolism of patients with
hypercatabolic ARF[3-6]. This report summarizes data concerning the
potential role of proteases in enhanced protein breakdown in
patients with acute and chronic renal failure.

MATERIAL AND METHODS

Proteolytic activity in plasma, dialysate and urine fractions
was measured, as previously described, using as substrates phosphorylase kinase (from rabbit skeletal muscle), azocasein or a
particular fraction of hepatocytes[3-7]. Polyacrylamide gel electro-

405

phoresis in the presence of sodium dodecylsulfate was carried out
according to Weber and Osborn[8]. The inhibitory activity of alpha$_1$-
protease inhibitor (alpha$_1$-antitrypsin) was determined with a commercial test system (Boehringer, Mannheim, FRG). Plasma concentrations of alpha$_2$-macroglobulin (α_2M) and alpha$_1$-protease inhibitor
(α_1PI) were evaluated by a radial immunodiffusion technique with
standardized immunodiffusion plates (Behringwerke, Marburg, FRG).
Quantitative estimation of the plasma levels of the elastase-alpha$_1$-
protease inhibitor (E-α_1PI) complex was carried out with a highly
sensitive enzyme-linked immunoassay[9]. We report here the results
obtained in: 12 healthy subjects aged from 22 - 39 years (seven
males, five females), 12 chronically uremic patients aged from 22 -
66 years undergoing regular dialysis therapy (RDT) for 41.3 + 11.6
month (nine males, three females) and 10 patients with posttraumatic
ARF aged from 19 - 45 years. Every 4th day venous blood was collected immediately before and after dialysis. Furthermore, dialysate
and urine fractions were collected weekly on ice in the presence of
0.01 % azide.

RESULTS

     Ultrafiltrated plasma fractions, obtained from healthy sub-
jects and RDT patients and prepared by ultrafiltration with an Amicon XM 50 filter, had no effect on degradation of phosphorylase kinase. However, the ultrafiltrated plasma fractions, obtained from
patients with posttraumatic ARF, were proteolytic. Figure 1 shows
the digestion of the three subunits of phosphorylase kinase after
incubation with ultrafiltrated plasma for 1 min as well as 10 and
24 hours. One minute after incubation, the typical subunit structure of this enzyme with the three polypeptide chains: alpha (MW
135,000), beta (MW 120,000) and gamma (MW 42,000) was observed. Ten
hours later, the alpha subunit was partially degraded, whereas the
gamma subunit was completely digested. After 24 hours the whole enzyme was destroyed. Phosphorylase kinase was also digested by plasma ultrafiltrates of the other nine patients with posttraumatic ARF
within 5 - 36 hour incubation period at 37 $^\circ$C.

     Free proteolytic activity in the plasma may be a consequence
of the low alpha$_2$-macroglobulin values, whereas the concentrations
of alpha$_1$-protease inhibitor (alpha$_1$-antitrypsin) were markedly enhanced. Figure 2 shows the trypsin binding capacity of plasma taken
from patients with posttraumatic ARF and from RDT patients. The
levels of proteolytic activity (after the addition of increasing
amounts of trypsin) are significantly higher in patients with posttraumatic ARF compared with RDT patients indicating a lower trypsin
binding capacity of the plasma in both groups of patients compared
with healthy subjects.

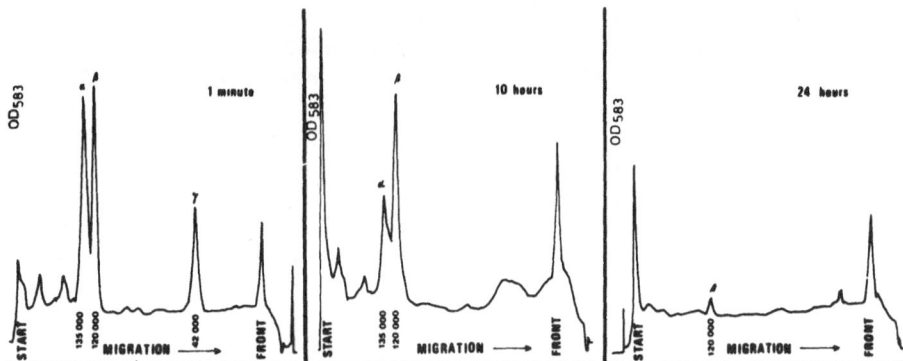

Fig. 1: Effect of an ultrafiltrate of plasma from patient H.H. on phosphorylase kinase. 0.1 ml ultrafiltrate (2.1 mg/ml), obtained by ultrafiltration of plasma through an Amicon XM 50 filter, was added to 0.1 ml phosphorylase kinase (1.5 mg/ml). Samples (1 min, 10 and 24 hr after incubation) were subjected to polyacrylamide (5 %) gel electrophoresis in the presence of sodium dodecylsulfate[8].

The effect of hemodialysis therapy on plasma proteinase activity, leukocyte counts, plasma elastase-alpha$_1$-protease inhibitor (E- $_1$PI) complex as well as $_1$PI activity and plasma concentration is shown in table 1. The proteolytic activity of the plasma fractions was measured using azocasein as a substrate. Before hemodialysis we found a significant increase in RDT patients (+ 244 %; p< 0.01) compared with plasma samples of healthy subjects (0.052 + 0.004 U/mg protein). The highest plasma proteolytic activity was observed in patients with posttraumatic ARF (0.255 + 0.044 U/mg). During hemodialysis therapy, however, there was a permanent decrease of plasma proteinase activity (0.127 + 0.019 vs. 0.037 + 0.007 U/mg; -71 %; p 0.001). Leukocyte counts decreased 10 min (21.1 %; n.s.) and 30 min (41.1 %; p<0.001) after initiation of hemodialysis therapy. We observed a maximal increase of the plasma E- α_1PI complexes after three hours (+ 409 %; p<0.001). In contrast, plasma α_1PI activity and α_1PI concentration were unchanged during hemodialysis therapy.

A parallel digestion of the subunits alpha, beta and gamma of
phosphorylase kinase was observed after incubation of urine fractions
of patients with posttraumatic ARF for 12 hours. However, when
approx. 100-times concentrated dialysates (pore size 10,000 dalton)
were incubated with phosphorylase kinase the alpha polypeptide chain
of the enzyme was completely degraded within 1 hour. After 4 hr of
incubation the gamma subunit was also totally destroyed, whilst
approx. 2/3 of the beta chain was digested.

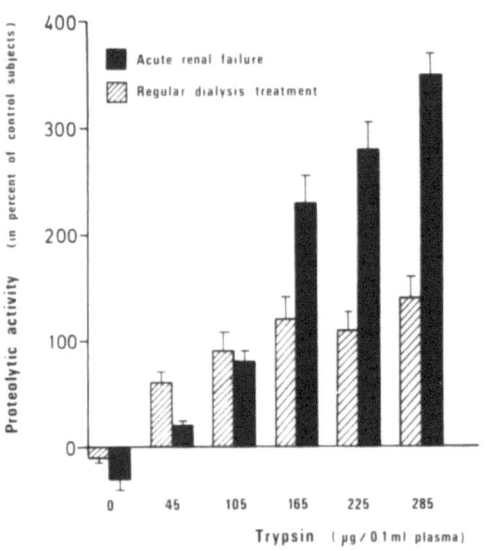

Fig. 2: Proteolytic inhibitory capacity in the plasma of RDT
patients and patients with posttraumatic ARF expressed in percent
of healthy subjects. 0.1 ml of plasma from experimental subjects
were incubated for 60 min at 37 °C with increasing amounts of
bovine trypsin. Proteolytic activity was measured using azocasein as a substrate[7]. Means ± SEM from five experiments.

Table 1

Effect of hemodialysis therapy on plasma proteinase activity, leukocyte counts, plasma elastase-alpha$_1$-proteinase inhibitor (E- α_1PI) complex as well as α_1PI activity and plasma concentration

	0	10	30	60	120	180 minutes
Proteinase activity (U/mg protein)	0.127 ± 0.019	0.103 ± 0.014	0.095 ± 0.014	0.087 ± 0.013	0.059 ± 0.010[b]	0.037 ± 0.007[c]
Leukocytes (cells/mm^3)	7,133 ± 582	5,625 ± 696	4,200 ± 251[c]	6,658 ± 441	5,950 ± 607	7,192 ± 824
E-α_1PI (ng/ml)	188 ± 20	196 ± 25	326 ± 56[a]	373 ± 32[c]	602 ± 84[c]	769 ± 128[c]
α_1PI (U/ml)	1.98 ± 0.11	1.89 ± 0.11	2.07 ± 0.10	2.06 ± 0.09	2.01 ± 0.09	2.08 ± 0.11
α_1PI (mg/dl)	254 ± 14	257 ± 13	268 ± 13	258 ± 12	260 ± 11	271 ± 13

Mean values ± SEM from 12 experiments before and during hemodialysis therapy
([a]p<0.05; [b]p<0.01; [c]p<0.001)

DISCUSSION

The object of our study was to investigate the role of proteases on the hypercatabolic state associated with ARF.

Figure 1 shows the proteolytic digestion of phosphorylase kinase by ultrafiltrated plasma from a hypercatabolic patient with
posttraumatic ARF. Comparable data were obtained with plasma ultrafiltrates from all patients with posttraumatic ARF.

Plasma ultrafiltrates from normal subjects and patients with
ARF after drug overdosage (barbiturate, flurazepam, carbromal,
different analgesics, aminoglycosides or cytostatics) did not
affect the subunit structure of phosphorylase kinase within 24
hours incubation[4]. Furthermore, plasma ultrafiltrate of patients
with posttraumatic ARF also digests hepatocyte membrane fractions,
whereas control ultrafiltrate only demonstrates very faint proteolytic degradation within 24 hours of incubation at 37 $^{\circ}C$[6].

Normally, digestion of protein in the plasma is effectively
limited by inhibitors of proteases. Alpha$_1$-antitrypsin (alpha$_1$-
protease inhibitor) and alpha$_2$-macroglobulin accounted for more
than 90 % of the total protease inhibiting capacity of plasma[10]. It
has been shown that plasma alpha$_1$-antitrypsin level increases in
patients with ARF particularly after multiple traumatic injuries.
In contrast to alpha$_1$-antitrypsin, the values of plasma alpha$_2$-mac-
roglobulin were markedly decreased[4]. Free proteolytic activity in
the plasma may be a consequence of the relative low alpha$_2$-macro-
globulin values. However, these inhibitors were measured by immunological techniques and their concentration shows no relationship to
the biological activity of these inhibitors. Trypsin binding capacity was significantly lower in RDT patients and dramatically reduced in patients with posttraumatic ARF compared to healthy controls.

We have previously shown that proteolytic activity of the administered plasma protein solutions are lower than proteolytic activity of plasma fractions of healthy controls[4]. Thus we can conclude that no proteolytic activity was infused into our patients
with ARF. However, alpha$_2$-macroglobulin was not detectible in the
available plasma protein solutions in contrast to relative low concentrations of alpha$_1$-antitrypsin[4]. Free proteolytic activity in
the plasma of a patient with posttraumatic ARF was observed due to
protease-antiprotease imbalance. After an initial determination of
80 mg/dl the alpha$_2$-macroglobulin values were too low to be detected. Addition of purified alpha$_2$-macroglobulin to the ultrafiltrates
resulted in complete inhibition of phosphorylase kinase digestion
in vitro[5]. According to our studies we favour the application of
alpha$_2$-macroglobulin or fresh frozen plasma to hypercatabolic

patients. Balldin et al. have demonstrated that the availability of alpha-macroglobulin is of vital importance for protection against proteinases[11]. Elastase[12] and trypsin[13] are bound by alpha-macroglobulin. Increased catabolism of alpha-macroglobulins after intravenous infusion of trypsin-alpha$_1$-antitrypsin complexes has been observed in dogs. More than half of the trypsin molecules were taken up by alpha-macroglobulins during the first hour. Subsequent to saturation of alpha-macroglobulins dogs became shocked[11].

Proteolytic degradation of the subunits alpha, beta and gamma of purified phosphorylase kinase was obtained by urine fractions of patients with posttraumatic ARF[6], whereas urine fractions of patients with nephrotic syndrome led to a prevalend digestion of the alpha polypeptide chain[14].

When dialysates of RDT patients were concentrated approx. 100-times we also observed rapid digestion of phosphorylase kinase[4]. From these experiments we may assume that proteinase may also be involved in the protein catabolism of RDT patients. Furthermore, it may be possible to remove in part proteolytic enzymes during dialysis treatment.

Hemodialysis therapy has been reported to be a catabolic event. It has been shown that glucose in the dialysate[15] as well as continuous amino acid infusion[16] are ineffective to prevent this catabolic state. The pathogenesis of hemodialysis induced protein catabolism remains unclear. One possibility could be the release of granulocyte proteinases after starting hemodialysis therapy. Craddock et al. have demonstrated hemodialysis-induced leukopenia and pulmonary vascular leukostasis resulting from complement activation by dialyzer cellophane membranes[17].

Neutrophil granulocytes contain a broad variety of agents that are involved in the defence and digestion of invading microorganisms[18]. These include elastase[18,19], cathepsin G, proteinase-3, cathepsin B, cathepsin D and collagenase[20-25]. Lysosomal proteinases are not restricted to the intracellular compartment. They are readily released extracellularly during cell death, phagocytosis, exposure to antigen antibody complexes, complement components and toxic substances such as endotoxins[26,27]. Under pathological conditions massive protease release may occur. This results in tissue injury[28,29] and degradation of plasma proteins, if the activities of the controlling proteinase inhibitors in plasma and tissues are insufficient.

Recently, Aasen et al.[30] observed in an experimental study (lethal endotoxin shock in canines) a clear relation of the initial drop of leukocytes (probably combined with degranulation of these cells) to the appearance of the leukocyte elastase-α_1-proteinase in-

hibitor complex in plasma. Shortly thereafter, the well-known disturbances of the blood systems arose indicating that consumption
of their components might be due, at least in part, to the action
of liberated leukocytic proteinases.

Granulocyte elastase may be released during hemodialysis
treatment probably due to the contact of blood cells with the blood
lines and the dialyzer membrane (cuprophane) system. This contact
may result in a "frustrated phagocytosis" and a subsequent proteinase release. However, leukopenia during hemodialysis therapy does
not parallel the increase of E-α_1PI. Pulmonary vascular leukostasis
resulting from complement activation by dialyzer cellophane membranes may be involved in this process[17]. Plasma α_1PI activity and
α_1PI concentration were unchanged during hemodialysis therapy
(Table 1).

Alpha$_1$-protease inhibitor is a dominating protease inhibitor
in plasma. It forms complexes with virtually all serine proteinases
[31,32] and some bacterial proteinases. A variety of oxidants including myeloperoxidase produced by neutrophils as well as cigarette
smoke are capable of inactivating alpha$_1$-protease inhibitor[33-35].

With decreased alpha$_1$-antiproteinase concentration inhibition
of bacterial and leukocyte proteinases in the bronchial secretion
may be insufficient or absent resulting in alveolar damage and
emphysema[36]. In idiopathic pulmonary fibrosis progression of the
disease is assumed due to a proteinase/antiproteinase imbalance.
Active collagenase has been observed in the lavage fluid of the
lower respiratory tract[37].

Our results offer the possibility that the hemodialysis-induced release of elastase may enhance the risk for development of
destructive lung disease in long-term RDT patients.

Proteolytic activity of the plasma fractions using azocasein
as a substrate was higher in RDT patients (+ 244 %; $p < 0.01$) compared with healthy controls (0.052 + 0.004 U/mg protein). However,
proteinase activity decreased permanently during hemodialysis
therapy indicating an activation of the plasma inhibitory capacity.
In agreement with this hypothesis we observed a further decrease
of plasma proteinase activity during hemodialysis therapy after the
addition of pork pancrease elastase to the plasma samples obtained
[38].

SUMMARY

1) Proteolytic enzymes exist in ultrafiltrated plasma, concentrated dialysates and urine fractions of patients with posttraumatic renal failure. Differences in digestion pattern of phosphory-

lase kinase suggest the existance of different proteases in patients with hypercatabolic renal failure.

2) Trypsin binding capacity is reduced in RDT patients and markedly lower in patients with posttraumatic ARF.

3) Protein catabolism is inhibited in vitro by $alpha_2$-macroglobulin. From our in vitro studies we favour the application of fresh frozen plasma instead of the available plasma protein solutions to hypercatabolic patients.

4) Hemodialysis may enhance proteinase inhibitory capacity of the plasma.

5) Hemodialysis therapy induces the increase of plasma E-α_1 PI. The continuous release of granulocyte elastase during hemodialysis therapy may enhance the risk for the development of destructive lung disease.

ACKNOWLEDGEMENT

This work was supported by the Deutsche Forschungsgemeinschaft, Ho 781/3-2 and Sonderforschungsbereich 51, München. The excellent technical assistance of Mrs. M. Röder and Mrs. U. Hof as well as the secretarial help of Miss M. Eiring is greatly appreciated.

REFERENCES

1. R. M. Abel, C. H. Beck, Jr., W. M. Abbott, J. A. Ryan, Jr., G. O. Barnett, and J. E. Fischer, Improved survival and acute renal failure after treatment with intravenous essential L-amino acids and glucose, N. Engl. J. Med. 288:695 (1973).
2. M. J. Blumenkrantz, J. D. Kopple, A. Koffler, A. K. Kamdar, M. D. Healy, E. I. Feinstein, and S. G. Massry, Total parenteral nutrition in the management of acute renal failure, Am. J. Clin. Nutr. 31:1831 (1978).
3. W. H. Hörl, and A. Heidland, Enhanced proteolytic activity - cause of protein catabolism in acute renal failure, Am. J. Clin. Nutr. 33:1423 (1980).
4. W. H. Hörl, J. Stepinski, C. Gantert, and A. Heidland, Evidence for the participation of proteases on protein catabolism during hypercatabolic renal failure, Klin. Wochenschr. 59: 751 (1981).
5. W. H. Hörl, C. Gantert, I. O. Auer, and A. Heidland, In vitro inhibition of protein catabolism by $alpha_2$-macroglobulin in plasma from a patient with posttraumatic acute renal failure, Am. J. Nephrol. 2:32 (1982).

6. W. H. Hörl, J. Stepinski, and A. Heidland, Further evidence for
   the participation of proteases in protein catabolism during
   hypercatabolic renal failure, in: "Acute Renal Failure," H. E.
   Eliahou, ed., John Libbey, London (1982).
7. W. H. Hörl, R. M. Schäfer, and A. Heidland, Role of urinary
   alpha$_1$-antitrypsin in Padutin (kallikrein) inactivation, Eur.
   J. Clin. Pharmacol. 22:541 (1982).
8. K. Weber, and M. Osborn, The reliability of molecular weight determinations by dodecylsulfate polyacrylamide gel electrophoresis, J. Biol. Chem. 244:4406 (1969).
9. S. Neumann, N. Hennrich, G. Gunzer, and H. Lang, Enzyme-linked
   immunoassay for human granulocyte α_1-proteinase inhibitor complex, in: "Progress in Clinical Enzymology - II,"D.M.Goldberg,
   M. Werner, eds., Masson Publ., New York (1983), in press.
10.C. B. Laurell, and I. O. Jeppson, Protease inhibitors in plasma,
   in: "The Plasma Proteins," F. W. Putnam, ed., Academic Press,
   New York, San Francisco, London (1975).
11.G. Balldin, C. B. Laurell, and K. Ohlsson, Increased catabolism
   of α-macroglobulins after intravenous infusion of trypsin-α_1-
   antitrypsin complexes in dogs, Hoppe-Seyler's Z. Physiol.
   Chem. 359:699 (1978).
12.E. L. Gustavsson, K. Ohlsson, and A. S. Olsson, Interaction between human pancreatic elastase and plasma protease inhibitors,
   Hoppe-Seyler's Z. Physiol. Chem. 361:169 (1980).
13.G. Balldin, K. Ohlsson, and A. S. Olsson, Studies on the influence of Trasylol on the partition of trypsin between the
   human plasma protease inhibitors in vitro, Hoppe-Seyler's Z.
   Physiol. Chem. 359:691 (1978).
14.W. H. Hörl, J. Stepinski, R. M. Schäfer, and A. Heidland, Role
   of proteases in hypercatabolic patients with renal failure,
   Kidney int. in press (1983)
15.P. Farrell, and P. Hone, Dialysis-induced catabolism, Am. J.
   Clin. Nutr. 33:1417 (1980).
16.F. Gotch, M. Borah, M. Keen, and J. Sargent, The solute kinetics of intermittent dialysis therapy (IDT), Proc. Ann. Con-
   tractors Conf. Artif. Kidney Program NIAMDD 10:105 (1977).
17.P. Craddock, J. Fehr, A. Dalmasso, K. Brigham, and H. Jacob,
   Hemodialysis leukopenia. Pulmonary vascular leukostasis resulting from complement activation by dialyzer cellophane
   membranes, J. Clin. Invest. 59:879 (1977).
18.J. Blondin, and A. Janoff, The role of lysosomal elastase in the
   digestion of Escherichia coli proteins by human polymorphonuclear leukocytes, J. Clin. Invest. 58:971 (1976).
19.A. Janoff, and J. Scherer, Mediators of inflammation in leukocyte lysosomes. IX. Elastinolytic activity in granules of human polymorphonuclear leukocytes. J. Exp. Med. 128:1137 (1968).
20.R. Rindler, F. Schmalzl, und H. Braunsteiner, Isolierung und
   Charakterisierung der chymotrypsinähnlichen Protease aus

neutrophilen Granulozyten des Menschen, Schweiz. Med. Wschr. 104:132 (1974).

21. M. Baggiolini, U. Bretz, and B. Dewald, Subcellular localization of granulocyte enzymes, in: "Neutral Proteases of Human Polymorphonuclear Leukocytes," K. Havemann, A. Janoff, eds., Urban and Schwarzenberg, Baltimore/Munich (1978).

22. G. Lazarus, J. Daniels, R. Brown, H. Bladen, and H. Fullmer, Degradation of collagen by a human collagenolytic system, J. Clin. Invest. 47:2622 (1968).

23. K. Ohlsson, and J. Olsson, The neutral proteases of human granulocytes. Isolation and partial characterization of two granulocyte collagenases, Europ. J. Biochem. 36:473 (1973).

24. U. Bretz, and M. Baggiolini, Biochemical and morphological characterization of azurophil and specific granules of human neutrophilic polymorphonuclear leukocytes. J. Cell. Biol. 63: 251 (1974).

25. G. Murphy, S. Reynolds, U. Bretz, and M. Baggiolini, Collagenase is a component of the specific granules of human neutrophil leukocytes, Biochem. J. 162:195 (1977).

26. A. Janoff, J. Blondin, R. Sandhaus, A. Mosser, and C. Malemud, Human neutrophil elastase: in vitro effects on natural substrates suggest important physiological and pathological action, in: "Proteases and Biological Control," E. Reich, ed., Cold Spring Harbor Laboratory, Cold Spring Harbor (1975).

27. J. Smolen, and G. Weissmann, The granulocyte: Metabolic properties and mechanisms of lysosomal enzyme release, in: "Neutral Proteases of Human Polymorphonuclear Leukocytes," K. Havemann, A. Janoff, eds., Urban and Schwarzenberg, Baltimore/Munich (1978).

28. A. Janoff, and J. Zeligs, Vascular injury and lysis of basement membrane in vitro by neutral protease of human leukocytes, Science 161:702 (1968).

29. C. Cochrane, and A. Janoff, The Arthus reaction. A model of neutrophil and complement mediated injury, in: "The Inflammatory Process," B. W. Zweifach, L. Grant, R. T. McCluskey, eds., Academic Press, New York, San Francisco, London (1974).

30. A. O. Aasen, and K. Ohlsson, Release of granulocyte elastase in lethal canine endotoxin shock, Hoppe-Seyler's Z. Physiol. Chem. 359:683 (1978).

31. J. Travis, P. Giles, L. Porcelli, C. Reilly, R. Baugh, and J. Powers, Human leukocyte elastase and cathepsin G: structural and functional characteristics, in: "Protein Degradation in Health and Disease," Ciba Foundation Symposium 75, Excerpta Medica, Amsterdam, Oxford, New York (1980).

32. G. Francis, S. Knowles, and F. Ballard, Inactivation of cytosol enzymes by a liver membrane protein, in: "Protein Degradation in Health and Disease," Ciba Foundation Symposium 75, Excerpta Medica, Amsterdam, Oxford, New York (1980).

33. D. Johnson, and J. Travis, Structural evidence for methionine at the reactive site of human α_1-proteinase inhibitor, J. Biol. Chem. 253:7142 (1978).

34. H. Carp, and A. Janoff, In vitro suppression of serum elastase inhibitory capacity by reactive oxygen species generated by phagocytosing polymorphonuclear leukocytes, J. Clin. Invest. 63:793 (1979).

35. A. Cohen, The effects in vivo and in vitro of oxidative damage to purified alpha$_1$-antitrypsin and to the enzyme inhibiting activity of plasma. Am. Rev. Respir. Dis. 119:953 (1979).

36. N. Matheson, P. Wong, and J. Travis, Enzymatic inactivation of human alpha$_1$-proteinase inhibitor by neutrophil myeloperoxidase. Biochem. Biophys. Res. Commun. 88:402 (1979).

37. J. Gadek, G. Fells, R. Zimmerman, S. Rennard, and R. Crystal, Antielastases of the human alveolar structures. Implications for the protease-antiprotease theory of emphysema, J. Clin. Invest. 68:889 (1981).

38. W. H. Hörl, M. Jochum, A. Heidland, and H. Fritz, Release of granulocyte proteinases during hemodialysis, Am. J. Nephrol. in press (1983).

RELEASE OF GRANULOCYTE NEUTRAL PROTEINASES IN PATIENTS WITH ACUTE AND CHRONIC RENAL FAILURE[+]

August Heidland[1], Walter H. Hörl[1], Norbert Heller[1],
Harmut Heine[2], Siegfried Neumann[3], and Ekkehard Heidbreder[1]

[1]Department of Medicine, University of Würzburg,
[2]Department of Anatomy, University of Würzburg,
[3]Biochemical Research, E. Merck, Darmstadt, FRG

INTRODUCTION

The pathogenesis of altered protein metabolism in hypercatabolic renal failure is not fully understood. There is growing evidence that the degradation of proteins is markedly enhanced in such conditions, but the rise in protein synthesis is inadequate[1,2]. Accelerated degradation of endogenous proteins may be predominantly intracellular, particularly within lysosomes. Extracellular protein degradation may also occur due to an imbalance in the circulating proteinase - antiproteinase system. Hörl et al.[3,4] have demonstrated free proteolytic activity in plasma ultrafiltrate of patients with hypercatabolic renal failure. The first data suggesting enhanced proteolytic activity were reported by Richet and Ardaillou[5], who observed increased tripeptidase activity in the plasma of hypercatabolic renal failure patients.

The origin of the circulating proteinases in uremia is as yet unknown. Potential sources are the striated muscle in rhabdomyolysis, kidneys (particularly the cortex[4]), lungs, pancreas (uremic pancreopathy[6,7]), blood cells and a variety of bacterial strains[8,9,10,11]. The present study was undertaken to examine the potential release of proteinases from polymorphonuclear (PMN) neutrophils in patients with acute and chronic renal failure (on dietary or regular hemodialysis treatment, RDT). The results obtained are in accordance with an enhanced release under these conditions.

[+]This work was supported by the Deutsche Forschungsgemeinschaft

417

MATERIALS AND METHODS

The cytochemical test of Klessen[12] was used to evaluate neutral granulocyte proteinases. He demonstrated, that the incubation of formalin-sublimate fixed blood smears of normal healthy subjects with 0.25 MNaCl borate buffer (pH 8.5) for two to three hours results in a ring-shaped degradation of erythrocytes and plasma around each PMN neutrophil (so called halo formation). Staining was performed by colloidal iron reaction. 10 - 20 granulocytes per blood smear were evaluated and the halo size graded using arbitrary units of 0 to 3. Two blood smears were performed for each patient, one after 2 and the other after 3 hours of incubation. The plasma elastase-α_1 proteinase inhibitor complex (elastase-α_1PI) was determined using the immunoassay of Neumann et al.[13]. Assays of the most important proteinase inhibitors ($\alpha_1$proteinase inhibitor, α_1PI) and $\alpha_2$macroglobulin (α_2M) were performed by a radial immunodiffusion technique with standardized immunodiffusion plates (Behringwerke, Marburg, FRG). White cell counts were determined by a Coulter Counter-Model B. Serum creatinine, blood urea nitrogen, total protein, sodium and potassium were determined by SMAC, Technicon.

PROBANDS AND PATIENTS

Control group: The control group included 13 normal healthy subjects (8 females, 5 males) with ages ranging between 20 and 36 years, $\bar{x}$ 26.2 years.

Patient group 1 (Table 1) consisted of 12 patients (8 males, 4 females) with chronic renal failure (CRF) treated with a low protein (30 g) diet combined with an oral supplement of 8 - 10 g essential amino acids. Diagnoses included chronic glomerulonephritis (7), polycystic kidney disease (3), and chronic pyelonephritis (1). The serum creatinine ranged from 6.9 - 14.2 mg/dl, $\bar{x}$ 9.9 $\pm$ 0.8 mg/dl.

Patient group 2 (Table 2) included 6 patients with acute renal failure (4 females, 2 males) of various origin (postoperative (3), septicemia (1), acute pancreatitis (1), and dye-induced renal failure (1) in a diabetic patient). Serum creatinine: 3.0 - 9.8 mg/dl, $\bar{x}$ 6.2 $\pm$ 1.1 mg/dl.

Patient group 3 (Table 3): 13 patients with chronic renal failure undergoing maintenance hemodialysis 3 x/week (7 males, 6 females) were studied. Diagnoses: chronic glomerulonephritis (9), polycystic kidney disease (1), amyloidosis (1), diabetic nephropathy (1), and analgesic nephropathy (1). Dialysis of all patients was performed with the single-use dialyser Gambro Lundia for 4 - 6 hours. Anticoagulation within the dialyser was achieved with a low dose of heparin infused into the afferent blood line. Blood samples were drawn immediately prior to the initiation of hemodialysis and in 6 patients at 2 hours intervals during dialysis treatment.

RESULTS

Neutral PMN proteinases could be identified in the blood smears of the 13 healthy subjects by a marked halo formation around each neutrophil. After a 2 hour incubation, 11 probands showed grade 2 - 3 halos, in 2 persons halo size was grade 1 to 2. After 3 hours, the halo size reached grade 3 in 8 probands and grade 2 to 3 in 4 (Figure 1). The serum concentration of elastase $\alpha_1$proteinase inhibitor complex in normal subjects is 83.1 $\pm$ 24.2 ng/ml[13].

In the 12 patients with chronic renal failure on dietary treatment (Table 1) blood smears revealed a reduction or absence of halo formation after a 2 hour incubation period in 8 patients (grade 0 to 1) (Figure 2) and normal in 4 patients (grade 1 to 2). After a 3 hour incubation halo formation was reduced in 5 patients and normal in 7 subjects. The elastase $\alpha_1$proteinase inhibitor complex was normal in 9 patients. In 3 cases, enhanced concentrations of 177, 185 and 370 ng/ml were observed (Table 1). In these patients the serum creatinine exceeded 13 mg/dl. $\alpha_2$macroglobulin, $\alpha_1$proteinase inhibitor and serum transferrin were normal.

In the six acute renal failure patients, halo formation was markedly reduced (Table 2). After a 2 hour incubation, the halo effect averaged 0 to 1 in 5 patients and in 1 patient grade 1 to 2, after 3 hours halo size averaged grade 1 to 2. The reduced proteolytic activity was basically independent of disease pathogenesis. We observed a comparable reduction of halo size in patients with acute pancreatitis, shock induced renal failure (rupture of an aortic aneurysm and postoperatively) Figure 3). Halo formation was even suppressed in the diabetic with dye induced renal failure.

The elastase $\alpha_1$proteinase inhibitor complex was simultaneously increased to 204.9 $\pm$ 32 µg/l (Table 2). A normal value was only obtained in one patient (R. I.) with postoperative renal failure after Billroth-I resection. Halo formation was clearly reduced in this patient. The degree of uremic intoxication in the ARF patients was not severe. Their creatinine averaged 6.2 $\pm$ 1.1 mg/dl and BUN 80.2 $\pm$ 10.6 mg/dl. $\alpha_1$proteinase inhibitor concentration was in the upper normal range or increased, $\alpha_2$macroglobulin was normal.

Patients undergoing chronic hemodialysis treatment (RDT) showed a markedly reduced halo formation after 2 and 3 hours of incubation (Table 3). A normal halo occured only in one patient. No patient showed signs of infectious disease (pneumonia, endocarditis or shunt infection).

Halo size was either unchanged or slightly reduced after 2 hours of hemodialysis. After 4 hours of dialysis, subnormal blood smear patterns were observed (Figure 4).

The elastase α_1 proteinase inhibitor complex was normal in 6 and slightly increased in 7 patients. Transferrin was in the lower normal range: α_1 proteinase inhibitor, and α_2 macroglobulin were normal.

DISCUSSION

Neutrophil granulocytes contain a broad variety of proteolytic enzymes that are involved in the defence and digestion of invading microorganisms[14]. These include the neutral proteinases elastase[14,15], cathepsin B[16], and collagenase[17,18] and the two acid proteinases, cathepsin B and D. Lysosomal proteinases are released extracellularly during cell death, phagocytosis, exposure to antigen antibody complexes, activated complement components (C3a and C5a) and toxic substances (endotoxins)[19,20]. Under pathological conditions, massive release of proteinases may occur resulting in plasma protein degradation and tissue injury, if the concentration or function of the controlling inhibitors is inadequate.

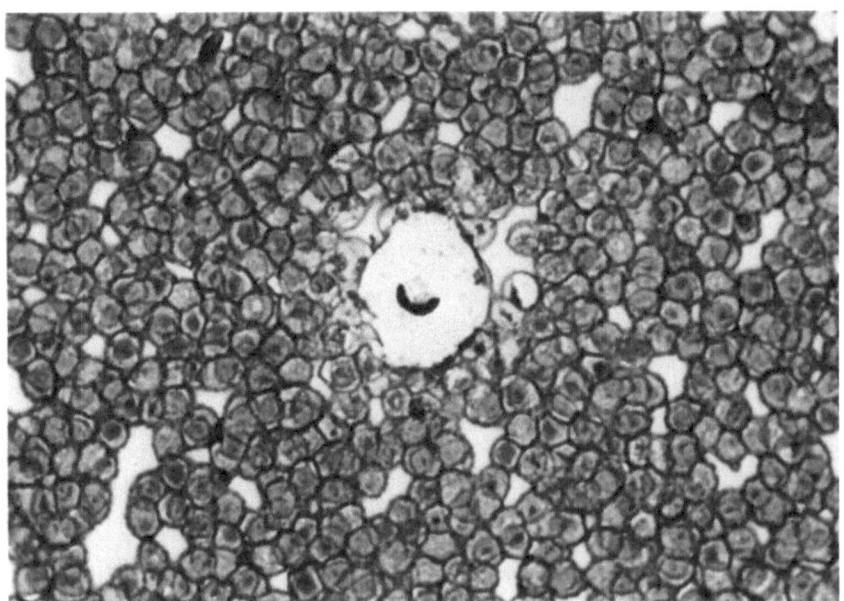

Fig. 1: Ringshaped degradation of erythrocytes and plasma around a PMN leukocyte in a blood smear of a normal subject. Incubation time with 0.25 M NaCl borate buffer 2 hours.

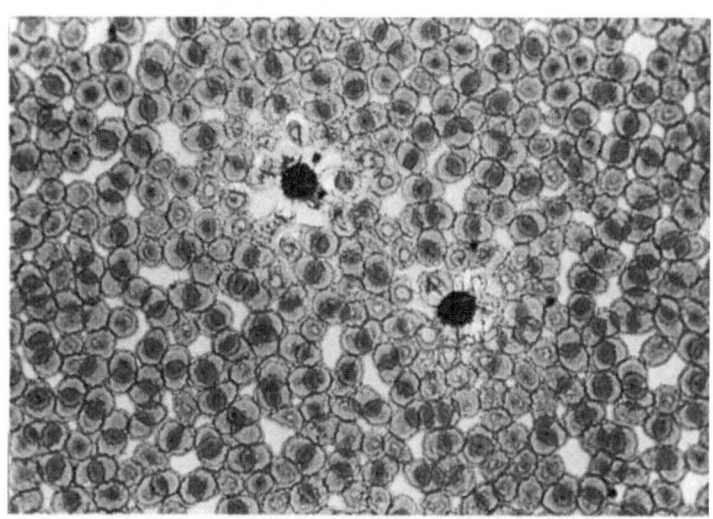

Fig. 2: Decreased halo formation with small spots of erosion around
2 leukocytes in the blood smear of a 74 years old patient (K.A.) with
chronic renal insufficiency treated with a low protein diet.
Incubation time with 0.25 M NaCl borate buffer 2 hours.

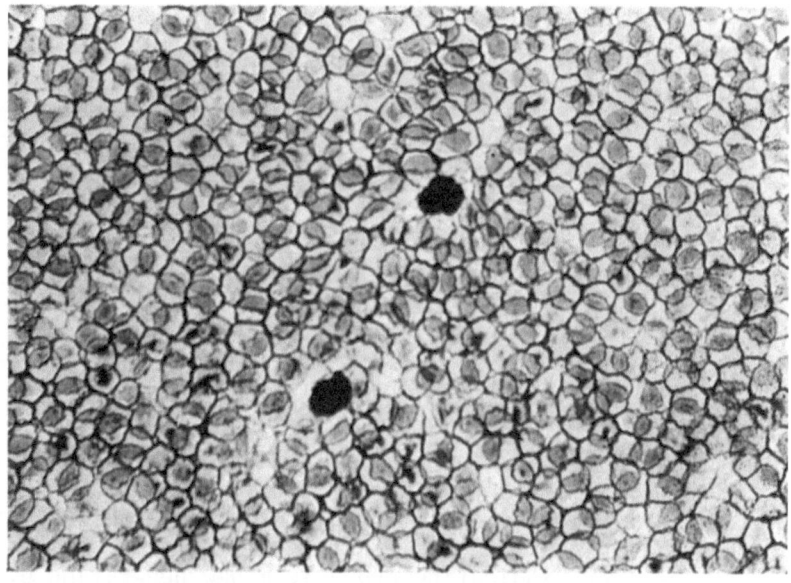

Fig. 3: Reduced halo formation in the blood smear of a 70 years old
patient with acute renal failure due to rupture of an aortic aneurys-
ma (M.M.). Incubation time with 0.25 M NaCl borate buffer 3 hours.

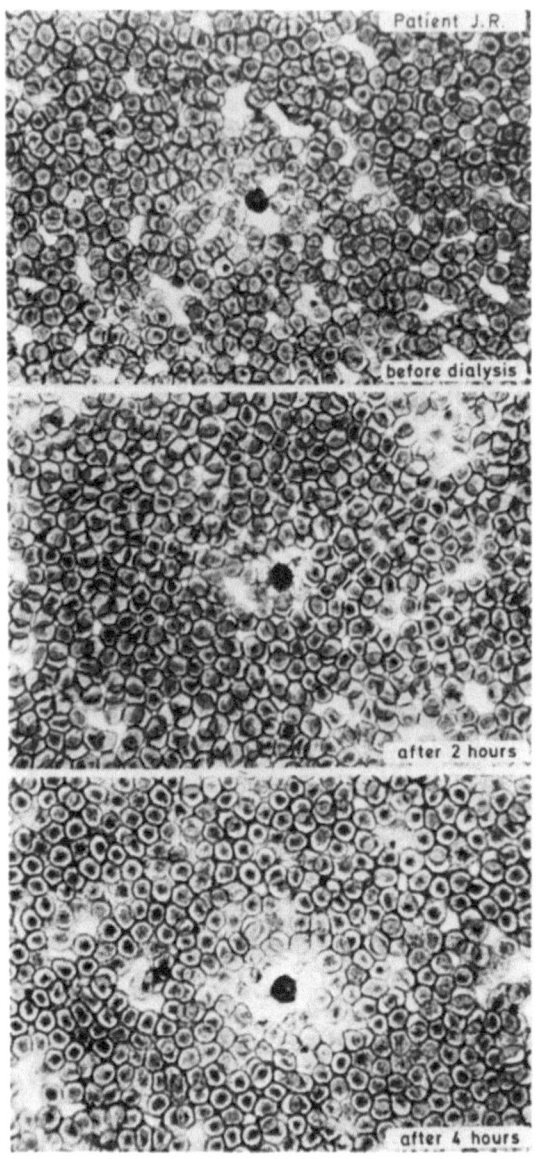

Fig. 4: Blood smears of a 25 years old patient on regular hemodialysis treatment (R.J.). Prior to the initiation and after 2 hours
of hemodialysis halo formation is markedly reduced or absent. After
4 hours of hemodialysis a subnormal proteolytic activity of PMN
leukocytes can be seen, indicating probably a new generation of
leukocytes. Incubation time 3 hours.

Table 1

Data of patients with chronic renal insufficiency on low protein diet (n = 12) Single and mean values $\pm$ SEM

Patient	Sex	Age years	Diagnosis	Serum creat. mg/dl	BUN mg/dl	Serum elast.-α_1PI μg/l	Serum α_1PI mg/dl	Serum α_2M mg/dl	Leuko-cytes x10^9/l	PMN leukocyte proteolytic activity halo effect 2 h	3 h
K.A. [x]	♀	74[x]	Chronic glomerulonephritis	14.2	122	185[x]	375	332	6.8	1	1-2
C.G.	♂	17	Chronic glomerulonephritis	13.6	104	177	290	472	4.0	1	2-3
S.M.H.	♂	48	Chronic pyelonephritis	13.5	81	370	280	266	6.8	0-1	1
B.L.	♂	54	Uninephrectomy, glomerulosclerosis	6.9	90	97	250	194	5.2	1	1-2
S.H.	♂	78	Chronic glomerulonephritis	10.3	94	73	380	284	8.5	0-1	1
G.E.	♂	47	Polycystic kidney disease	8.0	66	71	280	180	6.0	2	2
F.C.	♂	22	Chronic glomerulonephritis	10.3	78	85	290	276	4.1	1	1
B.E.	♂	56	Polycystic kidney disease	7.1	71	115	250	194	6.9	0-1	1
F.H.	♂	51	Chronic glomerulonephritis; autonomic hyperparathyroidism	8.0	62	116	290	186	7.0	1-2	1
K.W.	♂	47	Polycystic kidney disease	7.0	71	37	170	194	4.2	2	
K.M.	♀♂	51	Pyelonephritis atrophic kidney	10.2	104	92	250	194	6.1	2-3	2-3
H.H.	♀♂	87	Chronic glomerulonephritis	10.0	110	57	280	224	5.9	1-2	1-2
Mean values		52.7 $\pm$ 5.9		9.9 $\pm$ 0.8	87.8 $\pm$ 5.6	127.1 $\pm$ 24.9	282.1 $\pm$ 16.1	249.7 $\pm$ 24.7	6.0 $\pm$ 0.4	1-2	$<$ 2
Normal values (n = 13)				0.90 $\pm$ 0.11	14 $\pm$ 3	83.8[+] $\pm$ 24.2	283 $\pm$ 31	305 $\pm$ 38	6.4 $\pm$ 0.6	2	2-3

x refused dialysis treatment

+ Neumann et al., 1981

Table 2

Data of patients with acute renal failure (n = 6). Single and mean values ± SEM

Patient	Sex	Age years	Diagnosis	Serum creat. mg/dl	BUN mg/dl	Serum elast.- α_1PI µg/l	Serum α_1PI mg/dl	Serum α_2M mg/dl	Leuko-cytes x10^9/l	PMN leukocyte proteolytic activity halo effect 2 h	3 h
G.E.	♂	57	Renal failure postoperatively (prostatectomy)	3.0	75	245	455	224	14.1	1-2	1-2
B.E.	♀	79	Acute renal failure, sepsis, diabetic nephropathy	9.8	113	219	520	260	15.2	0-1	1
R.I.	♀	50	Renal failure postoperatively (Billroth's first method), septicemia	3.9	70	83.5	390	116	1.2	1	1
M.M.	♂	70	Renal failure postoperatively (rupture of an aortic aneurysma)	8.7	111	172.5	390	208	10.2	0-1	1-2
K.J.	♀	74	Acute renal failure, diabetes, pancreatitis, cholelithiasis	5.6	50	317	320	280	29.1	0-1	1
N.B.	♀	65	Dye-induced renal failure, diabetes mellitus	6.2	62	192.5	360	408	6.6	0-1	1-2
Mean values		65.8 ± 4.4		6.2 ± 1.1	80.2 ± 10.6	204.9 ± 31.8	405.8 ± 29.1	244.3 ± 39.3	12.7 ± 3.9	1	1.25
Normal values (n = 13)				0.90 ± 0.11	14 ± 3	83.8+ ± 24.2	283 ± 31	305 ± 38	6.4 ± 0.6	2	2-3

+Neumann et al., 1981

Table 3

Data of patients on regular hemodialysis treatment (RDT) (n = 13)   Single and mean values ± SEM

Patient	Sex	Age	Diagnosis	Serum creat.	BUN	Serum elast.- α_1PI	Serum α_1PI	Serum α_2M	Leukocytes	PMN leukocyte proteolytic activity halo effect	
		years		mg/dl	mg/dl	μg/l	mg/dl	mg/dl	x10^9/l	2 h	3 h
M.M.	♀	61	Chronic glomerulonephritis	8.9	62	132	350	240	9.5	0-1	1
M.Z.	♀♂	61	Analgetic nephropathy	7.0	54	174	260	280	10.2	0-1	0-1
K.K.	♂	64	Chronic glomerulonephritis	9.3	90	238	330	408	10.2	0-1	1
B.A.	♂	54	Chronic glomerulonephritis	9.3	36	71	220	260	9.2	1	1
B.H.	♂	54	Chronic glomerulonephritis	11.6	94	101.5	250	280	9.1	1	1
H.J.	♂	42	Amyloid kidney	11.1	71	211	220	492	7.2	0	0-1
S.F.	♂	52	Chronic glomerulonephritis	14.0	83	153.5	110	296	9.2	0-1	0-1
S.B.	♂	35	Chronic glomerulonephritis	13.3	91	161.5	70	312	7.7	1	1
W.W.	♀♂	51	Polycystic kidney disease	11.5	68	80	220	280	10.2	1	1
K.S.	♂	20	Chronic glomerulonephritis	7.7	94	90	230	604	7.0	0-1	0-1
H.W.	♂	53	Chronic glomerulonephritis	11.5	118	51	170	260	6.9	2	2
R.J.	♀	25	Chronic glomerulonephritis	8.0	75	91.5	250	280	5.1	1	1
S.J.	♀♀	41	Diabetic nephropathy	6.8	116	124.5	290	328	8.5	1	1
Mean values		47.2 ± 3.8		10.0 ± 0.6	81.1 ± 6.4	129.2 ± 15.6	225.5 ± 21.7	332.3 ± 29.6	8.5 ± 0.4	< 1	< 1
Normal values (n = 13)				0.90 ± 0.11	14 ± 3	83.8[+] ± 24.2	283 ± 31	305 ± 38	6.4 ± 0.6	2	2-3

[+]Neumann et al., 1981

Decreased PMN neutrophil proteolytic activity has been described in febrile infections[21]. According to our investigations, similar alterations of leukocyte function may occur in patients with acute and chronic renal insufficiency. More than half of our patients with chronic renal failure on dietary treatment showed a decreased halo formation around neutrophils. This phenomenon could not be attributed to overt or latent infection. All patients were afebrile and showed no leukocytosis or leukopenia. Chronic pyelonephritis was the cause of renal insufficiency in only one patient. Elastase$\alpha_1$proteinase inhibitor complex was normal in 9 patients, 3 patients showed a marked rise to 177, 185 and 370 ng/ml, respectively. The serum creatinine concentration was extremely high in these patients (>13 mg/dl).

In the 6 acute renal failure patients, leukocyte halo formation was dramatically reduced or absent in spite of less pronounced renal insufficiency. The elastase$\alpha_1$proteinase inhibitor complex was simultaneously elevated in 5/6 patients (rise by a factor of 2 to 4). The increased serum concentration of α-elastase proteinase inhibitor complex was not limited to patients with septicemia and postoperative renal failure, but also occured in 1 patient with dye induced ARF.

In RDT patients, leukocyte neutral proteolytic activity was extremely reduced. The halo effect was unchanged after 2 hours of dialysis. After 4 hours of hemodialysis a subnormal halo formation could be seen.

Half of the RDT patients showed a simultaneous rise of elastase $\alpha_1$proteinase inhibitor complex. A further increase of elastase$\alpha_1$proteinase inhibitor complex occurs during each hemodialysis, and persists for the whole treatment period[22,23].

The decreased halo formation of leukocytes in uremia may either be due to a defect in enzyme synthesis in the promyelocyte or due to an enhanced enzyme release into the circulation. The observation of an increased serum concentration of elastase$\alpha_1$proteinase inhibitor complex in, at least, some of the uremic patients supports the hypothesis of an increased release of this proteinase. There is probably a concomitant rise of other proteinases such as cathepsin B and collagenase.

The mechanisms of proteinase release in uremia is unknown. The role of "uremic toxins" has to be considered seriously in the absence of an active immunologic disease. Additional factors such as trauma, shock and septicemia are important in ARF patients. In patients on hemodialysis treatment another factor is involved. There is a tremendous rise of the elastase$\alpha_1$proteinase inhibitor complex due to the blood contact with the dialyser (Cellophan, resp. Cuprophane) membrane.

The use of cuprophane or cellophane membranes for dialysis results in a pronounced granulocytopenia during the first 30 min of treatment[26]. This is followed by tendency towards an overshoot of the neutrophil count. Recent studies have implicated an activation of the complement system by the alternative pathway: C5a seems to be largely responsible for this neutropenia. C5a induces increased adhesiveness of neutrophiles with subsequent leukoembolization particularly in the lung[27,28]. Leukocyte activation or cell death is associated with a release of lysosome, lactoferrin[29,30] and according to our studies of elastase. Enzymes may be released from circulating granulocytes, aggregated cells in the lung or other tissues. Dying or activated granulocytes are entrapped within the dialyser and release their granular constituents. It was recently shown, that leukocytes leaving the cellophane dialyser had a significantly decreased lysosyme content[29]. Hemodialysis associated complement activation is probably not solely due to the simple plasma cellophane (cuprophane) interaction. The cleavage of inactive C5 complement component by products released from the PMN specific granules could also play an important role.

The observation of leukocytes with a subnormal halo formation at the end of each dialysis can only be explained by a new leukocyte generation released from the bone marrow in an environment which is less toxic than at the beginning of the hemodialysis treatment. According to Hörl et al.[23] the inhibitory capacity improves at the end of hemodialysis treatment by an unknown mechanism.

The findings of a decreased PMN neutrophil proteolytic activity in uremia and the increased serum concentration of elastase $\alpha_1$proteinase inhibitor complexes does not prove that there is free proteolytic activity in the serum. The prerequisites for free proteolytic activity in the serum are

1. no inhibition of proteinase inhibitors,
2. inadequate inhibitor concentration or activity and
3. excessive proteinase release.

In our patients with chronic renal failure on dietary or hemodialysis treatment, the concentrations of the most important inhibitors ($\alpha_2$macroglobulin and $\alpha_1$proteinase inhibitor) were normal, as determined by immunological methods. The concentration of $\alpha_1$proteinase inhibitor was slightly increased in patients with acute renal failure, while that of $\alpha_2$macroglobulin was lower. A functional defect of these inhibitors cannot be excluded. Hörl et al.[31] have shown, that the inhibitory capacity of these proteins is reduced in uremia.

We seriously have to consider whether or not some complications of acute and chronic uremia, particularly of longterm hemodialysis treatment (anemia, muscle wasting, altered protein metabolism, lung disease) are consequences of a protease-antiprotease imbalance.

SUMMARY

In uremic intoxication proteolytic activity in plasma and striated muscle is enhanced. To get further insight into the underlying mechanisms the neutral proteinases of polymorphonuclear (PMN) leukocytes were investigated in patients with acute and chronic renal failure. The following studies were performed:
1. Neutral proteolytic activity of PMN neutrophils in blood smears (according to Klessen, 1978).
2. Serum levels of elastase α_1 proteinase inhibitor complex (Neumann et al., 1981).

In about half of the patients with chronic renal insufficiency on dietary treatment the proteolytic activity of PMN leukocytes (halo formation are due to digestion of erythrocytes and plasma) was reduced. The serum concentration of elastase α_1 proteinase inhibitor complex was normal in most subjects, but increased in 3 patients with the highest serum creatinine levels (>13 mg/dl). In the patients with acute renal failure (ARF) of various origin (postoperatively, septicemia, pancreatitis or dye induced) halo formation was either reduced or absent. Serum elastase α_1 proteinase inhibitor was increased in 5/6 patients by a factor of two to four. Also in the 15 patients on regular hemodialysis treatment halo formation was substantially reduced, while the serum levels of elastase α_1 proteinase inhibitor complex was slightly increased.

The finding of reduced proteolytic activity of PMN neutrophils in uremia is probably due to an enhanced release of proteinases into the circulation as indicated by the elevated serum levels of elastase α_1 proteinase inhibitor complex in some patients. The release of proteinases might be in part due to the effect of "uremic toxins". In the RDT patients the contact of the blood with the dialyzer (cuprophane) membrane might be an additional factor. In the patients with ARF the underlying disease (infection, shock, trauma) contributes to the release of proteinases. These disturbances may be harmful for the patient, if the blood concentration or function of the most important proteinase inhibitors (α_1 proteinase inhibitor, α_2 macroglobulin) is reduced.

REFERENCES

1. C. L. Long, M. Jeevanandam, B. M. Kim, and J. M. Kinney, Whole-body protein synthesis and catabolism in septic man, Am. J. Clin. Nutr. 30:1340 (1977).
2. R. H. Birkhahn, C. L. Long, D. Fitkin, M. Jeevanandam, and W. S. Blakemore, Whole-body protein metabolism due to trauma in man as estimated by L-^{15}N alanine, Am. J. Physiol. 241:E64 (1981).

3. W. H. Hörl, and A. Heidland, Enhanced proteolytic activity -
    cause of protein catabolism in acute renal failure, Am. J.
    Clin. Nutr. 33:1423 (1980).
4. W. H. Hörl, J. Stepinski, and A. Heidland, Further evidence for
    the participation of proteases in protein catabolism during
    hypercatabolic renal failure, in: "Acute Renal Failure," H.
    E. Eliahou, ed., John Libbey, London (1982).
5. G. Richet, and R. Ardaillou, L'activité tripeptidasique due
    plasma au cours des affections severes contribution a l'étude
    de l'hypercatabolisme protidique, Presse Medicale 30:1229
    (1959).
6. E. Heidbreder, W. Ralla, and A. Heidland, The exocrine pancreas
    in uremia: clinical and experimental investigations, in:
    "Renal Insufficiency," A. Heidland, H. Hennemann, J. Kult,
    eds., Georg Thieme Verlag, Stuttgart (1976).
7. A. Borgström, and K. Ohlsson, Studies on the turnover of endo-
    genous cathodal trypsinogen in man, Europ. J. Clin. Invest.
    8:379 (1978).
8. J. Y. Tai, A. A. Kortt, T. Y. Lia, and S. D. Elliott, Primary
    structure of streptococcal proteinase, III. Isolation of
    cyanogen bromide peptides: complete covalent structure of
    the polypeptide chain, J. Biol. Chem. 251:1955 (1976).
9. St. P. Riepe, J. Goldstein, and D. H. Alpers, Effect of secreted
    bacteroides proteases on human intestinal brush border hydro-
    lases, J. Clin. Invest. 66:314 (1980).
10. A. L. Goldberg, J. D. Kowit, J. D. Etlinger, and Y. Klemes, Selec-
    tive degradation of abnormal proteins in animal and bacterial
    cells, in: "Protein Turnover and Lysosome Function," H. L.
    Segal, D. Doyle, eds., Academic Press, New York (1978).
11. K. Morihara, and H. Tsuzuki, Production of protease and elastase
    by Pseudomonas aeruginosa strains isolated from patients,
    Infect. Immun. 15:679 (1977).
12. C. Klessen, On testing the activity of proteases from human poly-
    morphonuclear neutrophils on blood smears, J. Histochem. Cy-
    tochem. 26:759 (1978).
13. S. Neumann, G. Gunzer, N. Hennrich, M. Jochum, K. H. Duswald,
    and H. Fritz, Enzyme-linked immunoassay for human granulo-
    cytic elastase alpha-1-proteinase inhibitor complexes. De-
    termination in plasma of patients after major surgery, Poster
    Symposion on Inflammation Markers, Lyon, France (1981).
14. J. Blondin, and A. Janoff, The role of lysosomal elastase in the
    digestion of Escherichia coli proteins by human polymorpho-
    nuclear leukocytes, J. Clin. Invest. 58:971 (1976).
15. A. Janoff, and J. Scherer, Mediators of inflammation in leuko-
    cyte lysosomes. IX. Elastinolytic activity in granules of
    human polymorphonuclear leukocytes, J. Exp. Med. 128:1137
    (1968).

16. R. Rindler, F. Schmalzl und H. Braunsteiner, Isolierung und Charakterisierung der chymotrypsinähnlichen Protease aus neutrophilen Granulozyten des Menschen, Schweiz. Med. Wschr. 104:132 (1974).

17. G. S. Lazarus, J. R. Daniels, R. S. Brown, H. A. Bladen, and H. M. Fullmer, Degradation of collagen by a human collagenolytic system, J. Clin. Invest. 47:2622 (1968).

18. K. Ohlsson, and J. Olsson, The neutral proteases of human granulocytes. Isolation and partial characterization of two granulocyte collagenases, Europ. J. Biochem. 36:473 (1973).

19. A. Janoff, J. Blondin, R. A. Sandhaus, A. Mosser, and C. J. Malemud, Human neutrophil elastase: in vitro effects on natural substrates suggest important physiological and pathological action, in: "Proteases and Biological Control," E. Reich et al., eds., Cold Spring Harbor Laboratory, Cold Spring Harbor (1975).

20. J. E. Smolen, and G. Weissmann, The granulocyte: Metabolic properties and mechanisms of lysosomal enzyme release, in: "Neutral Proteases of Human Polymorphonuclear Leukocytes," K. Havemann, A. Janoff, eds., Urban and Schwarzenberg, Inc. Baltimore-Munich (1978).

21. C. Klessen, and W. Tekolf, Cytochemical investigation of neutral proteases in polymorphonuclear (PMN) neutrophiles in acute inflammatory diseases, Histochemistry 69:307 (1980).

22. A. Heidland, W. H. Hörl, N. Heller, H. Heine, S. Neumann, and E. Heidbreder, Proteolytic enzymes and catabolism - enhanced release of granulocyte proteinases in uremic intoxication and during hemodialysis, Kidney Int. in press (1983).

23. W. H. Hörl, M. Jochum, A. Heidland, and H. Fritz, Release of granulocyte proteinases during hemodialysis, Am. J. Nephrol. in press (1983).

24. R. Egbring, W. Schmidt, G. Fuchs, and K. Havemann, Demonstration of granulocyte proteases in plasma of patients with acute leukemia and septicemia with coagulation defects, Blood 49: 219 (1977).

25. A. O. Aasen, and K. Ohlsson, Release of granulocyte elastase in lethal canine endotoxin shock, Hoppe-Seyler's Z. Physiol. Chem. 359:683 (1978).

26. L. W. Kaplow, and J. A. Goffinet, Profound neutropenia during the early phase of hemodialysis, JAMA 203:1135 (1968).

27. P. R. Craddock, J. Fehr, A. P. Dalmasso, K. L. Brigham, and H. S. Jacob, Hemodialysis leukopenia. Pulmonary vascular leukostasis resulting from complement activation by dialyzer cellophane membranes, J. Clin. Invest. 59:879 (1977).

28. P. R. Craddock, D. E. Hammerschmidt, J. G. White, A. P. Dalmasso, and H. S. Jacob, Complement (C5a)-induced granulocyte aggregation in vitro. A possible mechanism of complement-mediated leukostasis and leukopenia, J. Clin. Invest. 60:260 (1977).

29. H. Wysocki, R. Czarnecki, B. Wierusz-Wysocka, A. Wykretowicz, K. Wysocki, and K. Baczyk, The selective polymorphonuclear neutrophils (PMN) degranulation as the probable additional mechanism of the hemodialysis (HD)-induced complement activation, Int. J. Art. Org. 4:174 (1981).

30. R. Hällgren, P. Venge, and B. Wilström, Hemodialysis-induced increase in serum lactoferrin and serum eosinophil cationic protein as signs of local neutrophil and eosinophil degranulation, Nephron 29:233 (1981).

31. W. H. Hörl, J. Stepinski, C. Gantert, M. Hörl, and A. Heidland, Evidence for the participation of proteases on protein catabolism during hypercatabolic renal failure, Klin. Wschr. 59: 751 (1981).

32. J. D. Conger, W. S. Hammond, A. C. Alfrey, S. R. Contiguglia, W. E. Huffer, and R. E. Stanford, Pulmonary calcification in chronic dialysis patients: Clinical and pathological studies, Am. Intern. Med. 83:330 (1975).

CHANGES IN COMPONENTS OF THE PLASMA KALLIKREIN-KININ AND FIBRINOLYTIC SYSTEMS INDUCED BY A STANDARDIZED SURGICAL TRAUMA

A.O. Aasen, J. Stadaas, T.E. Ruud and P. Kierulf

Central Laboratory and Surgical Department
Ullevaal University Hospital, University of
Oslo, Oslo 1, Norway

INTRODUCTION

Trauma might lead to pathological plasma proteolysis. In recent studies we found that several components of the plasma proteolytic enzyme systems becomes markedly reduced in patients with multiple trauma (1). Furthermore,these studies revealed significantly more pronounced decreases within the first week after injury in the fatal cases than in the patients that survived. In order to further evaluate these changes, components of the plasma kallikrein-kinin and fibrinolytic systems were studies in patients with a standardized surgical trauma.

MATERIALS AND METHODS

Eight patients admitted to our surgical department to have surgery for obicity were studied. All patients had a standardized surgical procedure with an extensive gastric exclusion, leaving only a small poach of the ventricle to be anastomosed to the small intestine. All operations were performed by the same surgical team. The clinical course was uneventful in all cases.

Citrated plasma samples were prepared preoperatively and daily for 7-8 days after surgery. Venous blood was drawn from a peripheral vein, deep frozen (-70°C), thawned and tested en bloc.

Chromogenic peptide substrate assays were used to

determine plasma prekallikrein (PKK), functional plasma
kallikrein inhibition (KKI), plasma plasminogen (Plg)
and functional plasma fast antiplasmin (AP). The assays
were performed on an automated centrifugal analyzer
(Cobas-Bio, Hoffman La Roche, Basel, Switzerland)
adapting previously published manual methods (2).

STATISTICS

     Non parametric statistical methods were used to
measure centrality and distribution of the data,
determining median values and 25 and 75 percentiles.
Wilcoxon rank sum test was used to test statistical
significance. The limit of significance was set to
p< 0.05.

RESULTS

     Plasminogen values were significantly reduced on the
first day after surgery (Fig.1) when compared with preoperative values. Thereafter this parameter gradually
increased reaching normal values within the first week
(Fig.1).

     Plasma prekallikrein values also revealed the same
pattern of changes (Fig.2) beeing significantly reduced
on the first day after surgery and thereafter normalizing
within the first week after the operation.

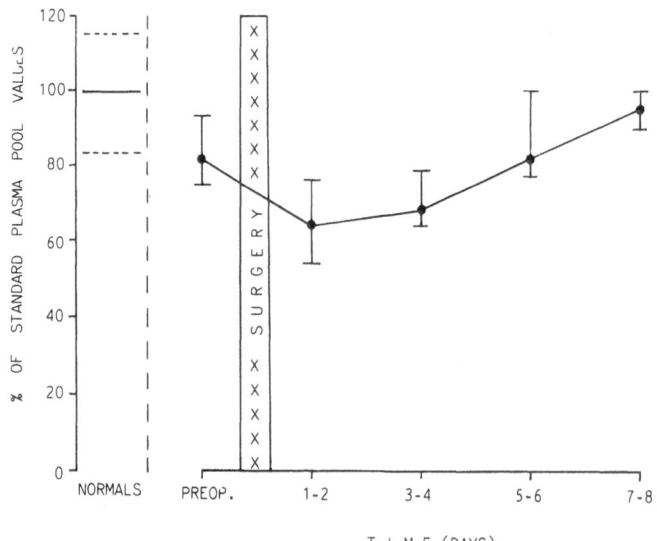

Fig.1.   Plasminogen in patients with gastric exclusion.

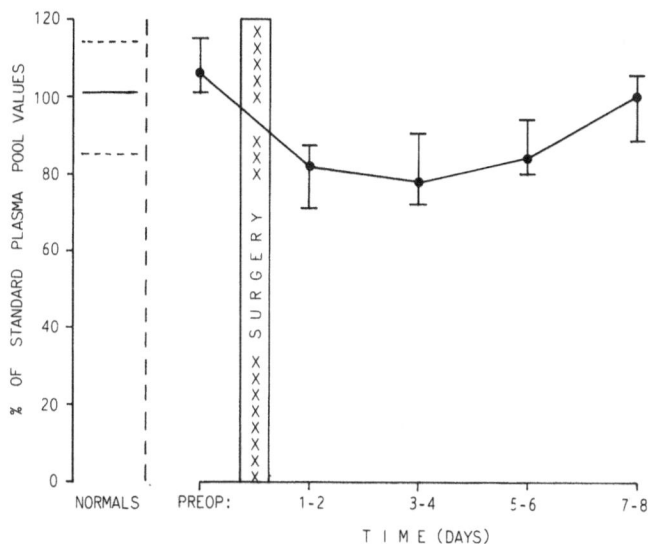

Fig.2. Prekallikrein in patients with gastric exclusion.

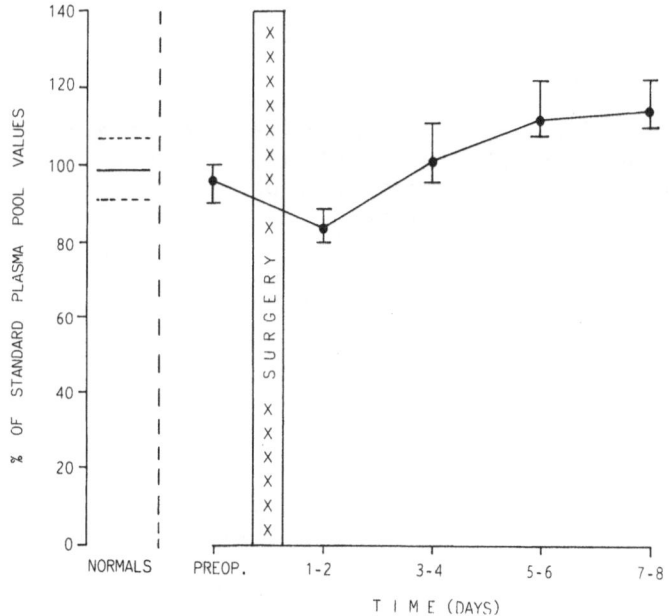

Fig.3. Functional plasma kallikrein inhibition in patients with gastric exclusion.

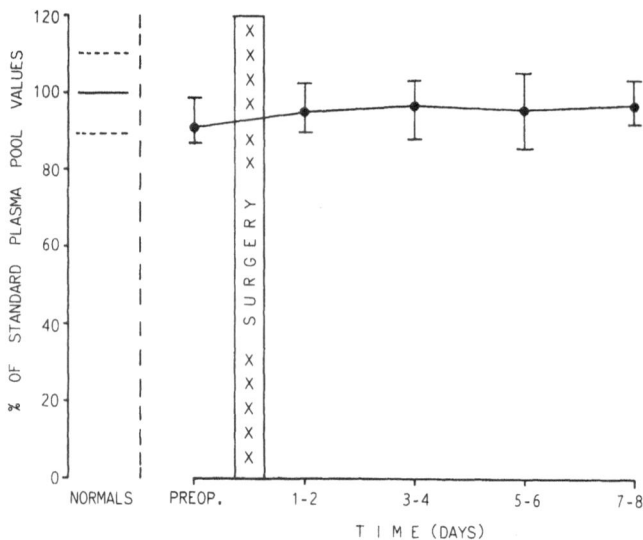

Fig.4.   Functional plasma fast antiplasmin activity in
         patients with gastric exclusion.

     Functional plasma kallikrein inhibition was
significantly reduced on day 1-2 after surgery when
compared with preoperative values (Fig.3).
For the rest of the observation period this parameter
was within the normal range or raised (Fig.3).

     Functional plasma fast antiplasmin activity was
unchanged and within the normal range for the whole
observation period (Fig.4).

DISCUSSION

     The present study in patients having a standardized
surgical trauma revealed significantly reduced values
on the first few days postoperatively for several of the
paramethers studied (PKK, Plg and KKI). All parameters
were normalized within the first week after surgery.
Functional plasma antiplasmin activity was unchanged for
the whole observation period. Several factors including
plasma dilution can be responsible for the changes seen.
Proteolytic process i.e. coagulation, fibrinolysis, and
kinin generation should also be considered.

The present observations extend and underline the clinical significance of our previous findings in multi-traumatized patients (1). In multitraumatized patients surviving the injury the reductions of several components of the plasma proteolytic enzyme systems also were normalized within the first week after the injury. In fatal case, however, values were kept low or declined during the same time. The cours in fatal cases were complicated with septicemia and lung insufficiency.

The present study thus suggest that monitoring plasma prekallikrein, plasminogen and functional inhibitor capacities of plasma protease systems appears to be important in evaluating surgical patients post-operatively.

REFERENCES

1. P.Kierulf, A.O.Aasen, S.Aune, H.C.Godal, T.E.Ruud and J.Vaage, Chromogenic peptide substrate assays in patients with multiple trauma, Acta Chir. Scand. Supp. 509: 69 (1982).
2. A.O.Aasen, P.Kierulf and J.H.Strømme, Methodological considerations on chromogenic peptide substrate assays and application on automated analyzers, Acta Chir. Scand. Supp. 509: 17 (1982)

ACKNOWLEDGEMENT

The financial support by Norwegian Research Council for Sciences and Humanities is acknowledged.

ENDOTOXINS AND COAGULATION PARAMETERS IN PATIENTS WITH TRAUMATIC HAEMORRHAGIC- AND BACTERIOTOXIC SHOCK

A. Stemberger, F. Strasser, G. Blümel, B.v. Hundelshausen*
S. Jelen*, O. Schmidt* and G. Tempel*
Institute for Experimental Surgery and *Anaesthesiology of
The Technical University Munich, Ismaninger Str. 22
D-8000 München 80, F.R.G.

INTRODUCTION

Endotoxins trigger the pathogenesis of gramnegative septice-mia.Lipopolysaccharides like lipid A initiate the endotoxin in-duced changes in blood coagulation,e.g. complement, fibrinolysis, the kallikrein-kinin system and the fibrile response (2,5,6,9,10, 13,15,16,18,22,23,24).

Determination via the endotoxin induced gelation reaction of the hemolymph of the horse-shoe crab (limulus amoebocyte lysate, lal) in aqueous solution has been established by Levin and Bang (11,12). Japanese investigators have shown that chromogenic sub-strates which were originally developed for coagulation assays. are susceptible to the so called proclotting enzyme (8).

The aim of this investigation was to monitor septic markers in patients suffering from hemorrhagic and/or bacteriotoxic shock. This includes endotoxin and parameters of coagulation and fibrino-lysis such as antithrombin III (at III), factor X, plasminogen, antiplasmin and fibronectin.

MATERIALS AND METHODS

Test procedures for the endotoxin assay were performed accor-ding to a previously published paper (20). Heparinized plasma sam-ples of patients were diluted with pyrogen free water (1:3) and boiled in a water bath for 10 minutes.

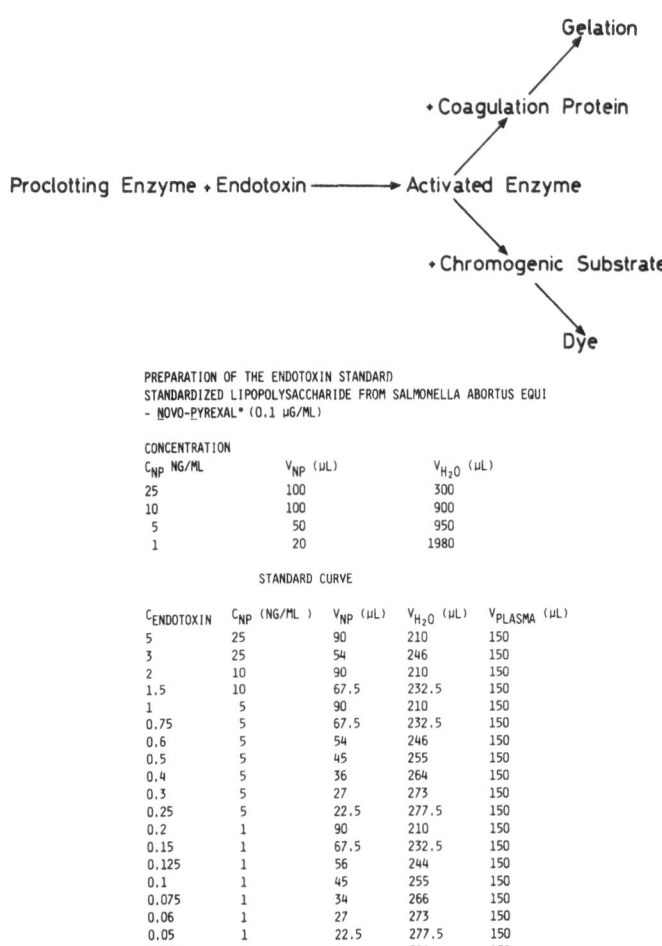

Fig. 1: Scheme of endotoxin induced l.a.l. reactions

```
TESTSYSTEM  MUNICH

INITIAL RATE METHOD                      SAMPLE         BLANK
                                         CUVETTE        CUVETTE
_____

BUFFER                                   400 µL         450 µL
MGCL₂                                     60 µL          60 µL
PLASMA ( DILUTED 1:3 )                    20 µL          20 µL
MIX AND INCUBATE ( 37°C)                        2 MIN
LAL ( ON ICE )                            50 µL          ---
MIX AND INCUBATE (37°C) EXACTLY                30 MIN
SUBSTRATE                                400 µL         400 µL


MIX IMMEDIATELY AND READ   0, 10, 20, 30, 45, 60, SECONDS OR
CONNECT SPECTROPHOTOMETER TO A RECORDER
_____


BUFFER               0.1 MOL TRIS PH 8.0
MGCL₂ SOLUTION       0.5 MOL
LAL DISSOLVED WITH AQUADEST
SUBSTRATE            0.6 MMOL S-2423 IN BUFFER
```

The following solutions were used: 0.1 M Tris-HCL buffer pH 8.0; 0.5 M $MgCl_2$; chromogenic substrate (2423, donated from Kabi Diagnostika) was dissolved in buffer; the l.a.l. was purchased from local distributors (Fa.Concept, Fa.Pyrotel) and dissolved in water. Endotoxin dilution were prepared from a standardized endotoxin preparation of salmonella abortus equi (Novo Pyrexal, Hermal Chemie). For properties see Galanos et al.(4). Pyrogen free test tubes were purchased from Byk Mallinckrodt.

Parameters for coagulation and fibrinolysis were determined as follows: at III and antiplasmin with test kits of Kabi Diagnostika GmbH (17), factor X using the X a-sensitive substrate S-2222 (1). Plasminogen was measured after streptokinase activation using the plasmin sensitive substrate S-2251 (3). Fibronectin was measured with a commercially available immunological method (Tinaquant[R], Boehringer Mannheim).

RESULTS AND DISCUSSION

In a group of 30 polytraumatized patients endotoxin was monitored over a period of 7 days after onset of septicemia (n=13) or in the non septic group (n=17) after hospitalization. Heparinized platelet poor plasma samples were immediately analyzed or stored deep frozen until determination. Analyses were performed with a recently developed chromogenic substrate assay as shown in materials and methods. A typical standard curve is seen in figure 2.

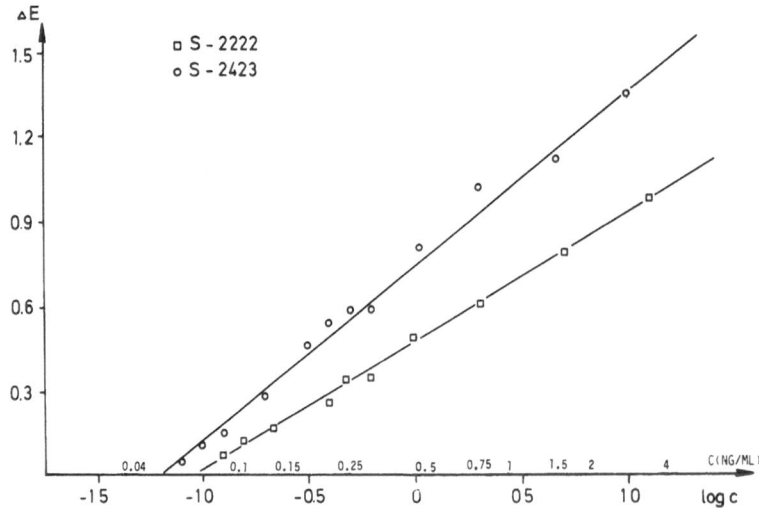

Fig. 2:   Calibration curve of the l.a.l. Chromogenic substrate
          endotoxin assay (S-2423 was only used in these inves-
          tigations).

These investigations have shown that the addition of polyvalent proteinase inhibitors to the anticoagulant solution decreases
the test sensitivity. This could be demonstrated both for drugs like
aprotinin (Trasylol) or Gabexate mesilate (Foy). It is noted that
these drugs are often administered during septic shock (fig. 3) (21).

Endotoxin could be demonstrated in polytraumatized patients
with septicemia. However in most cases neither positive blood culture nor positive endotoxin levels were found even in the septic
group (fig. 4).

In man endotoxins initiate different biological changes. Endotoxins stimulate the leucocyte dependent liberation of thromboplastin like materials followed by activation of the coagulation
system. Fibrin formation causes in conjunction with aggregated
platelets microthrombus formation. As parameters for activated
coagulation we used at III the key inhibitor for coagulation and
factor X as the key factor of the extrinsic and intrinsic pathway.

In the early post-traumatic period and after onset of septicemia a decrease at III was demonstrated with a slight improvement
during the following days. However as seen in fig. 5 a statistically
significant difference between both groups could not be shown.

INFLUENCE OF GABEXAT MESILAT (FOY) AND APROTININ ON THE LAL – CHROMOGENIC SUBSTRATE REACTION

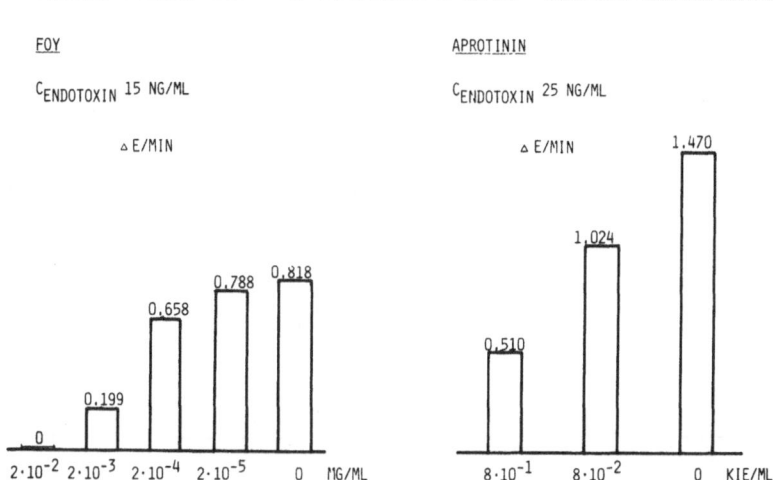

Fig. 3: Influence of polyvalent proteinase inhibitors on the l.a.l. chromogenic substrate endotoxin assay

	day after onset of septic complications						
	1	2	3	4	5	6	7
bloodcult.pos Endotoxin pos bloodcult.neg	·· ····	· ···	· ··	· ··	· ··	· ·	· ·
bloodcult.pos Endotoxin neg bloodcult neg	··· ·····	··· ······ ·	··· ······ ··	··· ······ ··	··· ······ ··	··· ······ ···	··· ··· ··

	day after hospitalization of polytrauma						
	1	2	3	4	5	6	7
bloodcult.pos Endotoxin pos bloodcult.neg	····	·	··	···			
bloodcult.pos Endotoxin neg bloodcult neg	····· ····· ·	······ ····· ··	······ ······ ···	······ ······ ··	····· ····· ··	····· ·····	······ ······

Fig. 4: Endotoxins in patients with traumatic hemorrhagic and/or bacteriotoxic shock

AT III

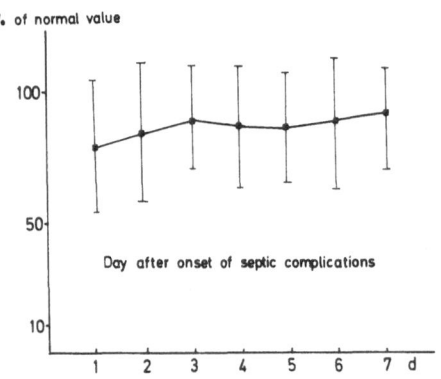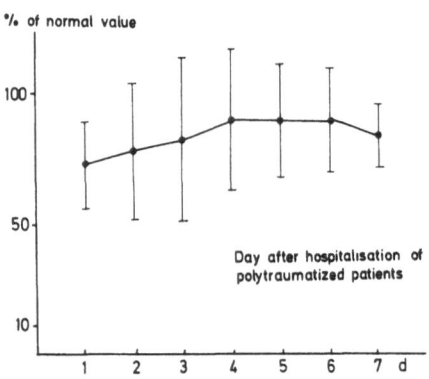

Fig. 5: At III levels in patients with traumatic hemorrhagic and/
or bacteriotoxic shock

FACTOR X

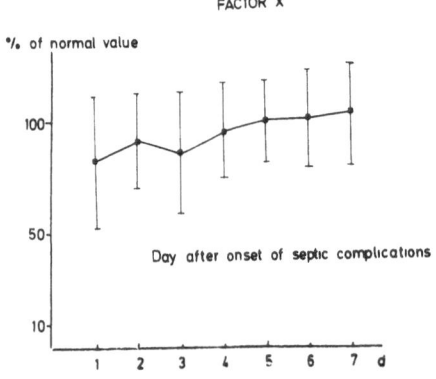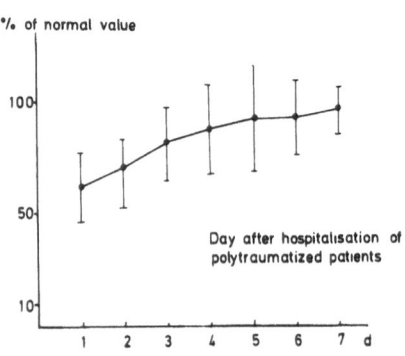

Fig. 6: Factor X activities in patients with traumatic hemorrhagic
and / or bacteriotoxic shock

Similar results were obtained for factor X (see fig. 6). A decrease in the factor X values was seen in both groups.But an improvement was noticed during the following seven days.

Concerning the fibrinolytic system in polytraumatized patients a slight activation of the fibrinolytic system was noted. The acute phase proteins plasminogen and antiplasmin as the sum of the different inhibitors like α_2-plasmin inhibitor are elevated during recovery period. In septic patients this observation was less obvious. Statistically significant differences could not be shown (fig. 7).

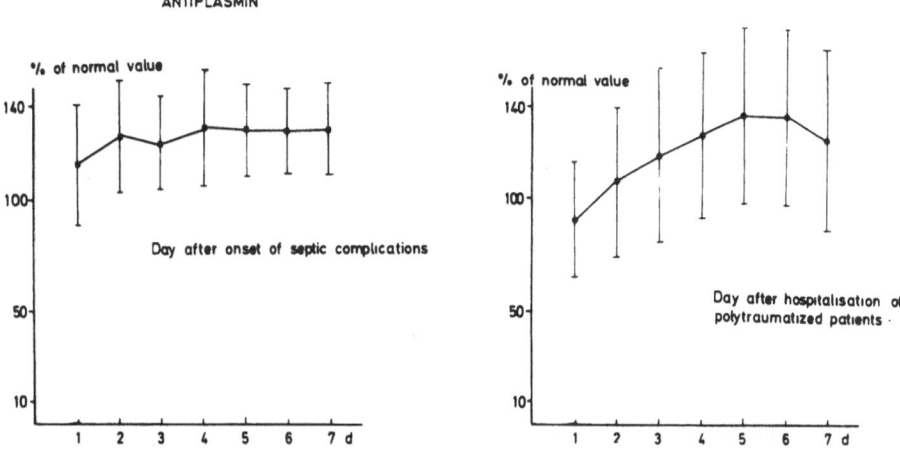

Fig. 7: Antiplasmin activities in patients with traumatic hemorrhagic and/or bacteriotoxic shock

Fibronectin as a marker of the RES clearance function (14,19) was not significantly altered. Abnormally low values were observed in both groups (fig. 8).

These data indicate that the early posttraumatic period is characterized by a slight activation of the coagulation and fibrinolytic system. However during the following seven days a tendency to normalization of these parameters is detected. This is the result of shock therapy such as tranfusion of fresh blood, fresh frozen plasma and surgical intervention for blood staunching. Endotoxins were found in only some of the investigated septic patients. In polytraumatized non septic patients low levels of endotoxin were noted during hemorrhagic shock. Endotoxin induced disturbances in blood coagulation have not been observed and confirm previous findings (7,20). However a severe septic shock during peritonitis* resulted in high endotoxin levels and a decrease in coagulation and fibrinolysis parameters.

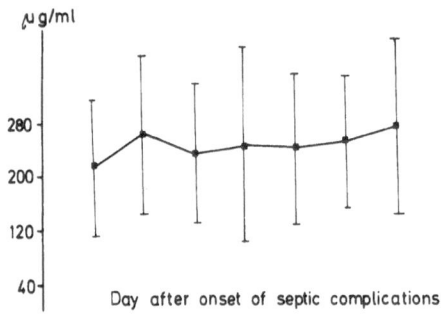

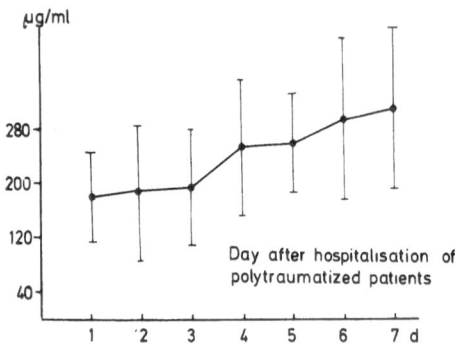

Fig. 8: Fibronectin values in patients with traumatic hemorrhagic and/or bacteriotoxic shock

    In conclusion these investigations have shown that in treated patients endotoxin is not a sensitive marker in the early diagnosis of gram negative infections.

REFERENCES

1. L. Aurell, P. Friberger, G. Karlsson, and G. Claeson, A new sensitive and highly specific chromogenic peptide substrate for factor X a, Thromb. Res. 11:595 (1977).
2. E. Bohn, G. Müller-Berghaus, The effect of leucocyte and platelet transfusion on the activation of intravascular coagulation by endotoxin in granulocytopenic and thrombocytopenic rabbits, Amer. J. Pathol. 84:239 (1976).
3. P. Friberger, M. Knoes, S. Gustavsson, L. Aurell, and G.Claeson, Methods for determination of plasmin, antiplasmin and plasminogen by means of substrate S-2251, Haemostasis 7:138 (1978).

*unpublished data in cooperation with Duswald,Jochum

4. C. Galanos, O. Lüderitz, and O. Westphal, Preparation and pro-
    perties of a standardized lipopolysaccharide from salmonella
    abortus equi (Novo pyrexal), Zbl. Bakt. Hyg. I. Abt. Orig. A.
    243:226 (1979).
5. D. L. Heene, Therapie der sepsisbedingten Hämostaseopathien, in:
    "Infektion, Blutgerinnung und Hämostase," R. Marx, H. A.
    Thies, eds., F. K. Schattauer Verlag, Stuttgart-New York
    (1977).
6. D. L. Heene und W. Kirschstein, Nachweis von Fibrinogenderivaten
    im Plasma bei Sepsis, in: "Fibrinolyse, Thrombose, Hämostase,"
    E. Deutsch, K. Lechner, eds., F. K. Schattauer Verlag, Stutt-
    gart-New York (1979).
7. B. v. Hundelshausen, G. Tempel, S. Jelen, A. Stemberger, G.
    Blümel, S. Haas, H.-M. Fritsche und G. Vogel, Zum Problem der
    Verbrauchskoagulopathie im unmittelbaren posttraumatischen
    Verlauf, in: "25 Jahre DGAI, Anaesthesiologie und Intensiv-
    medizin," Bd. 130, K. H. Weis, G. Cunitz, eds., Springer Ver-
    lag, Berlin-Heidelberg-New York (1980).
8. S. Iwanaga, T. Morita, T. Harada, S. Nakamura, M. Niwa, K. Taka-
    da, T. Kimura, and S. Sakakibara, Chromogenic substrates for
    horseshoe crab clotting enzyme: its application of the assay
    of bacterial endotoxins, Haemostasis 7:183 (1978).
9. M. A. Kane, J. E. May, and M. M. Frank, Interactions of the
    classical and alternate complement pathway with endotoxin
    lipopolysaccharide. Effect on platelets and blood coagulation,
    J. Clin. Invest. 52:370 (1973).
10. G. J. Kociba, W. F. Loeb, and R. F. Wall, Development of pro-
    coagulant (tissue thromboplastin) activity in cultured leuko-
    cytes, J. Lab. Clin. Med. 79:778 (1972).
11. J. Levin, and F. B. Bang, The role of endotoxin in the extra-
    cellular coagulation of limulus blood, Bul. Johns. Hopkins
    Hosp. 115:265 (1964).
12. J. Levin, T. E. Poore, N. S. Young, S. Margolis, N. P. Zauber,
    A. S. Townes, and W. R. Bell, Gram-negative sepsis: detection
    of endotoxemia with the Limulus test with studies of asso-
    ciated changes in blood coagulation, serum lipids and comp-
    lement, Ann. Intern. Med. 76:1 (1972).
13. R. C. Mechanic, E. III. Frei, M. Landy, and W. W. Smith, Quanti-
    tative studies of human leucocytic and febrile response to
    single and repeated doses of purified bacterial endotoxin,
    J. Clin. Invest. 41:167 (1961).
14. D. F. Mosher, and E. M. Williams, Fibronectin concentration is
    decreased in plasma of severely ill patients with dissemi-
    nated intravascular coagulation, J. Lab. Clin. Med. 91:729
    (1978).
15. G. Müller-Berghaus, and E. Lohmann, The role of complement in
    endotoxin-induced disseminated intravascular coagulation.
    Studies in congenitally C6-deficient rabbits, Br. J. Haema-
    tol. 28:403 (1974).

16. G. Müller-Berghaus, Aktivierung der Blutgerinnung bei Sepsis durch gramnegative Bakterien, in: "Fibrinolyse, Thrombose, Hämostase," E. Deutsch, K. Lechner, eds., F. K. Schattauer Verlag, Stuttgart-New York (1979).

17. O. R. Ødegärd, M. Lie, and U. Abildgaard, Heparin cofactor activity measured with an amidolytic method, Thromb. Res. 6: 287 (1975).

18. L. Roka, Infektion und Blutgerinnung aus biochemischer Sicht, in: "Infektion, Blutgerinnung und Hämostase," R. Marx, H. A. Thies, eds., F. K. Schattauer Verlag, Stuttgart-New York (1977).

19. Th. M. Saba, and E. Jaffe, Plasma fibronectin: its synthesis by vascular endothelial cells and role in cardiopulmonary integrity after trauma as related to reticuloendothelial function, Amer. J. Med. 68:577 (1980).

20. A. Stemberger, F. Strasser, B. v. Hundelshausen, G. Tempel, and G. Blümel, Detection of plasma endotoxin with a limulus amoebocyte lysate (LAL)-chromogenic substrate assay, Acta Chir. Scand. Suppl. 509 (1982) in press.

21. S. Sumida, Experimental studies on the effect of prophylactic proteinase inhibitor therapy on endotoxin shock, in press, available upon request.

22. E. Thaler und G. Kleinberger, Sepsis und Blutgerinnung, Intensivmed. 16:54 (1977).

23. E. Thaler, Verhalten der Inhibitoren der Blutgerinnung und Fibrinolyse bei Sepsis, in: "Fibrinolyse, Thrombose, Hämostase," E. Deutsch, K. Lechner, eds., F. K. Schattauer Verlag, Stuttgart-New York (1979).

24. S. M. Wolff, M. Rubenstein, J. H. Mulholland, and D. M. Alling, Comparison of haematologic and febrile response to endotoxin in man, Blood 26:190 (1965).

ACKNOWLEDGEMENTS

The authors appreciate the technical assistance of Miss C. Kaufer and G. Triandafillides and the secretary work of Miss S. Ryska.

STUDIES ON PATHOLOGICAL PLASMA PROTEOLYSIS IN SEVERELY
BURNED PATIENTS USING CHROMOGENIC PEPTIDE SUBSTRATE
ASSAYS: A PRELIMINARY REPORT

T.E. Ruud, P. Kierulf, H.C. Godal, S. Aune
and A.O. Aasen

Surgical department, medical department 9 and
Central laboratory, Ullevaal University
Hospital, Oslo 1, Norway

INTRODUCTION

Severe burns may lead to disturbances in the plasma
proteolytic enzyme systems. These changes may occur both
in the early stages after the injury and later during
the reconstruction phase, particularly if severe infections occur. Using chromogenic peptide substrate assay
technique, the kallikrein-kinin, the fibrinolytic and
the coagulation systems were closely monitored in this
study. Of particular interest was to see if changes of
components in these systems could be of any prognostic
or therapeutic value.

MATERIALS AND METHODS

Patients

Nine patients, age 17 - 80 years, with more than
30 % third degree combustions, were included in the
study. The observation time was the first four weeks
after the injury. All. the patients were critically ill
during the observation period, and five patients died
due to complications.

Citrated plasma (1/10 vol 0,13 mol/1 citrate) was
prepared regularly from venous blood, snap frozen in
liquid nitrogen, and stored in small aliquots at -70 °.

Assays

    Chromogenic peptide substrate assays was performed
by using the automated centrifugal analyzer (Cobas-Bio,
Hoffman La Roche, Basle, Switzerland) (Aasen et al. 1982).
Plasma prekallikrein (PKK) (Amundsen et al. 1978), plasma functional kallikrein inhibition (KKI) and plasma
kallikrein activity (KK) (Aasen et al. 1980) were all
studied using the chromogenic peptide substrate S-2302
(Kabi diagnostica AB, Stockholm, Sweden). Plasma prothrombin (Pthr) was performed according to Bergstrøm and
Egeberg, 1978, using Ecarin (Pentapharm, Basle, Switzerland) as activator. Plasma plasminogen (Plg) and plasmin
activity (Pl) was determined (according to Friberger and
Knös, 1979) using substrate S-2251 (Kabi diagnostica AB,
Stockholm, Sweden). Platelet count was performed on a
coulter Thrombocounter[R]. Serum fibrinogen-fibrin degradation products (FDP) were determined immunochemically
by an agglutination test (Thrombo Welco Test, Reagent
Ltd, Beckenham, England). The ethanol gelation test was
performed according to Godal and Abildgaard, 1966.

    PKK, Plg, Pthr, and KKI values are expressed as per
cent of a human standard plasma pool. Results were analyzed statistically by Willcoxon rank sum test. P values
<0,05 are considered significant. Median values with
25 and 75 fractiles are given.

RESULTS

The kallikrein-kinin system

    PKK values were significantly and markedly depress-
ed during the initial two weeks after injury (fig. 1).
KK activity and KKI values were within normal range, except for a few occasions under critical episodes with
extensive plasma proteolysis (table 1).

The coagulation system

    Pthr values were significantly lowered during the
first week (fig. 2). A high frequency (65 %) of positive
ethanol gelation tests was found during the observation
period and the platelet levels were low during the
initial days after injury (fig. 3).

The fibrinolytic system

Significantly depressed Plg values were found during the initial two weeks (fig. 4). Pl values were occassionally elevated(table 1). A rise in fibrinogen and FDP levels were regulary seen, indicating activation of the fibrinolytic system.

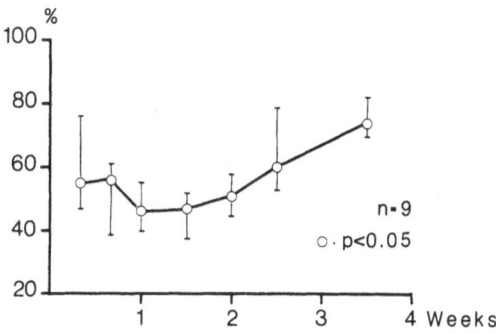

Fig. 1: Plasma prekallikrein values found during the first weeks after admission in 9 patients with severe burns.

Table 1: Changes in components of the plasma kallikrein-kinin and fibrinolytic systems during extensive proteolysis in patients with severe burns.

KK U/l	112	117	124	159	N < 40
PKK %	45	60	60	85	N=100
KKI %	60	61	50	75	N=100
Plasminogen %	54	82	70	96	N=100
Plasmin U/l	200	48	72	73	N < 20

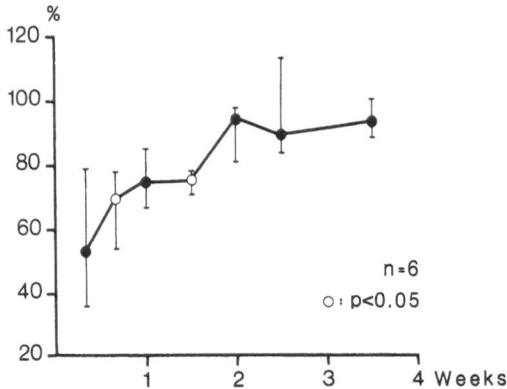

Fig. 2: Plasma prothrombin values found during the first
four weeks after admission in 6 patients with severe
burns.

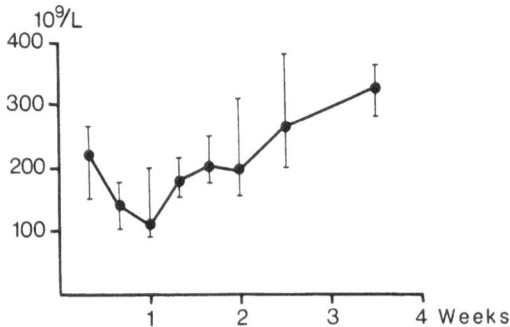

Fig. 3: Platelet values found during the first weeks
after admission in 8 patients with severe burns.

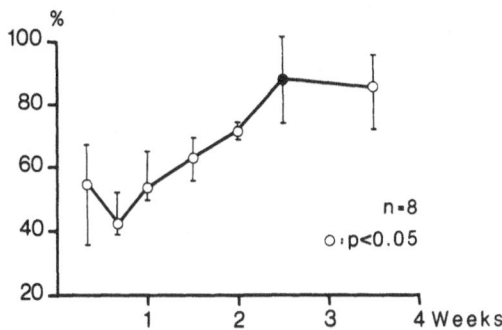

Fig. 4: Plasma plasminogen values found during the first four weeks after admission in 8 patients with severe burns.

DISCUSSION

The present study confirms that the proteolytic enzyme systems i.e kallikrein-kinin, coagulation and fibrinolytic systems, becomes activated in severely burned patients. Episodes with extensive plasma proteolysis (elevated KK and plasmin activity, decreased KKI, PKK and Plg values) occurred during critical illness among patients who later had a fatal outcome. Monitoring these parameters could therefore be of prognostic and therapeutic value in patients with severe burns. Chromogenic peptide substrate assay technique applied in the study is easy to perform and can be fully automated (Aasen et al. 1982). Therefore, such assays are very suitable for clinical use.

ABSTRACT

Changes in components of the plasma proteolytic enzyme systems were studied in nine patients with severe burns using chromogenic peptide substrate assay technique. All patients were critically ill and five died during a four week observation period. Plasma prekallikrein, plasminogen, prothrombin and platelet values decreases regularly in all patients after the injury. These changes were seen together with a high frequency of increases serum FDP values and presence of soluble fibrin. During critical illness extensive plasma proteolysis occured, indicated by elevated plasma kallikrein and plasmin activities combined with lowered functional plas-

ma kallikrein inhibition, prekallikrein and plasminogen
values. The study confirms that plasma pathological proteolysis occurs in patients with severe burns, and indicates that close monitoring of components of the plasma
proteolytic enzyme systems can give information of prognostic and therapeutic value in the severely burned
patient.

REFERENCES

Aasen, A. O., Smith-Erichsen, N., Gallimore, M. J., and
        Amundsen, E., 1980, Studies of plasma kallikrein-
        kinin system in plasma samples from normal indi-
        viduals and patients with septic shock, in: "Ad-
        vances in Shock Research," W. Schumer, J. J.
        Spitzer, B. E. Marshall, eds., Liss. A.R. Inch.,
        New York.
Aasen, A. O., Kierulf, P.,and Strømme,J.H.,1982, Methodo-
        logical considerations on chromogenic peptide sub-
        strate assays and application on automated analy-
        zers, Acta Chir. Scand., supp. 509:17.
Amundsen, E., Gallimore, M. J., Aasen, A. O., Larsbraaten,
        M., and Lyngaas, K., 1978, Activation of human plas-
        ma prekallikrein: influence of activators, acti-
        vation time and temperature and inhibitors, Thromb.
        Res.,13:625.
Bergstrøm, K., and Egberg, N., 1978, Determination of
        Vitamin K sensitive coagulation factors in plasma,
        Thromb. Res.,12:531.
Friberger, P., and Knös, M., 1979, Plasminogen determin-
        ation in human plasma, in: "Chromogenic Peptide
        Substrate," M. F. Scully, V. V. Kahhar, eds.,
        Churchill Livingstone.
Godal, H. C., and Abildgaard, U., 1966, Gelation of solu-
        ble fibrin in plasma by ethanol, Scand. J. Haematol.,
        3:342.

CHANGES IN COMPONENTS OF THE PLASMA PROTEASE SYSTEMS RELATED TO

COURSE AND OUTCOME OF SURGICAL SEPSIS

Nils Smith-Erichsen, Ansgar O. Aasen and
Egil Amundsen

Department of Anaesthesia, Akershus Central Hospital
1474 Nordbyhagen, Norway
Central Laboratory, Ullevaal Hospital, Oslo 1, Norway
Institute for Surgical Research, Rikshospitalet, Oslo 1,
Norway

INTRODUCTION

Septicemia in the surgical patient is a serious complication with high mortality[1]. Aggressive surgery and early intervention with mechanical ventilation in addition to conventional treatment is essential to reduce mortality of this ailment[2]. Computed tomography has greatly improved the possibilities of diagnosing and localizing intraabdominal septic foci, but a lag in time of 14 days from onset of symptoms until the computed tomography was performed have been reported[3]. Accordingly, still new and simpler tools seem necessary to improve the daily evaluations and decision making in these critically ill patients.

Several clinical studies have shown that extensive proteolytic activity occurs in plasma during septicemia and septic shock [4,5,6, 7,8,9,10]. Recent studies by us [7,8,9,11] and others [12,13,14] have indicated that determination of components in the plasma proteolytic enzyme systems using assays based on chromogenic peptide substrates may give valuable information about pathophysiology, course and prognosis of septic shock. Furthermore, these components in relation to surgery and trauma may serve as early warning signs of emerging septicemia.

In order to further elucidate our previous findings and to test the clinical usefulness of chromogenic peptide substrate assays, we have studied proteolytic activity and its inhibition in 18 surgical patients with complicating septicemia.

PATIENT MATERIAL AND METHODS

The patients were selected into the study when they fulfilled four of the following criteria of septicemia: Temperature above 38^0 C., need of respiratory support, leucocytes above $15 \times 10^9/1$ or below $5 \times 10^9/1$, thrombocytes below $100 \times 10^9/1$, positive blood culture or an obvious septic focus.

Patients were considered to be in septic shock when the use of a vasopressor agent was necessary to restore hypotension and oliguria.

The shock lasted as long as the need of vasopressor support was present.

Eighteen surgical patients (14 men and 4 women, age 12-83 years) with complicating septicemia treated in the intensive care unit of Akershus Central Hospital were studied. There were 9 fatal cases and 9 survivors with a similar age distribution between the two groups. The median observation time was 8 days (range 4-20 days) in the fatal cases, and 37 days (range 13-54 days) in the survivors. Septic shock occurred in 6 fatal cases and 3 survivors. Gram-negative microbes were cultured from blood in 7 patients, and gram-positive bacteria in blood from one. A variety of gram-negative microbes were cultured from the septic focus in the remaining patients.

In 10 patients sepsis was due to intra-abdominal infections complicating major trauma and surgery. Four patients underwent surgery because of primary intra-abdominal infections with septic symptoms (cholecystitis,perforated colon, pelvic abscess). Serious postoperative wound infections and intrapulmonary abscesses after major trauma were considered to be the septic focus in the remaining 4 patients.

The 9 survivors underwent 19 reoperations on the suspicion of a persistent septic focus whereas only 7 reoperations were performed on this indication in the fatal cases.

At the time of death clinical and laboratory signs of sepsis were present in all the fatal cases, and autopsy confirmed a persistent septic focus in all but one of the patients. In this patient only microscopic changes after an intra-abdominal abscess evacuated 14 days prior to death was found.

The first plasma sample was made from citrated blood as soon as the diagnosis was ascertained and immediately stored at -70^0. Thereafter plasma samples were made at 24 h minimum interval time until death or recovery, stored at -70^0, and thereafter analyzed and retrospectively compared to clinical course and outcome.

Plasma prekallikrein (PKK), Functional kallikrein inhibition (KKI), Plasminogen (Plg), and Functional antiplasmin activity (AP) were determined as described previously [7], [15], [16], [17].

Functional antithrombin III activity was measured using a quantichrome AT-III kit (Abbott Diagnostics Ltd., North Chicago, Ill., USA) according to the manufacturers' description.

All values were expressed as the percentage of values obtained with a standard plasma pool of healthy donors.

To describe centrality and distribution of data median values with 25 and 75 percentiles were used. Statistical significance was tested with Wilcoxon rank sum test and P-values < 0.05 were considered significant.

RESULTS

PKK-values were markedly reduced in both groups at day 1 and remained low until death in the patients with persistent sepsis. In the survivors low values were also encountered during the first two weeks of treatment, thereafter a gradual increase towards the normal range was observed. Recovery values for this parameter were in the lower normal range. The PKK-values in the fatal cases were significantly lower during the whole course compared to the survivors. (Day 1 P = 0.01, First week P < 0.05, Second week P < 0.01, Third week P < 0.005) (Table 1). Plg-values were reduced in both groups at day 1. In the survivors Plg increased during the first week of therapy and had returned to the normal range within two weeks. In the fatal cases with persistent sepsis Plg remained low, and a marked reduction was observed in the third week. No significant difference in Plg-values occurred between the two groups until the second week. (Table 1). AP-values varied widely in both groups, but remained in the normal range during the whole observation period in the survivors. In the patients with persistent sepsis a significant reduction in AP-values were observed in the second and third week compared to the survivors (P < 0.025)   (Table 1).

During septic shock AP-values were in the normal range and no significant differences between values obtained during non-fatal septic shock and fatal septic shock were observed (Table 2). Marked reductions in AT-III values were found in both groups at day 1. They remained low until death in the patients with per-sistent sepsis, whereas normal values were obtained within 2 weeks after start of therapy in the survivors. No significant differences between the two groups were obtained until the second week (Table 1). AT-III values obtained during fatal and non-fatal septic shock revealed no significant differences (Table 2). Changes in KKI was associated with the development of septic shock. Marked reduction in KKI values were observed in patients dying in septic shock (Fatal septic shock = FSS), whereas this para-meter remained in the normal range in patients who survived septic shock (Resuscitated septic shock = RSS)   (Table 2).

    In patients with sepsis without shock KKI-values were in the
upper normal range (Median 105%, 25 percentile 99%, 75 percentile
115%), and these values were significantly higher than those
obtained during RSS (P < 0.005).

Table 1

	Day 1	1st week	2nd week	3rd week	Recovery
PKK Survivors	67(54-75)	55(47-62)	58(51-87)	74(58-98)	91(78-101)
PKK Fatal	47(40-55)	44(33-50)	50(42-57)	41(32-45)	
Plg Survivors	72(63-75)	79(71-87)	109(98-114)	107(98-118)	120(111-132)
Plg Fatal	65(47-89)	67(52-81)	66(60-80)	25(20-53)	
AP Survivors	97(80-105)	92(83-109)	91(80-97)	100(92-105)	103(100-110)
AP Fatal	90(74-107)	91(72-103)	73(59-87)	74(56-87)	
AT-III Surv.	63(58-70)	69(62-77)	95(72-107)	95(76-116)	114(96-135)
AT-III Fatal	51(45-67)	67(50-75)	61(53-71)	54(47-57)	

Median values for plasma prekallikrein (PKK), Plasminogen (Plg),
Functional antiplasmin activity (AP), and Functional antithrombin
III activity (AT-III) with 25 and 75 percentiles in parenthesis
for the survivors and fatal cases with persistent sepsis.

    Day 1 are values from the first plasma sample collected, first,
second and third week show values obtained during the first, second
and third week of therapy.

    The recovery values were obtained at the time of discharge
from the intensive care unit or immediately after.

    All values are expressed as the percentage of values obtained
with a standard plasma pool of healthy donors.

Table 2

	FSS	RSS
KKI	74 (61-87)	97 (87-107)
AT-III	63 (54-72)	66 (59-69)
AP	93 (74-105)	94 (80-113)

Median values with 25 and 75 percentiles in parenthesis for
Functional kallikrein inhibition (KKI), Functional antithrombin
III activity (AT-III) and Functional antiplasmin activity (AP)
during fatal septic shock (FSS) and non-fatal septic shock (RSS).
All values are expressed as the percentage of values obtained with
a standard plasma pool of healthy donors.

DISCUSSION

This study further confirms that extensive proteolytic activity occurs in plasma during sepsis. Of particular clinical importance is the finding that persistent sepsis caused by an unrecognized septic focus was associated with continuous low values for several components of the plasma protease systems whereas these components rapidly returned towards normal values upon adequate therapy. Also, our previous findings of functional kallikrein inhibition as a critical determinant of survival in septic shock [7], [11], is further confirmed by the present study. Determination of components in the plasma-protease system using chromogenic peptide substrate assays are easy and fast, especially when automated and, therefore, suited for clinical use.

Another practical clinical aspect of this study is the observation that the survivors underwent nearly three times as many re-operations in order to evacuate a suspected septic focus as the fatal cases. This is in agreement with Ledingham and McArdle[2] who found that aggressive surgery and early intervention with mechanical ventilation significantly reduced mortality in septic shock.

Major trauma and surgery also induce proteolytic activity in plasma [12], [13], [14], and low values for PKK, Plg and AT-III have been encountered in these patients. However, in uncomplicated cases these parameters returned to normal values within one week, therefore, these authors also claim that persistent low values for these components are strongly suggestive of complicating sepsis or adult respiratory distress syndrome.

Activation of the kallikrein-kinin system has been implicated in the pathophysiology of septic shock [4,7,10,11,18]. The preponderance of septic shock among the fatal cases combined with significantly lower PKK-values compared to the survivors and significant reductions in KKI-values during septic shock support this hypothesis. Furthermore, these findings suggest a more extensive activation of the kallikrein-kinin system in the fatal cases than in the survivors. This study also support our previous findings [7], [11] that normal functional anti-kallikrein activity during septic shock is essential for survival. It may also be concluded from the present findings that the degree of reduction in PKK during sepsis may serve as a prognostic indicator which is in agreement with earlier observations [7], [11].

In contrast, the values obtained during septic shock for the functional activities of the two other major inhibitors examined, gave no prognostic information although the AT-III values were markedly reduced in both fatal and non-fatal septic shock. This further emphazises the role of the kallikrein-kinin system in septic shock.

Activation of the coagulation system with consumption of AT-III during sepsis is well documented [6,9,12,13,19]. Reduction in this parameter may also partly be due to unspecific proteolytic degradation by released granulocyte elastase from destroyed leucocytes [19,20,21]. The rapid return of AT-III to normal values in the survivors makes it a useful guide in monitoring therapy of sepsis, whereas no prognostic conclusions can be made from the degree of reduction in this parameter according to the data obtained in this study. The observed reductions in Plg during sepsis suggests its conversion to plasmin as a result of activation of the fibrinolytic system [8,19]. The wide variation in Plg-values in early stages of sepsis made any prognostic conclusions from this parameter impossible.

The marked reduction in Plg observed in the third week of persistent sepsis combined with the reduction in AP during the second and third week, may implicate a pronounced fibrinolytic activity in longstanding sepsis. The rapid return of Plg to normal values in the survivors makes this parameter to a useful tool in evaluating the course of persistent sepsis.

The finding of marked reduced AT-III values with simultaneously well preserved functional antiplasmin activity may play a central role in developing the microembolism syndrome [22] by allowing formed fibrin to deposit in the pulmonary vessels and other organs, and thus contribute to the respiratory failure [23] and liver damage [24] commonly observed in sepsis.

The very low levels of PKK, AT-III and Plg observed in the long-standing cases of sepsis may thus be due to a combined effect of sustained consumption and reduced synthesis.

Another problem still to be solved is the influence of increased capillarypermeability and dilution occurring during sepsis on the measured parameters. Increased disappearance rate from plasma for albumin has been observed [25], but if this is valid for proteases has not yet been proved.

We, therefore, conclude that by serial determinations of several key components in the plasma protease systems using chromogenic peptide substrate assays, it is possible to obtain valuable information about course and prognosis of surgical sepsis.

REFERENCES

1. R.D. Weisel, L. Vito, R.C. Dennis, C.R. Valeri and H.R. Hectman: Myocardial depression during sepsis. Am J Surg 133:512 (1977)

2. I.McA. Ledingham and C.S. McArdle: Prospective study of the treat-
    ment of septic shock. Lancet June 3, 1194 (1978)
3. P. Trunet, J.R Le Gall, P.-L. Fagniez, D. Larde, N. Vasile, J.
    Carlet and M. Rapin: Computed tomography and post-laparotomy
    intra-abdominal abscesses. Int Care Med 8:193 (1982)
4. J.W. Mason, U. Kleeberg, P. Dolan and R.W. Coleman: Plasma
    kallikrein and Hageman factor in gram-negative bacteremia.
    Ann Int Med 73:545 (1970)
5. J. Haviger: Disseminated intravascular coagulation in patients
    with infections: Diagnostic and therapeutic approach.
    Mater Med Pol 8:206 (1976)
6. H. Fritz: Proteinase inhibitors in severe inflammatory
    processes (septic shock and experimental endotoxemia):
    Biochemical and therapeutic aspects, in: Protein Degradation
    in Health and Disease, D. Evered and J. Whelan eds. Excerpta
    Medica, Amsterdam (1980)
7. A.O. Aasen, N. Smith-Erichsen , M.J. Gallimore and E. Amundsen:
    Studies on components of the plasma kallikrein-kinin system
    in plasma from normal individuals and patients with septic
    shock, in: Advances in Shock Research, W. Skinner, J.J.
    Spitzer, B.E. Marshall eds., Alan R. Liss Inc. New York (1980)
8. M.J. Gallimore, A.O. Aasen, N. Smith-Erichsen, M. Larsbraathen,
    K. Lyngaas and E. Amundsen: Plasminogen concentration and
    functional activities and concentrations of plasmin inhibitors
    in plasma samples from normal subjects and patients with
    septic shock. Thromb Res 18:601 (1980)
9. N. Smith-Erichsen, A.O. Aasen, M.J. Gallimore and E. Amundsen:
    Studies of components of the coagulation system in normal
    individuals and septic shock patients, Circ Shock (in press)
10. E.S. Kalter, A. Timmermans and B.N. Bouma: The kallikrein
    generating system during sepsis and bacterial shock, in
    Recent Progress on Kinins, H. Fritz, G. Dietze, F. Friedler,
    G.L. Haberland eds., Birkäuser Verlag, Basel (1982)
11. N. Smith-Erichsen, A.O. Aasen and E. Amundsen: The functional
    inhibition of plasma kallikrein, a critical factor in septic
    shock. Adv Exp Med Biol (in press)
12. H.G. Schipper, J. Roos, F.v.d. Meulen and J.W. ten Cate:
    Antithrombin III deficiency in surgical intensive care
    patients. Thromb Res 21:73 (1981)
13. H.R. Büller, C. Bolwek, J. ten Cate, J. Roos, L.H. Kahle and
    J.W. ten Cate: Postoperative hemostatic profile in relation
    to gram-negative septicemia. Crit Care Med 5:311 (1982)
14. A.O. Aasen, P. Kierulf, J. Vaage, H.C. Godal and S. Aune:
    Determination of components of the plasma proteolytic enzyme
    systems give information of prognostic value in patients with
    multiple trauma. Adv Exp Med Biol (in press)

15. E. Amundsen, M.J. Gallimore, A.O. Aasen, M. Larsbraathen and
    K. Lyngaas: Activation of plasma prekallikrein: Influence
    of activators, activation time and temperature and inhibitors.
    Thromb Res 13:625 (1978)
16. A.O. Aasen, O.D. Saugstad, B. Lium and I. Guldvog: Plasma
    antiplasmin activities in experimental lung insufficiency.
    Acta Chir Scand suppl 499, 113 (1980)
17. M.J. Gallimore, E. Amundsen, A.O. Aasen, M. Larsbraathen,
    K. Lyngaas and L. Svendsen: Studies on plasma antiplasmin
    activity using a new specific chromogenic tripeptide
    substrate. Thromb Res 14:51 (1979)
18. J.A. Robinson, M.L. Klodnycky, H.S. Loeb, M.R. Racie and
    R.M. Gunnar: Endotoxin, prekallikrein, complement and
    systemic vascular resistance - Sequential measurements in
    man. Am J Med 59:61 (1975)
19. J. Witte, M. Jochum, R. Scherer, W. Schramm, K. Hochstrasser
    and H. Fritz: Disturbances of selected plasmaproteins in
    hyperdynamic septic shock. Int Care Med 8:215 (1982)
20. A.O. Aasen and K. Ohlsson: Release of granulocyte elastase
    in canine endotoxin shock. Hoppe-Zeyler's Z.
    Physiol Chem 359:683 (1978)
21. R. Egbring, W. Schmidt, G. Fuchs and K. Havemann: Demonstra-
    tion of granulocytic proteases in plasma of patients with
    acute leukemia and septicemia with coagulation defects.
    Blood 49:219 (1977)
22. T. Saldeen: The microembolism syndrome. Microvasc Res
    11:227 (1976)
23. C.H.A. Clowers Jr.: Pulmonary abnormalities in sepsis.
    Surg Clin North Am 5:993 (1974)
24. I. Suten: The viscera in shock, in: Shock, Pathology,
    Metabolism, Shock cell, Treatment. I. Suten, T. Baudilla,
    A. Cafrita, A. Bucur and V. Candea eds. Abacus Press,
    Turnbridge Wells, England (1979)
25. R.C. Birk, R.F. Wilson: Central venous pressure and blood
    volume determination in clinical shock. Circulation
    38:42 (1968)

ROLE OF PROTEASES IN THE DEVELOPMENT OF ACUTE PANCREATITIS

M. Wanke

Pathologisches Institut

2370 Rendsburg, FRG

Morphologically, acute pancreatitis is marked by adipose tissue necrosis, pancreatic juice edema, parenchymal necrosis and hemorrhage.

In the individual case, the prognosis depends on the extent of the necrotizing process and the proportions of these four chief components.

We can go on to differentiate morphologically between a primary nonautodigestive biliary type (BP), and a primary autodigestive lipolytic-proteolytic type (LPP)[8].

ACUTE BILIARY PANCREATITIS (BP)

Ductogenic acute pancreatitis has resisted attempts to bring its etiology, pathogenesis, and morphological substrate together in a consistent picture, and this has led to general misunderstanding of it. The morphological picture of BP is determined by the concentration of unconjugated bile acids [8, 13]. Intraductal instillation of 0.5 - 5 % sodium taurocholate solution into a dogs pancreas, for example, has the same effect as the detergent Triton X-100. Instillation of the noxa immediately causes coagulation necrosis of the duct epithelium and acinar complexes, which is the primary goal of the infusate. The intramural vessels of the pancreatic ducts are hyperemic. Vascular edema and hemorrhages cause mechanical ruptures of the acini and lobuli. The hypoxic acinar epithelium undergoes vacuolate degeneration. The acinar epithelium is destroyed; initially no autodigestion occurs. During the hypoxic phase of BP, the pancreatic enzymes are activated. The time factor is important in assessing the morpholo-

gy of BP. The hemorrhagic components can complicate any type of pancreatitis in it's late phase, and is especially characteristic of
the BP and is already dominant in the initial phase. In addition
to the direct vessel damage due to bile, a histamine type increased vessel permeability also occurs. Generally the vessel factor has biliary, lipolytic, and proteolytic-elastolytic causes;
pathogenetically it is many sided.

MIXED FORMS OF ACUTE PANCREATITIS (MP)

        During chyme reflux the effect of the three components bile,
substrate, and activated enzymes accumulates. Therefore the
following factors are pathogenetically important for the course
and type of reflux pancreatitis:

        1. Composition of the reflux

            a) Duodenum contents, high/low fat contents, fine/coarsely
               emulsified
            b) Contents of active proteolytic/lipolytic enzymes
            c) Ratio of conjugated to unconjugated bile acids
            d) Bile plus duodenum contents with active proteolytic
               and lipolytic enzymes

        2. Quantity of the reflux

        3. Time of contact

        4. Reason for reflux

            a) Papillary stones
            b) Papillary spasm
            c) Afferent loop syndrome after B II stomach resection
            d) Parasites in the pancreatic duct system

        In practice, mixed forms of pancreatitis result morphologi-
cally in a high procentage (BP/LPP or LPP/BP); known under the
problematic terminus "cholecystopancreatitis" (figure 1).

        The morphological picture of "cholecystopancreatitis" depends
upon the localisation of the concrements, the anatomy of the biliary and pancreatic ductal system, and the anatomy of papilla Vateri[5];
so we find BP in the sense of Opie and Halsted, mixed forms MP as
well as LPP e.g. "pseudocholecystopancreatitis".

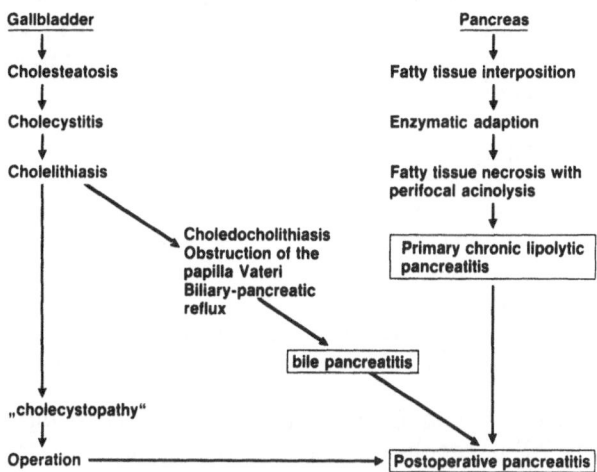

Figure 1:   Aetio-Pathogenesis of "Cholecystopancreatitis." Hyper-
           cholesterinemia/Hyperlipemia as common underlying factors.

ACUTE LIPOLYTIC-PROTEOLYTIC PANCREATITIS (LPP)

        Regardless of the cause, course, and primary aggression of
the noxa on the parenchymal and mesenchymal tissue, an early
disturbance causes edema and enzymatic parapedesis, so that the
organ-specific interchangeability of metabolism is stopped. This
initial edema, Zoepffel edema, is a common morphological substrate
of all kinds of acute pancreatitis. In our opinion, the main difference between acute and chronic (relapsing) pancreatitis is the
length of time and the volume of parapedesis.

     The etiology

            1. biliary tract diseases
            2. obstruction - hypersecretion
            3. processes in duodenum - duodenal reflux
            4. toxic factors - e.g. alcoholism
            5. infection
            6. vascular processes
            7. endocrine factors
            8. metabolic changes
            9. immunological factors
           10. traumatic
           11. neurogenic factors
           12. hereditary factors

determines the morphological picture (BP, MP, LPP), and the form
and pace of parenchyma destruction[9]. The noxa invades the organelles either of cell respiration (hypoxic type) or of cell metabolism
(metabolic type). If we don't take in count the opinion of Opie
and Halsted on acute biliary reflux pancreatitis (BP), we can distinguish various stages and phases in the course of primary autodigestive pancreatitis[12]:

1. preliminary phase
2. minimal lesions   - severity grade 0
3. edematous pancreatitis   - severity grade I
4. acute LPP without   - severity grade II
   with hemorrhages   - severity grade II/III
5. acute LPP with biliary
   componente and hemorrhages   - grade II/III
6. afterphase with sequelae of enzyme diversion

   If the preliminary phase is humorally or morphologically realized, the phases of parenchyma destruction up to autodigestion
are:

1. Dyscholia
2. Edema
3. Circulatory disturbance
4. Necrobiosis
5. Autodigestion

Four pathogenetic factors are of particular importance:

1. Obstruction to outflow of pancreatic juice
2. Extreme stimulation of pancreatic secretion
3. Activation of juice enzymes
4. Loss of protection against digestion of glandular
   parenchyma due to metabolic decline.

PRELIMINARY PHASE

   The preliminary phase of acute LPP is characterized by interposition of adipose tissue in internal adiposity, or by adipose
tissue proliferation in cachexia, also by vascular sclerosis and
fibrosis around the pancreatic ducts. As a result, outflow of
lymph is impeded and recurrent juice edema develops - fat deposits
pave the way to acute lipolytic-proteolytic pancreatitis. Figure
2 provides an overview of the primary disease site:

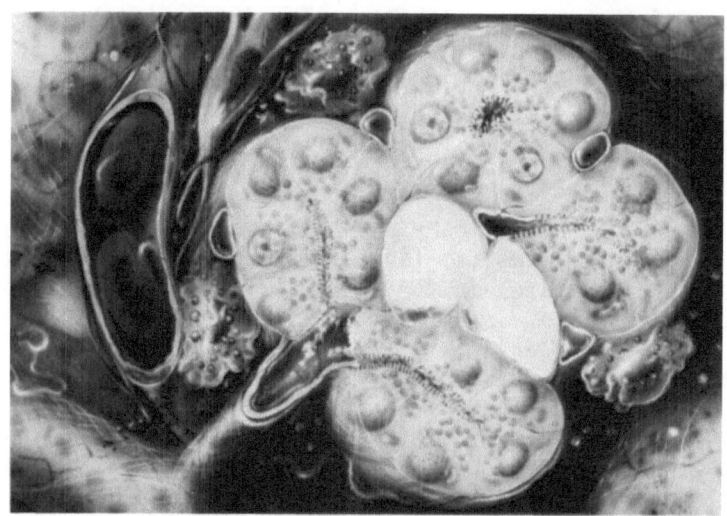

Figure 2: Overview of the primary disease site. Fat deposits
          pave the way to acute autodigestive pancreatitis.

    In figure 2 four acini with their epithelia are grouped around
two centrally located fat cells. The major portion of the juice is
led via initial pancreatic passages bordered by isthmic epithelial
cells into larger excretory ducts. Accessory secretion passes via
the bases of the acinar epithelial cells into the interstitium by
parapedesis and is drained primarily via lymphatic vessels as demonstrated in figure 3.

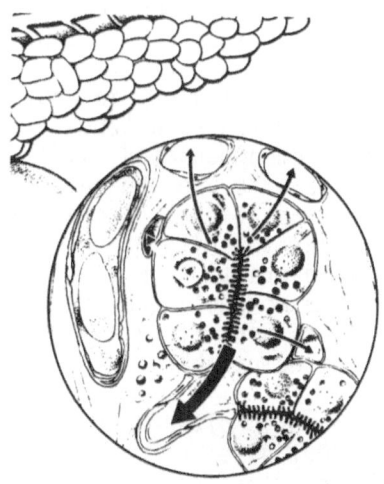

Figure 3

The dense capillary network supplies the necessary high oxygen demand of the acinar epithelium. The interstitial tissue contains numerous perivascular and periacinar mast cells (11, 14). The perivascular space is rich in collagenous and elastic fibers as well as fibrocytes (compare tab. 6: summary of the lipolytic and proteolytic enzymes that take part in autodigestion and their cellular substrate).

If the stage "preliminary phase" is morphologically realized, disseminated minimal lesions (autodigestive) develop, and upon that base further phases in the course of primary autodigestive pancreatitis:

DYSCHOLIA

The morphological signs of acinar dyscholia are hydropic vacuolic degeneration, indicating the acinar cells' acute oxygen insufficiency (see below, shock); loss of basal basophilia, indicating the incipient destruction of the ergastoplasm; acidophilic degeneration, and dyscholic atrophy or necrosis[9].

EDEMA

Edematous pancreatitis - severity grade I

The most important early morphological change is edema. The physiological secretion process of the gland is already marked by increased parapedesis through its inclination to edema production; here we see the Achilles' heel of the pancreas and the key to understand acute pancreatitis. Every extra burden on the secretion process tends, through chronic inveterate edema, to produce a gradual organ alteration. Vascular and acinar edema have to be differentiated in their biological aggression.

From a general pathological view, in Rössle's opinion, pancreatic edema can be interpreted as a serous inflammation. Compensated, it is unimportant; decompensated, it is a component of necrosis. The moderate variants are: edematous pancreatitis, acute pancreatic edema, acute transitional pancreatitis, and interstitial pancreatitis - in all cases severity grade I - compare CT results! Nevertheless, the enzyme disruption is biochemically, and clinically, always present! The morphological changes are limited to discrete dyscholic lesions. Here clinical and pathological statistics are as a rule not comparable.

Acute pancreatic edema is often seen in association with acute episodes of bile duct disease (compare "Cholecystopancreatitis"). Enlargement of the intrapancreatic secretion duct with juice retention and reflux into the interstitial space is also often seen as a precipitating factor in acute pancreatitis. The

capacity of the lymph drainage determines whether a biochemical
pancreatitis will become morphologically manifest. Thus the organ's
own lymph system acts as an overflow and safety valve. On the one
hand the edema can join the acute pancreatitis; on the other hand,
it can stand as an intermediate position of the BP and LPP varie-
ties. It causes the lipolytic-proteolytic late phase of BP and the
subacute exacerbation of LPP. Depending on the preliminary phase
components, and on the local inhibitor capacity, e.g. in mast cells[4]
and on the functional capacity of the lymph system, acute pancreatitis can finish in the following stages:

> 1. minimal lesions   severity grade 0/I
> 2. edematous pancreatitis   I
> 3. acute LPP   II
> 4. acute LPP with hemorrhages
>   and massive autodigestion   II/III

CIRCULATORY DISTURBANCE

In addition to the secretion against an obstruction by

> 1. Pancreatic lithiasis
> 2. Papillary folding of the duct epithelium
> 3. Squamous epithelial metaplasia of duct epithelium
> 4. Scar stenosis of the pancreatic duct
> 5. Concrements
> 6. Papillitis stenosans Vateriana
> 7. Juxtapapillary duodenal diverticulae
> 8. Papillary cancer
> 9. Parasite invasion

and edema[3] decreased arteria flow, obstructed venous return, and
increased capillary permeability (due to mast cell enzyme liberation), are potential factors in the precipitation of acute pancreatitis. If the arterial circulation cannot adept to the necessary
increase, e.g. in arteriosclerosis (cf. the age distribution of
acute pancreatitis[12]), then work hypoxia and cell necrosis result.

In shock the pancreas occupies a special position since it
can either itself produce shock or be secondarily involved in it[14].
One has to differentiate between pancreatitis as a consequence of
shock and postpancreatic shock. Circulatory disorders are of essential significance in both modes. Since 1965 we have repeatedly
emphasized the central significance of local acidosis[7] in the
pancreas and its responsibility for the onset of acute pancreatitis, a view which is based on clinicopathological and experimental
investigations[12]. This is confirmed by an unexpected frequency of
immediately noticeable lipolytic-proteolytic pancreas necrosis in
shock - up to 47 %.

The LPP goes hand in hand with autodigestive acinar epithelial necrosis and in initial fat necrosis (compare later), whose genesis is attributed to acidosis. While the absolute hypoxia causes secretion to cease, the partial ischemia induces extrusion of zymogen granules, vacuolig degeneration, nuclear edema, and monocellular necrosis and is thus the equivalent of acinar dyscholia.

Arteriosclerosis of the vessels of the pancreas on the one hand causes extensive organ fibrosis and reduction of the excretory parenchyma; on the other hand there are significant relationships with acute pancreatitis (5 -: compare age distribution of acute LPP: "Ranson-signs" of prognosis!).

The relative rarity of pancreatitis during childhood[10] can be attributed to the fact that the preliminary phase is not yet morphologically realized. The morphological substrate of this prephase comprises:

1. Periductal fibrosis
2. Lymph blockade - intramural/cisterna chyli
3. Saliva blockade
4. Arteriosclerosis
5. Fatty tissue interposition

If the preliminary phase is humorally realized we find acute pancreatitis in childhood in the sense of "cortisone pancreatitis". Intrapancreatic fatty tissue is of increasing importance as an intraorganic substrate for the already actively present pancreas lipase. Beyond the age of 25 we found a marked interstitial proliferation of fatty tissue in the pancreas, and beyond the age of 50 up to 63 %.

Considering the change in body weight of patients who died of acute autodigestive pancreatitis, these relations are significant. Only 15 % of 1165 cases had normal weight according to the Geigy tables, while the rest showed intrapancreatic proliferation of fatty tissue in cachexia or increased fatty tissue interposition corresponding to obesity - 55 % of our cases. The frequency of the lipolytic-proteolytic foci - minimal lesions - which initially were not clinically detectable ran parallel to the aging process.

NECROBIOSIS

Necrobiosis is defined as the intravital death of a cell with nucleus and cell-membrane under pathological condition. The nuclear atrophy, or karyolysis, is proof of an increased proteolytic activity and is pH-dependent. In nuclear atrophy and chromatolysis the protein components are splitt off the nucleoprotein by proteases, while polynucleotidases take care of the nuclein. Therefore, it is legitimate to describe this specific destruction of the pancreas with inflammation as autodigestion (1, 2, 3); lipolytic-proteolytic foci develop. Their number depends upon the extent of the interstitial fat tissue. Thus the picture of chronic pancreatitis develops and in nearly 85 % of our cases paves the way for the preliminary phase of acute LPP. The low frequency of acute autodigestive pancreatitis in childhood is, as mentioned above, thought to be caused by the lack of development of the preliminary phase.

AUTODIGESTION

Thus the morphological substrate of acinar dyscholia in acute autodigestive pancreatitis is distinct and determinate. An autodigestive lipolytic and proteolytic necrosis develops out of the dyschylic necrobiosis. Increased interposed adipose tissue and contiguous acinar complexes represent the trigger site for acute autodigestion.

To begin with, there is interacinar adipose tissue necrosis. Depending upon the composition of triglycerides, different fatty acids are released. The local acidosis activates proteolytic proenzymes, which results in perifocal acidolysis (demonstrated by fibrinolysis autographs method of Todd and Warren by 2) and triggers a chain reaction known as acute LPP - figure 4:

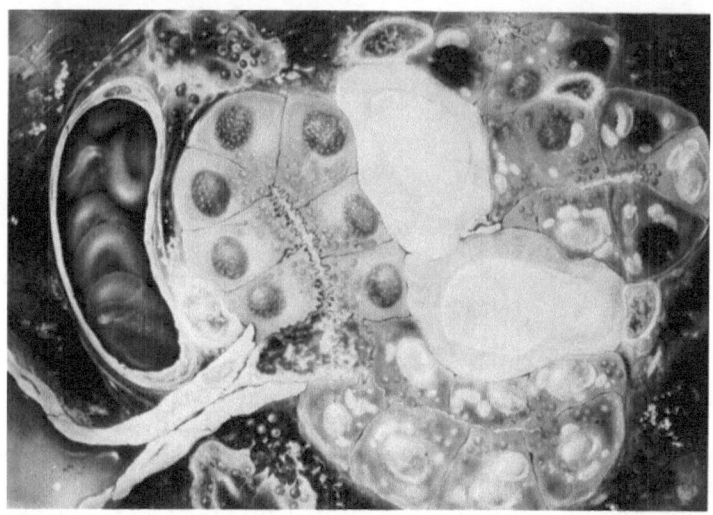

Figure 4:   Early stage of acute lipolytic-proteolytic pancreatitis.

In addition, circulatory disturbances like hemorrhagic edema
and hemorrhages result. The edema causes the structure of the lobu-
lar unit to rupture. The intramural lymph system decompensates the
flood of dtritus, necrotic acinar epithelial components, as well
as juice and vascular edema. The acute exacerbation is triggered:

1. Haematogenically, through direct damage to the acinar epithelium with invasion in the organelles of the structure and function circulation by the metabolic form, as well as in the organelles of the cell respiration by shock and hypoxia.
2. Lymphogenically, from adjoining areas or metastatically.
3. Mesenchymally, through liberation of the mast cells' own
enzymes as well as indirect hematogenic and metastatic-septic excretion inflammation.
4. Ductogenically, through bile or chyme reflux as well as
bacteriogenic ascending and parasitally.
The figure 5 shows the full-blown picture of acute lipolytic proteolytic pancreatitis with hemorrhages:

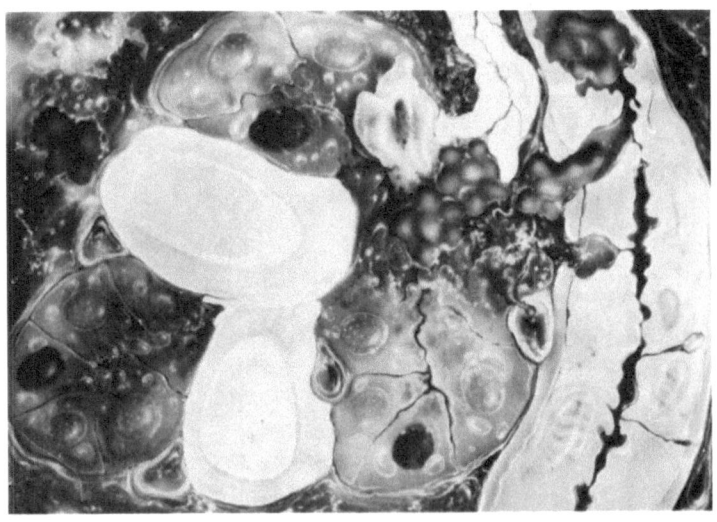

Figure 5: Late stage of acute lipolytic-proteolytic pancreatitis with hemorrhages.

Necrobiotic acini cluster around two central necrotic fat
cells. Individual cells are in different stages of necrosis. Owing
to isthmic and lymphatic blockade there is interstitial edema from
pancreatic juice rich in enzymes and protein. The tissue mast cells
rupture and become degranulated. The capillary wall thickens as a
result of edema and becomes permeable as a result of acidosis -
capillary leak. The terminal and preterminal vasculature is decom-

pensated. Red blood cells penetrate the interstitium and take on a thorn apple shape. The two phenomena bring about hemorrhagic edema. Because of the edema the transit zone between the capillary lumen and the interstitium is enlarged and the tissue can no longer be adequately nourished and supplied with oxygen. A vicious cycle begins.

Table 6.   Summary of the lipolytic and proteolytic enzymes that take part in autodigestion and their particular cellular substrate.

Enzyme	Substrate	Effect
Lipase	Triglyceride intracellular-extracellular	Fat tissue necrosis, local acidosis after triglyceride fissure, intraacinar lipolysis, loosening of the cell membrane
Phospho-lipase A	Cell membrane phosphatide	Lysophosphatide formation membrane destruction, dyshoria
Trypsin	Activation from trypsinogen chymotrypsinogen prephospholipase A–B preelastase prekallikrein denatured protein frame	Coagulation necrosis, dyshoria
Chymotrypsin, carboxypeptidases	Denatured protein frame	Coagulation necrosis, dyshoria
Elastase proteolytic component	Protein frame	Coagulation necrosis
elastolytic component	Elastic and kollagenic fibers	Elastocollageno-lysis, dyshoria
Kallikrein	Kinin	Dyshoria

The common feature of the proteolytic enzymes that take part in autodigestive tissue destruction is that their effect spreads from the periphery to the lobuli, after activation in an acidic milieu during the phase of increased parapedesis. Hemorrhages can complicate every type of pancreatitis in its late stage; but they are especially characteristic of the biliary form, where they are already dominant in the very initial phase as mentioned above.

The tables summarize morphogenesis and clinical pathology of pathology of acute pancreatitis, local and systemic;

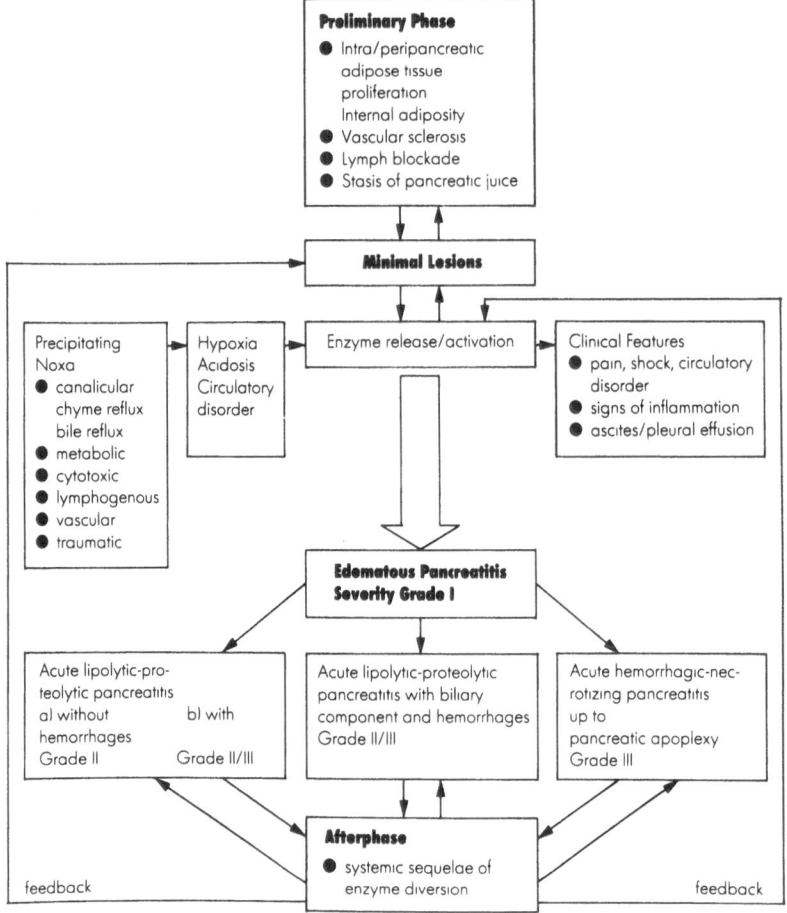

Preliminary Phase
- Intra/peripancreatic adipose tissue proliferation Internal adiposity
- Vascular sclerosis
- Lymph blockade
- Stasis of pancreatic juice

Minimal Lesions

Precipitating Noxa
- canalicular chyme reflux bile reflux
- metabolic
- cytotoxic
- lymphogenous
- vascular
- traumatic

Hypoxia Acidosis Circulatory disorder

Enzyme release/activation

Clinical Features
- pain, shock, circulatory disorder
- signs of inflammation
- ascites/pleural effusion

Edematous Pancreatitis Severity Grade I

Acute lipolytic-proteolytic pancreatitis
a) without hemorrhages
Grade II
b) with
Grade II/III

Acute lipolytic-proteolytic pancreatitis with biliary component and hemorrhages
Grade II/III

Acute hemorrhagic-necrotizing pancreatitis up to pancreatic apoplexy
Grade III

Afterphase
- systemic sequelae of enzyme diversion

feedback   feedback

Clinical pathology of acute pancreatitis

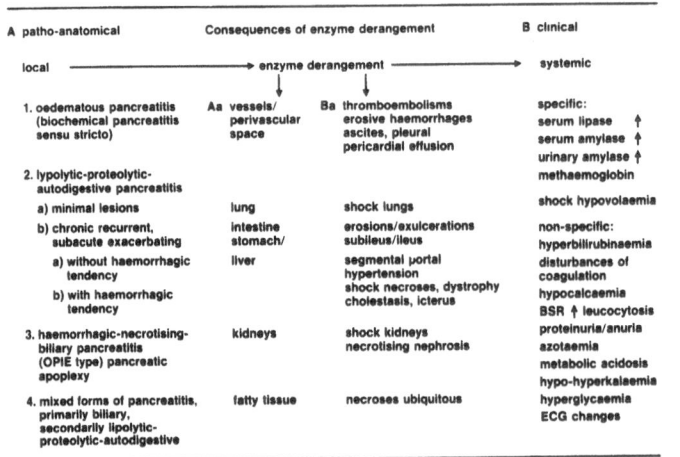

A patho-anatomical	Consequences of enzyme derangement		B clinical
local ———→	——→ enzyme derangement ——→		systemic
	Aa vessels/ perivascular space	Ba thromboembolisms	
1. oedematous pancreatitis (biochemical pancreatitis sensu stricto)		erosive haemorrhages ascites, pleural pericardial effusion	specific: serum lipase ↑ serum amylase ↑ urinary amylase ↑
2. lypolytic-proteolytic-autodigestive pancreatitis			methaemoglobin
a) minimal lesions	lung	shock lungs	shock hypovolaemia
b) chronic recurrent, subacute exacerbating	intestine stomach/	erosions/exulcerations subileus/ileus	non-specific: hyperbilirubinaemia
a) without haemorrhagic tendency	liver	segmental portal hypertension	disturbances of coagulation
b) with haemorrhagic tendency		shock necroses, dystrophy cholestasis, icterus	hypocalcaemia BSR ↑ leucocytosis
3. haemorrhagic-necrotising-biliary pancreatitis (OPIE type) pancreatic apoplexy	kidneys	shock kidneys necrotising nephrosis	proteinuria/anuria azotaemia metabolic acidosis hypo-hyperkalaemia
4. mixed forms of pancreatitis, primarily biliary, secondarily lipolytic-proteolytic-autodigestive	fatty tissue	necroses ubiquitous	hyperglycaemia ECG changes

REFERENCES

1. V. Becker und W. Wilde, Pankreasschäden durch Trypsin in vitro, Klin. Wschr. 41:73 (1963).
2. U. Bleyl, K. H. Grözinger, W. Nagel und M. Wanke, Histotopochemie aktiver proteolytischer Enzyme bei der experimentellen autodigestiven Pankreatitis, Virchows Arch. (Path. Anat.) 342:26 (1967).
3. W. Doerr, P. B. Diezel, K. H. Grözinger, H. G. Lasch, W. Nagel, J. R. Rossner, M. Wanke und F. Willig, Pathogenese der experimentellen autodigestiven Pankreatitis, Klin. Wschr. 43: 125 (1965).
4. H. Fritz, J. Kruck, I. Rüsse, and H. G. Liebich, Immunofluorescence studies indicate that the basic trypsin-kallikrein-inhibitor of bovine organs (Trasylol) originates from mast cells, Hoppe-Seyler's Z. Physiol. Chem. 360:437 (1979).
5. L. Hollender, P. Lehnert, and M. Wanke, Acute pancreatitis. Interdisciplinary synopsis, Urban & Schwarzenberg, Munich/Baltimore (1983).
6. A. S. Todd, The histological localization of fibrinolysin activator, J. Path. 78:281 (1959).
7. M. Wanke, Isthmusblockade und Hypoxie als Ursachen chronisch rezidivierender wie akuter tryptischer Pankreatitis, Gastroenterologia 103:103 (1965).
8. M. Wanke, Experimentelle Pankreatitis. Proteolytische, lipolytische und biliäre Form, Thieme, Stuttgart (1968).
9. M. Wanke, Experimental acute pancreatitis, in: "Current Topics in Pathology," vol. LII, E. M. Grundmann, W. H. Kirsten, eds., Springer, Berlin/Heidelberg/New York (1970).
10. M. Wanke, Significance of lipolytic pancreatitis in childhood in the development of acute relapsing pancreatitis, Beitr. Pathol. 146:272 (1972).
11. M. Wanke, Akute Pankreaserkrankungen, in: "Pankreas," (Handbuch der inneren Medizin, 5. neubearb. Aufl., Bd.III/6), M. M. Forell, ed., Springer, Berlin/Heidelberg/New York (1976).
12. M. Wanke, Acute and chronic pancreatitis, in: "Topics in Acute and Chronic Pancreatitis," L. A. Scuro, A. Dagradi, eds., Springer, Berlin/Heidelberg/New York (1981).
13. M. Wanke, W. Nagel und F. Willig, Formen der experimentellen Pankreatitis patho-anatomisch gesehen, Frankf. Z. Pathol. 75:207 (1966).
14. M. Wanke und G. Schumann, Patho-anatomisches Bild des Schocks und seiner verschiedenen Formen, Chirurg 45:97 (1974).

ON THE POTENTIAL ROLE OF TRYPSIN AND TRYPSIN INHIBITORS
IN ACUTE PANCREATITIS

Åke Lasson and Kjell Ohlsson

Departments of Surgery, Clinical Chemistry
and Experimental Research, Malmö General
Hospital, S-214 01 Malmö, Sweden

INTRODUCTION

Acute pancreatitis is still a disease with many
questions concerning both etiology and pathophysiology
in spite of all the research in this field. Many so
called "triggers" have been investigated, but none of
them have so far been convincingly shown to be the one
"trigger". Trypsin has long been designated as a key-
enzyme[1] capable of converting all other pancreatic
enzymes into their active forms, and many investigators have shown active trypsin both in peritoneal
fluid and in blood in acute pancreatitis.[2 3 4 5] Fig. 1
schematically shows the possible events, when activated
trypsin comes free in acute pancreatitis. Within the
gland trypsin is directly complexed with PSTI (=pan-
creatic secretory trypsin inhibitor), but this complex
dissociates when it comes in contact with α_2-macroglo-
bulin (α_2-M) and α_1-antitrypsin (α_1-AT).[6] Most of the
trypsin forms complex with α_2-M, and so is not immunoreactive, and is discharged via the reticuloendothelial system (RES) in a few minutes. A small part of
the trypsin complexes with α_1-AT and this complex is
immunoreactive, as is also trypsinogen. Active trypsin
in the circulation is quickly complexed mostly by α_2-M
and a small part with α_1-AT. Fig. 2 shows 45 patients
with acute pancreatitis, who all have immunoreactive
trypsin along with their raised amylase. With a further
development of the radioimmunoassay[3], it is possible to
separate between trypsinogen and active trypsin bound
to α_1-AT, and table 1 shows the values for these 45

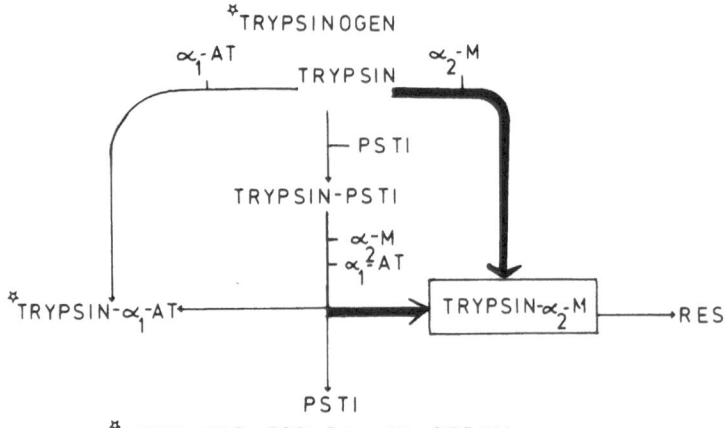

Fig. 1.   A schematic illustration of the interaction
           between trypsin and main endogenous trypsin
           inhibitors.

patients . The high trypsinogen (normally below 25 µg/l)
is a result of leakage. Most patients have rather high
proportion of their immunoreactive trypsin as trypsin-
α_1-AT, some of them almost 100 %.

     The results of our earlier studies on dogs on
bile-induced pancreatitis correspond closely to the
results of both *in vivo* and *in vitro* experiments using
trypsin.[7] This applies both to changes in the complement and kinin systems and to changes in the protease-
antiprotease pattern. This resemblance between *in vivo*
and *in vitro* results also exists in man, both concerning the complement and the kinin systems.[8] As an
example, fig. 3 a and b shows the crossed immunoelectrophoresis pattern with antiserum against C3 and
against high molecular weight kininogen (HMW-kininogen)
from patient samples and from *in vitro* experiments
with and without trypsin. On the basis of this identity, we have made a series of experiments adding
human cationic trypsin to human serum and plasma in
order to see, if there is any correlation between the
cleavage of C3 and kininogen and the degree of trypsin
saturation of the two main antiproteases α_2-M and α_1-AT.
We have also evaluated Aprotinin, a commercially available trypsin inhibitor, known as Trasylol®, in this
model.
     The addition of human cationic trypsin to human
serum or plasma followed by incubation at 37 °C for
30 or 60 min., gives a gradual saturation of first
α_2-M and then, to a lesser extent, α_1-AT (Fig. 4).

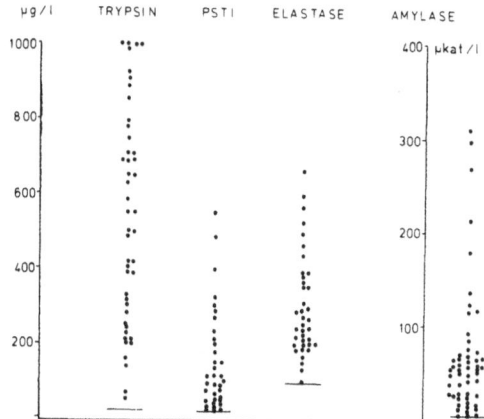

Fig. 2.   Immunoreactive trypsin, elastase, PSTI and
           amylase levels in blood in 45 patients with
           acute pancreatitis.

Table 1.   Immunoreactive cationic trypsin in sera of
           patients with severe acute pancreatitis (N=45)

Trypsinogen	659 µg/l	98 - 1000 µg/l
Trypsin bound to α_1-AT	29.8 %	2 - 100 %

Values for α_2-M were calculated using the low molecular
weight substrate BAPNA,[9] while α_1-AT complexation was
calculated from crossed immunoelectrophoresis analyses.[10] Identical reaction mixtures using ^{125}I-labeled
trypsin were also analyzed with gel chromatography on
Sephadex G-200. Similar results were then obtained.
Nearly all the radioactivity eluted in the α_2-M peak
and only about 30 % in the α_1-AT peak, until α_2-M
approached saturation.
      The next question is, whether anything else hap-
pens when, for instance α_2-M, the most important antiprotease, is saturated. As in our earlier dog experiments, we have used C3, the third complement factor,
and kininogen, the precursor to bradykinin, as indica-
tors of activation of the complement and kinin systems.
      C3 can be activated both by classical and alterna-
tive complement activation and changes in C3-levels
have been found to parallel changes in both total hemo-
lytic activity and in C4 in acute pancreatitis.[11] Trypsin cleaves C3 as shown in fig. 3 a. In serum without

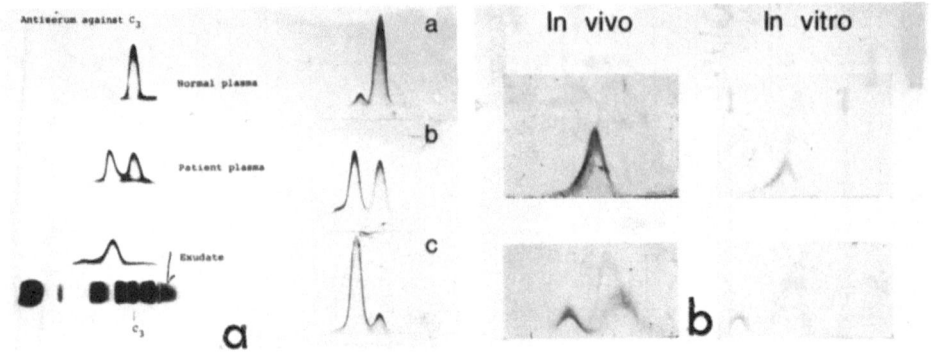

Fig. 3 a.   Crossed immunoelectrophoresis patterns
            using antiserum against human C3 with nor-
            mal plasma (top, left and right), plasma
            from a patient with acute pancreatitis
            (middle, left) peritoneal exudate from this
            patient (bottom, left) and serum with in-
            creasing amounts of trypsin (middle and
            bottom right).
Fig. 3 b.   Crossed immunoelectrophoresis patterns
            with antiserum against human HMW-kininogen
            with normal serum (top, left and right),
            serum from a patient with acute pancreati-
            tis (bottom, left) and serum with 5.2 µM
            of trypsin (bottom, right).

trypsin there is only little spontaneous cleavage. With
increasing amounts of trypsin, C3 is degraded and with
high trypsin concentrations there is no native C3 left.
This cleavage occurs promptly, when nearly all the
α_2-M is saturated and it occurs in spite of at least
90 % free and reactive α_1-AT (Fig. 5). Preliminary
results in human acute pancreatitis seem to favour a
direct enzymatic effect of trypsin on C3 rather than
the usual activation sequences, since C2, C4 and
factor D seem unaffected.[12]
        There are many difficulties in measuring altera-
tions in the kinin system[13] and especially in measuring
the active end-product, bradykinin, which is very quickly degraded. Assays of kininogen are often used for this
purpose and we measured kininogen both by crossed immunoelectrophoresis and by bioassay on the esterous
rat uterus.[14] Trypsin activates kininogen (Fig. 3 b)
and, as seen in Fig. 5, this activation takes place,
when most of the α_2-M is complexed with trypsin,
while most of the α_1-AT is still free and reactive.
        To summarize this part of the experimental series,

there is a gradual saturation of first α_2-M and then
α_1-AT, when human cationic trypsin is added to human
serum or plasma. When α_2-M is saturated to 70 %, a
prompt and nearly complete activation of both the complement and kinin systems occurs with possible clinical
consequences. α_1-AT does not stop this activation.
The activation pattern is identical for C3 and kininogen in plasma and serum but occurs more easily in
serum, where the trypsin needed for full activation is
about 2/3 of that needed in plasma. This is probably
explained by some consumption of α_2-M during the clotting process in serum. Our ongoing studies on human
pancreatitis in 18 patients requiring peritoneal lavage
and in 12 patients requiring only food restriction for
a few days, show that these low concentrations of
active α_2-M really exist both in blood and especially
in the peritoneal exudate during the first 2-6 days
in patients with severe acute hemorrhagic pancreatitis[12]
(Fig. 6 and table 2). Correspondingly, we found a
total cleavage of both C3 and kininogen in the peritoneal exudate, while we found about 70-80 % cleavage
in blood.

     Aprotinin has been tried for about 20 years as a
treatment of acute pancreatitis, but while there are
many reports on good effects in dogs[16][17], most reports on
treatment in humans fail to show any benefit at all.[18][19][20]
To try to elucidate a possible species difference, we
have tested Aprotinin in our human *in vitro* model.
Fig. 7 shows the effect of increasing amounts of
Aprotinin on trypsin-induced C3-cleavage, using a
constant amount of human cationic trypsin. A 60 µmolar

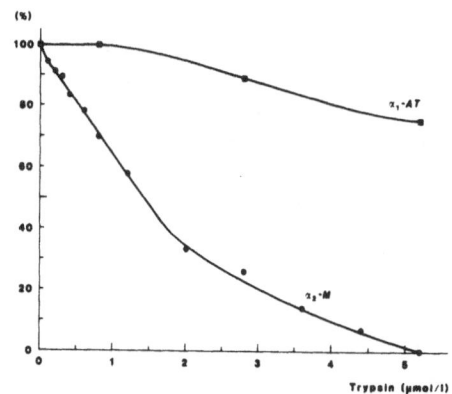

Fig. 4.   Free α_2-M (●—●) and α_1-AT (■-■), respec-
           tively, in per cent with increasing amounts
           of human cationic trypsin added to 500 µl of
           human serum. Final volume 1000 µl.

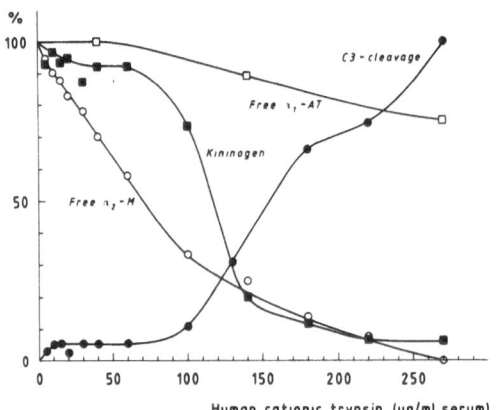

Fig. 5.   The relationship between the amount of free
           α₂-macroglobulin (α₂-M), α₁-antitrypsin
           (α₁-AT), kininogen and C3-cleavage in reac-
           tion mixtures of 0.5 ml human serum and in-
           creasing amounts of human cationic trypsin.
           Final volume was made 1.0 ml with saline.

concentration of Aprotinin blocks trypsin-induced C3-
cleavage almost completely, but this concentration is
about 15 times higher than the one ever used clinically
(Fig. 8). Total blockage occurs with a 20 μmolar concentration of Aprotinin when plasma instead of serum
is used. The reason for this difference needs further
elucidation. This concentration is about 5 times the
one ever used in patients anyhow, and it is also much
higher in the human model than in our corresponding
dog model. Fig. 8 shows that similar concentrations
of Aprotinin are needed to block trypsin-induced
kininogen consumption and again Aprotinin is more
effective in plasma than in serum. Fig. 8 also shows
that Aprotinin saves much of both α₁-AT and α₂-M.
      To summarize this part of the series, Aprotinin
can block the trypsin-induced cleavage of C3 and
kininogen in human serum and plasma, probably resulting
in abolished activation of these systems, provided that
sufficiently high concentrations are used. Maybe clinical trials with these high concentrations will prove
Aprotinin to be as good a remedy in human acute pancreatitis, as it is in dog pancreatitis.

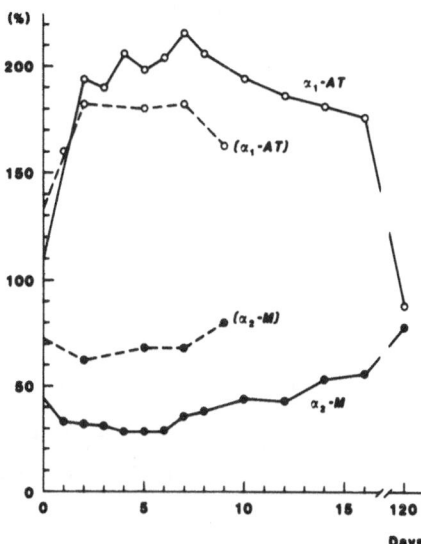

Fig. 6.   Values for α₂-M and α₁-AT in blood from two
          patients with acute pancreatitis. ●—● Rep-
          resents a very sich patient treated by peri-
          toneal lavage at the intensive care unit,
          while ●---● represents a patient needing only
          food restriction for 3 days. Values were ob-
          tained by immunoelectrophoresis (ref. 15).

Table 2.   Concentrations of the main inhibitors in
           the peritoneal exudate of patients with
           severe acute pancreatitis (n=18)

	mean value	range
α_1-antitrypsin	0.65 g/l	(0.38 - 1.3)
α_2-macroglobulin	0.38 g/l	(0.22 - 0.76)

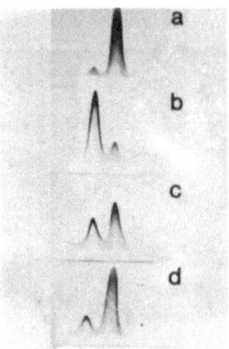

Fig. 7.   Patterns obtained on crossed immunoelectro-
           phoresis of human serum using anti-human C3
           of (a) human serum, (b) reaction mixture
           of 500 µl of human serum and 5.0 µM of human
           trypsin, (c) 500 µl of human serum containing
           7.5 µM of Aprotinin followed by 5.0 µM of
           human trypsin, (d) = (c) but the serum con-
           taining 60 µM of Aprotinin. Final volume was
           made 1000 µl with saline.

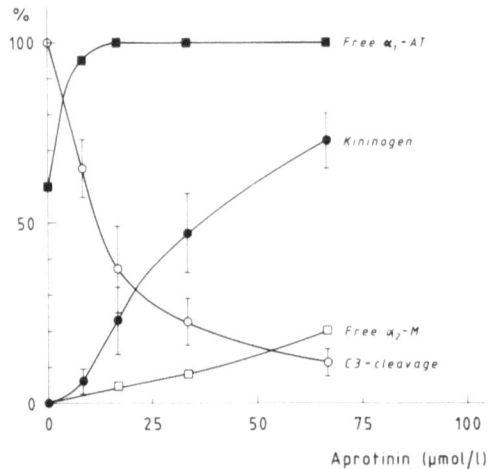

Fig. 8.   Mean values of trypsin-induced C3-cleavage
           (spontaneous cleavage withdrawn), kininogen,
           free α_2-M and free α_1-AT in reaction mixtures
           of 500 µl serum and incrasing amounts of
           Aprotinin followed by a constant amount of
           human trypsin. Final volume 1000 µl.

ABSTRACT

The protective role of α_2-macroglobulin, α_1-anti-trypsin and Aprotinin against trypsin-induced effects on C3 and kininogen was studied in a human *in vitro* model.

When human cationic trypsin was added to human serum or plasma, there was a gradual saturation of α_2-macroglobulin and later of α_1-antitrypsin. When α_2-macroglobulin was 70 % saturated, there was a prompt cleavage of both C3 and kininogen, in spite of 80 % free and active α_1-antitrypsin. These biochemical changes and antiprotease levels are identical to our findings in patients with acute pancreatitis, especially in their peritoneal exudate.

Very high concentrations of Aprotinin, 5-15 times higher than ever used clinically, blocked the cleavage of both C3 and kininogen, while doses commonly used clinically were without significant effect.

The clinical implications are:
1. A trypsin-induced activation of both the complement and kinin system with clinical consequence is possible in patients with acute pancreatitis because of very low α_2-macroglobulin levels.
2. Aprotinin in adequate doses, 5-15 times higher than ever used clinically, seems to protect against activation of two systems.

REFERENCES

1. H. Chiari, Über selbst verdauung des menschlichen Pancreas. Z Heilk 17:69 (1896).
2. G. Balldin and K. Ohlsson, Demonstration of pancreatic protease-antiprotease complexes in the peritoneal fluid of patients with acute pancreatitis, Surgery 85:451 (1979).
3. A. Borgström and K. Ohlsson, Immunoreactive trypsin in serum and peritoneal fluid in acute pancreatitis, Hoppe-Seyler's Z Physiol Chem 359:677 (1978).
4. D. Vercaigne, C. Morcamp, C. Martin, J. P. Joly, B. Hillemand, and J. P. Raoult, "Tryptic-like" activity in sera of patients with pancreatitis, Clin Chim Acta 106:269 (1980).
5. J. W. Brodrick, M. C. Geokas, C. Largman, M. Fassett, and J. H. Johnson, Molecular forms of immunoreactive pancreatic cationic trypsin in pancreatitis patient sera, Am J Physiol 237:E474 (1979).

6. A. Eddeland and K. Ohlsson, Studies on the
    pancreatic secretory trypsin inhibitor in
    plasma and its complex with trypsin in vivo
    and in vitro, Scand J Clin Lab Invest 38:
    507 (1978).
7. G. Balldin, E. L. Gustafsson, and K. Ohlsson,
    Influence of plasma protease inhibitors and
    Trasylol on trypsin-induced bradykinin-release
    in vitro and in vivo, Europ Surg Res 12:260
    (1980).
8. G. Balldin, A. Eddeland, and K. Ohlsson,
    Studies on the role of the plasma protease in-
    hibitors on C3-activation in vitro and in
    acute pancreatitis, Scand J Gastroent 16:
    603 (1981).
9. P. O. Ganrot, Partition of trypsin between
    alpha$_2$-macroglobulin and the other trypsin
    inhibitors of serum, Arkiv för kemi 26:577
    (1967).
10. P. O. Ganrot, Crossed immunoelectrophoresis,
    Scand J Clin Lab Invest 29, suppl 124:39 (1972).
11. I. M. Goldstein, D. Cala, A. Radin, H. B.
    Kaplan, J. Horn, and J. Ranson, Evidence of
    complement catabolism in acute pancreatitis,
    Am J Med Sci 275:257 (1978).
12. A. Lasson and K. Ohlsson, Studies on acute
    human pancreatitis, Unpublished.
13. C. Regoli and J. Barabé, "Pharmacology of
    bradykinin and related kinins", Pharm Rev
    32:1 (1980).
14. E. Ofstad, Formation and destruction of plasma
    kinins during experimental acute hemorrhagic
    pancreatitis in dogs. Scand J Gastroent
    5:9 (1970).
15. C. B. Laurell, Electroimmunoassay, Scand J
    Clin Lab Invest 29, suppl 124:21 (1972).
16. G. Balldin and K. Ohlsson, Trasylol prevents
    trypsin-induced shock in dogs, Hoppe-Seyler's
    Z Physiol Chem 360:651 (1979).
17. C. W. Imrie and M. Mackenzie, Effective Apro-
    tinin therapy in canine experimental bile-
    trypsin pancreatitis, Digestion 22:32 (1981).
18. C. W. Imrie, I. S. Benjamin, J. C. Ferguson,
    A. J. McKay, I. Mackenzie, J. O'Neill, and
    L. H. Blumgart, A single-centre double-blind
    trial of Trasylol therapy in primary acute
    pancreatitis, Br J Surg 65:337 (1978).
19. M. R. C. Multicentre Trial of Glucagon and
    Aprotinin, Lancet 24:632 (1977).

20. M. R. C. Multicentre Trial, Morbidity of acute pancreatitis: the effect of aprotinin and glucagon, Gut 21:334 (1980).

ACKNOWLEDGEMENTS

This investigation was supported by grants from the Swedish Medical Research Council (project no B83-17X-03910-11B), Thorsten and Elsa Segerfalks Foundation for Medical Research and Education, and Malmö General Hospital Foundation against Cancer.

STUDIES ON THE KALLIKREIN-KININ SYSTEM IN PLASMA AND

PERITONEAL FLUID DURING EXPERIMENTAL PANCREATITIS

T.E. Ruud, A.O. Aasen, P. Kierulf, J. Stadaas
and S. Aune

Surgical department and Central laboratory
Ullevaal University Hospital, Oslo 1, Norway

ABSTRACT

Using chromogenic peptide substrate assay technique,
components of the plasma kallikrein-kinin system and
trypsin activity were studied in plasma and peritoneal
exudate during acute pancreatitis in pigs. In the plasma
no significant changes occured, but increased kallikrein
activity was found in the peritoneal exudate. This finding was paralleled by a reduction in prekallikrein levels
and functional kallikrein inhibition values. The trypsin
activity in peritoneal exudate, however, increased inconstantly. These results emphasize the significance of
peritoneal protease-antiprotease imbalance during acute
pancreatitis.

INTRODUCTION

A variety of ethiological and pathogenetic factors
have been proposed in acute pancreatitis. In 1970
Creutzfeldt and Schmidt proposed the following sequence
of events leading to acute pancreatitis: Injury of acinar cells, liberation or activation of trypsin, activation of prekallikrein and appearance of edema and shock.
Using immunochemical methods, proteases have been found
in the peritoneal exudate during both experimental and
acute pancreatitis in humans (Balldin 1980). The proteases appears initially in complex with inhibitors, then
in free forms after saturation of the inhibitor capacity. Ofstad (1970) reported kinin formation accompanied

by kininogen consumption in peritoneal exudate during
experimental pancreatitis. Within few hours about 1/3 of
the total plasma volume is lost from the circulation in
experimental acute pancreatitis (Ryan et al. 1965, Carey
and Rodgers 1966, Anderson et al. 1967). This fact may
partly explain the shock development seen during this
condition. This process is likely influenced by generation of vasoactive agents (Amundsen et al. 1968). Chromogenic substrate peptide assay technique has made close
monitoring of components of the plasma proteolytic enzyme systems possible. The aim of this study, using this
technique, was to see if components of the plasma kallikrein-kinin system exists in normal peritoneal fluid,
and if changes occurs in the peritoneum during experimental acute pancreatitis in pigs.

MATERIALS AND METHODS

        Acute pancreatitis was induced in 8 laparotomized,
anesthetized juvenile pigs, weighing 15-25 kg, by retrograde injections of 10 ml 10 % Na-Taurocholeate into
the pancreatic duct (Elliot et al. 1959, Schön et al.
1963, Doerr et al. 1965). The observation time was 6
hours, 2 sham operated pigs served as controls. Trypsin
activity (TRY) was determined using the chromogenic
peptide substrate S-2222 according to the producer (KABI
Diagnostica, Stockholm, Sweden). Plasma prekallikrein
(PKK) (Amundsen et al. 1978), plasma functional kallikrein inhibition (KKI) and plasma kallikrein activity
(KK) (Aasen et al.1978) were all studied in plasma and
peritoneal exudate before and during acute experimental
pancreatitis, using the chromogenic peptide substrate
S-2302. Median values with 25 and 75 fractiles are
given. PKK and KKI are expressed as per cent of a human
standard plasma pool value. Results were analyzed statistically by Willcoxon rank sum test. P values less
than 0,05 were considered significant.

RESULTS

        All animals developed serious acute pancreatitis.
Five of the pigs died during the observation period.
The PKK, KK and KKI values found before the induction
of pancreatitis are summarized in table 1. During the
observation time no significant changes in plasma PKK,
KK or KKI values occured. In the peritoneal exudate,
however, PKK values were reduced to about 60 % of the
initial levels, whereas the KKI activity was reduced by
about 20 % (fig. 1,2). These changes were paralleled by

a significant increase in KK activity (fig. 3). Trypsin
activity in the peritoneal exudate during acute pancreatitis increased inconstantly. In two pigs, both succomb-
ing, TRY increases from less than 20 U/l to more than
4000 U/l. In the others, the increases were less pronounced or TRY activity was unaltered. In the controls no
change in PKK, KK, KKI or TRY occured.

Table 1: Plasma and peritoneal fluid PKK, KK and KKI
activities before induction of acute experimental
pancreatitis.

n = 18	Peritoneal fluid	Vena porta	Vena cava
PKK %	70 (62–81)	64 (33–100)	74 (32–100)
KK u/l	27 (19–36)	24 (16–34)	27 (16–33)
KKI %	55 (47–69)	93 (83–107)	93 (80–104)

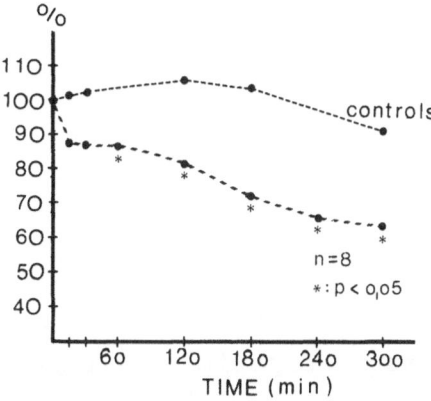

Fig. 1: Plasma prekallikrein values in peritoneal exudate during acute experimental pancreatitis.

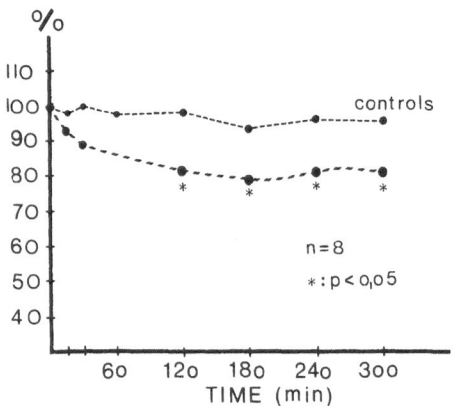

Fig. 2: Functional plasma kallikrein inhibition in
peritoneal exudate during experimental pancreatitis.

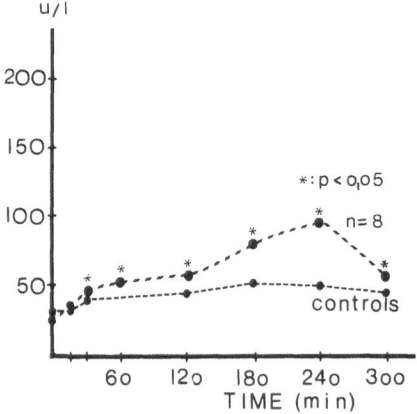

Fig. 3: Plasma kallikrein activity in peritoneal
exudate during acute experimental pancreatitis.

DISCUSSION

The present study, using chromogenic peptide substrate assay technique, shows that the plasma kallikrein-
kinin system exists in the peritoneal fluid in pigs. The
KK activity and PKK values were similar in plasma and
peritoneal fluid. KKI values in peritoneal fluid, however,
were about 60 % of the plasma levels. After induction of
acute pancreatitis marked changes were found in the
kallikrein-kinin system in the peritoneal exudate, whereas no changes occured in plasma. Release of proteolytic
enzymes and in particular trypsin, is considered to be
of great importance for the pathophysiological changes
seen during acute pancreatitis (Ofstad 1970, Ohlsson and
Eddeland 1975, Amundsen 1976 and Balldin 1980). Trypsin
is known to be an activator of the kallikrein-kinin
system, and is likely an important mediator for the
changes in this system reported here. Following complete
saturation with inhibitors like alpha-2 macroglobulin or
alpha-1 antiprotease, trypsin will appear in free, active form (Balldin 1980). Also trypsin in complex with
alpha-2 macroglobulin can be detected with chromogenic
peptide substrates, because these complexes retain the
ability to split several substrates (Nagasawa et al.
1970, Harpel and Mosesson 1973, Largemann et al. 1977).
The study shows that activation of the kallikrein-kinin
system and release of trypsin occurs locally during experimental acute pancreatitis in pigs. This may be of
importance for shock development seen during this condition.

REFERENCES

Aasen, A. O., Smith-Erichsen, N., Gallimore, M. J., and
    Amundsen, E., 1980, Studies of the plasma kalli-
    krein-kinin system in plasma samples from normal
    individuals and patients with septic shock, in:
    "Advances in Shock Research," W. Schumer, J. J.
    Spitzer, B. E. Marshall, eds., Liss, A.R., Inch,
    New York.
Amundsen, E., Ofstad, E., and Hagen, P. O., 1968, Ex-
    perimental acute pancreatitis in dogs, 1. Hypoten-
    sive effect induced by pancreatic exudate, Scand.
    J. Gastroent., 3:659.
Amundsen, E., 1976, Pathophysiologische Problematik der
    akuten Pancreatitis, Leber Magen Darm, 6:199.
Amundsen, E., Gallimore, M. J., Aasen, A. O., Lars-
    braaten, M., and Lyngaas, K., 1978, Activation
    of human plasma prekallikrein: Influence of

activators, activation time and temperature and inhibitors, Thromb. Res., 13:625.

Anderson, M. C., Schoenfeld, F. B., Lams, W. B., and Suwa, M., 1967, Circulatory changes in acute pancreatitis, Surg. clin. N. Amer., 47:127.

Balldin, G., 1980, On protease-antiprotease imbalance with special reference to the protective role of protease inhibition in acute pancreatitis, Malmø.

Carey, L. C., and Rodgers, R. E., 1966, Pathophysiological alterations in experimental pancreatitis, Surgery, 60:171.

Creutzfeld, W., and Schmidt, M., 1970, Aetiology and pathogenesis of pancreatitis (current concepts), Scan. J. Gastroent., 5, suppl. 6.

Doerr, W., Diezel, P. B., Grozinger, K. H., Nagel, W., Rossner, J. A., Wanke, M. und Willig, F., 1965, Pathogenese der experimentellen autodigestiven Pancreatitis, Klin. Wschr., 43:125.

Elliot, D. W., Williams, R. D., and Steward, W. R. C., 1959, The role of trypsin and bile salts in the pathogenesis of acute pancreatitis, Surg. forum, 9:533.

Harpel, P. C., and Mosesson, M. W., 1973, Degradation of human fibrinogen by plasma α_2-macroglobulin-enzyme complexes, J. Clin. Invest., 52:2175.

Largman, C., Johnson, J. H., Brodrick, J. W., and Geokas, M. C., 1977, Proinsulin conversion to desalanyl insulin by α_2-macroglobulin-bound trypsin, Nature, 269:168.

Nagasawa, S., Han, B. H., Sugihara, H., and Suzuki, T., 1970, Studies on α_2-macroglobulin in bovine plasma. II Interaction of α_2-macroglobulin and trypsin, J. Biochem., 67:821.

Ofstad, E., 1970, Formation and destruction of plasma kinins during experimental acute hemorrhagic pancreatitis in dogs, Scan. J. Gastroent. 5, suppl. 1.

Ohlsson, K., and Eddeland, A., 1975, Release of proteolytic enzymes in bile induced pancreatitis in dogs, Gastroenterology, 69:668.

Ryan, J. W., Moffat, J. G., and Thompson, A. G., 1965, Role of bradykinin system in acute hemorrhagic pancreatitis, Arch. Surg., 91:14.

Schön, H., Meinkel, K. und Henning, N., 1963, Über das Verhalten von Pancreassystemen bei der experimentellen Pancreatitis der Ratte, Klin. Wschr., 41:612.

THE INFLUENCE OF THE KALLIKREIN-KININ SYSTEM IN THE DEVELOPMENT OF THE PANCREATIC SHOCK

H. Kortmann, E. Fink, and G. Bönner[+]

Chir. Klinik der Universität München
Klinikum Großhadern
[+]II. Med. Klinik der Universität Köln

Kallikrein and other kininogenases like trypsin and plasmin release kinins by limited proteolysis from high (HMWK)- and low molecular weight kininogen (LMWK). The kinins are vasoactive peptides with fairly unknown physiology.

The kallikrein-kinin system has long been supposed to be involved in the pathogenesis of the severe course of acute hemorrhagic pancreatitis (AHP). However, missing any specific test method former attempts at demonstrating such a role have been inconclusive. We developed specific and highly sensitive immunoassays for studying the influence of the kallikrein-kinin system on the process of pancreatic shock in an animal model.

MATERIALS AND METHODS

Studies were performed on piglets on either sex with an average body weight of 19,3 kg. They were deprived of food but not water for 24 hours. Anesthesia was induced by intraperitoneal injection of 1 ml Azaperon (Stresnil[R], Janssen GmbH, West Germany) together with 5 ml Metomidat-Hydrochlorid (Hypnodil[R], Janssen GmbH, West Germany), and was maintained with small amounts of intravenous sodium pentobarbital (Nembutal[R], Ceva GmbH, West Germany). Catheters were inserted into the left a. carotis com. for continuous registration of blood pressure and arterial blood gas analysis and into the left v. jugularis using for blood sampling and infusion of Ringer's solution. A 7 F-Thermistor Swan-Ganz catheter (Gould Medical Products Division, West Germany) was inserted through the right external jugular vein and guided into the pulmonary artery. Pulmonary artery pressure (PAP), central venous pressure (CVP) and pulmonary artery wedge pressure

(PCWP) were continuously monitored. Cardiac output (CO) was computed by the thermodilution method. Plasma volume was calculated with the Evans-Blue method as described by Linderkamp[1].

An acute hemorrhagic-necrotizing pancreatitis (AHP) was induced by retrograde intraductal injection of 10 % Na-Taurocholic-Acid (Sigma Chemie GmbH, West Germany) in 15 piglets. The 6 control pigs were only sham operated without any other difference in the treatment. They were killed after a postoperative observation period of at least 14 hours. To analyse the data of all animals, independent of their survival time, comparable time periods were chosen. Analysed data were taken at the end of the periods shown in the diagrams.

Glandular kallikrein concentration was determined by radioimmunoassay as described earlier[2]. Native kininogen concentration was measured by a radioimmunoassay of bradykinin[3]. The differentiation in HMWK and LMWK was essentially carried out as described by Uchida and Katori[4]. Plasmakallikrein was determined by chromogenic peptide substrate assay (substrate: S-2302 Kabi Vitrum AB, Sweden)[5]. Results were analysed statistically by Wilcoxon rank sum test. P values $< 0,05$ were considered significant.

RESULTS

The glandular kallikrein increased in plasma immediately after TCA injection that means after the onset of AHP. The highest values were obtained prefinally (Fig. 1).

Quantitatively most glandular kallikrein was excreted in the ascites. Compared to plasma the concentration was 10 to a hundredfold higher. With the propagation of pancreatitis the values enlarged. Because of the large variations significantly higher values compared with those after sixty minutes were only seen during the last two hours (Fig. 2).

The plasma kallikrein, a kininogenase quite different from glandular kallikrein in respect to molecular size and physiologic function, showed in the AHP animals as well as in the controls a continuous fall in activity. Significant differences between the two groups were only seen in the last state.

The total kininogen concentration in plasma decreased continuously after TCA injection. The highest loss was found in the prefinal period. Although the control values decreased too, there was a highly significant difference between AHP animals and the controls. The differentiation into H- and LMWK showed with 86 % loss a higher consumption of the HMWK compared with only 61 % loss of LMWK. The relation of L- to HMWK altered from 2:1 before setting the AHP to 6:1 in the prefinal stage. Nevertheless most of the liberated kinins were released from the LMWK (Fig. 4).

The continuous production of ascitic fluid yielded a strong hemoconcentration with a 30 % increase of hematocrite and a 50 % reduction in plasma volume as against only 20 % with the controls in 14 hours (Fig. 5).

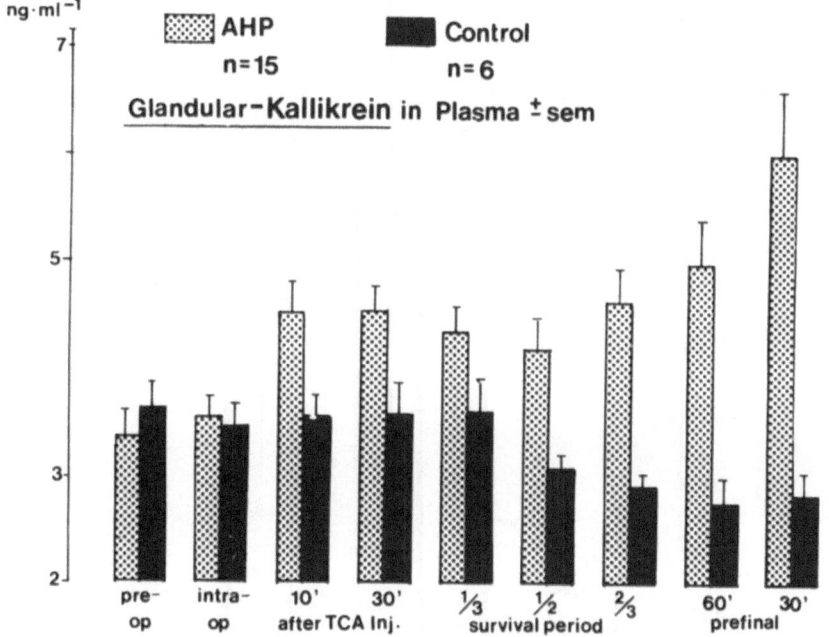

Fig. 1: Glandular kallikrein in plasma after Na-Taurocholic-Acid (TCA) injection

As a consequence arterial pressure, cardiac output, and central venous pressure decreased continuously (Fig. 6). All the animals with AHP died of hypovolemic shock.

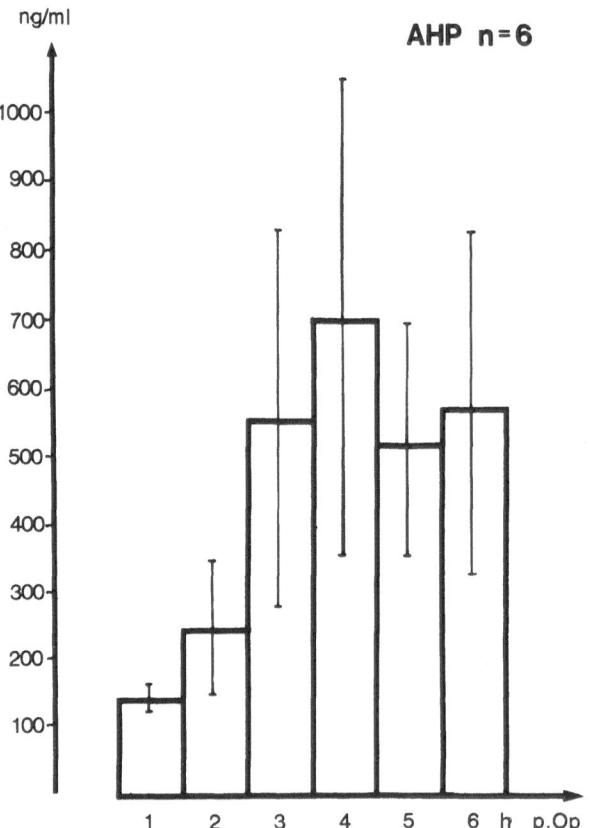

Fig. 2: Tissue kallikrein concentration in ascitic fluid

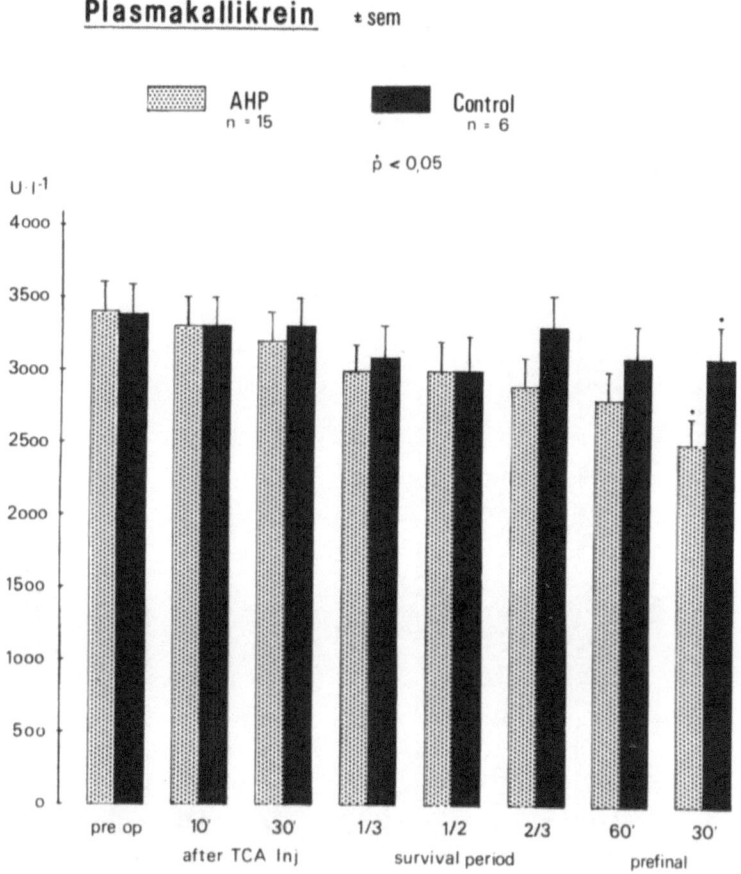

Fig. 3: Plasmaprekallikrein concentration expressed in plasma-kallikrein activity in blood plasma in the course of experimental AHP

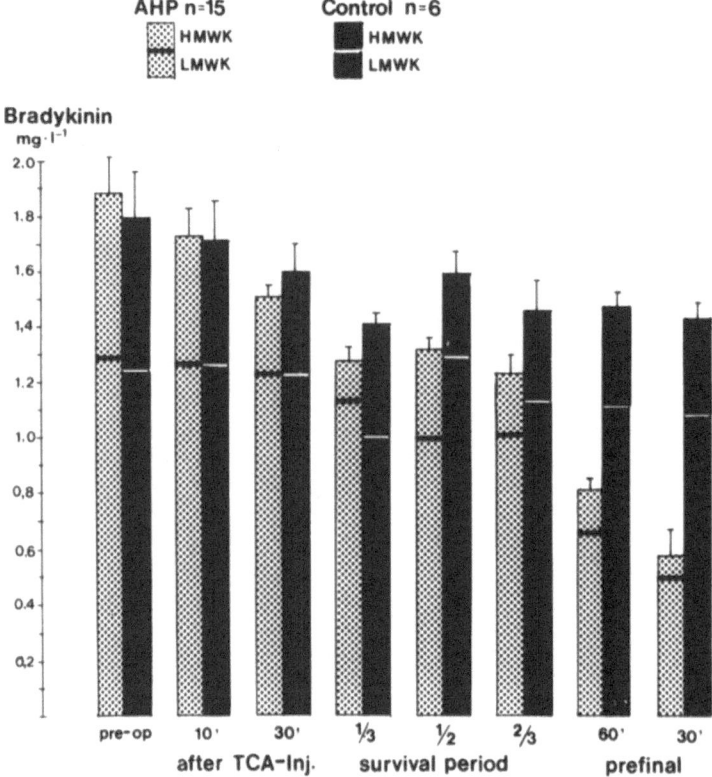

Decrease of Total Kininogen Concentration in Plasma
(HMWK & LMWK)
(Bradykinin – RIA)

Fig. 4: Native kininogen concentration and differentiation in HMWK and LMWK in the course of experimental AHP. Native kininogen is expressed in bradykinin equivalents.

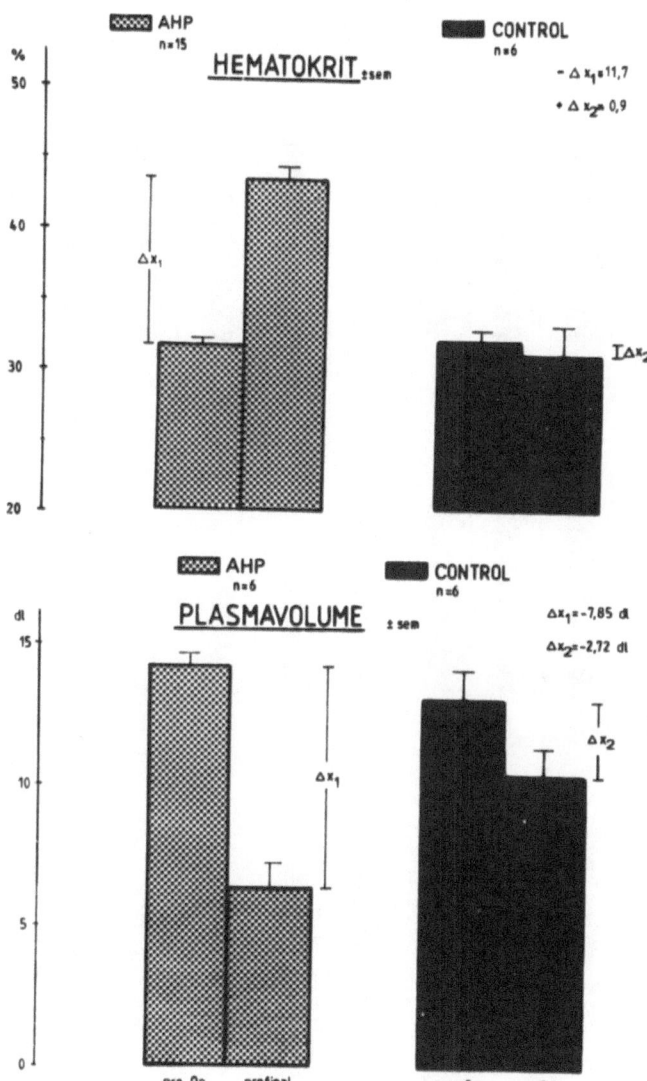

Fig. 5: Hemoconcentration and loss of plasma volume in experimental AHP

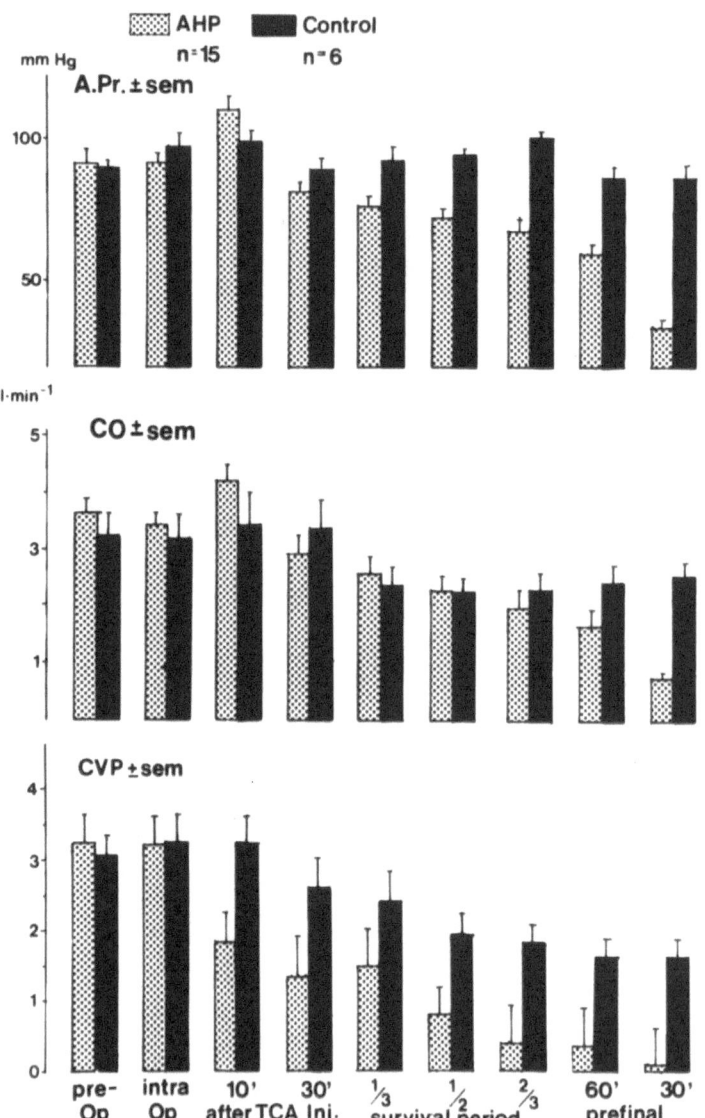

Fig. 6: Alterations of hemodynamic parameters in the course of experimental AHP

DISCUSSION

With specific immunoassays we could show that there is indeed a high loss of total kininogen in experimental AHP, which means a strong liberation of kinins. These potent vasoactive peptides increase the local capillary permeability resulting in continuous ascitic fluid production and final hypovolemic shock[6,7,8]. HMWK is an important factor in solid phase activation of the intrinsic coagulation system[9]. Although we didn't find global coagulopathy the high consumption of HMWK might have caused the microthrombolies seen in lung and kidney on histological examinations.

REFERENCES

1. O. Linderkamp, T. Mader, O. Butenandt, and K. P. Riegel, Plasma volume estimation in severely ill infants and children using a simplified Evans Blue method, Pediatrics 125:135 (1977).
2. E. Fink, and Ch. Güttel, Development of a radioimmunoassay for pig pancreatic kallikrein, J. Clin. Chem. Clin. Biochem. 16: 381 (1978).
3. G. Bönner, M. Marin-Grez, D. Beck, M. Deeg, and F. Gross, Measurement of kallikrein activity in urine of rats and man using a chromogenic tripeptide substrate. Validation of the amidolytic assay by means of bradykinin radioimmunoassay, J. Clin. Chem. Clin. Biochem. 19:165 (1981).
4. Y. Uchida, and M. Katori, Differential assay method for high molecular weight and low molecular weight kininogen, Thrombosis Res. 15:127 (1979).
5. E. Ammundsen, M. J. Gallimore, A. O. Aasen, M. Labraaten, and K. Lyngaas, Activation of human plasma prekallikrein: Influence of activators, activation time and temperature and inhibitors, Thrombosis Res. 13:625 (1978).
6. B. Haraaldsson, B. Rippe, B. J. Moxham, and B. Folkow, Permeability of fenestrated capillaries in the isolated pig pancreas, with effects of bradykinin and histamin, as studied by simultaneous registration of filtration and diffusion capacities, Acta Physiol. Scand. 114:67 (1982).
7. J. W. Ryan, J. G. Moffat, and A. G. Thompson, Role of bradykinin system in acute hemorrhagic pancreatitis, Arch. Surg. 91:14 (1965).
8. J. A. Barrowman, M. A. Perry, P. R. Kvietys, M. Ulrich, and D. N. Granger, Effects of bradykinin on intestinal fluid exchange, Can. J. Physiol. Pharmacol. 59:786 (1981).
9. K. D. Wuepper, and C. Cochrane, Effect of kallikrein on coagulation in vitro, Proc. Soc. Exp. Biol. Med. 141:271 (1972).

PROTEOLYTIC ENZYMES AND ENHANCED MUSCLE

PROTEIN BREAKDOWN

B. Dahlmann, L. Kuehn, and H. Reinauer

Biochemical Department
Diabetes Forschungsinstitut
Auf'm Hennekamp 65   4000 Düsseldorf F.R.G

For studies of intracellular protein breakdown very often tissues like liver are used. In hepatocytes, the major part of intracellular protein degradation probably results from the degradation of more or less soluble cytoplasmic proteins. In contrast, when studying the intracellular protein breakdown in muscle cells, a very different cellular organization must be considered. Thus, only a minor fraction of muscle cellular proteins is localized in the soluble cytoplasmic compartment while 60 % of the intracellular proteins are constituents of the contractile myofilaments[1]. Additionally, it is known that a major part of the cytoplasmic proteins are tightly bound or attached to myofibrils[2]. As shown for the enzyme aldolase[3], this may also influence the degradative rate of these proteins.

In muscle tissue similar to other tissues, the basal protein turnover is characterized by a heterogeneous degradation rate of cellular proteins. There are both short-lived proteins with a high turnover rate and long-lived proteins with a low turnover rate. Not only do the average turnover rates of cytoplasmic proteins and contractile proteins differ by a factor of two[4], but there is also considerable heterogeneity in the turnover rate within each the group of sarcoplasmic and myofibrillar proteins, respectively[5].

While this heterogeneity is a characteristic feature of normal cells, the loss of heterogeneity, on the other hand, is equally typical for a pathologically enhanced

protein breakdown. Millward, for instance, reported that
during 'enhanced degradation of starvation the average
degradation rates of sarcoplasmic and contractile
proteins are very similar'[6]. Furthermore, Dice and
coworkers[7] have shown that normally long-lived cytoplasmic proteins are degraded at the same rate as are
short-lived cytoplasmic proteins, during starvation and
diabetes mellitus.

The question is: what may be the mechanism responsible for this qualitative and quantitative change in
muscle protein breakdown? Theoretically, there are
several possibilities: It may be a consequence of either
a modification of substrate proteins, rendering them
more susceptible to proteolytic attack or an alteration
of the proteolytic capacity or specificity in the cell.
A third possibility would be that alterations occur both
at the substrate and at the enzyme level.

To our knowledge, no examples are known for post-
translational modifications of proteins resulting in the
transformation of a long-lived into a short-lived protein.
Nevertheless, several post-translational modifications
of proteins have been suggested to initiate the degradative process. These are, for instance, carbamylation,
phosphorylation, glycosylation, reductive cleavage of
disulphide bridges, loss of coenzyme or substrate[6].
All these mechanisms have been investigated using soluble
cytoplasmic proteins and it would not be surprising if
the rate of one of these post-translational protein
modifying reactions would also increase during conditions
of enhanced protein catabolism to initiate the degradation of proteins.

It is presently unknown whether such mechanisms are
also responsible for initiating the degradation of myofibrillar proteins. However, a few cases of protein modifications influencing the digestibility of contractile
proteins have been reported. For example, phosphorylation
of myosin light chain 2 inhibits the digestion of this
molecule and concomitantly determines the site of cleavage by chymotrypsin in the myosin heavy chain[8]. Recently,
Borejdo[9] has shown that binding of MgADP to myosin-ATPase
active site increases the digestibility of myofibril-
bound myosin. Although these experiments were performed
in vitro, phosphorylation of light chain 2 as well as
actin-myosin-ADP interactions are physiologically
occurring phenomena[10,11] and thus could well be involved
in the triggering of myofibrillar protein degradation.

Table I Endopeptidases isolated from striated and non-striated muscle tissue

ENZYME	PH OPTIMUM	MOL.WT.	SOURCE	
CATHEPSIN B	~5.0	25 000	SKELETAL MUSCLE	RAT [13]
CATHEPSIN D	3.0 - 4.0	42 000	SKELETAL MUSCLE	RAT [13]
			SKELETAL MUSCLE MYOMETRIUM	RABBIT [22] PIG [23]
			HEART MUSCLE	RABBIT [24]
CATHEPSIN L	3.5 - 7.0	24 000	SKELETAL MUSCLE	RABBIT [14]
CA-ACTIVATED NEUTRAL PROTEINASE (CAF)	~7.0	80 000 (+ 30 000)	SKELETAL MUSCLE HEART MUSCLE	RABBIT,PIG,CHICKEN,MAN [25,26] BOVINE,PORCINE [25,26]
ALKALINE SERINE PROTEINASE	8.5 - 9.5	25 000	SKELETAL MUSCLE HEART MUSCLE	RAT,MICE,MAN [25,26] RAT,HAMSTER [25,26]
HYDROLASE H	7.5 - 8.0	340 000	SKELETAL MUSCLE	RABBIT [27]
HIGH MOLECULAR WEIGHT PROTEINASE	~7.2	550 000	SKELETAL MUSCLE	MAN [28]
INSULIN DEGRADING PROTEINASE	7.0 - 7.5	80 000 135 000	SKELETAL MUSCLE SKELETAL MUSCLE	RAT [29] PIG [30]
ATP-DEPENDENT PROTEINASE	~ 7.8		SKELETAL MUSCLE	RABBIT [31]
NEUTRAL TRYPSIN-LIKE PROTEINASE	7.0 - 8.0	33 000	INTESTINAL SMOOTH MUSCLE	RAT [26]
ALKALINE CYSTEINYL PROTEINASE	7.5 - 8.5		UTERUS SMOOTH MUSCLE	RAT [38]

Whatever the initiating reactions of the degradative process may be, the following step will be the proteolytic cleavage of proteins and this cleavage will probably be an endo-proteolytic one. Therefore, not only protein transformation but also a change in activity or specificity of these enzymes could be responsible for the enhanced, non-heterogeneous protein breakdown in muscle observed during several pathological conditions. Thus, the question arises what type of proteinase could be responsible for myofibrillar protein degradation.

As summarized in Table I, about 13 different proteinases have been extracted from muscle tissue. One group of proteinases has optimum activity in the acidic pH range while another group is active at neutral or even alkaline pH. If one of these proteinases participates in the proteolytic cleavage of myofibrillar proteins, it should be able to attack these proteins in vitro. Unsurprisingly therefore, numerous reports are concerned with the ability of these enzymes to degrade myofibrillar proteins in vitro. But, data are lacking for hydrolase H, for the high molecular weight proteinase and the ATP-dependent proteinase. The hydrolytic activity of the so-called insulin-degrading enzyme is reported to be specific towards insulin and glucagon[12].

With the exception of the Ca-activated proteinase the enzymes summarized in Table II degrade the purified components of the thick filament myosin heavy chain and myosin light chain. A similar susceptibility to the proteolytic enzymes has been found for purified actin. Furthermore, all enzymes degrade troponin T and troponin I in vitro, whereas troponin C is more resistant to proteolytic attack. Tropomyosin, the other major component of the thin filament, is degraded by cathepsin B, Ca-activated proteinase and the alkaline serine proteinase, but not by cathepsin L or the trypsin-like proteinase from the smooth muscle. The main Z-line protein, α-actinin, is reported to be degraded only by cathepsin L and by the trypsin-like proteinase from smooth muscle.

In general, these data show that the digestibility of myofibrillar proteins by the proteolytic enzymes is very similar. Only the Ca-activated proteinase has a more restricted specificity. Therefore, no clearcut information can be derived from these data as to which proteinases are able to attack myofibrillar proteins and which are unable to do so.

Table II Degradability of myofibrillar proteins by muscle proteinases

	CATH.B 13,32		CATH.D 13,22,24		CATH.L 14,33		CAF 25,26		ALK.SERINE PROT. 25,26 SKELETAL		ALK.SERINE PROT. 34,35 HEART		NEUTR.TRYP.PROT. 35,37		ALK.CYST.PROT. 38	
	SP	MP	SP	MP	SP	MP	SP	MP	SP	MP	SP	MP	SP	MP	SP	MP
THICK FILAMENT																
MYOSIN																
- HEAVY CHAIN	+		+	+	+	+	-	-	+		+	+	+	+		
- LIGHT CHAIN LC₁	(+)		+	+	+		-	-			-				+	
LC₂	(+)		+		+	+	-	-	+		+	+	+			
LC₃							-	-								
C-PROTEIN			+		+		+	+	+		+	+				
THIN FILAMENT																
ACTIN	+		+		+		-		+		-	-	+			
TROPOMYOSIN	+				-		+	(+)	+		+		+	-		
TROPONIN TN-C	-				-		-		-		+		+	+		
TN-I	+				+	+	+	+	+		+		+			
TN-T	+				+	+	+	(+)	+		+		+	+		
M-LINE																
M-PROTEIN		+					-		+	-	+	+				
Z-LINE																
α-ACTININ	-		+			-	-		-		-	-	+			

+ DEGRADED   SP: SINGLE PROTEIN
- NOT DEGRADED MP: MYOFIBRILLAR-BOUND PROTEIN

Another prerequisite for an involvement of a proteolytic enzyme to participate in myofibrillar protein breakdown is that the enzyme and the substrate must be in
physical proximity. Cathepsin B, D and L are known to be
localized intralysosomally [13,14]. Therefore, if these
enzymes are to be involved in myofibrillar protein degradation, myofibrillar proteins would have to enter the
lysosome. Until now, however, neither have fractions of
myofibrils or myofilaments been detected in lysosomes,
nor has a release of lysosomal enzymes into the cyto-
plasma been observed except under conditions of cell
necrosis[15]. As a consequence, single myofibrillar proteins or protein subunits would have to enter the lysosomes to be degraded by the lysosomal proteinases. Such
a mechanism would be in accordance with findings by
Gerard and Schneider[16], who detected degradation products
of actomyosin in a lysosomal fraction from rat gastrocnemius muscle. If these observations are correct and if an
increased autophagy may be responsible for the pathologically enhanced, non-heterogeneous protein breakdown,
a question still remains to be answered: How are myofibrils disassembled into monomeric proteins which can
then be enclosed in autophagic vacuoles?

A conceivable mechanism would be that proteinases
localized in the cytoplasm depolymerize myofilaments by
limited proteolysis. With regard to their extra-lysoso-
mal origin, the neutral and alkaline proteinases appear
to be good candidates for such a function. Accordingly,
myofibrillar proteins still bound within the myofibrillar structure should be accessible to these proteinases. As shown in Table II, the susceptibility of
myofibrillar bound contractile proteins to proteolytic
attack is indeed the same as that of purified contractile proteins. These data are from digestion experiments where whole myofibrils were incubated with purified proteolytic enzymes and where digestion has been
monitored by SDS-polyacrylamide gel electrophoresis.

By means of light and electron microscopy, some
additional information can be obtained on the degradation of myofibrils. As summarized in Table III, it is
interesting to see that fragmentation of myofibrils into
single sarcomeres was observed in the presence of cathepsin B, Ca-activated proteinase or the alkaline serine
porteinase from heart muscle, although these enzymes do
not degrade α-actinin, the main component of the Z-disk.
In these cases, Z-line destruction is the result of
α-actinin being removed from the Z-line, without further
degradation of the protein molecule.

Table III Degradation of myofibrillar structures by muscle proteinases

	CATHEPSIN [32] B D L	CAF [25,26]	ALKAL.SERINE PROT. [25,26] SKELETAL	ALKAL.SERINE PROT. [34,35] HEART	NEUTRAL TRYPSIN LIKE PROT. [37]
Z-LINE					
- DESTRUCTION	+	+	+	+	+
- α-ACTININ RELEASE	+	+	+	+	
FRAGMENTATION OF MYO-FIBRILS INTO SACROMERES	+	+	-	+	
M-LINE					
- DESTRUCTION	+	±		+	+
THICK FILAMENT					
- A BAND DESTRUCTION	+	-	+		+
THIN FILAMENT					
- I BAND DESTRUCTION		-			+

In accordance with the susceptibility of purified contractile proteins, destruction of A- and I-bands as well as destruction of M-lines has been observed due to the action of cathepsin B, alkaline serine proteinase and intestinal trypsin-like proteinase. On the other hand, Ca-activated proteinase does not degrade myosin or actin and thus cannot destroy the A- and I-band within the myofibrils. Its degradative action on troponin only leads to a loss of periodicity of the thin filament[17].

If, in addition to the ability to attack whole myofibrils, the activity of these non-lysosomal proteinases is increased during certain conditions of enhanced muscle protein breakdown, their involvement in disassembling myofibrils and myofilaments could well be envisaged. As concerns the ATP-dependent proteinase, the high molecular weight proteinase, hydrolase H and also the smooth muscle trypsin-like and alkaline cysteinyl proteinase, no data on their possible change in activity during protein catabolic conditions have been published.

Table IV   Alterations of non-lysosomal proteinase activity during various conditions of enhanced muscle protein breakdown

PROTEINASE	INCREASE OF ENZYME ACTIVITY	
CA-ACTIVATED PROTEINASE [25,26]	DUCHENNE DYSTROPHY	MAN
	BECKER DYSTROPHY	MAN
	VITAMIN E DEFICIENCY	RABBIT
	HEREDITORY DYSTROPHY	MICE,HAMSTER
ALKALINE SERINE PROTEINASE [25,26,39]	DUCHENNE DYSTROPHY	MAN
	BECKER DYSTROPHY	MAN
	HEREDITORY DYSTROPHY	MICE
	DIABETES MELLITUS	RAT
	STARVATION	RAT
	TESTOSTERONE DEFICIENCY	RAT
	GLUCOCORTICOID TREATMENT	RAT
	ENDURANCE TRAINING	RAT

In contrast, and as summarized in Table IV, during several types of muscle dystrophy, an increased activity is observed for the Ca-activated proteinase as well as the alkaline serine proteinase. However, there are some situations of enhanced muscle protein breakdown like starvation, diabetes mellitus or after glucocorticoid treatment, where the Ca-activated proteinase is unchanged. Importantly, during all these conditions, the alkaline serine proteinase activity is increased.

Since, in conditions like diabetes mellitus and starvation, qualitative and quantitative alterations in the degradation of sacroplasmic proteins have been found, it was of interest to study whether the alkaline serine proteinase has the enzymatic properies to be responsible for this altered protein degradation.

As previously mentioned, Dice and coworkers[7] have found that in muscle tissue of diabetic and starved rats, normally long-lived proteins are degraded at the same rate as are normally short-lived proteins. In general, long-lived proteins are molecules with a basic isoelectric point, small subunits, low surface hydrophobicity and a low carbohydrate content. Conversely, short-lived cytoplasmic proteins are acidic molecules, have large subunits and a considerable surface-hydrophobicity and carbohydrate content[18]. Thus, in muscle tissue of diabetic rats the turnover rate of sarcoplasmic proteins with neutral and basic isoelectric point is increased dramatically[7].

On the basis of these findings cytoplasmic proteins isolated from rat skeletal muscle tissue were fractionated into one group comprising proteins with isoelectric points predominantly above pH 6 and into another group of proteins with isoelectric points essentially lower than pH 6. Further analysis of these groups of cytoplasmic proteins revealed that the neutral and basic proteins are composed of subunits with both a smaller average molecular weight and a significantly lower surface hydrophobicity when compared with the acidic cytoplasmic proteins. When subjecting the two protein groups to in vitro degradation by the alkaline serine proteinase, we observed that digestion of the neutral and basic proteins is at a significantly faster rate than digestion of the acidic proteins[19]. Since the activity of this alkaline serine proteinase is increased during several protein catabolic conditions and since the enzyme preferentially degrades small, basic, non-hydrophobic cytoplasmic proteins, it is not unlikely that this proteolytic enzyme is involved in the pathologically enhanced degradation of normally long-lived sarcoplasmic proteins.

However, a point of uncertainty bears on this notion, concerning the localization of the enzyme within muscle tissue. We have recently shown that the enzyme is localized in mast cells distributed throughout the muscle tissue[20]. Although preliminary results by Stauber (personal communication) indicate that the number of mast cells is increased in muscle tissue of diabetic rats and that mast cells are degranulating at various times after induction of diabetes, it is an open question whether the mast cell proteinase enters the muscle cell. It is furthermore unknown which proteinase assumes the function of this alkaline serine proteinase in species like rabbit, whose muscle tissue is reported to contain no mast cells.

At present it is unknown whether proteolytic enzymes from other cells within muscle tissue like fibroblasts, macrophages, epithelial cells, take part in the degradation of muscle cell proteins. Although these cells are probably not involved in the turnover of normal muscle cells, it cannot be excluded that proteinases originating from these cells are taken up by muscle cells under pathological conditions[21].

In conclusion, our present understanding of the mechanism of intramuscular protein degradation, both under normal conditions and during enhanced catabolism, is still unsatisfactory.

ACKNOWLEDGEMENT

This study was supported by the Ministerium für Wissenschaft und Forschung des Landes Nordrhein-Westfalen, Düsseldorf, by the Ministerium für Jugend, Familie und Gesundheit, Bonn and by the Deutsche Forschungsgemeinschaft (SFB 113), Bonn, F.R.G..

REFERENCES

1. E. Helander: On quantitative muscle protein determination. Acta Physiol.Scand. Vol. 41,Suppl.141 (1957).
2. F.M. Clarke and C.J. Masters: Interactions between muscle proteins and glycolytic enzymes. Minireview. Int.J.Biochem. 7 : 359 (1976).
3. J.R.Dedman, D.M.Payne and B.G.Harris: Increased proteolytic susceptibility of aldolase induced by actin binding. Biochem.Biophys.Res.Commun. 65:1170 (1975).
4. D.J.Millward: Protein turnover in skeletal muscle. I. Measurement of rates of synthesis and catabolism

of skeletal muscle protein using $[^{14}C]$ Na_2CO_3 to label protein. Clin.Sci.39: 577 (1970).

5. J.C. Waterlow, P.J. Garlick and D.J. Millward: Protein turnover in mammalian tissues and in the whole body. North-Holland Publishing Company, Amsterdam (1978).

6. D.J. Millward: Protein degradation in muscle and liver. in: Comprehensive Biochemistry. (M.Florkin, A.Neuberger and L.L.M.Van Deenan, eds.)Vol. 19 B (1), p. 153, Elsevier/North-Holland. Amsterdem(1980).

7. J.F.Dice, C.D.Walker, B.Byrne, and A.Cardiel: General characteristics of protein degradation in diabetes and starvation. Proc.Natl.Acad.Sci.USA 75:2093(1978).

8. C.J.Ritz-Gold, R.Cooke, D.K.Blumenthal and J.T.Stull: Light chain phosphorylation alters the conformation of skeletal muscle myosin. Biochem.Biophys.Res.Commun. 93: 209 (1980).

9. J.Borejdo: Magnesium adenosine 5'-diphosphate influences proteolytic susceptibility of myosin in myofibrils. Biochemistry 21 : 234 (1982).

10. K.Bárány, M.Bárány, J.M.Gillis and M.J.Kushmerick: Phosphorylation-dephosphorylation of the 18 000 dalton light chain of myosin during the contraction-relaxation cycle of frog muscle. J.Biol.Chem. 254: 3617 (1979).

11. R.S. Adelstein and E. Eisenberg: Regulation and kinetics of the acto-myosin-ATP interaction. Ann. Rev.Biochem. 49 : 921 (1980).

12. F.K.Baskin, W.C.Duckworth and A.E.Kitabchi: Site of cleavage of glucagon by insulin-glucagon protease. Biochem.Biophys.Res.Commun. 67 : 163 (1975).

13. W.N.Schwartz and J.W.C.Bird: Degradation of myofibrillar proteins by cathepsin B and D. Biochem.J. 167 : 811 (1977).

14. A.Okitani, U.Matsukura, H,Kato and M.Fujimaki: Purification and some properities of a myofibrillar protein-degrading protease, cathepsin L, from rabbit skeletal muscle. J.Biochem. 87 : 1133 (1980).

15. A.R.Poole and J.S.Mort: Biochemical and immunological studies of lysosomal and related proteinases in health and disease. J.Histochem.Cytochem.29: 494 (1981).

16. K.W.Gerard and D.L.Schneider: Evidence for degradation of myofibrillar proteins in lysosomes. J.Biol. Chem. 254: 11798 (1979).

17. W.R.Dayton, D.E.Goll, M.G.Zeece, R.M.Robson and W.J.Reville: A Ca^{2+}-activated protease possibly involved in myofibrillar protein turnover. Purification from porcine muscle. Biochemistry 15: 2150 (1976).

18. F.J.Ballard: Intracellular protein degradation. Essays Biochem. 13 : 1 (1977).

19. B.Dahlmann, L.Kuehn and H.Reinauer: Susceptibility of muscle cytosolic proteins to muscle alkaline proteinase (chymase). Hoppe Seyler's Z.physiol.Chem. 363 : 979 (1982).

20. W.T.Stauber, B. Dahlmann, V.Fritz and H.Reinauer: Immunohistochemical localization of two myofibrillar degrading proteinases in skeletal muscle. In preparation.

21. R.Libelius, I.Lundquist, S.Tågerud and S.Thesleff: Endocystosis and lysosomal enzyme activities in dystrophic muscle: The effect of denervation. Acta Physiol.Scand. 113 : 259 (1981).

22. A.Okitani, T.Matsumoto, Y.Kitamura and H.Kato: Purification of cathepsin D from rabbit skeletal muscle and its action towards myofibrils. Biochim.Biophys. Acta 662 : 202 (1981).

23. E.G.Afting and M.L.Becker: Two-step affinity-chromatographic purification of cathepsin D from pig myometrium with high yield. Biochem.J. 197 : 519 (1981).

24. E.F.Ogunro, R.B.Lanman, J.R.Spencer, A.G.Ferguson and M. Lesch: Degradation of canine myosin and actin by cathepsin D isolated from homologous tissue. Cardiovasc.Res. 13: 621 (1979).

25. for references see: T.Obinata, K.Maruyama, H.Sugita, K.Kohama and S. Ebashi: Dynamic aspects of structural proteins in vertebrate skeletal muscle. Muscle & Nerv 4 : 456 (1981).

26. for references see: J.W.C.Bird and J.H.Carter: Proteolytic enzymes in striated and non-striated muscle. in:Degradative processes in heart and skeletal muscle (K.Wildenthal, ed.) Chapt 3, p. 51, Elsevier/North-Holland, Amsterdam (1980).

27. A.Okitani, T.Nishimura and H.Kato: Characterization of hydrolase H, a new muscle protease possessing aminoendopeptidase activity. Eur.J.Biochem. 115 : 269 (1981).

28. M.F.Hardy, D.Mantle, T.Edmunds and R.J.T. Pennington: A high molecular-weight enzyme from skeletal muscle which hydrolyses chymotrypsin substrates. Biochem. Soc.Trans. 9 : 218 (1981).

29. W.C.Duckworth, M.A.Heinemann and A.E.Kitabchi: Purification of insulin-specific protease by affinity-chromatography. Proc.Natl.Acad.Sci.USA 69:3698 (1972).

30. K.Yokono, Y.Imamura, K.Shii, H.Sahai and S.Baba: Purification and characterization of insulin-degrading enzyme from pig skeletal muscle. Endocrinology 108 : 1527 (1981).

31. J.D.Etlinger, S.Speiser, E.Wajnberg, M.J.Glucksman: ATP-dependent proteolysis in erythroid and muscle cells. Acta biol.med.germ. 40 : 1285 (1981).

32. T.Noda, K.Isogai, H.Hayashi and N. Katunuma: Susceptibilities of various myofibrillar proteins to cathepsin B and morphological alterations of isolated myofibrils by this enzyme. J. Biochem. 90 : 371 (1981).

33. U.Matsukura, A.Okitani, N.Nishimuro and H. Kato: Mode of degradation of myofibrillar proteins by an endogenous protease, cathepsin L. Biochim.Biophys. Acta 662 : 41 (1981).

34. U.Murakami and K.Uchida: Degradation of rat cardiac myofibrils and myofibrillar proteins by a myosin-cleaving protease. J.Biochem. 86 : 553 (1979).

35. T.Kuo, F.Giacomelli, K.Kithier and A.Malhotra: Biochemical characterization and cellular localisation of serine protease in myopathic hamster. J.Mol.Cell. Cardiol. 13 : 1035 (1981).

36. J.Kay, R.F.Siemanowski, L.M.Siemanowski and D.E.Goll: Degradation of smooth muscle myosin by trypsin-like serine proteinases. Biochem.J. 201 : 267 (1982).

37. J.Kay, L.M.Siemanowski, R.F.Siemanowski,J.A.Greweling and D.E.Goll: Degradation of myofibrillar proteins by trypsin-like serine proteinases. Biochem.J. 201 : 279 (1982).

38. E.G.Afting and M.Roth: A novel alkaline proteinase associated with the actomyosin complex from pig uterine myometrium. Hoppe Seyler's Z.physiol.Chem. 362 : 453 (1981).

39. for references see: B. Dahlmann, L. Kuehn, M. Rutschmann and H. Reinauer: Studies on the alkaline proteinase system in rat skeletal muscle. Acta biol. med.germ. 40 : 1349 (1981).

Ca^2 – ACTIVATED PROTEINASES, PROTEIN DEGRADATION AND MUSCULAR DYSTROPHY

John Kay

Department of Biochemistry, University College
P.O.Box 78,
Cardiff CF1 1XL, Wales, U.K.

Progressive muscular dystrophies in man are characterised by muscle degeneration and weakness. A gradual loss of muscle tissue takes place until, in the late stages of most dystrophies (and particularly in the X-linked Duchenne type), very little functional muscle remains. During necrosis, there is a marked loss of both myofibrillar and soluble proteins from the muscle fibres. While the underlying causes may in part be related to membrane defects (1), alterations in protein turnover in the muscle tissue must be involved in the development of the condition to some extent. The level of protein in any tissue is determined, under steady state conditions, as a balance between the rate of protein synthesis and its rate of degradation. The observed loss of muscle mass in wasting disorders would appear then to be explained most readily by an increase in the rate of protein degradation and, indeed, such measurements have been made in patients with Duchenne dystrophy (2). However, the same net effect i.e. muscle degeneration, can be achieved as a result of a diminished rate of protein synthesis in muscle and very recently, measurements made in vivo with dystrophic boys have suggested that in all types of dystrophy, muscle protein synthesis is depressed and that there is little change in protein breakdown from the rates observed with unafflicted male patients (3).

Nevertheless, a great deal of attention has been focussed on the degradative aspects of muscle protein turnover especially since the mechanisms by which protein breakdown is regulated in skeletal muscle are poorly understood. The highest concentration of proteinases in most cell types is found within the lysosomes and these hydrolytic enzymes are also present within the lysosomes of myocytes. However, skeletal muscle is composed of at least five different types of cell and the lysosomal content of some of these (e.g. fibroblasts and macrophages) is considerably higher than that of the myocytes. Thus, the increased levels of lysosomal cathepsins B & D that have been measured in dystrophic tissue may be due largely to infiltrating non-myocytic cells. In polymyositis and other inflammatory myopathies, the non-myocytic cells are relatively numerous and probably do make a significant contribution to the total hydrolytic activity in muscle affected with these disorders.

However, ultrastructural studies have indicated that the lysosomes endogenous to myocytes do not appear to engulf whole myofibrils nor intact thick or thin filaments even during rapid muscle atrophy. Individual myofibrillar proteins have considerably different half-lives (4) and a non-uniform turnover of these proteins is apparently inconsistent with a lysosomal route of catabolism of intact myofibrils. Recent evidence (5,6,7) has been obtained with specific lysosomotropic agents to indicate that while the degradation of "general" proteins and certain cell-surface receptor proteins (e.g. the acetyl-choline receptor (8)) is inhibited to some extent by these compounds, the breakdown of myofibrillar proteins is not affected. Thus, it seems unlikely that the lysosomal proteinases are involved in the initial, rate-limiting steps of myofibrillar protein degradation. It is more probable that these all-important reactions are initiated within the sarcoplasm by neutral, i.e. non-lysosomal proteinases, and then the partially-degraded fragments can be completely broken down to amino acids after uptake into the lysosomal compartment(4).

A number of tissue endopeptidases active at neutral pH have been characterised in vitro and implicated in protein catabolism in living tissues. These include:-
a) the ATP-dependent proteinases of reticulocytes, liver, heart and skeletal muscle (9,10,11)
b) the chymotrypsin-like proteinase from rat mast cells - this contributes most of the neutral-alkaline proteolytic activity measured in rat muscle homogenates (12) but since it has been shown to originate from tissue mast cells (13), its role in myofibrillar protein degradation must be restricted to conditions where there is an increased penetration of mast cells into muscle.
c) the trypsin-like serine proteinase type of activity (14). This is particularly active towards native proteins (15,16) and does not originate within mast cells.
d) the Ca^{2+} - activated cysteine proteinases.

Since many biological responses are known to be triggered by an influx of Ca^{2+}, including an increase in cellular protein degradation in isolated muscle fibres (made permeable in vitro with the ionophore, A23187-17,18), considerable interest has been aroused in a proteolytic system that was apparently able to be switched on only in response to increased levels of calcium. Increased leakiness of the muscle membranes to Ca^{2+} might, then, permit a role for such an enzyme in enhancing the disassembly of myofibrillar proteins in dystrophic conditions as well as in post-mortem autolysis. It has been suggested that elevated levels of this type of activity can be detected in (human) dystrophic and (rabbit) atrophying muscle (19,20).

A Ca^{2+}-activated proteinase was first purified completely from pig skeletal muscle (21) and was found to have little effect on most of the contractile proteins of muscle. Its action seemed to be restricted to digesting only troponins I & T and releasing α-actinin (intact) from the Z-lines of muscle (22). Similar enzymes have now been isolated from virtually every cell type known e.g. brain, liver, heart, kidney, skeletal, smooth and cardiac muscles, blood cells including erythrocytes and platelets, oviduct, nerve tissue, spinal cord and even tumour cells, and several names have been coined to describe this Ca^{2+}-activated type of activity, including CAF (21,23), CANP (24) and calpain(25).

Within the last two years, a second form of the enzyme has been detected (26) and this has also now been purified from a variety of tissues including cardiac (23) and skeletal (27) muscles, liver (28), brain and enythrocytes (25,29). The major operational distinction between the two forms of the enzyme is that, while the original form required high (millimolar) concentrations of Ca^{2+} for maximal activity, the new form is activated by much lower (i.e. micromolar) concentrations of Ca^{2+}.

Both forms of the enzyme appear to be present in most tissues although the relative amounts vary considerably (25) and some cells such as human erythrocytes seem to contain only the low (Ca^{2+}- requiring) enzyme (30,31) while it has been reported that the Ca^{2+}- dependent activity from human skeletal muscle can be resolved into four peaks of proteolytic activity (32).

The high and low Ca^{2+}- requiring forms are readily extracted from the tissues by low ionic strength buffers without the need to resort to detergents and so they cannot be truly intrinsic membrane proteins. Nevertheless, both forms are adsorbed strongly to hydrophobic matrices such as phenyl-Sepharose (23)and so it seems likely that they will be strongly associated with membranes in the tissues.

The two forms have similar properties (Table 1):- both are active optimally at pH 7.5, both are inhibited by sulphydryl modifying reagents and exogenous inhibitors of cysteine proteinases and both appear to consist of two subunits - a large 80,000 dalton subunit which contains the active site cysteine residue (23,33) and a second catalytically inactive 30,000 dalton subunit.

Table 1. The properties of Ca^{2+} - activated proteinases

Effector or Property	Low Ca^{2+} Form	High Ca^{2+} Form
pH optimum	7.5	7.5
Subunits	80,000 & 30,000	80,000 & 30,000
Charge on electrophoresis at pH 7.5	Negative	More Negative
EDTA	Inhibited	Inhibited
Ca^{2+}(50μM)	Activated	No effect
Ca^{2+}(5mM)	Activated	Activated
Iodoacetamide	Inhibited	Inhibited
p-OH-Mercuribenzoate	Inhibited	Inhibited
Leupeptin	Inhibited	Inhibited
Mersalyl	Inhibited	Inhibited
E-64-C	Inhibited	Inhibited
Ep 475	Inhibited	Inhibited

The function of the latter is not clear at present although it has been suggested to be essential for the expression of the full activity intrinsic to the 80,000 subunit (24,34). The enzymes purified from chicken and rabbit skeletal muscles have been reported to contain only the 80,000 subunit (35,36,37). While this subunit is undoubtedly responsible for the catalytic activity, it is possible that the 30,000 subunits became diss-ociated during chromatographic purification steps that were carried out at high salt concentrations. With this reservation, it is noteworthy that both forms of the enzyme always appear to have identical subunit constructions irrespective of the tissue or species of origin. An antiserum raised against the 80,000 subunit from the high form from pig muscle was shown to cross-react with the 80,000 subunit from the low form (27). The 30,000 subunit appears to be identical in both forms (38).

Apart from the altered Ca^{2+} requirements, the only apparent difference between the high and low forms of the enzyme is that the intact high form has a greater negative charge on electro-phoresis of the native proteins at pH 7.5. The biochemical nature of these differences is not known but the possibility has been considered that post-translational modification e.g. phosphorylation/dephosphorylation, might permit interconversion between the forms and account for the different electrophoretic mobilities and Ca^{2+}-requirements. In addition, a number of authors have observed that limited autolysis of the high form of the enzyme in the presence of millimolar concentrations of Ca^{2+},

generates an altered form of the enzyme that is activated by micromolar Ca^{2+} (34,39,40,41). However, it now seems clear that this
"autolysed" form is not the same as the low form of the enzyme
which occurs naturally and is isolated from the tissues and so this
form may be an artefact resulting from the in vitro procedure. At
present, then, there is no known means whereby high form can be
converted into the low-Ca^{2+} requiring form and, indeed, it may be
that the two enzymes are derived from different cellular locations
(see below) and need not be interconvertible in vivo. Recently, it
has been suggested on the basis of peptide mapping and immunoauto-
radiography (38) that while both forms do have an identical 30,000
subunit, the two catalytically active subunits, although both of
apparently identical molecular weight, in fact are substantially
different. This new proposal merits further investigation,
particularly since it has frequently been suggested that the high
-Ca^{2+} requiring enzyme is an "off" form (since the intracellular
Ca^{2+} concentration can never reach millimolar concentrations) and
that activation to the low form would switch on the enzyme
responsive to the micromolar levels of Ca^{2+} that can be attained
intracellularly.

This presumes, of course, that the enzymes are intracellular
but there is an additional factor which also has to be taken into
account. In common with most other proteolytic systems (42),
wherever a proteinase is found in nature, there too exists a
specific inhibitor of that enzyme to regulate its function. Thus,
an endogenous inhibitor (sometimes referred to as calpastatin (25))
of the Ca^{2+}-activated proteinase has been purified from a number
of tissues and characterised. In contrast to the uniformity found
among the proteinases, the inhibitor molecules vary considerably
(Table 2) depending on the tissue and species of origin. Overall
molecular weights differ and while some inhibitors are monomers,
others are oligomeric proteins which may undergo dissociation
on interaction with the proteinase (43). All, however, are
remarkably heat (and acid) stable proteins and contain few
hydrophobic amino acid residues but appear to be very susceptible
to degradation by other types of proteinase (e.g. trypsin).

Each inhibitor is specific, then, in interacting only with
Ca^{2+}-dependent proteinases and Ca^{2+} is essential for inhibition
to occur. Thus, the low form is inhibited at low concentrations
of Ca^{2+} while the high form is affected only at millimolar
concentrations of Ca^{2+}(23). The inhibitors do not act by
sequestration of Ca^{2+} but interact directly with the proteinase
once Ca^{2+} has been introduced. It has been suggested for the red
blood cell proteinase that binding of the tetrameric inhibitor
dissociates not only the inhibitor into 60,000 monomers but
dissociates the 30,000 subunit from the enzyme leaving an
inhibited enzyme complex of mol.wt. 140,000 (43).

Table 2   Endogenous Inhibitors of Ca^{2+}-activated
          proteinases

Tissue of Origin	Approx. Mol.wt.	Subunit size	Reference
Bovine Cardiac Muscle	100,000	100,000	23
Rabbit Skeletal "	67,000	?	44
" " "	68,000	34,000	45
Chicken Skeletal "	68,000	68,000	46
Rat Liver	280,000	70,000	25
" Brain	300,000	?	25
Human Erythrocytes	240,000	60,000	30,43

This implies that, if the enzyme and inhibitor have ready access to each other in vivo, the proteinase can never be active since as soon as Ca^{2+} is released, the interaction with inhibitor will be promoted (binding is fairly tight-K_i approx.10^{-8}M; and each inhibitor molecule may be able to bind more than one molecule of enzyme 23,44,45). The enzyme is then unable to begin its attack on substrate proteins. No information is, as yet, available on the cellular location of the inhibitor while conflicting reports have appeared describing immunofluorescent localisation studies on the enzyme. In one, it was concluded that the muscle enzyme was located on the extracellular face of the cell membrane (47) while in the other, the enzyme was considered to be located inside striated muscle cells at the Z-lines of myofibrils and at the cytoplasmic surface of the cell membrane (48). Clearly, this dichotomy of opinion requires further examination at the electron microscope level. However, the reports may not be incompatible in that it may be possible that the low form of the enzyme is located on the internal face of the membrane where micromolar levels of Ca^{2+} would be available periodically while the high Ca^{2+} - requiring form is situated on the external side of the membrane, facing into millimolar concentrations of Ca^{2+}. Indeed, it has been suggested previously that this enzyme may be involved in the processing of hormone receptor proteins (49,50).

Other substrate proteins that have been found to be susceptible to cleavage by Ca^{2+}-activated proteinases include neurofilament proteins, C-protein, desmin, filamin and vimentin (51,52,53,54,55) but not actin, myosin nor α-actinin (16). Thus, this enzyme has a very restricted proteolytic activity and even those proteins that are susceptible are cleaved only to large polypeptide fragments and not to small peptides or amino acids. The enzyme has little or no activity towards synthetic peptide substrates. It has been reported that the requirement for Ca^{2+} may not be absolute in some cases. The rabbit muscle enzyme is activated to a small extent by Mg^{2+} (37) while Mn^{2+} can apparently act synergistically to lower the Ca^{2+} requirement of the high form from chicken muscle(56).

These observations may involve interaction between the metal ion and the substrate proteins used rather than a direct stimulation of the protein itself.

The lack of "general" proteolytic activity is likely to be a reflection of the unusual and restricted specificity of this type of enzyme and suggests that Ca^{2+}-activated proteinases may act physiologically to accomplish a specific function rather than as general proteinases involved in initiating the degradation of groups of proteins during their metabolic turnover. The serine proteinases e.g. the trypsin-like enzyme (14,15,16) and the chymotrypsin-like enzymes (12) have been shown to be readily able to degrade most myofibrillar proteins including actin and myosin and so they (together with the lysosomal proteinases) would appear to be much better equipped to carry out this cellular function.

Indeed, metabolic investigations have indicated that the Ca^{2+}-stimulated increase in "general" protein degradation observed in intact muscles flooded with Ca^{2+} (courtesy of A 23187) in vivo (57) and in vitro (18) cannot be prevented by treatment with mersalyl- a good inhibitor of Ca^{2+}-activated proteinases (Table 1) which can apparently penetrate the muscle cell without entering the lysosomal compartment. Thus, while it must be borne in mind that these experiments examined total cell protein breakdown and not specifically that of the myofibrillar component (see earlier), nevertheless, the data serve as a warning that just because a proteinase that is activated by Ca^{2+} has been identified in a tissue in which Ca^{2+} also stimulates protein breakdown, it is not safe to assume that the one is causally related to the other.

It has been suggested that the increase in the breakdown of muscle proteins stimulated by elevated Ca^{2+} levels may be brought about indirectly by activation of a Ca^{2+}-sensitive phospholipase (18) although this proposal has not met with universal agreement (11). The phospholipase, in turn may act on membrane phospholipids to release free fatty acids including arachidonic acid which gives rise to elevated levels of the prostaglandin, PGE_2. This has been shown to stimulate lysosomal proteolysis (58). Thus, the balance of lysosomal/non-lysosomal protein breakdown in muscle (and in other tissues) is still a matter of controversy.

However, the involvement of Ca^{2+}-activated proteinases with their widespread distribution and poor "general" proteolytic activity in initiating turnover of myofibrillar proteins is becoming increasingly uncertain. The effects of the enzyme in platelets (59,60) and smooth muscle cells indicate that it may act to detach actin filaments from their intracellular attachments whether these are Z-lines in striated muscle or attachment plaques in other cell types. This, in turn, may have effects on cell shape, mobility, the secretion of specialised intracellular organelles, aggregation, etc. These alterations together with studies on changes in membrane proteins and neurofilament proteins are under investigation in many different laboratories (see 61). In terms of muscular dystrophy, it is not known currently whether the level of the proteinase is elevated or the amount of the inhibitor is diminished in dystrophic muscle in comparison to normal tissue. It has been observed that the relative proportions of the low and high forms of the enzyme and the inhibitor are considerably different in various rat tissues (25). The present evidence suggests that the expression of the Ca^{2+}-dependent activity can be controlled in three ways:-

   1) by its limited specificity which restricts its action
   2) by an endogenous inhibitor which is capable of removing the activity completely and
   3) by the concentration of Ca^{2+} that is available to the two forms of the enzyme.

Its role in the pathogenesis of wasting and other (e.g. neoplastic) diseases is probably still worthy of further investigation.

ACKNOWLEDGEMENTS

   Supported by grants from the SERC, ARC & Muscular Dystrophy Group of Great Britain. It is a pleasure to acknowledge the contributions made by Jane Greweling & Dorothy Fairhurst and the excellent secretarial assistance of Barbara Power.

REFERENCES

1. J. A. Lucy, Is there a membrane defect in muscle and other cells? Br. Med. Bull. 36:187 (1980).
2. R. O. McKeran, D. Halliday, and P. Purkiss, Increased myofibrillar protein catabolism in duchenne muscular dystrophy measured by 3-methyl histidine excretion in the urine, J. Neurol. Neurosurg. Psychiatr. 40:979 (1977).
3. M. J. Rennie, R. H. T. Edwards, D. J. Millward, S. L. Wolman, D. Halliday, and D. E. Matthews, Effects of duchenne muscular dystrophy on muscle protein synthesis, Nature 296:165 (1982).
4. J. Kay, Intracellular protein degradation, Biochem. Soc. Trans. 6:789 (1978).
5. J. F. Riebow, and R. B. Young, Effects of leupeptin on protein turnover in normal and dystrophic chicken skeletal muscle cells in culture, Biochem. Med. 23:316 (1980).
6. K. Wildenthal, J. R. Wakeland, J. M. Ord, and J. T. Stull, Interference with lysosomal proteolysis fails to reduce cardiac myosin degradation, Biochem.Biophys. Res. Commun. 96: 793 (1980).
7. P. J. Bates, G. A. Coetzee, and D. R. van der Westhuyzen, The degradation of endogenous and exogenous proteins in cultured

smooth muscle cells. Biochim. Biophys. Acta. 719:377 (1982).

8. P. Libby, and A. L. Goldberg, The control and mechanism of protein breakdown in striated muscle: studies with selective inhibitors, in: "Degradative Processes in Heart and Skeletal Muscle," K. Wildenthal, ed., Elsevier/North Holland (1980).

9. A. L. Goldberg, N. P. Strnad, and K. H. S. Swamy, Studies of the ATP dependence of protein degradation in cells and cell extracts, in: "Protein Degradation in Health and Disease," D. C. Evered, J. Whelan, eds., CIBA Found. Symp. 75, Excerpta Medica, Amsterdam (1980)..

10. A. L. Goldberg, and F. S. Boches, Oxidized proteins in erythrocytes are rapidly degraded by the adenosine triphosphate-dependent proteolytic system, Science 215:1107 (1982).

11. S. Rapoport, W. Dubiel, and M. Muller, Characteristics of an ATP-dependent proteolytic system of rat liver mitochondria, FEBS. Lett. 147:93 (1982).

12. B. Dahlmann, L. Kuehn, M. Rutschmann, I. Block, and H. Reinauer, Characterisation of two alkaline proteinases from rat skeletal muscle, in: "Proteinases and their Inhibitors, Structure, Function & Applied Aspects," V. Turk, Lj. Vitale, eds., Pergamon Press, Oxford (1981).

13. T. Edmunds, and R. J. T. Pennington, Mast cell origin of "Myofibrillar Protease" of rat skeletal and heart muscle, Biochim.Biophys. Acta 661:28 (1981).

14. R. J. Beynon, and J. Kay, The interactivation of native enzymes by a neutral proteinase from rat intestinal muscle, Biochem. J. 173:291 (1978).

15. J. Kay, R. F. Siemankowski, L. M. Siemankowski, and D. E. Goll, Degradation of smooth muscle myosin by trypsin-like serine proteinases, Biochem. J. 201:267 (1982).

16. J. Kay, L. M. Siemankowski, R. F. Siemankowski, J. A. Greweling, and D. E. Goll, Degradation of myofibrillar proteins by trypsin-like serine proteinases, Biochem. J. 201:279 (1982).

17. P. H. Sugden, The effects of calcium ions, ionophore A23187 and inhibition of energy metabolism on protein degradation in the rat diaphragm and epitrochlearis muscles in vitro, Biochem. J. 190:593 (1980).

18. H. P. Rodemann, L. Waxman, and A. L. Goldberg, The stimulation of protein degradation in muscle by Ca^{2+} is mediated by prostaglandin E_2 and does not require the calcium-activated protease, J. Biol. Chem. 257:8716 (1982).

19. N. C. Kar, and C. M. Pearson, A calcium-activated neutral protease in normal and dystrophic human muscle, Clin. Chim. Acta 73:293 (1976).

20. W. R. Dayton, J. V. Schollmeyer, A. C. Chan, and C. E. Allen, Elevated levels of a calcium-activated muscle protease in rapidly atrophying muscles from vitamin E-deficient rabbits, Biochim. Biophys. Acta 584:216 (1977).

21. W. R. Dayton, D. E. Goll, M. G. Zeece, R. M. Robson, and W. J. Reville, A Ca^{2+}-activated protease possibly involved in myofibrillar protein turnover. Purification from porcine muscle, Biochemistry 15:2150 (1976).

22. W. R. Dayton, W. J. Reville, D. E. Goll, and M. H. Stromer, A Ca^{2+}-activated protease possibly involved in myofibrillar protein turnover. Partial characterization of the purified enzyme, Biochemistry 15:2159 (1976).

23. A. Szpacenko, J. Kay, D. E. Goll, and Y. Otsuka, A different form of the Ca^{2+}-dependent proteinase activated by micromolar levels of Ca^{2+}, in: "Proteinases and their Inhibitors, Structure, Function and Applied Aspects," V. Turk, Lj. Vitale, eds., Pergamon Press, Oxford (1981).

24. S. Tsuji, and K. Imahori, Studies on the Ca^{2+}-activated neutral proteinase of rabbit skeletal muscle. The characterization of the 80K and the 30K subunits, J. Biochem. 90:233 (1981).

25. T. Murachi, K. Tanaka, M. Hatanaka, and T. Murakami, Intracellular Ca^{2+}-dependent protease (Calpain) and its high molecular weight endogenous inhibitor (Calpastatin), Adv. in Enz. Reg. 19:407 (1981).

26. R. L. Mellgren, Canine cardiac calcium-dependent proteases: Resolution of two forms with different requirements for calcium, FEBS Lett 109:129 (1980).

27. W. R. Dayton, J. V. Schollmeyer, R. A. Lepley, and L. R. Cortes, A calcium-activated protease possibly involved in myofibrillar protein turnover. Isolation of a low-calcium-requiring form of the protease, Biochim. Biophys. Acta 659:48 (1981).

28. G. N. de Martino, Calcium-dependent proteolytic activity in rat liver: Identification of two proteases with different calcium requirements, Arch. Biochem. Biophys. 211:253 (1981).

29. A. Kishimoto, N. Kajikawa, H. Tabuchi, M. Shiota, and Y. Nishizuka, Calcium-dependent neutral proteases, widespread occurrence of a species of protease active at lower concentrations of calcium, J. Biochem. 90:889 (1981).

30. T. Murakami, M. Hatanaka, and T. Murachi, The cytosol of human erythrocytes contains a highly Ca^{2+}-sensitive thiol protease (Calpain I) and its specific inhibitor protein (Calpastatin), J. Biochem. 90:1809 (1981).

31. E. Melloni, B. Sparatore, F. Salamino, M. Michetti, and S. Pontremoli, Cytosolic calcium dependent proteinase of human erythrocytes: Formation of an enzyme-natural inhibitor complex induced by Ca^{2+} ions, Biochem. Biophys. Res. Commun. 106:731 (1982).

32. M. F. Hardy, D. Mantle, and R. J. T. Pennington, Calcium-activated proteinases in human skeletal muscle, Biochem. Soc. Trans. 9:219 (1981).

33. S. Tsuji, S. Ishiura, M. Takahashi-Nakamura, T. Katamoto, K. Suzuki, and K. Imahori, Studies on a Ca^{2+}-activated neutral proteinase of rabbit skeletal muscle. II. Characterization of sulfhydryl groups and a role of Ca^{2+} ions in this enzyme, J. Biochem. 90:1405 (1981).

34. K. Suzuki, S. Tsuji, S. Kubota, Y. Kimura, and K. Imahori, Limited autolysis of Ca^{2+}-activated neutral protease (CANP) changes its sensitivity to Ca^{2+} ions, J. Biochem. 90:275 (1981).

35. S. Kubota, K. Suzuki, and K. Imahori, A new method for the preparation of a calcium activated neutral protease highly sensitive to calcium ions, Biochem. Biophys. Res. Commun. 100: 1189 (1981).

36. S. Ishiura, H. Murofishi, K. Suzuki, and K. Imahori, Studies of a calcium-activated neutral protease from chicken skeletal muscle. I. Purification and characterisation, J. Biochem. 84:225 (1981).

37. J. L. Azanza, J. Raymond, J-M. Robin, P. Cottin, and A. Ducastaing, Purification and some physico-chemical and enzymic properties of a calcium ion-activated neutral proteinase from rabbit skeletal muscle, Biochem. J. 183:339 (1979).

38. M. J. Wheelock, Evidence for two structurally different forms of skeletal muscle Ca^{2+}-activated protease, J. Biol. Chem. 257:12471 (1982).

39. K. Suzuki, S. Tsuji, S. Ishiura, Y. Kimura, S. Kubota, and K. Imahori, Autolysis of calcium-activated neutral protease of chicken skeletal muscle, J. Biochem. 90:1787 (1981).

40. D. R. Hathaway, D. K. Werth, and J. R. Haeberle, Limited autolysis reduces the Ca^{2+} requirement of a smooth muscle Ca^{2+}-activated protease, J. Biol. Chem. 257:9072 (1982).

41. R. L. Mellgren, A. Repetti, T. C. Muck, and J. Easly, Rabbit skeletal muscle calcium-dependent protease requiring millimolar Ca^{2+} purification, subunit structure, and Ca^{2+}-dependent autoproteolysis, J. Biol. Chem. 257:7203 (1982).

42. J. Kay, Proteolysis-A degrading business but food for thought, Biochem. Soc. Trans. 10:277 (1982).

43. E. Melloni, B. Sparatore, F. Salamino, M. Michetti, and S. Pontremoli, Cytosolic calcium dependent neutral proteinase of human erythrocytes: the role of calcium ions on the molecular and catalytic properties of the enzyme, Biochem. Biophys. Res. Commun. 107:1053 (1982).

44. P. Cottin, P. L. Vidalenc, and A. Ducastaing, Ca^{2+}-dependent association between a Ca^{2+}-activated neutral proteinase (CaANP) and its specific inhibitor, FEBS Lett. 136:221 (1981).

45. M. Takahashi-Nakamura, S. Tsuji, K. Suzuki, and K. Imahori, Purification and characterization of an inhibitor of calcium-activated neutral protease from rabbit skeletal muscle, J. Biochem. 90:1583 (1981).

46. S. Ishiura, S. Tsuji, H. Murofushi, and K. Suzuki, Purification of an endogenous 68000-dalton inhibitor of a Ca^{2+}-activated neutral protease from chicken skeletal muscle, Biochim. Biophys. Acta 701:216 (1982).

47. R. Barth, and J. S. Elce, Immunofluorescent localization of a Ca^{2+}-dependent neutral protease in hamster muscle, Am. J. Physiol. 240:E493 (1981).

48. W. R. Dayton, and J. V. Schollmeyer, Immunocytochemical localization of a calcium-activated protease in skeletal muscle cells, Exper. Cell. Res. 136:423 (1981).

49. G. A. Puca, E. Nola, V. Sica, and F. Bresciani, Estrogen binding proteins of calf thymus, molecular and functional characterization of the receptor transforming factor: a Ca^{2+}-activated protease, J. Biol. Chem. 252:1358 (1977).

50. W. V. Vedeckis, M. R. Freeman, W. T. Schrader, and B. W. O' Malley, Progesterone-binding components of chick oviduct: partial purification and characterization of a calcium-activated protease which hydrolyzes the progesterone receptor, Biochemistry, 19:335 (1980).

51. N. C. Collier, and K. Wang, Purification and properties of human platelet P235. A high molecular weight protein substrate of endogenous calcium-activated protease(s), J. Biol. Chem. 257:6937 (1982).

52. S. M. Tahara, and J. A. Traugh, Differential activation of two protease-activated protein kinases from reticulocytes by a Ca^{2+}-stimulated protease and identification of phosphorylated translational components, Eur. J. Biochem. 126:395 (1982).

53. T. Tashiro, and Y. Ishizaki, A calcium-dependent protease selectively degrading the 160 000 M_r component of neurofilaments is associated with the cytoskeletal preparation of the spinal cord and has an endogenous inhibitory factor, FEBS Lett. 141:41 (1982).

54. U. J. P. Zimmerman, and W. W. Schlaepfer, Characterization of a brain calcium-activated protease that degrades neurofilament proteins, Biochemistry 21:3977 (1982).

55. W. J. Nelson, and P. Traub, Purification and further characterization of the Ca^{2+}-activated proteinase specific for the intermediate filament proteins vimentin and desmin, J. Biol. Chem. 257:5544 (1982).

56. K. Suzuki, and S. Tsuji, Synergistic activation of calcium-activated neutral protease by Mn^{2+} and Ca^{2+}, FEBS Lett. 140:16 (1982).

57. K. W. Gerard, and D. L. Schneider, Protein turnover in muscle: Inhibition of the calcium activated proteinase by mersalyl without inhibition of the rate of protein degradation, Biochem.Biophys. Res. Commun. 94:1353 (1980).

58. H. P. Rodemann, and A. L. Goldberg, Arachidonic acid, prosta-
     glandin E_2 and $F_2\alpha$ influence rates of protein turnover in
     skeletal and cardiac muscle, J. Biol. Chem. 257:1632 (1982).
59. J. A. Truglia, and A. Stracher, Purification and characteriz-
     ation of a calcium dependent sulfhydryl protease from human
     platelets, Biochem. Biophys. Res. Commun. 100:814 (1981).
60. S. Rosenberg, A. Stracher, and R. C. Lucas, Isolation and
     characterization of actin and actin binding protein from
     human platelets, J. Cell. Biol. 91:201 (1981).
61. S. Ishiura, Calcium-dependent proteolysis in living cells,
     Life Sci. 29:1079 (1981).

MUSCLE CATHEPSIN D ACTIVITY, AND RNA, DNA AND PROTEIN CONTENT

IN MAINTENANCE HEMODIALYSIS PATIENTS

G. F. Guarnieri, G. Toigo, R. Situlin, L. Faccini,
R. Rustia, and F. Dardi
With the technical assistance of M. Crevatin

Institute of Medical Pathology, University of Trieste
Italy

INTRODUCTION

In man, skeletal muscle is the largest protein pool in the body. Muscle losses in response to undernutrition, inactivity, endocrine abnormalities, trauma, severe infection, fever etc. can be very extensive, as a consequence of reduced muscle protein synthesis and/or increased degradation[1-8]. The rate of muscle protein degradation can paradoxically increase or decrease during wasting, and its changes are accompanied by changes of proteinase activity in muscle[9].

Evidence of muscle protein wasting and abnormal muscle metabolism is common in uremia[10-22], including patients treated with maintenance hemodialysis[23-31]. In the present investigation we studied muscle metabolism in chronically uremic patients on regular hemodialysis treatment, by measuring in muscle biopsy specimens the concentration of the DNA, RNA and alkali-soluble (i.e. non-collagen) protein (ASP), and the activity of total and "free" cathepsin D. Besides, the nutritional status was assessed by means of standard methods[24]. Muscle DNA content is considered a reliable reference standard in malnourished men[32,33]. The RNA:DNA ratio is a measure of the capacity of the cell to synthesize proteins and the ASP:DNA ratio is an index of a hypothetical muscle cell size, i.e. the imaginary volume of cytoplasm managed by a single nucleus[1,3,33]. Cathepsin D is an acid lysosomal endopeptidase, which is capable of degrading myofibrillar proteins[34,35]. It seems to have a role in lysosomal proteolysis, but a great deal remains to be learnt about its physiological role[36]. It is present in "free" (non-sedimentable) and in bound forms, and its activity increases in denervation[35,37], in various myopathies[35,38], in cancer[39], in glucocorticoid[40,41] or D-thyroxine[42] administration, and in cardiac muscle in starvation[43]. Insulin, which is a known inhibitor of protein breakdown in muscle, lowers the activity of the free enzyme[44,45].

PATIENTS, MATERIALS AND METHODS

We examined 10 stable chronically uremic patients on maintenance hemodialysis, whose clinical and laboratory data are reported in table 1. Patients with systemic diseases were excluded from the study. Patients received vitamin supplements and phosphate binders; iron and other medications were administered as necessary.

Hemodialysis was performed 3 times per week for 4 to 6 hours, using standard dialysates containing glucose (1 g/1) and acetate, and 1.0 - 1.2 m^2 plate dialyzers.

Dietary intake was evaluated by an experienced dietician by means of dietary interviews and food tables. Methods for anthropometric measurements, blood and serum analysis, muscle biopsy studies and statistical calculations were previously reported[23,24]. For cathepsin D activity determination we used the method of Mycek[46], modified according to method I of Barrett and Heath[36]. About 20 mg of muscle were gently homogenized in a glass to glass homogenizer containing 0.25 M sucrose (1:19 w/v). An aliquot of this first homogenate was centrifuged at 48,000 g for 30 min at 2o-4°C and the supernatant was used for free activity determination. A second aliquot of the first homogenate was diluted with 2 volumes of 0.3 % Triton X-100 in 0.25 M sucrose and centrifuged at 1,200 g for 15 min. at 2o-4°C; the pellet was washed with 1 volume of 0.2 % Triton X-100 in 0.25 M sucrose and the combined supernatants were used for total activity determination. The incubation mixture contained 0.20 ml of a 2.5 % solution (w/v) of dialyzed hemoglobin, 0.05 ml of 1 M Na formate-formic acid buffer pH 3.3 and 0.05-0.10 ml of extract. Four assays were run, both for free and total activity determination and after 60-120-180-240 min of incubation at 37°C they were stopped with 0.50 ml of ice-cold 5 % (w/v) TCA. The clear supernatant was read at 280 nm against water. Blanks were prepared by adding the TCA solution to the incubation mixture before extract addition.

RESULTS

Protein and energy intake (table 1) was close to dietary requirements for hemodialysis patients[31,47]. Significant correlations were present between protein intake and SUN (table 2), and between dietary energy and dietary proteins.

Table 1 also shows that the mean blood lymphocyte count was low; it was under the normal range in 5 out of 10 patients.

Table 1.   Clinical and laboratory data of the patients, expressed
           as mean ± S.D. Ranges are reported in parentheses.

AGE, years	55 + 12 (30-67)
DIALYSIS TREATMENT, months	32 + 33 (4-120)
DIETARY PROTEINS, g/kg des. wt/day	1.38 + 0.20 (1.06-1.37)
DIETARY ENERGY, kcal/kg des. wt/day	34 + 7 (26-43)
UREA NITROGEN, mg/dl	93 + 9
CREATININE, mg/dl	13.9 + 1.3
TRIGLYCERIDES, mg/dl	220 + 72
CALCIUM, mg/dl	9.3 + 0.5
INORGANIC P, mg/dl	6.4 + 2.0
HEMOGLOBIN, g/l	8.0 + 1.5
LYMPHOCYTES/mm^3	1,483 + 475

Anthropometric indexes (table 3) were only slightly decreased.
Some significant correlations were present between anthropometric
indexes, and between the SST and the serum triglyceride concentration (table 2).

Table 2.   Significant correlations present in hemodialysis
           patients.

		r	p<
age	- total cathepsin D	0.82	0.005
age	- free cathepsin D	0.65	0.050
months on dialysis	- serum total proteins	0.71	0.030
dietary energy	- dietary proteins	0.67	0.040
dietary proteins	- serum urea N	0.73	0.020
dietary proteins	- total cathepsin D/DNA	- 0.76	0.020
dietary energy	- total cathepsin D/DNA	- 0.67	0.040
serum urea N	- serum calcium	- 0.67	0.040
serum urea N	- total cathepsin D/DNA	- 0.81	0.005
serum urea N	- free cathepsin D/DNA	- 0.76	0.020
PDW	- AFA	0.66	0.050
RBW	- SST	0.76	0.020
RBW	- AFA	0.76	0.020
SST	- serum triglycerides	0.64	0.050
albumin	- transferrin	0.70	0.025
DNA	- ASP	- 0.70	0.025
RNA/DNA	- ASP/DNA	0.84	0.005
total cathepsin D	- free cathepsin D	0.82	0.005
total cathepsin D/DNA	- free cathepsin D/DNA	0.93	0.001
total cathepsin D/DNA	- RNA/DNA	0.79	0.005
free cathepsin D/DNA	- RNA/DNA	0.83	0.005

Table 3.   Anthropometric measurements. Results are expressed
           as mean ± S.D. The percentage of values under the
           5th percentile[48] is reported in parentheses.

WEIGHT	
% ideal body wt. (PDW)	100 ± 10
% mean body wt. (RBW)	90 ± 8
SKINFOLD THICKNESS	
triceps (TST), mm	9.7 ± 4.0 (20 %)
subscapular (SST), mm	10.6 ± 3.3 (20 %)
ARM FAT AREA (AFA), mm^2	1,311 ± 552 (30 %)
ARM MUSCLE CIRCUMFERENCE (AMC), mm	253 ± 22 (0 %)
ARM MUSCLE AREA (AMA), mm^2	5,149 ± 876 (0 %)

Serum protein content (table 4) was often decreased. Serum
transferrin and albumin were significantly correlated (table 2).

The muscle DNA concentration was higher in hemodialysis patients than in controls, whereas the ASP content, and the RNA:DNA
and the ASP:DNA ratios were decreased (table 5). The two ratios
were significantly correlated.

Table 4.   Serum protein content. Results are expressed as
           mean ± S.D. The percentage of values under (↓) or
           above (↑) normal range is reported in parentheses.

	NORMAL RANGE	PATIENTS
TOTAL PROTEINS, g/dl	6.3 - 7.8	7.0 ± 0.6 (↓ 10 %)
ALBUMIN, mg/dl	⩾ 3,500	3,213 ± 509 (↓ 70 %)
TRANSFERRIN, mg/dl	205 - 374	188 ± 98 (↓ 60 %)
PSEUDOCHOLINESTERASE, U/l	1,900 - 3,800	2,006 ± 624 (↓ 40 %)
C3, mg/dl	100 - 200	94 ± 13 (↓ 70 %)
C3 ACTIVATOR, mg/dl	18 - 40	10 ± 2 (↓ 100%)
IgM, mg/dl	45 - 145	117 ± 55 (↑ 0 %)

Table 5.   Muscle DNA, RNA, ASP (alkali-soluble proteins), water
           and fat content. Results are expressed as mean $\pm$ S.D.

	CONTROLS	PATIENTS	LEVEL OF SIGNIFICANCE
DNA, mg/100 g w. wt.	34 $\pm$ 8	50 $\pm$ 13	0.005
RNA, mg/100 g w. wt.	77 $\pm$ 23	68 $\pm$ 13	N.S.
ASP, g/100 g w. wt.	18.0 $\pm$ 2.1	15.4 $\pm$ 2.3	0.005
RNA/DNA, mg/mg	2.32 $\pm$ 0.59	1.48 $\pm$ 0.59	0.005
ASP/DNA, mg/mg	554 $\pm$ 109	337 $\pm$ 127	0.005
Water, g/100 g FFS	301 $\pm$ 56	323 $\pm$ 69	N.S.
Fats, g/100 g FFS	25 $\pm$ 17	13 $\pm$ 12	N.S.

     Both total and free cathepsin D activities (table 6) were
lower in hemodialysis patients than in controls; the percentage of
free cathepsin D was normal. Significant correlations (table 2)
were present between total and free cathepsin D activities, between cathepsin D activities and the RNA:DNA ratio, between dietary
intake and total cathepsin D activity (inverse correlations), and
between serum urea N and cathepsin D activities (inverse correlations).

Table 6.   Muscle cathepsin D activity. Results (as mean $\pm$ S.D.) are
           expressed in U/100 mg w. wt. or U/μg DNA.

	CONTROLS (N = 7)	PATIENTS (N = 10)	LEVEL OF SIGNIFICANCE
Total cathepsin D	98 $\pm$ 17	64 $\pm$ 16	0.005
Free cathepsin D	42 $\pm$ 8	29 $\pm$ 7	0.005
Free/total x 100	44 $\pm$ 10	47 $\pm$ 9	N.S.
Total cath. D/DNA	2.98 $\pm$ 0.68	1.33 $\pm$ 0.46	0.005
Free cath. D/DNA	1.28 $\pm$ 0.38	0.62 $\pm$ 0.23	0.005

DISCUSSION

Evidence for wasting and malnutrition is common in chronically uremic patients on conservative, hemodialysis and peritoneal dialysis treatment[10-31] and it has been ascribed to many causes[49]. Many abnormalities of protein metabolism have been reported in uremia such as dialysis induced protein catabolism, a less efficient protein utilisation, greater suppression of muscle protein synthesis in fast, decreased protein synthesis in rat skeletal muscle, "in vitro" inhibition of protein synthesis by uremic plasma, low protein turnover rates, resistance to insulin-mediated aminoacid uptake in skeletal muscle, increased gluconeogenesis, abnormalities of aminoacid metabolism, impaired polyamine formation, defective protein synthesis in liver[23].

In the patients of the present study abnormalities of different protein pools were present, because plasma protein concentration, blood lymphocyte count, muscle ASP and RNA content, and muscle cathepsin D activity were significantly decreased. Anthropometric indexes, instead, were less severely affected than in other uremic patients on a lower protein and energy intake[23-26]. The increase of muscle DNA concentration in hemodialysis patients suggests a reduced muscle "cell size"[1,3,33] and, in fact, DNA was inversely correlated to muscle ASP concentration. Muscle ASP concentration, and the ASP:DNA and RNA:DNA ratios were significantly lower in patients than in controls, and the two ratios were significantly correlated. Therefore, a parallel muscle depletion of non collagen proteins and ribosomes (about 80 % of RNA is ribosomal and its content is correlated to the intensity of protein synthesis[1-5]) was present in our patients. An abnormal muscle polysome profile and a decrease in muscle RNA concentration in chronically uremic rats, and a lower muscle protein content in acute and chronic uremia have been reported[23]. An insufficient dietary intake, uremic toxins, metabolic abnormalities, increased PTH levels, and resistance to insulin could affect muscle metabolism and muscle protein and RNA content in uremia[23].

Also muscle total and free cathepsin D activities were decreased in hemodialysis patients and they were significantly correlated. If cathepsin D has a role in the physiological and/or pathological degradation of myofibrillar proteins, this finding may suggest a reduced lysosomal proteolysis in these stable patients. Cathepsin D activity of the patients was about 44 % of controls and it was, therefore, more markedly decreased than muscle ASP (61 %). This may suggest that the lower enzyme activity of hemodialysis patients is not only a consequence of an overall muscle protein depletion, but points out a specifically reduced proteinase activity

in the muscle of these patients. On the other hand, in chronically
uremic adults, who are depleted of muscle proteins[23,26], we
previously found normal hexokinase levels[15] . The lower RNA:DNA
ratio(i. e. an index of muscle protein synthetic activity), and
the lower cathepsin D activity (which is possibly an index of
muscle protein degradation) suggest a reduced muscle protein
turnover in these patients. The two indexes were significantly
correlated. Low protein turnover rates were also reported by
Conley et al.[50] in hemodialysis children by means of ^{15}N-lysine
studies. Muscle cathepsin D activity was directly correlated with
age (as previously reported by Lundholm and Scherstén[51]) and
inversely correlated with protein and energy intake and with
serum urea nitrogen. Of course, a cause-effect relationship may
be either absent, or present only between cathepsin D activity
and one of the two variables, being protein intake and SUN
directly correlated. These findings, however, may suggest an inhibitory effect of toxins arising from dietary protein catabolism on enzyme activity, and/or that protein intake has a
regulatory effect on muscle cathepsin D activity. An increased
muscle protein breakdown has been sometimes reported in fasting
and in malnutrition, although muscle turnover in muscle seems
to be primarily regulated by protein synthetic activity[1,2].
Protein intake was quite high in our patients and also energy
intake may be considered sufficient. Nevertheless, muscle and
plasma protein synthesis was clearly reduced. Therefore, dietary
intake and/or protein metabolism were not adequate in these
patients for a normal protein synthesis in liver and in muscle.
The low cathepsin D activity, which was reduced concomitantely
to the high protein intake, could not compensate for the reduced protein synthesis in muscle. Alternatively, the level
of cathepsin D activity may be not relevant for the control
of protein turnover in muscle and it is to be considered merely
an epiphenomenon. In acutely uremic rats muscle proteinase activity is normal (J. D. Kopple, personal communication).

In conclusion, the hepatic synthesis of plasma proteins,
muscle protein and RNA content and muscle cathepsin D activity
were reduced in the hemodialysis patients. These findings
suggest a decreased muscle protein synthesis and, possibly,
degradation. The determination of protein and nucleic acid
content, and proteinase activity in muscle biopsy specimens
enables the clinician to directly study, "in vivo", muscle
protein synthesis and degradation.

REFERENCES

1. D. J. Millward, and J. C. Waterlow, Effect of nutrition on
   protein turnover in skeletal muscle, Fed. Proc. 37:2283
   (1978).
2. J. C. Waterlow, P. J. Garlick, and D. J. Millward, The
   effects of nutrition and hormones on protein turnover
   in muscle, in: "Protein Turnover in Mammalian Tissues
   and in the Whole Body," J. C. Waterlow, P. J. Garlick,
   D. J. Millward, eds., Elsevier/North-Holland Biomedical
   Press, Amsterdam (1978).
3. C. A. Spence, and F. M. Hansen-Smith, Comparison of the
   chemical and biochemical composition of thirteen muscles
   of the rat after dietary protein restriction, Br. J.
   Nutr. 39:647 (1978).
4. A. C. Bylund, J. Holm, J. Lundholm, and T. Schersten, Incorporation rate of glucose carbon into metabolites in
   relation to enzyme activities and RNA levels in human
   skeletal muscles, Enzyme 21:39 (1976).
5. D. B. Cheek, and J. E. Graystone, Insulin and growth hormone:
   regulators of growth with particular reference to muscle,
   Kidney Int. 14:317 (1978).
6. O. Z. Lernau, S. Nissan, B. Neufeld, and M. Mayer, Myofibrillar protease activity in muscle tissue from
   patients in catabolic conditions, Eur. J. Clin. Invest. 10:357 (1980).
7. M. S. Pote, and W. Altekar, Muscle aldolase: the stress-
   dependent modifications of catalytic and structural
   properties by rat muscle lysosomal cathepsin B, Biochim. Biophys. Acta 661:303 (1981).
8. H. N. Munro, and V. R. Young, Urinary excretion of Nτ-
   methylhistidine (3-methylhistidine): a tool to study
   metabolic responses in relation to nutrient and
   hormonal status in health and disease of man, Am. J.
   Clin. Nutr. 31:1608 (1978).
9. D. J. Millward, P. C. Bates, J. G. Brown, S. R. Rosochacki,
   and M. J. Rennie, Protein degradation and the regulation of protein balance in muscle, Ciba Found. Symp.
   75:307 (1979).
10. C. Delaporte, J. Bergström, M. Broyer, and A. M. Dartois,
    Variations in muscle cell protein of severely uremic
    children, Kidney Int. 10:239 (1976).
11. M. A. Holliday, C. Chantler, R. MacDonnel, and J. Keitges,
    Effect of uremia on nutritionally-induced variations
    in protein metabolism, Kidney Int. 11:236 (1977).

12. J. D. Kopple, Abnormal amino acid and protein metabolism in uremia, Kidney Int. 14:340 (1978).
13. J. Bergström, P. Fürst, L. O. Norée, and E. Ulnnass, Intracellular free amino acids in muscle tissue of patients with chronic uremia: effect of peritoneal dialysis and infusion of essential amino acids, Clin. Sci. Mol. Med. 54:51 (1978).
14. J. B. Li, and S. J. Wassner, Muscle degradation in uremia: 3-methyl-histidine release in fed and fasted rats, Kidney Int. 20:321 (1981).
15. G. F. Guarnieri, G. Toigo, S. De Marchi, R. Situlin, and L. Campanacci, Muscle hexokinase and phosphofructokinase activity in chronically uremic patients, in: "Uremia - Pathobiology of Patients Treated for 10 Years or more," C. Giordano, E. A. Friedman, eds., Wichtig Editore, Milano (1981).
16. W. C. Arnold, and M. A. Holliday, In vitro suppression of insulin-mediated amino acid uptake in uremic skeletal muscle, Am. J. Clin. Nutr. 33:1428 (1980).
17. B. F. Westervelt Jr., Uremia and insulin response, Arch. Intern. Med. 126:865 (1975).
18. V. Sahgal, J. Hughes, A. Shah, and A. Quintanilla, Muscle abnormalities in chronic renal failure, Kidney Int. 14:732 (1978).
19. A. J. Clark, M. Wong, J. D. Kopple, and M. E. Swendseid, Ribosome profiles and RNA, DNA content of muscle in uremia, Fed. Proc. 35:596 (1976).
20. J. D. Kopple, B. Cianciaruso, and S. G. Massry, Does parathyroid hormone cause protein wasting? Contr. Nephrol. 20:138 (1980).
21. P. Fürst, A. Alverstrand, and J. Bergström, Effects of nutrition and catabolic stress on intracellular amino acid pools in uremia, Am. J. Clin. Nutr. 33:1387 (1980).
22. T. Nakao, S. Fujiwara, K. Isoda, and T. Miyahara, Impaired lactate production by skeletal muscle with anaerobic exercise in patients with chronic renal failure, Nephron 31:111 (1982).
23. G. F. Guarnieri, G. Toigo, R. Situlin, L. Faccini, L. Campanacci, G. Bazzato, U. Coli, and S. Landini, Protein-calorie undernutrition in chronically uremic patients. Investigation by muscle biopsy, in Self care dialysis - The fifty-fourth Hahnemann International Symposium, Venezia, Italy, April 2 - 3, 1982, in press.
24. G. F. Guarnieri, L. Faccini, T. Lipartiti, F. Ranieri, F. Spangaro, D. Giuntini, G. Toigo, F. Dardi, F. Berquier-Vidali, and A. Raimondi, Simple methods for nutritional assessment in hemodialyzed patients, Am. J. Clin. Nutr. 33:1598 (1980).
25. G. F. Guarnieri, F. Ranieri, T. Lipartiti, F. Spangaro, D. Giuntini, L. Faccini, G. Toigo, F. Legnani, A. Raimondi, and L. Campanacci, Protein-calorie malnutrition in hemodialysis patients, Int. J. Art. Org. 3:143 (1980).

26. G. F. Guarnieri, G. Toigo, R. Situlin, L. Faccini, U. Coli, S.
    Landini, G. Bazzato, and L. Campanacci, Muscle biopsy studies
    in chronically uremic patients. Evidence for malnutrition,
    Kidney Int. in press (1983).

27. B. J. Thunberg, A. P. Swamy, and R. V. M. Cestero, Cross-sectional
    and longitudinal nutritional measurements in maintenance hemo-
    dialysis patients, Am. J. Clin. Nutr. 34:2005 (1981).

28. R. A. Ward, M. J. Shirlow, J. M. Hayes, G. V. Chapman, and P. C.
    Farrell, Protein catabolism during hemodialysis, Am. J. Clin.
    Nutr. 32:2443 (1979).

29. I. M. Cohen, J. Griffiths, R. A. Stone, and T. Leech, The crea-
    tine kinase profile of a maintenance hemodialysis population:
    a possible marker of uremic myopathy, Clin. Nephrol. 13:235
    (1980).

30. P. C. Farrell, and P. W. Hone, Dialysis-induced catabolism, Am.
    J. Clin. Nutr. 33:1417 (1980).

31. M. F. Borah, P. Y. Schoenfeld, F. A. Gotch, J. A. Sargent, M.
    Wolfson, and M. H. Humphreys, Nitrogen balance during inter-
    mittent dialysis therapy of uremia, Kidney Int. 14:491 (1978).

32. E. E. Gordon, K. Kowlasky, and M. Fritts, Muscle proteins and
    DNA in rat quadriceps during growth, Am. J. Physiol. 210:
    1033 (1966).

33. G. A. O. Alleyne, R. W. Hay, D. I. Picou, J. P. Stanfield, and
    R. G. Whitehead, Protein-energy malnutrition, Edward Arnold
    Publisher (1977).

34. J. W. Bird, J. H. Carter, R. E. Triemer, R. M. Brooks, and A. M.
    Spanier, Proteinases in cardiac and skeletal muscle, Fed. Proc.
    39:20 (1980).

35. W. N. Schwartz, and J. W. Bird, Degradation of myofibrillar
    proteins by cathepsins B and D, Biochem. J. 167:811 (1977)
    (quoted by Bird et al., refer. 34).

36. A. L. Barrett, and M. F. Heath, Lysosomal enzymes, in:"Lysosomes -
    Laboratory Handbook," J. T. Dingle, ed., North-Holland Pub-
    lishing Company, Amsterdam - New York - Oxford (1977).

37. L. C. Sellin, R. Libelius, I. Lundquist, S. Tagerud, and S. Thes-
    leff, Membrane and biochemical alterations after denervation
    and during reinnervation of mouse skeletal muscle, Acta
    Physiol. Scand. 110:181 (1980).

38. J. B. Li, Protein synthesis and degradation in skeletal muscle
    of normal and dystrophic hamsters, Am. J. Physiol. 239:E401
    (1980).

39. K. Lundholm, A. C. Bylund, J. Holm, and T. Scherstén, Skeletal
    muscle metabolism in patients with malignant tumor, Europ.
    J. Cancer 12:465 (1976).

40. A. F. Clark, and P. J. Vignos Jr., The role of proteases in ex-
    perimental glucocorticoid myopathy, Muscle Nerve 4:219 (1981).

41. B. G. Vernon, and P. J. Buttery, The effect of the growth pro-
    moter trenbolone acetate, dexamethasone and thyroxine on ske-
    letal muscle cathepsin D activity, Proc. Nutr. Soc. 40:13A
    (1981).
42. R. S. Decker, and K. Wildenthal, Lysosomal alterations in heart,
    skeletal muscle, and liver of hyperthyroid rabbits, Lab. In-
    vest. 44:455 (1981).
43. K. Wildenthal, A. R. Poole, and J. T. Dingle, Influence of star-
    vation on the activities and localization of cathepsin D and
    other lysosomal enzymes in hearts of rabbits and mice, J. Mol.
    Cell. Card. 7:841 (1975).
44. D. E. Rannels, R. Kao, and H. E. Morgan, Effect of insulin on
    protein turnover in heart muscle, J. Biol. Chem. 25:1694 (1975).
45. J. B. Li, S. R. Rannels, M. E. Burkart, and L. S. Jefferson,
    Effects of insulin on protein degradation and lysosomal
    cathepsin D in perfused skeletal muscle, Fed. Proc. 34:335
    (1975).
46. M. J. Mycek, Cathepsins, in: "Methods in Enzymology," G. E. Perl-
    mann, L. Lorand, eds., Academic Press, New York - London (1970).
47. J. D. Kopple, Dietary requirements, in: "Clinical Aspects of
    Uremia and Dialysis," S. G. Massry, A. L. Sellers, eds.,
    Charles C. Thomas, Springfield (1976).
48. A. R. Frisancho, New norms of upper limb fat and muscle areas
    for assessment of nutritional status, Am. J. Clin. Nutr. 34:
    2540 (1981).
49. J. D. Kopple, S. G. Massry, V. Bonomini, A.Heidland, eds., Pro-
    ceedings of the 2nd International Congress on Nutrition in Re-
    nal Disease,June 13-15,1979,Bologna,Am.J.Clin.Nutr. 33 (1980).
50. S. B. Conley, G. M. Rose, A. M. Robson, and D. M. Bier, Effects
    of dietary intake and hemodialysis on protein turnover in
    uremic children, Kidney Int. 17:837 (1980).
51. K. Lundholm, and T. Scherstén, Leucine incorporation into pro-
    teins and cathepsin D activity in human skeletal muscles;
    the influence of the age of the subject, Exp. Geront. 10:155
    (1975).

ENHANCED MUSCLE PROTEIN DEGRADATION AND AMINO ACID

RELEASE FROM THE HEMICORPUS OF ACUTELY UREMIC RATS

Reinhild M. Flugel-Link, Isidro B. Salusky,
Michael R. Jones and Joel D. Kopple

Division of Nephrology and Hypertension, Harbor-UCLA Medical
Center, Division of Pediatric Nephrology, UCLA Center for the
Health Sciences, and the Division of Nutritional and Environ-
mental Services, UCLA Schools of Medicine and Public Health

INTRODUCTION

Patients with acute renal failure are often severely catabolic (1-3)
and may manifest marked muscle protein wasting. Although muscle is the
major endogenous source of protein and amino acids for the organism,
there are little data concerning muscle protein and amino acid metabolism in acute uremia. Since hepatic uptake of amino acids, gluconeogenesis and urea synthesis are increased in rats with experimental acute
uremia (4-7), it is likely that muscle provides increased amounts of amino
acids for the enhanced metabolic activity of the liver.

There are only a few studies of muscle protein turnover in acute
renal failure (8-10) and it is not clear yet whether enhanced release of
muscle amino acids reflects increased degradation or decreased synthesis.
We therefore examined muscle protein synthesis and degradation, urea
nitrogen appearance, and amino acid uptake and release in acutely uremic
rats and in sham-operated control rats. Also, the activities of certain
muscle proteolytic enzymes and muscle ATP, creatine phosphate and
cyclic AMP were assayed.

METHODS

Male Sprague-Dawley rats (200-250 g) underwent either bilateral
nephrectomy or sham operation. After surgery the rats were kept in
metabolic cages and were fasted except for tube feeding twice with 3 ml
of water containing 0.6 mEq of sodium bicarbonate, at 8 hr and 24 hr
after surgery, to reduce acidosis and hyperkalemia. All rats were studied
30 hr after surgery.

Muscle protein synthesis and degradation was evaluated with the hemicorpus perfusion technique which was performed according to the method of Jefferson (11). Rats were given heparin intraperitoneally (2500 U/kg, for nephrectomized rats) and were anesthetized with sodium pentobarbital i.p., 50 mg/kg. The abdomen was opened, and the celiac, superior and inferior mesenteric, renal, adrenal and spermatic arteries and veins were ligated. The chest was opened, and the aorta was cannulated above the diaphragm. The perfusate was then infused into the aorta at a flow rate of 10 ml/min, and the vena cava was severed. The outflow from the vena cava was allowed to drain into a basin and was recirculated through the hemicorpus for a period of 120 min. The initial composition of the perfusate consisted of Krebs-Henseleit bicarbonate buffer (12), 3% bovine serum albumin, 30% washed bovine red blood cells, 15 mM glucose, and amino acids at the plasma levels of normal fasting rats (13). Phenylalanine was added at five times the normal plasma concentrations in order to facilitate equilibration of labeled phenylalanine. Phenylalanine was used for assessing protein turnover because it is neither synthesized nor degraded by muscle (13). Perfusate was sampled at 60 and 120 min of perfusion for measurement of potassium, phosphorus, calcium, amino acids, and radioactivity.

Gastrocnemius muscle was taken from one leg at 60 min and from the other leg at 120 min. Protein synthesis was calculated from the quantity of ^{14}C-phenylalanine incorporated into protein of this muscle between 60 and 120 min divided by the average intracellular specific activity. Protein degradation was assessed from dilution of ^{14}C-phenylalanine by unlabeled phenylalanine from protein breakdown (13). Urea nitrogen appearance was measured during the 30 hr period after surgery and was calculated as described earlier (14). In gastrocnemius muscle, ATP and creatine phosphate were measured in the perchloric acid homogenate by enzymatic analysis (15) and cyclic-AMP was determined in the 5% TCA extract using a double antibody radioimmunoassay (16).

In order to assay the activity of the proteolytic lysosomal enzymes, cathepsin Bl and D, gastrocnemius, psoas, and soleus muscles were finely minced with a scissors, and homogenized with 9 vol of ice cold 0.25 N sucrose containing 0.2% Triton X-100 in a glass homogenizer with intermittent cooling in ice. Total activity of cathepsin Bl was analyzed according to Barrett (17) with BANA (α-N-benzoyl-DL-arginine 2-naphthylamide hydrochloride) as substrate. Activity is expressed in nmol 2-naphthylamine/h/mg noncollagen protein (NCP). Cathepsin D activity was assayed by the method of Anson (18) with hemoglobin as substrate. Alkaline protease activity was measured at pH 9.1 in the myofibrillar fraction of the three muscles according to Mayer et al (19). Activities of cathepsin D and alkaline protease are expressed as μg tyrosine/h/mg NCP. Noncollagen protein in the muscle homogenate or myofibrillar fraction was determined as described by Pennington et al (20).

Plasma and muscle amino acids were determined in anesthetized uremic and sham rats; the specimens were obtained simultaneously from abdominal aortic blood and gastrocnemius muscle . Muscle intracellular amino acid concentrations were calculated as previously described (21). The amino acids were analyzed within eight days of collection with a Beckman 121MB amino acid analyzer using a lithium buffer system. Potassium was measured with an Instrumentation Laboratory 143 flame photometer, phosphorus with a Technicon AutoAnalyzer, calcium with a Perkin-Elmer (303) atomic absorption spectrophotometer, and urea with a Beckman BUN analyzer. Radioactivity was determined with a Beckman liquid scintillation counter, Model L5-3150.

RESULTS

In the two groups of rats the initial body weight was similar at the time of surgery. After 30 hours, the uremic rats had lost less weight compared to sham animals probably because of their inability to excrete urine (Table 1). Plasma urea nitrogen and plasma potassium concentrations were significantly increased in the uremic rats. Urea nitrogen appearance, an indicator of net protein degradation (14), was significantly greater in the uremic rats as compared to the sham animals (Table 1).

TABLE 1

Weight Change, Plasma Urea and Potassium, and Urea Nitrogen
Appearance in Acutely Uremic and Sham-Operated Rats

	Uremic n=10	Sham n=10
Weight loss (g)	$10 \pm 3^{a,c}$	20 ± 6
Plasma urea nitrogen (mg/100 ml)	184 ± 22^{c}	14 ± 3
Plasma potassium (mEq/liter)	7.7 ± 1.1^{c}	3.6 ± 0.3
Urea nitrogen appearance (mg N/30 h)	196 ± 50^{b}	151 ± 35

[a]Mean $\pm$ SD.
Probability that values in uremic and sham rats are not different; [b]p <0.05, [c]p < 0.001.

Most plasma amino acid concentrations at 30 hours after surgery were decreased in the uremic animals. These included alanine, glutamine, and total essential and total nonessential amino acids which were each significantly decreased (Table 2). Plasma histidine and phenylalanine were increased in the uremic rats. The only amino acids which were not different from sham rats were the three branched chain amino acids, leucine, isoleucine and valine, and aspartic acid, citrulline, and taurine. Similar findings were present with regard to many muscle intracellular amino acid concentrations. Many muscle amino acids were reduced in the acutely uremic rats (Table 2). These included histidine, lysine, threonine, total essential amino acids, tyrosine, arginine, glutamine, glycine, serine, taurine, total nonessential amino acids and total amino acids. Muscle intracellular methionine and glutamic acid were each significantly increased in the uremic rats.

TABLE 2

Plasma and Intracellular Muscle Amino Acid Concentrations in Acutely
Uremic and Sham-Operated Rats

	PLASMA (μmol/liter)		MUSCLE (μmol/liter intracellular water)	
	Uremic (n=6)	Sham (n=6)	Uremic (n=6)	Sham (n=6)
Alanine	$194 \pm 39^{a,c}$	395 ± 98	3274 ± 619	3802 ± 714
Glutamine	284 ± 37^{c}	549 ± 106	1156 ± 423^{c}	2920 ± 479
Total Essential[d]	888 ± 166^{b}	1220 ± 72	2135 ± 16^{b}	4394 ± 1200
Total Nonessential[e]	909 ± 67^{c}	2009 ± 224	14015 ± 1301^{b}	17956 ± 2084

[a]Mean $\pm$ SD.
Probability that values in uremic and sham rats are not different;
[b]$p < 0.01$, [c]$p < 0.001$.
[d]Calculated as the sum on the concentrations of histidine, isoleucine, leucine, lysine, methionine, phenylalanine, threonine, and valine.
[e]Calculated as the sum of the concentrations of alanine, arginine, asparagine, aspartic acid, glutamic acid, glutamine, glycine, ornithine, proline, and serine.

TABLE 3

Protein Synthesis and Degradation in the Perfused Rat Hemicorpus

	N	Protein Synthesis nmol/h/g muscle	Protein Degradation nmol/h/g hemicorpus
Uremic	9	37.8 ± 15.4^{a}	233 ± 42
Sham	10	51.7 ± 19.2	170 ± 36
p[b]		NS	< 0.001

[a]Mean $\pm$ SD.
[b]Probability that values in uremic and sham rats are not different; NS, not significantly different.

Protein synthesis and degradation rates are shown in Table 3. The rate of protein synthesis in gastrocnemius muscle, expressed as nmoles phenylalanine incorporated/g muscle/hr, tended to be lower in uremic rats, but the difference between the two groups was not statistically significant. However, protein degradation, expressed in nmoles phenyl-alanine released/g hemicorpus/hr, was significantly increased, by 30%, in the uremic rats. The finding of enhanced protein breakdown in muscle of uremic rats is supported by the increased net release of phenylalanine, tyrosine and alanine, total nonessential amino acids and total amino acids in the acutely uremic rats during the 60 min of perfusion (Table 4); net release of citrulline, an amino acid not found in protein, was greater in the sham operated controls.

TABLE 4

Net Release or Uptake of Amino Acids
During Perfusion of the Rat Hemicorpus[a]

	Uremic (n=7)	Sham (n=8)	p[b]
Tyrosine[d]	-3486 ± 992[c]	-2650 ± 411	<0.05
Alanine[d]	-56314 ± 17090	-31850 ± 12141	<0.01
Citrulline[d]	-733 ± 468	-1225 ± 225	<0.01
Total Nonessential[d,e]	-107729 ± 30603	-72530 ± 17619	<0.02
Total Amino Acids[d,e,f]	-153854 ± 42778	-114467 ± 21422	<0.05
Potassium (mEq/60 min)	-0.22 ± 0.02	-0.13 ± 0.05	<0.001
Phosphorus (mg/60 min)	-2.63 ± 0.43	-1.05 ± 0.65	<0.001
Calcium (mg/60 min)	-1.52 ± 0.72	-0.98 ± 1.23	NS

[a]Refers to quantity released into perfusate between 60 and 120 minutes after onset of perfusion study; a positive sign indicates net uptake; negative sign indicates net release.
[b]Probability that values in uremic and sham rats are not different; NS, not significantly different.
[c]Mean $\pm$ SD.
[d]Values are given in nanomoles released per 60 minutes of perfusion.
[e]Total essential and nonessential amino acids are calculated as described in Table 2.
[f]Calculated as the sum of net release or uptake of total essential amino acids, total nonessential amino acids and cystine, tyrosine, citrulline, and taurine.

There was also increased net release of potassium and phosphorus from the hemicorpus of the uremic animals between 60 and 120 min of perfusion (Table 4). However, calcium release was similar in the two groups of rats.

To evaluate the possibility that enhanced proteolysis might be due to altered energy levels or cyclic-AMP, we examined muscle concentrations of ATP, creatine phosphate and cyclic-AMP. The results obtained 30 hr after surgery are shown in Table 5. There was no difference in the muscle concentrations of these compounds between the uremic and sham animals. Since proteases are present in skeletal muscle, they also might be a cause for the enhanced protein breakdown. Therefore, cathepsin B1, cathepsin D and an alkaline protease (pH 9.1) were measured in homogenates of gastrocnemius, psoas, and soleus muscles. The activity of each of these enzymes did not differ in any of the three muscles in the uremic as compared to the sham animals (Table 6).

TABLE 5

Nucleotides and Creatine Phosphate in Muscle of Acutely
Uremic and Sham-Operated Rats 30 hours after Surgery

	ATP μmol/g	Creatine Phosphate μmol/g	cAMP pmol/g
N	6^a	5	9
Uremic	6.55 ± 0.86^b	17.7 ± 3.5	505 ± 94
Sham	6.40 ± 0.29	13.5 ± 1.4	491 ± 105
p^c	NS	NS	NS

[a]Refers to the number of rats in each of the two groups studied (uremic and sham).
[b]Mean $\pm$ SD.
[c]Probability that values in uremic and sham rats are not different; NS, not significantly different.

DISCUSSION

The results of these studies suggest that there is enhanced muscle protein wasting in acutely uremic rats which is due to increased protein degradation. Although muscle protein synthesis appeared to be reduced in these uremic animals, the values were not significantly different from sham-operated controls. This study also provides other evidence for enhanced protein wasting in acutely uremic animals. Thus, there was increased release from the hemicorpus of the amino acids phenylalanine, tyrosine, alanine, the sum of all nonessential amino acids, total amino acids, and potassium and phosphorus. Also, the lower plasma and intracellular muscle concentrations of most amino acids may reflect the catabolic state of the uremic rats. Although it is possible that enhanced net release of phosphorus from the hemicorpus in uremic rats could be due to wasting of bone, the finding that the release of calcium was similar to that of sham operated rats suggests that the additional phosphorus loss in uremic rats was derived from soft tissue.

TABLE 6

Muscle Protease Activities in Acutely Uremic and Sham-Operated Rats

	N^a	Uremic[b]	Sham

Cathepsin B_1 (nmol 2-Naphthylamine/h/mg NCP[c])

Gastrocnemius	7	4.53 ± 1.22^d	5.22 ± 0.86
Psoas	7	6.31 ± 1.39^e	5.65 ± 1.59
Soleus	4	$9.83 \pm 0.47^{f,h}$	$9.11 \pm 0.71^{f,g}$

Cathepsin D (μg tyrosine/h/mg NCP)

Gastrocnemius	7	11.02 ± 1.85	12.97 ± 3.57
Psoas	7	10.00 ± 1.95	10.65 ± 1.58
Soleus	7	11.08 ± 1.15	10.84 ± 1.57

Alkaline Protease (pH 9.1) (μg tyrosine/h/mg NCP)

Gastrocnemius	6	3.85 ± 0.51	4.81 ± 0.86
Psoas	6	3.75 ± 0.72	3.95 ± 1.35
Soleus	5	5.31 ± 1.71	$6.27 \pm 0.83^{e,g}$

[a]Refers to the number of rats in each of the two groups studied (uremic and sham).
[b]There were no significant differences between corresponding values in the uremic as compared to the control rats.
[c]NCP – non-collagen protein.
[d]Mean $\pm$ SD.
[e,f]Probability that enzyme activity is not different from that in gastrocnemius muscle in the same rats; $p < 0.05$, < 0.001.
[g,h]Probability that enzyme activity is not different from that in psoas muscle in the same rat; $p < 0.01$, < 0.001.

The present study does not demonstrate the mechanisms responsible for increased muscle protein degradation in the acutely uremic rats. Since the concentrations of muscle ATP, creatine phosphate, and cyclic-AMP were normal and not different in the uremic and sham animals, the enhanced muscle protein degradation in uremic rats does not appear to be due to abnormal levels of these compounds. Since proteases can catabolize muscle proteins, altered protease activity might enhance muscle protein breakdown in the uremic rats. Indeed, the in vitro activity of certain proteases has been observed to be increased in muscle of animals and humans with muscular dystrophy (22-24) and in patients with cancer (25). In addition, Horl and Heidland have reported increased proteolytic activity in plasma from hypercatabolic patients with acute renal failure (26). Moreover, in some studies, the administration of protease inhibitors in vitro decreased proteolytic activity (27).

Based upon these earlier reports, we examined the activity of two lysosomal proteases and an alkaline protease in each of three different muscles. Our observations that the activity of these proteases did not differ between the uremic and sham rats suggest that the enhanced muscle protein degradation in the uremic rats was either due to other proteolytic enzymes or was caused by other factors. Another possibility is that our methods for measuring proteolytic activity were not sensitive enough to detect existing alterations.

The perfused posterior hemicorpus of the rat contains bone, adipose tissue, and skin as well as muscle, and it is possible that the metabolism in these tissues may have affected our findings concerning amino acid, protein and mineral metabolism. In fact, adipose tissue has branched chain aminotransferase activity (28) which could have contributed to the catabolism of the branched chain amino acids. Several factors, however, provide evidence that alterations in the metabolism of bone, adipose tissue and skin did not make a major contribution to our findings. First, most of the free amino acids and protein in the posterior hemicorpus reside in muscle. In addition, the finding that there was no difference in the rate of calcium release from the perfused hemicorpus of the uremic and control rats suggests that calcium and possibly protein turnover in bone were not different in the two groups of animals. These considerations suggest that the present observations concerning synthesis and degradation of protein and release of total amino acids reflect metabolic processes occurring in muscle.

Despite our finding that there was increased release of phenylalanine, tyrosine, alanine, total nonessential amino acids, and total amino acids from the perfused hemicorpus of the acutely uremic rats, these aminals displayed, in vivo, low concentrations of tyrosine, alanine and many other amino acids in plasma and muscle (Table 4). This observation is consistent with the possibility that there was enhanced degradation of these amino acids in non-muscle tissues, most likely the liver. Indeed, our finding that urea nitrogen appearance was increased in the uremic rats suggests the likelihood of increased degradation of amino acids in the liver. Previous studies in the isolated perfused liver or in liver slices from acutely uremic rats indicate that there is increased hepatic uptake and catabolism of amino acids, enhanced gluconeogenesis, and elevated activity of gluconeogenic enzymes (5, 7).

The present findings also suggest that the enhanced muscle protein degradation and amino acid release in uremic rats was not the primary cause for increased catabolism of amino acids in non-muscle tissues, although it may have served a permissive role. If it were the primary cause, then one would expect high, or at least normal, plasma and intracellular amino acid concentrations. Thus, the enhanced catabolic processes in the liver may be the primary cause of the catabolic state in the uremic rats.

The foregoing experiments in rats may explain why patients with acute renal failure do not seem to benefit from intravenous infusions of large quantities of essential and nonessential amino acids (3, 29). These experimental rat studies suggest that enhanced hepatic gluconeogenesis may be a primary cause of the hypercatabolic state in patients. Infusion of greater quantities of amino acids in patients may, by increasing the concentration of amino acids flowing into liver, enhance urea synthesis and possibly generation of other toxic metabolites. These considerations suggest that investigation into methods for reducing catabolic processes in liver may be a fruitful line of research for reducing the hypercatabolic state in acute uremia.

SUMMARY

Protein synthesis and degradation and net uptake and release of amino acids and minerals were investigated in the perfused hemicorpus of acutely uremic and control Sprague-Dawley rats. Rats underwent bilateral nephrectomy or sham surgery and were studied 30 hr after surgery. The uremic rats displayed greater urea N appearance (net urea generation), lower plasma and muscle concentrations of most amino acids, and increased muscle protein degradation as compared to control rats. Muscle protein synthesis was slightly but not significantly decreased in the uremic animals. There was greater net release of phenylalanine, tyrosine, alanine, total nonessential amino acids, total amino acids, potassium and phosphorus from the perfused hemicorpus of uremic rats and greater release of citrulline from sham rats. Muscle ATP, creatine phosphate, cyclic-AMP, and activities of cathepsin B1, cathepsin D, and alkaline protease were not different in the uremic and sham rats. These data provide evidence that acutely uremic rats sustain increased muscle protein wasting which is due to enhanced protein degradation. The increased protein degradation does not appear to be due to enhanced activities of muscle cathepsin B1, cathepsin D or alkaline protease.

REFERENCES

1. Silva, H., J. Pomeroy, A.I. Rae, S.M. Rosen, and S. Shaldon. Daily haemodialysis in "hypercatabolic" acute renal failure. 1964. Br. Med. J. 2:407-410.
2. Leonard, C.D., R.G. Luke, and R.R. Siegel. 1975. Parenteral essential amino acids in acute renal failure. Urology 6:154-157.
3. Feinstein, E.I., M.J. Blumenkrantz, M. Healy, A. Koffler, H. Silberman, S.G. Massry, and J.D. Kopple. 1981. Clinical and metabolic response to parenteral nutrition in acute renal failure: a controlled double-blind study. Medicine 60:124-137.
4. Sellers, A.L., J. Katz, and J. Marmorston. 1957. Effect of bilateral nephrectomy on urea formation in rat liver slices. Am.J.Physiol. 191(2):345-349.
5. Lacy, W.W. 1969. Effect of acute uremia on amino acid uptake and urea production by perfused rat liver. Am. J. Physiol. 216:1300-1305.

6. Lacy, W.W. 1970. Uptake of individual amino acids by perfused rat liver: effect of acute uremia. Am. J. Physiol. 219:649-653.
7. Frohlich, J., J. Scholmerich, G. Hoppe-Seyler, K.P. Maier, H. Talke, P. Schollmeyer, and W. Gerok. 1974. The effect of acute uremia on gluconeogenesis in isolated perfused rat livers. Eur. J. Clin. Invest. 4:453-458.
8. Mitch, W.E. 1981. Amino acid release from the hindquarter and urea appearance in acute uremia. Am. J. Physiol. 241:E415-E419.
9. Shear, L. 1969. Internal redistribution of tissue protein synthesis in uremia. J. Clin. Invest. 48:1252-1257.
10. Pils, P., W. Jettmar, D. Adamiker, and K.H. Tragl. 1981. Insulin and the in vitro protein synthesis of liver and skeletal muscle ribosomes in experimental acute uraemia. Horm. Metab. Res. 13:89-91.
11. Jefferson, L.S. 1975. A technique for perfusion of an isolated preparation of rat hemicorpus. Methods Enzymol. 39:73-82.
12. Krebs, M.A., and K. Henseleit. 1932. Hoppe-Seylers. J. Physiol. Chem. 210:33-66.
13. Jefferson, L.S., J.B. Li, and S.R. Rannels. 1977. Regulation by insulin of amino acid release and protein turnover in the perfused rat hemicorpus. J. Biol. Chem. 252:1476-1483.
14. Kopple, J.D. 1981. Nutritional therapy in kidney failure. Nutr. Rev. 39:193-206.
15. Lamprecht, W., and I. Trautschold. 1974. In Methods of Enzymatic Analysis. H.V. Bergmeyer, editor. 2nd English Edition, Vol. 4. Academic Press, New York. 2101-2110.
16. Torikai, S., M.S. Wang, K.L. Klein, and K. Kurokawa. 1981. Adenylate cyclase and cell cyclic AMP of rat cortical thick ascending limb of Henle. Kidney Int. 20:649-654.
17. Barrett, A.J. 1972. A new assay for cathepsin B1 and other thiol proteinases. Anal. Biochem. 47:280-293.
18. Anson, M.L. 1938. The estimation of pepsin, trypsin, papain and cathepsin with hemoglobin. J. Gen. Physiol. 22:79-89.
19. Mayer, M., R. Amin, and E. Shafrir. 1974. Rat myofibrillar protease: Enzyme properties and adaptive changes in conditions of muscle protein degradation. Arch. Biochem. Biophys. 161:20-25.
20. Pennington, R.J., and J.E. Robinson. 1968. Cathepsin activity in normal and dystrophic human muscle. Enzym. Biol. Clin. 9:175-182.
21. Flugel-Link, R.M., I. B. Salusky, M. R. Jones, and J. D. Kopple. Protein and amino acid metabolism in the posterior hemicorpus of acutely uremic rats. Am. J. Physiol. (In press).
22. Kar, N. C., and C. M. Pearson. 1972. Acid, neutral and alkaline cathepsins in normal and diseased human muscle. Enzyme 13:188-196.
23. Li, J. B. 1980. Protein synthesis and degradation in skeletcal muscle of normal and dystrophic hamsters. Am. J. Physiol. 239:E401-406.
24. Mayer, M. R., R. Amin, R. J. Milholland, and F. Rosen. 1976. Possible significance of myofibrillar protease in muscle catabolism. Enzyme activity in dystrophic, tumor-bearing, and glucocorticoid-treated animals. Exper. Molec. Pathol. 25:9-19.

25. Lundholm, K., A.-C. Bylund, J. Holm, and T. Schersten.   1976. Skeletal muscle metabolism in patients with malignant tumor.   Euro. J. Cancer 12:465-473.
26. Horl, W. H., and A. Heidland. 1980. Enhanced proteolytic activity - cause of protein catabolism in acute renal failure.   Am. J. Clin. Nutr. 33:1423-1427.
27. Libby, P., and A. L. Goldberg.   1980.   Effects of chymostatin and other proteinase inhibitors on protein breakdown and proteolytic activities in muscle. Biochem. J. 188:213-220.
28. Tischler, M. D., and A. L. Goldberg. 1980. Leucine degradation and release of glutamine and alanine by adipose tissue.   J. Biol. Chem. 255:8074-8081.
29. Feinstein, E. I., J. D. Kopple, H. Silberman, and S. G. Massry. 1983. Parenteral nutrition with increased nitrogen intake in the treatment of acute renal failure. Kidney Int 23:123 (abstract).

ENHANCED MUSCLE PROTEIN CATABOLISM IN UREMIA

H.R. Harter, T.A. Davis and I.E. Karl

Department of Medicine, Divisions of Nephrology and
Metabolism, and the Chromalloy American Kidney Center
Washington University School of Medicine, St. Louis,
Missouri

Acute and chronic uremia are associated with variable degrees
of myopathy, a retardation in growth, and a reduction in lean body
mass (1-3). With time, both the upper and lower extremities may
demonstrate diffuse muscle atrophy and weakness (3). These data
suggest that muscle protein catabolism may be increased, synthesis
decreased, or both may coexist in uremia. Significant changes in
plasma amino acids occur in both acute and chronic uremia. The
plasma phenylalanine to tyrosine ratio is reduced and plasma nonessential amino acid levels are usually normal while the levels of
essential amino acids including leucine, isoleucine, valine, and
lysine are decreased (4, 5). These data suggest that significant
abnormalities in protein metabolism do exist in uremia. Uremic
rats fed identical diets exhibit significantly reduced growth
rates compared to pair-fed control animals (6). Despite increased
protein intake, growth was still impaired in the uremic animals.
Thus, these data demonstrate that protein efficiency is reduced in
the growing uremic animal. The mechanisms responsible for these
observations have not been definitively elucidated.

Clinical studies designed to evaluate the effect of uremia on
muscle protein metabolism are limited in number (2, 5). Recent
data has demonstrated that various proteolytic enzymes are normal
despite the enhanced releases of alanine, phenylalanine and glutamine in the perfused hindquarter of the acutely uremic rat (7, 8).
Since the hindquarter preparation is composed of skin and bone in
addition to muscle (9), rather than selecting this in situ model
to evaluate the specific mechanisms responsible for the changes in
protein metabolism in muscle, an in vitro muscle preparation was
developed which would limit variables, e.g. hormonal, vascular and
neurogenic, as much as possible. The epitrochlearis muscle, which

557

weighs 20-30 mg and is 0.2 mm thick in a 120-140 g rat, can be removed rapidly and intact from the foreleg of the rat. This muscle
maintains viability for a period well over six hours as noted by
measurements of ATP, phosphocreatine, ADP, AMP, glycogen, lactate
to pyruvate ratios and glucose uptake (10). Studies also demonstrated that the releases of amino acids follow a pattern similar
to that seen in vivo and that alanine, glutamine and glutamate
account for 70% of the total amino acid efflux from muscle (11).
Various hormonal and dietary influences which are associated with
increased rates of nitrogen loss and muscle wasting, such as that
seen in fasting, diabetes produced with streptozotocin, or treatment
of animals with glucocorticoids or thyroxin, also produce an increased rate of skeletal muscle proteolysis as reflected in the sum
of alanine, glutamine and glutamate releases (12). The increased
release of these amino acids was not related to augmented glucose
uptake, suggesting that the carbon sources for these amino acids
may be derived from sources other than glucose (10). It also was
demonstrated that the epitrochlearis muscle has an epinephrine responsive adenylate cyclase system (13). Addition of epinephrine to
the incubating media reduced the muscle releases of alanine, glutamine and glutamate in diabetic and thyroxine and cortisone treated
rats.

    Studies from our laboratory utilizing the epitrochlearis muscle
have recently been extended to uremia. Male Sprague-Dawley rats
were made uremic by a 75% left renal infarction followed by a con-

Table 1. Plasma levels of BUN, Creatinine, Insulin and Glucagon in
          Control (C) and Uremic (Nx) Rats.

DIET	BUN mg/dl	CREATININE mg/dl	INSULIN uU/ml	GLUCAGON pg/ml
10% Casein				
C (22)	11.2 + 1.2	0.35 + 0.02	12.8 + 3.8	167 + 21
Nx (22)	29.4 + 6.5*	0.60 + 0.03*	8.2 + 2.2	575 + 49*
20% Casein				
C (40)	13.3 + 0.9	0.37 + 0.02	50.0 + 7.7	167 + 21
Nx (37)	36.5 + 2.0*	0.86 + 0.06*	16.8 + 3.3*	496 + 20.5*
40% Casein				
C (24)	37.2 + 1.7	0.24 + 0.02	52.2 + 8.0	157 + 27
Nx (18)	106.8 + 2.8*	0.76 + 0.06*	29.4 + 2.5*	632 + 73*

Number of animals studied appear in parentheses.
* $P < 0.05$ compared with corresponding C values.
* $P < 0.001$ compared with corresponding C values.

tralateral nephrectomy or were sham operated. Sham-operated control
rats were pair-fed to uremic animals for 10 days prior to study.
Initial studies were designed to test the effects of various protein intakes on in vitro muscle metabolism (14). Animals were fed
10%, 20%, or 40% isocaloric casein diets. Table 1 describes the
plasma chemistries in these animals. Plasma creatinine levels were
increased approximately two-fold in all groups of uremic animals
compared to control values. Of interest, the BUN values increased
significantly in both the uremic and control animals as the protein
content of the diet was increased. Insulin levels in these non-
fasted rats were lower in uremic animals compared to control values
and increased as the protein content of the diet was increased.
Glucagon values were increased two fold above control values regardless of dietary protein intake.

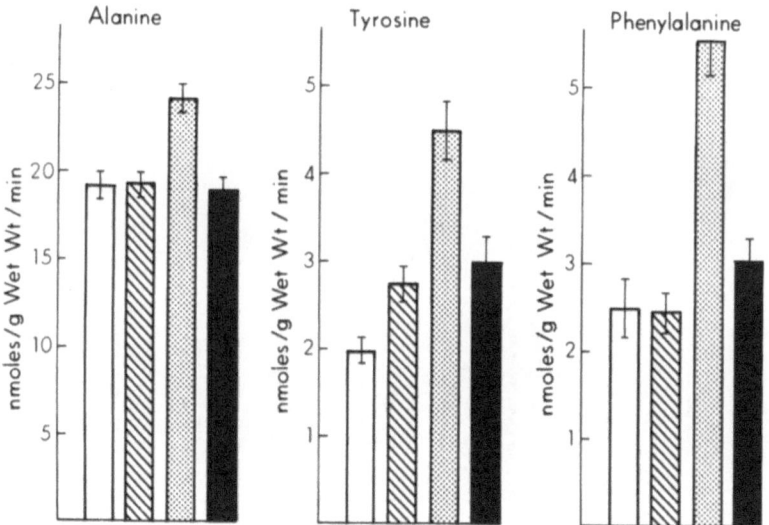

Fig. 1.   Effects of uremia and insulin on alanine, tyrosine, and
          phenylalanine release from muscles after one hour of incu-
          bation in 10% casein fed rats. ⊏⊐, muscles from control
          rats incubated without insulin; ⧄, muscles from control
          rats incubated with insulin; ⊡, muscles from uremic rats
          incubated without insulin; ■, muscles from uremic rats
          incubated with insulin.

The epitrochlearis muscles were removed and incubated for one hour in 500 ul of Kreb's Ringer bicarbonate buffer containing 5mM glucose and 5mM Hepes buffer. Where specified, 0.01u/ml of purified insulin was added. All incubations were conducted at 37°C in a metabolic shaker using continuous gassing with a 95% O_2 - 5% CO_2 mixture. The releases of alanine, tyrosine and phenylalanine were significantly increased in uremic rats fed 10% casein compared to values obtained from control rats (Fig. 1). Addition of insulin to the media reduced the release rates to control values. The releases of tyrosine and phenylalanine, both noncatabolizable amino acids at the level of muscle, decreased as dietary protein intake was increased in the uremic rats (Fig. 2). The addition of insulin suppressed these release rates to control levels. On the other hand,

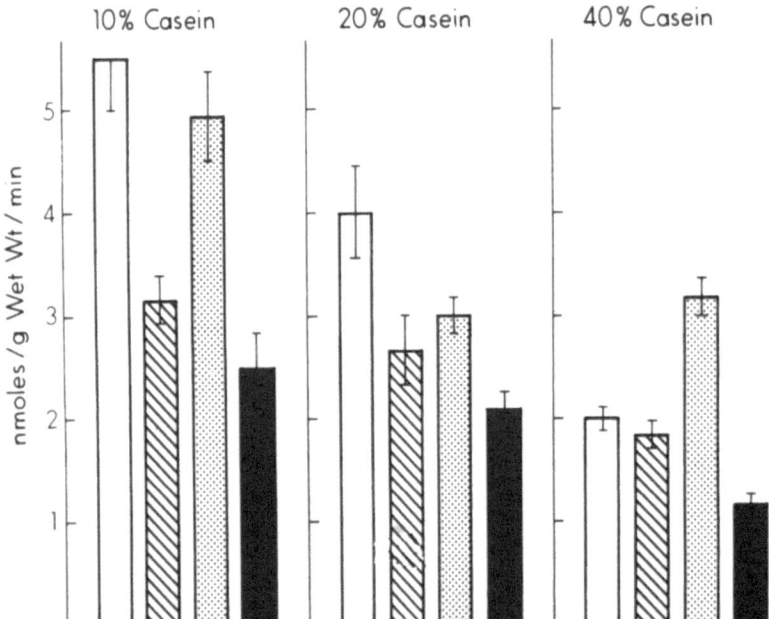

Fig. 2.   Effects of diet and insulin on phenylalanine and tyrosine release from muscle after one hour of incubation in uremic rats. ⊏⊐, phenylalanine release from muscles of uremic rats incubated without insulin; ▨, phenylalanine release from muscles of uremic rats incubated with insulin; ▦, tyrosine release from muscles of uremic rats incubated without insulin; ■, tyrosine release from muscles of uremic rats incubated with insulin.

alanine release was unaffected by changes in dietary protein and the effects of insulin were blunted. The releases of glutamine, gluta-mate, lactate and pyruvate were unaffected by uremia. Intracellular levels of alanine, glutamine, glutamate, ATP and glycogen were unaf-fected by uremia, diet or insulin. Thus, the changes in the observed release rates were not associated with changes in cellular energy or membrane permeability. These data demonstrate that uremia is associ-ated with increased muscle protein catabolism which is sensitive to dietary protein intake or to the addition of insulin to the incubating media. These data support the studies of DeFronzo et al which show that insulin resistance resides in peripheral tissues in uremia (15).

Since abnormalities in parathyroid hormone (PTH) and vitamin D metabolism exist in uremia and since PTH may produce toxic effects in various organ systems (16, 17), studies were carried out with each hormone. Parathyroid hormone did not affect in vitro metab-olism of muscles obtained from normal or uremic animals. Prior studies have shown that 25-OH cholecalciferol (25-OHD$_3$) stimulates phosphate uptake and increases cellular ATP levels in muscles ob-tained from rachitic rats (18). Furthermore, 25-OHD$_3$ but not 1,25 diOH cholecalciferol (1,25 diOHD3) has been shown to increase calcium uptake by the sarcolemma (19). These results are not sur-

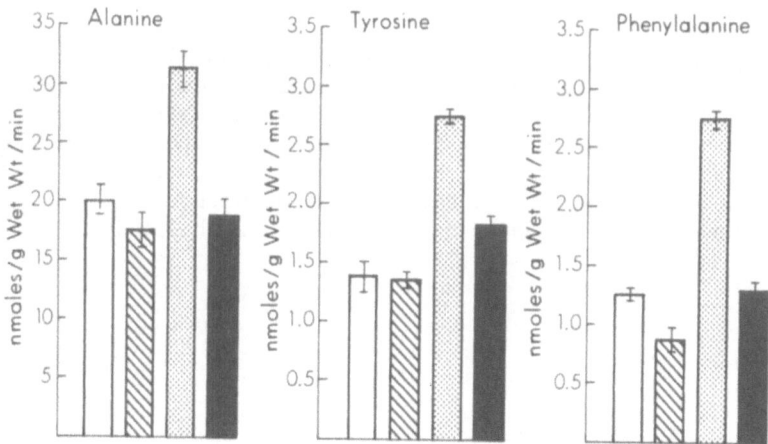

Fig. 3.   Effects of 25-OH cholecalciferol (25OHD$_3$) on alanine, tyrosine, and phenylalanine release from muscles from control and uremic rats after one hour of incubation. , muscles from control rats incubated without 25OHD$_3$; , muscles from control rats incubated with 25OHD$_3$; , muscles from uremic rats incubated without 25OHD$_3$; , muscles from uremic rats incubated with 25OHD$_3$.

prising since a cytosolic receptor protein has been isolated for
$25\text{-}OHD_3$, but not $1,25$ diOHD3 (20). We tested the effects of these
vitamin D metabolites on in vitro muscle metabolism in uremia (21).
Under baseline conditions, the releases of alanine, tyrosine and
phenylalanine from epitrochlearis muscles obtained from uremic
animals were increased 60-100% above control values (Fig. 3). The
addition of 100ng/ml of $25\text{-}OHD_3$ to the incubating media was asso-

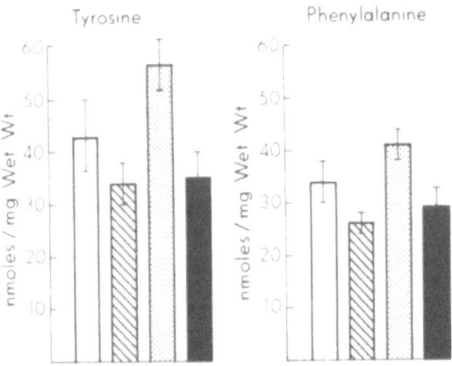

Fig. 4. Effects of 25-OH cholecalciferol (25OHD3) on muscle pool
         sizes of tyrosine and phenylalanine from muscles of uremic
         and control rats after one hour of incubation. ⊏⊐,
         muscles from control rats incubated without 25OHD3; ⧄,
         muscles from control rats incubated with 25OHD3; ⊡,
         muscles from uremic rats incubated without 25OHD3; ▬,
         muscles from uremic rats incubated with $25OHD_3$.

ciated with a significant reduction in the release of these amino
acids after one hour of incubation. $25\text{-}OHD_3$ had no effects on
amino acid release rates from muscle of normal rats. No additive
effects were seen when insulin and $25\text{-}OHD_3$ were added to the incubation media. Furthermore, the addition of lng/ml of $1,25$ diOHD3
did not affect these release rates from muscles obtained from normal
or uremic animals. ATP and glycogen levels were comparable in both

groups of muscles. Of interest, the intracellular levels of tyrosine and phenylalanine were significantly elevated in muscles from uremic animals and were returned to control levels by the addition of 25-OHD$_3$ (Fig. 4). On the other hand, tissue cyclic AMP levels and the incorporation of (^{3}H-) leucine into muscle protein were unaffected by the vitamin. These data suggest that resistance to the effects of 25-OHD$_3$ exists at the level of muscle.

Studies by Garber et al have demonstrated the presence of an epinephrine responsive adenylate cyclase system in the epitrochlearis muscle (13). Baseline activity of this adenylate cyclase system is reduced in epitrochlearis muscles obtained from uremic rats (22). Furthermore, the responses of this adenylate cyclase system to epinephrine and serotonin were blunted in uremia (23).

Thus, the increased muscle protein catabolism noted in uremia may be associated with in vitro resistance to various hormones including epinephrine, serotonin and insulin. Furthermore, there is a reduced cellular response to the effects of 25-OHD$_3$. Dietary protein intake also appears to have significant effects on these metabolic parameters. The mechanisms responsible for the apparent hormonal resistance in uremia are unknown but may relate to altered hormonal receptor binding or to a post-receptor defect (24).

Endurance exercise training has been shown to enhance the cellular response to insulin (25) and epinephrine (26). It is possible therefore, that exercise training may have a salutory effect on increased muscle protein catabolism in uremia. We have tested this hypothesis in our laboratory utilizing an exercise trained uremic

Table 2.   Plasma levels of BUN, Creatinine, Insulin and Glucagon in Sedentary and Exercise Trained Control and Uremic Rats.

	BUN mg/dl	CREATININE mg/dl	INSULIN uU/ml	GLUCAGON pg/ml
Sedentary Uremic (7)	$55.1 \pm 15.6^*$	$0.84 \pm 0.16^*$	21.0 ± 3.6	$323 \pm 45^{*}_{*}$
Exercise Uremic (5)	$24.9 \pm 5.4^{*}_{*}$	$0.57 \pm 0.07^{*}_{*}$	26.8 ± 2.3	$290 \pm 43^{*}_{*}$
Sedentary Control (10)	13.1 ± 1.0	0.43 ± 0.02	19.4 ± 2.5	233 ± 20
Exercise Control (10)	11.7 ± 0.8	0.42 ± 0.02	18.5 ± 2.2	194 ± 9

Number of animals studied appear in parentheses.
* P < .003 compared to control values.
* P < .05 compared to control values.

rat model (27). Female Sprague-Dawley rats weighing 80-90 gm were
made uremic by a 50% infarction followed by a contralateral nephrectomy. Sham-operated control animals were pair-fed to uremics within
both the sedentary and exercise trained groups. Two days after
surgery they were made to swim 15 minutes in a tank containing water
maintained at 35-37°C. The duration of swimming was slowly increased
until the animals could swim 2 hours daily five days per week. After

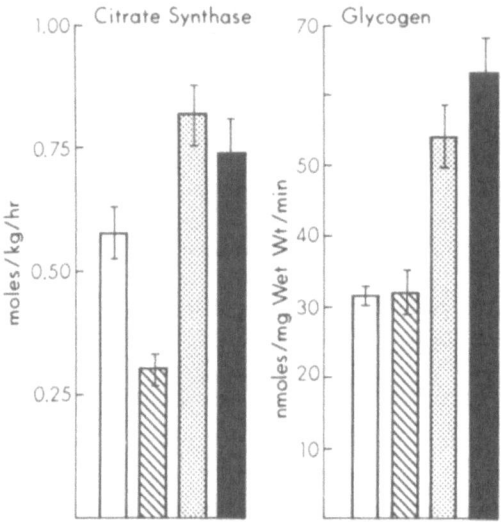

Fig. 5. Effect of exercise training on muscle citrate synthase
         activity and glycogen content in control and uremic rats.
         ⬜, sedentary control; ▨, sedentary uremic; ▦,
         exercise control; ■, exercise uremic.

four weeks, the animals were sacrificed and blood was obtained for
creatinine, BUN, glucose, insulin and glucagon determinations (Table
2). Plasma creatinine and BUN levels were significantly increased
in the sedentary uremic group. With exercise training, the BUN and
creatinine values fell in the uremic animals but were significantly
greater than controls. Insulin and glucagon values were unaffected
by exercise training.

Muscle glycogen content and citrate synthase activity were determined to document the degree of exercise conditioning in these animals (28; Fig. 5). Of interest, tissue citrate synthase levels were significantly decreased in epitrochlearis muscles obtained from sedentary uremic rats compared to control values. With exercise training significant increases in muscle citrate synthase and glycogen levels were noted in both control and uremic animals compared to corresponding sedentary values. Muscle weight and protein content were unaffected by uremia or exercise training.

Exercise training reduced the elevated tyrosine and phenylalanine releases from muscles obtained from uremic rats (Fig. 6). No changes were noted from muscles obtained from exercise trained control rats. The release of alanine was significantly increased from muscles obtained from sedentary uremic rats compared to control values. Of interest, no decrease in the release of this amino acid was noted with exercise training in uremic rats. Moreover, exercise training increased alanine release from muscles obtained from control rats.

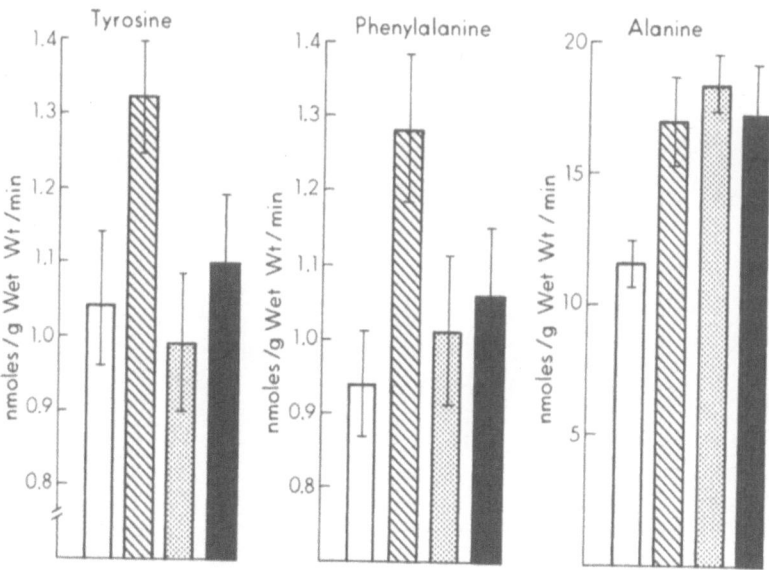

Fig. 6. Effect of exercise training on tyrosine, phenylalanine, and alanine release from muscles obtained from control and uremic rats after one hour of incubation. ▭, sedentary control; ▨, sedentary uremic; ▦, exercise control; ■, exercise uremic.

    To evaluate the maximal effects of insulin (0.01 u/ml) on these
in vitro metabolic parameters, glucose uptake and amino acid release
rates were determined after three hours of incubation. The effects
of insulin on glucose uptake by muscles obtained from exercised uremic
and control rats were increased compared to values obtained from the
corresponding sedentary animals (Fig. 7). Insulin significantly
reduced the releases of tyrosine and phenylalanine from muscles obtained from exercise trained uremic and control rats. These data
suggest that exercise training can reduce in vitro muscle protein
catabolic rates and these changes are associated with an enhanced
tissue response to the cellular effects of insulin. Whether these
observations reflect enhanced tissue binding of insulin or a post-
receptor phenomenon remains to be determined.

    In summary, chronic uremia is associated with enhanced muscle
protein catabolism. There appears to be resistance to the tissue
effects of at least three anticatabolic hormones including insulin,
epinephrine and serotonin. Furthermore, 25-OHD3 reduces the accelerated amino acid release rates from muscles of uremic rats to levels
comparable to those seen in control animals. Protein intake also
has significant effects on these release rates. As dietary protein
intake is increased the releases of tyrosine and phenylalanine are

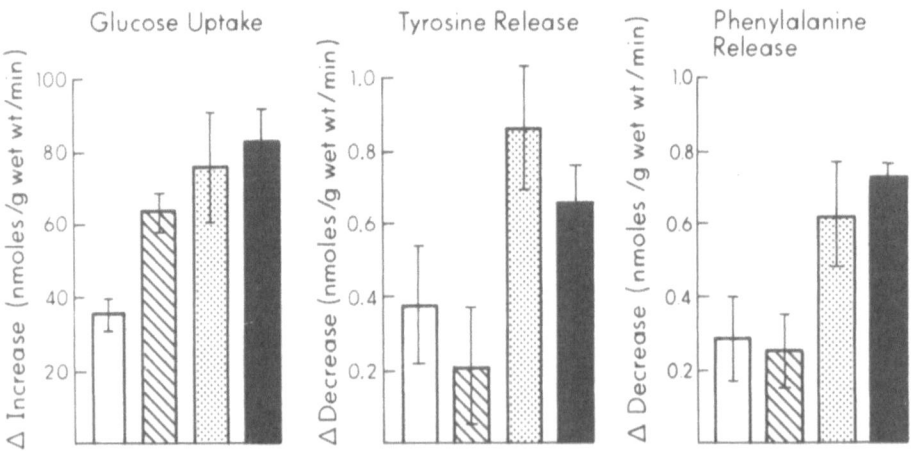

Fig. 7.   Effect of changes in glucose uptake and tyrosine and phen-
          yalanine release with the addition of insulin to the media
          during three hours of incubation in muscles obtained from
          sedentary and exercise trained control and uremic rats.
          ☐ , sedentary control;▨, sedentary uremic;▨, exercise
          control;■, exercise uremic.

decreased. Whether these observations reflect changes in cellular proteolytic enzyme activity is still not clear. One potential therapeutic modality that may help correct the increased proteolysis in uremic muscle is exercise training. Further studies are necessary to
document whether the improvement in muscle protein metabolism with
exercise training reflect only enhanced tissue sensitivity to insulin
or whether other hormones are involved as well.

ACKNOWLEDGEMENTS

     This work was supported in part by USPHS NIAMDD grants AM0997
and T32 HL07456-03. The technical assistance of Elise Tegtmeyer,
Dale Osborne, and Thomas Howard, and the secretarial assistance of
Brenda Lister are greatly appreciated.

REFERENCES

1. M. A. Holliday, and C. Chantler, Metabolic and nutritional fac-
    tors in children with renal insufficiency, Kidney Int. 14:
    306 (1978).
2. C. Delaporte, G. Jean, and M. Broyer,Free plasma and muscle ami-
    no acids in uremic children, Am. J. Clin. Nutr. 31:1647 (1978).
3. M. Floyd, D. D. Ayyar, P. Barwick, and D. Weightman, Myopathy
    in chronic renal failure, Q. J. Med. 172:509 (1974).
4. L. Shear, Internal redistribution of tissue protein synthesis
    in uremia, J. Clin. Invest. 48:1252 (1969).
5. J. D. Kopple, and M. R. Jones, Amino acid metabolism in patients
    with advanced uremia and in patients undergoing chronic dia-
    lysis, Adv. Nephrol. 8:233 (1979).
6. M. Wang, I. Vyhmeister, J. D. Kopple, and M. E. Swendseid,
    Effect of protein intake on weight gain and plasma amino
    acid levels in uremic rats, Am. J. Physiol. 230:1456 (1976).
7. R. M. Flugel-Link, I.B. Salusky, M.R. Jones, J.D. Kopple, En-
    hanced muscle protein degradation and amino acid release from
    the hemicorpus of acutely uremic rats, in: "Proteases: Poten-
    tial Role in Health and Disease," W.H. Hörl, A. Heidland,
    eds., Plenum Press, New York (1983).
8. W. E. Mitch, Amino acid release from the hindquarter and urea
    appearance in acute uremia, Am. J. Physiol. 241:E415 (1981).
9. V. R. Preedy, and P. J. Garlick, Rates of protein synthesis in
    skin and bones and their importance in the assessment of
    protein degradation in the perfused rat hemicorpus, Biochem.
    J. 194:373 (1981).
10. A. J. Garber, I. E. Karl, and D. M. Kipnis, Alanine and gluta-
    mine synthesis and release from skeletal muscle. I. Glyco-
    lysis and amino acid release, J. Biol. Chem. 251:826 (1976).
11. A. J. Garber, I. E. Karl, and D. M. Kipnis, Alanine and gluta-
    mine synthesis and release from skeletal muscle. II. The
    precursor role of amino acids in alanine and glutamine syn-
    thesis, J. Biol. Chem. 251:836 (1976).

12. I. E. Karl, A. J. Garber, and D. M. Kipnis, Alanine and gluta-
    mine synthesis and release from skeletal muscle. III. Die-
    tary and hormonal regulation, J. Biol. Chem. 251:844 (1976).
13. A. J. Garber, I. E. Karl, and D. M. Kipnis, Alanine and gluta-
    mine synthesis and release from skeletal muscle. IV. B-adre-
    nergic inhibition of amino acid release, J. Biol. Chem. 251:
    851 (1976).
14. H. R. Harter, I. E. Karl, S. Klahr, and D. M. Kipnis, Effects
    of reduced renal mass and dietary protein intake on amino
    acid release and glucose uptake by rat muscle in vitro. J.
    Clin. Invest. 64:513 (1979).
15. R. A. DeFronzo, A. Alvestrand, D. Smith, R. Hendler, E. Hendler,
    and J. Wahren, Insulin resistance in uremia, J. Clin. Invest.
    67:563 (1981).
16. E. Slatopolsky, K. Martin, and K. Hruska, Parathyroid hormone
    metabolism and its potential as a uremic toxin, Am. J.
    Physiol. 239:F1 (1980).
17. R. Gray, I. Boyle, and H. F. DeLuca, Vitamin D metabolism: The
    role of kidney tissue, Science 172:1232 (1971).
18. S. J. Birge, and J. G. Haddad, 25-hydroxycholecalciferol stimu-
    lation of muscle metabolism, J. Clin. Invest. 56:1100 (1975).
19. J. J. Pointon, M. J. O. Francis, and R. Smith, Effect of vita-
    min D deficiency on sarcoplasmic reticulum function and
    troponin C concentration of rabbit skeletal muscle, Clin.
    Sci. 57:257 (1979).
20. J. G. Haddad, and S. J. Birge, Widespread, specific binding of
    25-hydroxycholecalciferol in rat tissues, J. Biol. Chem.
    250:299 (1975).
21. H. R. Harter, S. J. Birge, K. J. Martin, S. Klahr, and I. E.
    Karl, The effects of vitamin D metabolites on the protein
    catabolism of muscle from uremic rats, Kidney Int. in press
    (1983).
22. A. J. Garber, Skeletal muscle protein and amino acid metabolism
    in experimental chronic uremia in the rat. Accelerated ala-
    nine and glutamine formation and release, J. Clin. Invest.
    62:623 (1978).
23. A. J. Garber, The regulation of skeletal muscle alanine and
    glutamine formation and release in experimental chronic
    uremia in the rat. Subsensitivity of adenylate cyclase and
    amino acid release to epinephrine and serotonin, J. Clin.
    Invest. 62:633 (1978).
24. D. Smith, and R. A. DeFronzo, Insulin resistance in uremia me-
    diated by postbinding defects, Kidney Int. 22:54 (1982).
25. V. Koivisto, V. Soman, E. Nadel, W. V. Tamborlane, and P. Fe-
    lig, Exercise and insulin: Insulin binding, insulin mob-
    ilization and counter-regulatory hormone secretion, Fed.
    Proc. 39:1481 (1980).
26. C. S. Liang, R. R. Tuttle, W. B. Hood, and H. Gavras, Con-
    ditioning effects of chronic infusions of dobutamine: Com-
    parison with exercise training, J. Clin. Invest. 64:613 (1979).

27. T. A. Davis, I. E. Karl, A. P. Goldberg, and H. R. Harter,
     Effects of exercise training on muscle protein catabolism
     in uremia, Kidney Int. in press (1982).
28. J. O. Holloszy, and F. W. Booth, Biochemical adaptation to
     endurance exercise in muscle, Annu.Rev.Physiol.38:273 (1976).

CATABOLIC STRESS ON INTRACELLULAR AMINO ACID POOL

Peter Fürst

Institute for Biological Chemistry and Nutrition
University of Hohenheim, Garbenstrasse 30
7000 Stuttgart 70, W. Germany

Many studies have documented abnormal plasma concentrations of the free amino acids in patients with different catabolic disorders. The interpretation and the pathogenesis of these abnormal patterns, however, remain speculative. Certain abnormalities appear to be caused by the catabolic stress per se, whereas others, resembling those observed in subjects with low intake of protein and energy, may be related to malnutrition. During catabolic stress the distribution of free amino acids showed to be altered between the extracellular and intracellular compartments[1-6]. Therefore the plasma concentrations do not necessarily reflect the intracellular concentration. Since skeletal muscle contains the largest pool of free amino acids, determination of the free amino acid concentration in muscle might be of particular interest in the study of amino acid and protein metabolism during catabolic stress and subsequent repletion.

METHODS

In the past 20 years a relatively simple and safe technique for taking muscle biopsies has been developed in Stockholm[7] and subsequently introduced and applied at Columbia University, New York[2-4]. In this procedure, a 0.5 cm incision in the thigh is made under local anaesthesia, permitting a needle biopsy of the quadriceps femoris muscle amounting to 30-50 mg of tissue[7,8]. Multiple biopsies can be taken in almost any situation. With the use of modern microanalytical techniques this biopsy can be analyzed for water[7], electrolytes[9,10], lipids[7] and free amino acids[11,12]. Similar measurements are made on a plasma sample and the distribution of intracellular and extracellular water and the concentration of intracellular amino acids and electrolytes may then be calculated[7],[10-12].

THE INTRACELLULAR MUSCLE FREE AMINO ACID PATTERN

One important finding is that there are considerable species differences in muscle amino acid concentrations[6]. This is important while it is frequently possible to make qualitative conjectures about human metabolism from animal studies, a quantitative under-standing, thus, requires direct measurement on man.

Another important finding is that the concentrations of muscle free amino acids change in response to changes in diet and physio-logical or pathological state, such as uremia[1],[5],[13-15], diabetes mellitus[16], postoperative trauma[3],[6],[17], injury[4],[18], sepsis[4], starvation[19], semistarvation and immobilisation[2],[3]. This is shown for selected amino acids in figure 1.

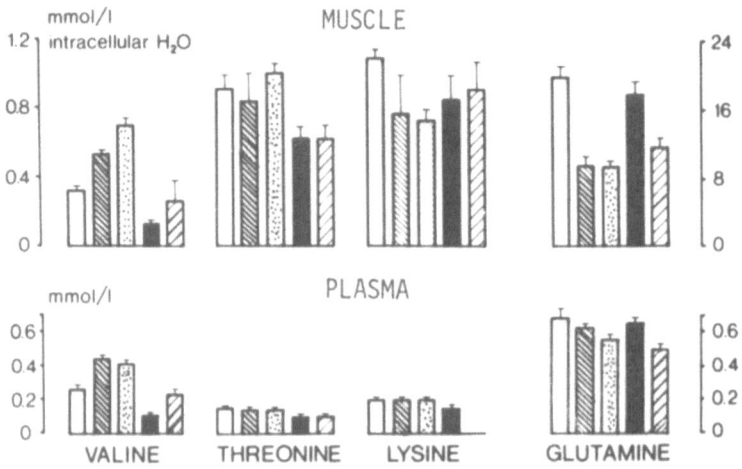

Fig. 1. Free amino acids in different catabolic states in muscle
         and plasma.
             □ normal   ■ uremia-untreated
             ▨ 11 day fast   ▧ diabetes-untreated
             ▦ after total hip replacement

The figure illustrates some typical changes in certain catabolic conditions. As shown each catabolic condition appears to have its own unique and reproducible intracellular pattern. Valine, like the other Branched-Chain Amino Acids (BCAA), increases uniformly in trauma and is unchanged in diabetes. Glutamine by contrast decreases considerably in trauma, starvation and diabetes but is unchanged in uremia. Changes in plasma free amino acid concentrations may parallel those in muscle, but frequently differ either quantitatively or qualitatively. These results indicate that the intracellular free amino acid pattern of muscle is characterized by a "unique pattern of catabolism" distinctive for the disease or condition studied.

Typical changes in intracellular muscle free and plasma free concentrations of BCAA, aromatic amino acids (AAA) and methionine (Met) are summarized in 6 different catabolic situations in fig. 2. Bedrest, starvation, semistarvation, postoperative trauma, severe injury and onset of sepsis are conditions associated with negative nitrogen balance which is presumed to originate in large part from net protein degradation in skeletal muscle[20]. In the above conditions the observed negative nitrogen balance varies in extent from - 1 g N per day on bedrest up to-20-40 g N or more per day in severe injury and sepsis. This means that in each of these conditions there is a net transport of amino acids out of muscle to other tissues.

The pattern of changes in amino acid concentrations in muscle and plasma shows many similarities during catabolism. In all cases there is an increase in BCAA, AAA and Met and a decrease in glutamine (Gln) and in basic amino acids. There is, however, a gradual response to the catabolic stress in the alterations for BCAA, AAA and Met, with minimal changes seen in bedrest and maximal changes in sepsis as illustrated in fig. 2. Obviously, the alterations observed in these conditions tend to follow a pattern according to the specifications of the different amino acid transport system. The BCAA together with AAA and Met are transported by the leucine preferring system (L-system)[21]. As shown in fig. 2 in the milder conditions increases of muscle BCAA are proportionally greater than those of the AAA and Met, while in severe injury and sepsis, the increases for muscle AAA and Met are greater than for BCAA. It is also notable that in the milder conditions, plasma concentrations of BCAA rise proportionally to intracellular conditions, but in severe catabolic state increased intracellular BCAA concentrations are not accompanied with similar rises in plasma.

A uniform reduction of about 50 % in muscle free Gln seems to be one of the most typical features of the response to starvation and trauma[2-4,6,17-19]. This consistent decrease in the intracellular Gln pool occurs irrespective of the degree of trauma or the mode of nutrition, suggesting that the observed depletion is virtually obligatory.

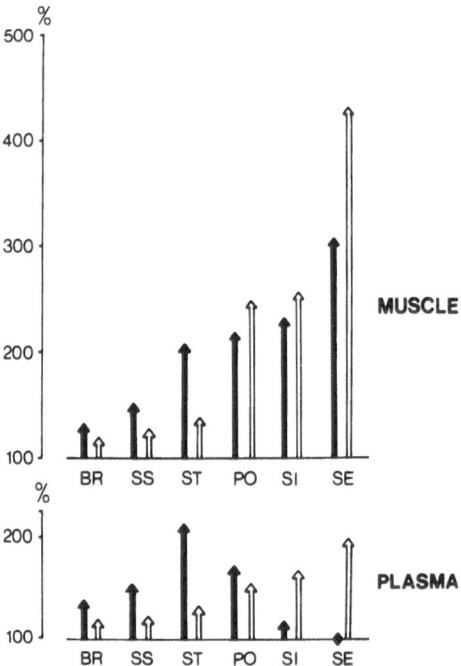

Fig. 2. Branched-chain amino acids (↑) and aromatic amino acids
and methionine (⇧) as affected by catabolism in muscle
and plasma. % changes from normal. BR = Bedrest; SS =
semistarvation; ST = starvation; PO = post-operative;
trauma; SI = severe injury; SE = sepsis.

HYPOTHESES AND INTERPRETATIONS

Increased tissue catabolism alone cannot explain the observed intracellular amino acid pattern, since certain amino acids show no change or reveal a decreased concentration. However, the selected changes seen for BCAA, AAA and Met may be accounted for, if it is postulated that the concentration of the L-system in muscle is rate limiting and that higher efflux requires higher intracellular concentrations. This helps to explain the finding that intracellular concentrations as well as intra-extracellular transmembrane gradients increase with increasingly negative nitrogen balance. In most severe catabolic states net proteolysis in muscle is much greater and there are fourfold increases in muscle concentrations of the AAA and Met[4,18]. The liver's ability to handle this outflow will be taxed, leading to increase in plasma concentrations. If the L-transport system is limiting in severe catabolic states, the high muscle concentrations of AAA and Met will competitively reduce the outflow of BCAA, thus plasma concentrations of BCAA will decrease with respect to those in muscle. This speculative explanation postulates: 1) that no adaptive changes occur in the muscle L-system, but rather that it becomes rate-limiting in severe catabolic states, 2) that the differences in behaviour of the BCAA and other amino acids are due to their different sites of catabolism, and 3) that marked increase in rates of catabolism of methionine, tyrosine and phenylalanine in the liver may occur with little increase in concentrations, whereas increased rates of catabolism of BCAA in muscle are more concentration dependent.

The lack of the most important non-essential amino acid Gln (more than 50 % of the total free intracellular amino acid pool), might be a serious limiting factor for optimum protein synthesis. Increased Gln transport, due to accelerated gluconeogenesis in the liver may serve as a possible explanation. An additional factor might be a relative inability to re-utilize available amino groups for Gln synthesis. While the majority of the intracellular amino acid concentrations returned to normal levels or tended toward normal during late convalescence[4], intracellular Gln continued to be reduced by 50 %. This finding indicates that this particular abnormality is very difficult to correct and seems to be a consistant quantitative response of the catabolic state. In late convalescence, when the patients are in the anabolic phase, the low Gln could be due to inadequate nutrition. Inadequate nutrition in the late convalescence phase may also explain the decreased intracellular arginine and lysine concentrations, since these amino acids showed very similar responses during starvation[19] and immobilisation plus semistarvation[2,3].

NEW THERAPEUTICAL APPROACHES IN INJURY AND IN CHRONIC UREMIA

It is slowly becoming clear that nutrition at height of the flow phase after injury requires further detailed studies of the influence of nutrients on the altered metabolism at this time. The imbalance of cellular amino acids in postoperative catabolism may be one of the fundamental inadequacies in cell nutrition adversely affecting normal protein synthesis. This indicates that proportions and minimum requirements based on data from healthy subjects cannot be applied directly in postoperative catabolism.

None of our results, however, support the recent suggestions[22,24] that administration of BCAA only may specifically effect protein synthesis in severe trauma and sepsis. The combination of tissue energy deficit[25,26] and intracellular accumulation of BCAA[3,4,6,18] suggests an impairment of substrate use rather than lack of fuel availability of the cell, further suggesting that alterations in BCAA do not mediate but rather result from changes in protein synthesis. These findings, thus, do not support the beneficial use of BCAA-enriched solutions[27,28] in severe trauma and sepsis.

Intracellular Gln depletion is virtually obligatory in trauma, whatever the nitrogen intake. Technical difficulties make inclusion of free Gln impossible in presently available commercial solutions. Recently, a Gln-containing peptide has been synthetised, highly soluble, stable and cellularly cleavable[29,30]. Future studies will determine whether inclusion of such a peptide in commercial preparations will promote nitrogen utilization.

A better understanding of the factors affecting protein synthesis and breakdown is needed not only from theoretical standpoint, but has important practical implication if it can lead to reduction of catabolic losses and optimisation of protein gains during repletion therapy.

Recent experiences in conservative treatment of chronic uremia may serve as an example. In the untreated uremic patients there exists a unique intracellular pattern of uremia, characterized with low concentrations of threonine, valine, tyrosine, lysine and histidine and high levels of arginine, ornithine and citrulline[5,13-15]. The distribution of the BCAA is abnormal in chronic uremia, showing low intra- and extracellular valine concentrations but normal or high intracellular, and low plasma levels of leucine and isoleucine[5,13,15]. Abnormalities in the BCAA pattern resembling those observed in rats fed a low protein diet containing an excess of leucine[31] and suggest the existence of an BCAA antagonism in chronic uremia. Antagonism of BCAA, with other words, excess or deficiency of one of the BCAA might change the distribution of the others and this phenomenon is associated with deterioration of growth and diminished amino acid utilization[32]. The abnormal distribution of BCAA persisted in patients treated with low protein diet supplemented with an amino acid mixture of essential amino acids in the proportions recommended by Rose even though the regimen provided 3-4 times the minimum requirement of valine[5,14]. Based on information of free amino acids in plasma and muscle a new amino acid formula (NAF) was, therefore, composed and tested with changed proportions between the essential amino acids and with addition of tyrosine. Long-term treatment with this new formulation resulted in improved nitrogen balance and correction of abnormalities in muscle and plasma AA[5,33,34]. The improved nitrogen balance was conceivable achieved by correction of BCAA antagonism and by repletion of the free amino acid pools which were low and possible limiting for protein synthesis. The results, thus, suggest that normalization of amino acid pools may improve nitrogen utilization. Although the observed difference in nitrogen balance may be small when expressed in grams of nitrogen per day, the difference in the cummulative balance may become great and of obvious clinical importance in long-term treatment.

We believe that determination of intracellular patterns before and during therapy provides a powerful tool in developing better nutritional means during catabolic stress. It is also to hope that the data presented will positively contribute to protease research, assuming that intracellular findings can be related to a corresponding alteration in protease activities.

REFERENCES

1. J. Bergström, P. Fürst, L-O. Norée and E. Vinnars, Intracellular amino acids in muscle tissue of patients with chronic uremia. The effect of peritoneal dialysis and infusion of essential amino acids, Clin. Sci. Mol. Med. 54:51 (1978).
2. J. Askanazi, D.H. Elwyn, J.M. Kinney, F.E. Gump, C.B. Michelsen, F.E. Stinchfield, P. Fürst, E. Vinnars and J. Bergström, Muscle and plasma amino acids after injury: the role of inactivity Ann. Surg. 188:797 (1978).
3. J. Askanazi, P. Fürst, C.B. Michelsen, D.H. Elwyn, E. Vinnars, F.E. Gump, F.E. Stinchfield and J.M. Kinney, Muscle and plasma amino acids after injury: hypocaloric glucose vs amino acids infusion, Ann. Surg. 191:465 (1980).
4. J. Askanazi, Y.A. Carpantier, C.B. Michelsen, D.H. Elwyn, P. Fürst, L.R. Kantrowitz, F.E. Gump and J.M. Kinney, Muscle and plasma amino acids following injury. Influence of intercurrent infection, Ann. Surg. 192:78 (1980).
5. A. Alvestrand, J. Bergström and P. Fürst, Plasma and muscle free amino acids in uremia: influence of nutrition with amino acids, Clin. Nephrol (In press) (1982).

6. P. Fürst, J. Bergström, J.M. Kinney and E. Vinnars, Nutrition in postoperative catabolism, in:"Nutritional Aspects of Care of the Critically Ill," J.E. Richards and J.M. Kinney, eds., Churchill Livingstone, Edinburgh, London (1977).

7. J. Bergström, Muscle electrolytes in man, Scand. J. Clin. Lab. Invest. 68:7 (1962).

8. J. Bergström, Percutaneous needle biopsy of skeletal muscle in physiological and clinical research, Scand. J. Clin. Lab. Invest. 35:609 (1975).

9. J. Bergström and A-M. Fridén, The effect of hydrochlorothiazide and amiloride administerd together on muscle electrolyte in normal subjects, Acta Med. Scand. 197:415 (1975).

10. J. Bergström, P. Fürst, B. Holmström, E. Vinnars, J. Askanazi, D. Elwyn, C.B. Michelsen and J.M. Kinney, Influence of injury and nutrition on muscle water and electrolytes, Ann. Surg. 193: 810 (1981).

11. J. Bergström, P. Fürst, L-O. Norée and E. Vinnars, Intracellular free amino acid concentration in human muscle tissue, J. Appl. Physiol. 36:693 (1974).

12. J. Bergström, A. Alvestrand, P. Fürst, E. Hultman, K. Sahlin, E. Vinnars and A. Widström, Influence of severe potassium depletion and subsequent repletion with potassium on muscle electrolytes, metabolites and amino acids in man, Clin. Sci. Mol. Med. 51:589 (1976).

13. A. Alvestrand, J. Bergström, P. Fürst, G. Germanis, U. Widstam, Effect of essential amino acid supplementation on muscle and plasma free amino acids in chronic uremia, Kidney Intern. 14:323 (1978).

14. P. Fürst, M. Ahlberg, A. Alvestrand and J. Bergström, Principles of essential amino acid therapy in uremia, Am. J. Clin. Nutr. 31:1949 (1978).

15. P. Fürst, A. Alvestrand and J. Bergström, Effects of nutrition and catabolic stress on intracellular amino acid pools in uremia, Am. J. Clin. Nutr. 33:1387 (1980).

16. A. Roch-Nordlund, A. Alinder, M. Ahlberg, P. Fürst and G. Werner, Nitrogen metabolism in diabetic patients, Acta Endocrinol. 77:190 (1974).

17. E. Vinnars, J. Bergström and P. Fürst, Influence of postoperative state in the intracellular free amino acids in human muscle tissue, Ann. Surg. 163:665.

18. P. Fürst, J. Bergström, L. Chao, J. Larsson, S-O. Liljedahl, M. Neuhauser, B. Schildt and E. Vinnars, Influence of amino acid metabolism in severe trauma, Acta Chir. Scand 494:136 (1979).

19. D. Elwyn, P. Fürst, J. Askanazi and J.M. Kinney, Effect of fasting on muscle concentrations of branched chain amino acids, in:"Metabolism and Clinical Implication of Branched Chain Amino and Keto-Acids," M. Walser and J.T. Williamson, eds., North Holland, Elsevier, New York (1981).

20. J.M. Kinney, Surgical diagnosis, patterns of energy, weight and tissue changes, in:"Metabolism and the Response to Injury," A.W. Wilkinson and E.P. Cuthbertson, eds., Portman Medical, Kent, (1976).

21. H.N. Christensen, Some special kinetic problems of transport, Adv. Enzymol. 32:1 (1969).

22. T.F. O'Donnell Jr., G.H.A. Clowes Jr. and G.L. Blackburn, Proteolysis associated with a deficit of peripheral energy fuel substrates in septic man, Surgery 80:192 (1976).

23. H.R. Freund, A. Lapidot and J.E. Fischer, The use of branched chain amino acids in the injured-septic patient, in:"Metabolism and Clinical Implications of Branched Chain Amino and Ketoacids," M. Walser and J.R. Williamson, eds., North-Holland/Elsevier, New York, Amsterdam, Oxford (1981).

24. B. Bistrian, Improvement in amino acid utilization in the critically ill with BCAA enriched parenteral nutrition, in:" Advances in Clinical Nutrition," I.D.A. Johnston, ed., MTP-Press, Lancaster (In press) (1982).

25. K.Y. Liaw, J. Askanazi, C.B. Michelsen, L.R. Kantrowitz, P. Fürst and J.M. Kinney, Effect of injury and sepsis on high-energy phosphates in muscle and red cells, J. Trauma 20:755 (1980).

26. K.Y. Liaw, J. Askanazi, C.B. Michelsen, P. Fürst, D.H. Elwyn and J.M. Kinney, Effect of postoperative nutrition on muscle high energy phosphates, Ann. Surg. 195:12 (1982).

27. H.R. Freund, J.A. Ryan and J.E. Fischer, Amino acid derangements in patients with sepsis: treatment with branched-chain amino acid rich infusion, Ann. Surg. 188:423 (1978).

28. H.R. Freund, N. Yoshimura, L. Lunetta and J.E. Fischer, The role of the branched chain amino acids in decreasing muscle catabolism in vivo, Surgery 83:611 (1978).

29. P. Stehle, B. Kühne, W. Kubin, P. Fürst and P. Pfaender, Synthesis and characterization of tyrosine and glutamine containing peptides, J. Appl. Biochem. (In press) (1982).

30. P. Stehle, B. Kühne, P. Pfaender and P. Fürst, Isotachophoretic separation of two synthetic peptides, J. Chromatogr. (In press) (1982).

31. F.L. Shinnic and E.A. Harper, Effects of branched-chain amino acid antagonism in the rat on tissue amino acid and ketoacid concentrations, J. Nutr. 107:887 (1977).

32. A.E. Harper, Amino acid toxicities and imbalances, in:" Mammalian Protein Metabolism," H.N. Munro and J.B. Allison, eds., Academic Press, New York (1964).

33. A. Alvestrand, M. Ahlberg, J. Bergström and P. Fürst, The effect of nutritional regimens on branched-chain amino acid (BCAA) antagonism in uremia, in:"Metabolism and the Clinical Implications of Branched-Chain Amino and Ketoacids, M. Walser and J.R. Williamson, eds., North-Holland/Elsevier, New York (1981).

34. A. Alvestrand, M. Ahlberg, J. Bergström and P. Fürst, Clinical results of long-term treatment with low protein diet and a new amino acid preparation in chronic uremic patients, Clin. Nephrol. (In press) (1982).

RHABDOMYOLYSIS: A CLINICAL ENTITY FOR THE STUDY OF ROLE OF PROTEASES

Shaul G. Massry

Division of Nephrology
University of Southern California
School of Medicine
Los Angeles, California 90033

INTRODUCTION

Many of the readers of these proceedings may be surprised to find the clinical subject of rhabdomyolysis included in a symposium dealing with the biochemistry and biology of proteases. However, a deeper look into the issue provides a clear justification.

We have heard during the previous two days of this meeting a great deal of information on the biochemical characteristics of various proteases and on their biological features. However, the clinical implications of the derangements in the function of proteases and the clinical expressions associated with increased release of proteases into the circulation are lagging behind the biochemical progress of the field. It is time that a closer collaboration be developed between the biochemist, the biologist and the clinician in an effort to advance the potential importance of the proteases in health and disease. Thus, disease states associated with increased release of proteases into the circulation should be studied. Two such classical diseases are acute pancreatitis and non-traumatic rhabdomyolysis. My task is to provide a brief description of the latter entity and to speculate on the potential role of proteases in the pathogenesis of the various complications associated with rhabdomyolysis.

THE CLINICAL ENTITY OF NON-TRAUMATIC RHABDOMYOLYSIS

Rhabdomyolysis by definition is muscle necrosis. Therefore, rhabdomyolysis must be associated with the release of various skeletal proteases into the circulation. Thus, this pathologic entity

can serve a unique human model for the study of the effects of proteases on various body functions.

The major cause of non-traumatic rhabdomyolysis is drug abuse
(1,2). It is not infrequently encountered after an alcohol binge
(2), heroin overdose (1) or phencyclidine abuse (3). Under such
circumstances the patient is usually obtunded or may even be comatose. He remains immobilized for a prolonged period of time, and
the pressure exerted by his body weight against the floor damages
the muscle and results in myonecrosis. Figure 1 provides schematic
presentation of the sequence of events that follows the injection
of a large dose of heroin and the consequences of rhabdomyolysis.

The clinical setting of rhabdomyolysis is that of a patient
known to use heroin or other drugs and brought to the hospital obtunded or comatose. Occasionally, the patient may be awake but
gives a history of recent obtundation or coma. He may complain of
muscle tenderness or may display overt evidence of muscle injury
involving one or more limbs or other parts of his body. The muscle
damage may be mild or extensive and associated with marked swelling.
Those with extensive muscle injury may also complain of neuropathy
with motor and/or sensory deficits. Many of these patients display
clinical evidence of dehydration. A decrease in urine output and
the passing of red-brown urine may be reported by the patient or
oliguria and urine discoloration are detected in the hospital.
Acute renal failure is not uncommon in these patients and could be
of the oliguric or non-oliguric variety. Thus, these patients have
clinical features of myopathy, nephropathy and neuropathy (1-5).

The laboratory findings in these patients include features of
rhabdomyolysis and of acute renal failure. Rhabdomyolysis causes
myoglobinuria, elevated blood levels of muscle enzymes, dispropor-
tionate rise in serum creatinine and leukocytosis (1-3). Myoglobinuria is responsible for the urine discoloration (red-brown
urine). The urine is strongly positive for blood by Dipstick but
only few or no red blood cells are found. Myoglobin in urine
should be measured to help confirm the diagnosis. The measurement
should be made early in the course of the disease since myoglobin
may not be detected several days after the myonecrosis. A red-
brown urine could be produced by other disease entities such as
hematuria, hemoglobinuria, porphyria and following the ingestion
of exogenous pigments.

The blood levels of muscle enzymes [creatine phosphokinase
(CPK) and aldolase] rise rapidly and reach high values (CPK >
24,000 U/L) to be followed by gradual decline. If the blood levels
of these enzymes do not fall, one must suspect continued myonecrosis
which is usually encountered in patients with extensive swelling of
limbs with marked compression of arteries and muscles. Such patients require fasciotomy of the affected limbs to relieve pressure.

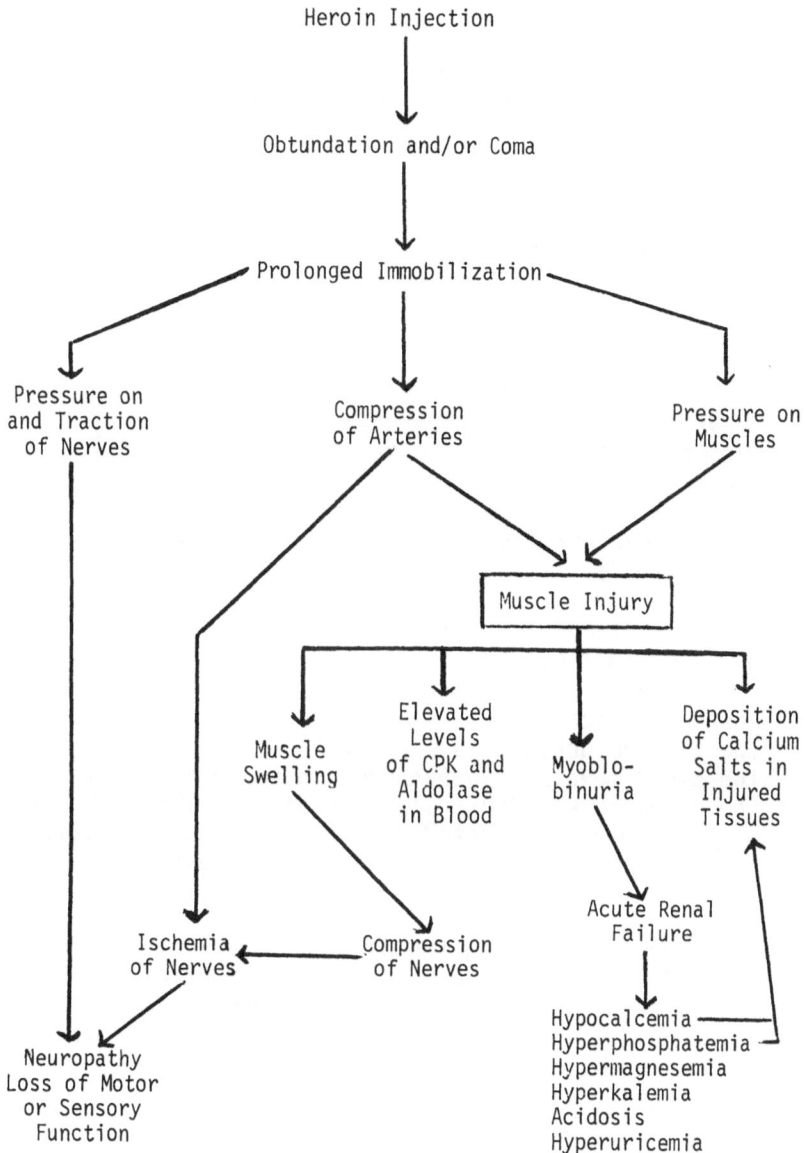

Figure 1.   A schematic presentation of the sequence of events fol-
            lowing an injection of a large dose of heroin. CPK =
            creatinphosphokinase.

The levels of serum creatinine are usually disproportionately higher than those of urea nitrogen. Almost all patients with rhabdomyolysis have moderate to marked leukocytosis (11,000-25,000 cells/mm^3) in the absence of infection. No data are available on the activity of the various proteases in the blood of these patients.

The laboratory features of acute renal failure in these patients are similar to those seen in patients with other causes of acute renal failure but are more pronounced in the patients with rhabdomyolysis (1,2). The duration of oliguria may range from 1-20 days followed by diuresis and recovery of renal function. About 20% of the patients have non-oliguric renal failure. The patients with acute renal failure have marked hyperphosphatemia and hypocalcemia. Occasionally, the concentration of serum phosphorus is not elevated and this is seen in patients who were phosphate depleted and hypophosphatemic prior to the development of rhabdomyolysis. Profound hyperuricemia is a usual finding in these patients. The mechanisms of the hyperuricemia include increased production of uric acid from the precursors released from the damaged muscle as well as decreased urinary excretion by the failing kidneys. These patients are usually hypercatabolic as manifested by negative nitrogen balance (6), hyperkalemia, severe acidosis and rapid rise in blood levels of urea nitrogen (1,2).

Deposition of calcium in damaged muscles during the oliguric phase is common in these patients. These calcium deposits can be detected by conventional x-rays, electron radiography or technetium 99m diphosphonate scan (7). The latter provides the best diagnostic tool. The calcium deposits disappear with recovery of renal function. The blood levels of parathyroid hormone are usually elevated during the oliguric and diuretic phase of the illness. The blood levels of 25OHD may be elevated during the oliguric and diuretic phase (8,9), and in a few patients the blood levels of 1,25(OH)$_2$D were found elevated during the diuretic phase (10).

Hypercalcemia occurs in the diuretic phase of acute renal failure in about 30% of the patients (9). The hypercalcemia usually appears shortly after the onset of diuresis and has been noticed during the 3rd-11th day of the diuretic phase. Occasionally it is delayed and in one patient it occurred 55 days after the onset of diuresis (9). It is usually mild, of short duration and relatively benign. Occasionally, it is severe, prolonged and symptomatic. It is generally accepted that the mechanism of this hypercalcemia is the remobilization into the extracellular space of calcium previously deposited in the damaged muscle. Release of vitamin D from the damaged muscle may also play a role. On rare occasions, hypercalcemia may develop during the oliguric phase of the acute renal failure and at a time when serum phosphorus levels are also markedly elevated (8). This situation is very hazardous since it could precipitate acute widespread calcium deposition in soft tissues including vital organs such as lungs and heart and lead to severe cardiopulmonary insufficiency (8).

POTENTIAL ROLE OF PROTEASES IN THE COMPLICATIONS OF RHABDOMYOLYSIS

It is not apparent at the present time whether rhabdomyolysis is associated with increased activity of proteases in the blood of these patients. However, it is most likely that such an event does occur. Hörl et al. (11) reported increased proteolytic activity in the serum of a patient with acute renal failure due to multiple trauma and muscle injury. Their observation is consistent with the postulate that muscle injury could be associated with increased proteolytic activity in the blood. Careful and detailed studies ar. needed to confirm or refute such a postulate and to character-ize the nature of such a proteolytic activity. If proteases are releused in the circulation, one can make a case for their potential role in the pathogenesis of the various complications discussed above.

The mechanisms of acute renal failure in patients with non-traumatic rhabdomyolysis are not known. Is it possible that pro-teolytic proteases released from damaged muscles can adversely af-fect the function and integrity of the renal cells? A similar hy-pothesis was advanced to explain the acute renal failure associated with acute pancreatitis (12). Moreover, patients with acute renal failure due to non-traumatic rhabdomyolysis are very catabolic and supplementation of large amounts of nitrogen were not successful to abate the negative nitrogen balance (6). It is tempting to spe-culate that increased proteolytic activity in the blood of these patients participates in the genesis of the hypercatabolic state. The data of Hörl and Heidland (13) provide support for such a con-tention. Finally, Kaptein et al. (14) reported that the binding of various thyroid hormones to their carrier proteins is reduced in patients with acute renal failure due to non-traumatic rhabdomyoly-sis. It is possible that proteases released from injured muscle may alter the configuration or the surface charges of the carrier pro-teins and alter their binding characteristics. Such a disturbance may affect other hormones and drugs and as such may have far-reaching consequences.

It is evident that non-traumatic rhabdomyolysis is a pathologic entity that provides a unique opportunity for the study of the po-tential role of proteases in the pathogenesis of a multitude of de-rangements that may have wide range clinical implications.

REFERENCES

1. R.A. Grossman, R.W. Hamilton, B.M. Morse, A.S. Penn, and M. Goldberg, Non-traumatic rhabdomyolysis and acute renal fail-ure, N. Engl. J. Med. 291:807 (1974).

2.   A. Koffler, R.M. Friedler, and S.G. Massry, Acute renal failure
      due to non-traumatic rhabdomyolysis, Ann. Intern. Med. 85:23
      (1976).
3.   M. Akmal, J.R. Valdin, M.M. McCarron, and S.G. Massry, Rhabdo-
      myolysis with and without acute renal failure in patients
      with phencyclidine intoxication, Am. J. Nephrol. 1:91 (1981).
4.   A.S. Penn, L.P. Rowland, and D.W. Fraser, Drugs, coma and myo-
      globinuria, Arch. Neurol. 26:336 (1972).
5.   M. Akmal and S.G. Massry, Peripheral nerve damage in patients
      with non-traumatic rhabdomyolysis, Arch. Intern. Med. (1983,
      in press).
6.   E.I. Feinstein, M.J. Blumenkrantz, M. Healy, A. Koffler, H.
      Silberman, S.G. Massry, and J.D. Kopple, Clinical and meta-
      bolic responses to parenteral nutrition in acute renal fail-
      ure: A controlled double-blind study, Medicine 60:124 (1981).
7.   M. Akmal, D.A. Goldstein, N. Telfer, E. Wilkinson, and S.G.
      Massry, Resolution of muscle calcification in rhabdomyolysis
      and acute renal failure, Ann. Intern. Med. 89:928 (1978).
8.   E.I. Feinstein, M. Akmal, D.A. Goldstein, N. Telfer, and S.G.
      Massry, Hypercalcemia and acute widespread calcifications
      during the oliguric phase of acute renal failure due to
      rhabdomyolysis, Miner. Elect. Metab. 2:193 (1979).
9.   E.I. Feinstein, M. Akmal, N. Telfer, and S.G. Massry, Delayed
      hypercalcemia in patients with acute renal failure asso-
      ciated with non-traumatic rhabdomyolysis, Arch. Intern. Med.
      141:753 (1981).
10.   F. Llach, A.J. Felsenfeld, and M.R. Haussler, The pathophysiol-
      ogy of altered calciu metabolism in rhabdomyolysis-induced
      acute renal failure: Interactions of parathyroid hormone,
      25-hydroxycholecalciferol and 1,25-dihydroxycholecalciferol,
      N. Engl. J. Med. 305:117 (1981).
11.   W.H. Hörl, C. Gantert, I.O. Auer, and A. Heidland, In vitro
      inhibition of protein catabolism by alpha$_2$-macroglobulin in
      plasma from a patient with post-traumatic acute renal fail-
      ure, Am. J. Nephrol. 2:32 (1982).
12.   G.T. Simon and J.P. Giacobino, Pathogenesis of the glomerular
      lesions in acute pancreatitis, Lancet 2:669 (1970).
13.   W.H. Hörl and A. Heidland, Proteolytic digestion: Cause of
      protein catabolism in acute renal failure, Kidney Int. 17:
      407 (1980).
14.   E.M. Kaptein, D. Levitan, E.I. Feinstein, J.T. Nicoloff, and
      S.G. Massry, Alterations of thyroid hormone indices in
      acute renal failure and in acute critical illness with and
      without acute renal failure, Am. J. Nephrol. 1:138 (1981).

INDEX

Actin, 507, 525
Acute renal failure, 405, 417
Adult respiratory distress
  syndrome, 319
Agonists
  of steroids, 129
  of steroid receptors, 130
Alanine
  aminopeptidase, 179
  release, 559, 560, 561, 565
Aldolase, 582
Alkaline
  cysteinyl proteinase, 507
  phosphatase, 215, 380
  protease, 551
  serine proteinase, 507
Alpha-D-glucosidase, 214
Alpha$_1$-
  antichymotrypsin, 5, 31, 74,
      336
  antitrypsin, 24, 74, 300, 336,
      406, 477
  proteinase inhibitor, 5, 89,
      104, 345, 357, 379, 392,
      406, 419
Alpha$_2$-
  antiplasmin, 30, 74, 400
  macroglobulin, 5, 28, 74, 78,
      99, 263, 300, 336, 345,
      363, 371, 392, 406, 477
  plasmin inhibitor, 4, 253, 392
  thiol proteinase inhibitor, 74
Alternative pathway
  of activation for complement,
      227, 235

Amino acids
  branched-chain, 574
  muscle pattern, 572, 574
  plasma pattern, 572, 574
  pool, 571
Aminopeptidase, 143, 193, 209,
      222
Amplification
  C3-convertase, 237
  heparin inhibition, 237
Amylase, 477
Anaphylatoxin, 229
  biological responses, 229
  carboxypeptidase, 229
  COOH-terminal arginyl residue,
      229
Androgen, 129
  receptors, 131
Antagonists
  of steroids, 129
  of steroid receptors, 130
Antichymotrypsin, 300
Antileukoprotease, 300, 303, 336
Antipain, 139
Antiplasmin, 278, 401, 434, 445,
      457
Antithrombin III, 4, 26, 74, 263,
      392, 457
Aorta
  arteriosclerosis, 111
  elastase-type proteases, 113
  smooth muscle cells, 113
Aprotinin, 69, 125, 287, 307,
      401, 443, 478
Arteriosclerosis, 111

Association rate constant, 98
  irreversible inhibitor, 100
  reversible inhibitor, 100
Autodigestion, 471
Azocasein, 405

Bacteriotoxic shock, 439
Beta$_1$-anticollagenase, 74, 300
Blood
  coagulation, 287
  platelets, 241
  smears, 422
Bowman Birk Inhibitor, 401
Bradykinin, 43, 70, 479
Brain cortical synaptosomes, 165
Bronchial
  inhibitor, 104
  mucous inhibitor, 5
Brush border membrane, 182, 209
Burned patients, 449

Calcium, 160
  activated neutral proteinase,
    507
  dependent kinase, 407
  dependent neutral proteinase,
    241
Calpain, 521
Calpastatin, 523
Captopril, 173
Carboxypeptidase B, 143
Casein, 558
Catabolic stress, 571
Catabolism, 413, 557
Cathepsin, 2, 345, 393, 507, 520,
  B, 143, 159, 193, 222, 551
  D, 143, 534, 551
  G, 79, 90, 313
Chymotrypsin, 2, 77, 90, 105, 133,
    138, 193, 227
  like proteinase, 315
Cigarette smoke, 93
  alpha$_1$PI oxidation, 93
C$_1$-inactivator, 29, 74
Classical pathway
  of activation for complement,
    227, 235

Clotting, 393
Coagulation, 235, 273, 436, 439
Collagen, 379
Collagenase, 3, 80
Complement system, 227, 235, 393,
    479
Concanavalin A, 181
C3-convertases, 235
Corticosteroid binding globuline,
    131
Corticosterone, 135
C-protein, 524
Creatine phosphokinase, 582
Cystine, 79
Cytochemistry, 193
Cytoskeleton, 123

Defibrination, 274
Desmin, 524
Dialysate, 405
  proteolytic activity, 407
Dialysis
  treatment, 407, 408, 409, 422,
    423, 425, 535
Dipeptidylpeptidase, 222
Disseminated intravascular
    coagulation, 331
Dissociation rate constant, 99
Duchenne
  muscular dystrophy, 519, 526
Dyscholia, 468

ε-aminocaproic acid, 134
Edema
  acute pancreatic edema, 468
  edematous pancreatitis, 468,
    469
Elastase, 2, 77, 90, 105, 111,
    193, 222, 253, 319, 345,
    355, 393, 406, 419, 479
Elastin, 4, 92, 113, 379
Elastolytic activity, 313
Emphysema, 12, 111, 300, 313
Endotoxin, 394, 439
Enkephalin, 165

Estradiol, 132
Estrogen, 129
  receptors, 131

Factor
  B, 236
  D, 236
  Xa, 237, 288
  XIII, 395
Fc-receptor, 241
Ferritin, 180
Fibrin, 274
  degradation products, 278, 450
  derivatives, 278
  split products, 279
Fibrinogen, 274
Fibrinolysis, 273, 393, 436, 441
Fibroblast, 124, 520
Fibronectin, 4, 124, 446
Filamin, 524
FITC-Thrombin, 269
Fluorescamin, 266
Fluorescence detection, 225

α-glutamyltranspeptidase, 179, 222
Gastrin, 143
Glomerulonephritis, 273
Glucagon, 143, 558
Glucocorticoid, 129
  receptors, 131
Glucose
  uptake, 67, 68, 566
Glycocalicin, 241
Glycocalyx, 123
Glycogen
  myocardial, 70
  muscle, 564
Granulocyte, 6
  elastase, 379
  proteases, 299, 335
Haemorrhagic shock, 439
Hageman factor, 235
Halo formation
  in blood smears 419
Heart
  steroid receptor components,
    132
Hemodialysis, 407, 539

Heparin, 238, 263, 273
  cofactor II, 263
High molecular weight proteinase,
    507
Hip replacement, 287
Hydrogen peroxide, 80
Hydrolase, 2, 507
Hypercoagulability, 331

Immunoassay
  for determination of elastase-
    α_1PI, 379
Indomethacin, 68
Inflammation, 13, 391
Insulin, 70, 143, 558
  degrading proteinase, 507
Inter-α_1-trypsin inhibitor, 31, 74
Iodoacetamide, 522

Kallidin, 42
Kallikrein, 24, 41, 143, 147, 193,
    222, 235, 393, 450, 457,
    489, 495
  inhibition, 1
  inhibitor, 69
Kidney
  steroid receptor components,
    132
Kidney cortex
  membrane-bound proteases, 179
Kinetic constants, 97
Kinin, 42, 436, 495
  systems, 479, 489
Kininase, 41
Kininogen, 41, 479, 495
Kininogenases, 495

Lactase, 215
Lactate
  blood level, 287
Leukocyte, 379
  elastase, 313
Leukodiapedesis, 374
Leupeptin, 134, 139, 522
Lipoproteins, 115
Liver
  steroid receptor components,
    132
Lysosomes, 520

Macrophages, 402, 520
Mast cells, 402
Mercuribenzoate, 522
Mersalyl, 522
Methylamine, 375
Mineralocorticoid, 129
  receptors, 131
Multipolar model
  of hormone receptors, 138
Muscle
  alkali soluble proteins, 537
  amino acid release, 525
  cathepsin D, 537
  DNA, 537
  energy metabolism, 63
  fats, 537
  injury, 583
  intracellular amino acids, 572, 574
  necrosis, 581
  protein catabolism, 557
  protein degradation, 525, 545
  RNA, 537
Muscular dystrophy, 519
Myeloperoxidase, 2, 94
Myocytes, 520
Myoglobulinuria, 582
Myonecrosis, 509
Myosin, 509, 525

Na-taurocholic acid, 496
Necrobiosis, 470
Neurofilament proteins, 524
  C-protein, 524
  desmin, 524
  filamin, 524
  vimentin, 524
Neutral trypsin-like proteinase, 507
Neutrophil, 402
  elastase, 2, 89

Osteoarthrosis, 345
Oxidation
  of α_1 proteinase inhibitor, 89
Pancreatic elastase, 316, 320, 339, 345, 355, 379, 395, 409, 423

Pancreatic shock, 495
Pancreatitis, 463, 477, 489, 496
Papain, 10, 133, 137, 180
Papaverin, 68
Parathormon, 143
Parathyroidhormone, 153
Peroxidase, 80, 302
Phenylalanine
  release, 559
Phosphorylase kinase, 321, 405
Plasmin, 24, 90, 102, 143, 193, 222, 227, 236, 278, 495
Plasminogen, 236, 254, 275, 401, 434, 450, 457
  activator, 193, 222
Platelet, 287
  adhesiveness, 288
  aggregability, 288
  aggregates, 288
Pleural empyema, 231
P-nitrophenyl phosphate, 381
Polymorphonuclear leukocytes, 229
Postproline, 193
Prekallikrein, 434, 450, 457, 490
Preprohormone, 141
Prepropth, 159
Progestagens, 129
Progesterone, 132
Prohormone, 141
Pronase, 227
Propth, 159
Prostaglandins, 67
Proteinase, 129, 227, 235, 379, 391, 519
  inhibitors, 73, 97, 263
Proteoglycan, 123, 379
Proteolysis, 49, 129, 138, 393
Prothrombin, 236, 264, 401, 450
Pulmonary artery pressure, 321
Pulmonary capillary perfusion, 321
Pulmonary diseases, 299
Pulmonary leucostasis, 331
Puromycin, 173

Receptor multiplicity, 130
Renal failure, 405, 417
Renin, 147

Respiratory resistance, 321
Rhabdomyolysis
  after alcohol binge, 582
  after drug abuse, 582
  after heroin overdose, 582
  after phencyclidine abuse, 582
Rheumatoid arthritis, 12, 335,
    345, 355

Section biochemistry, 193, 215
Seminal plasma inhibitor, 5
Sepsis, 455, 574
Septicemia, 391
Serine esterase, 236
Shock
  bacteriotoxic, 439
  endotoxin, 411
  haemorrhagic, 439
  septic, 393
  traumatic, 439
Skeltal muscle, 507
  proteases, 581
Skin fibroblast, 114
Solid phase antibody, 364, 381
Soybean trypsin inhibitor, 364
Stability time
  of the trypsin-interα-trypsin
  inhibitor complex, 106
Starvation, 575
Steroid receptors, 130
Synaptosomes, 174
Synovial
  fluids, 338, 345, 367
  membrane, 375
  tissue inhibitor, 5
Synovitis, 12

Tamoxifen, 132
Taurocholeate, 490
Testosterone, 138
Thiol-disulfide interchange, 79
Thrombin, 23, 90, 102, 193, 222,
    236, 263, 319
Thromboelastogram, 288
Thromboplastin time, 288
Total thoracic compliance, 321
Transcortin, 131

Trauma, 574
Traumatic shock, 439
Tropomyosin, 509
Troponin, 509
Trypsin, 90, 104, 133, 135, 143,
    156, 193, 227, 366, 406,
    408, 477, 495
  binding capacity, 364
  inhibitors, 134, 135, 477
  interα-trypsin inhibitor
  complex, 106
Trypsinogen, 79, 477
Tyrosine release, 559

Ultracytochemistry, 193
Urea nitrogen appearance, 547
Uremia, 557
Urokinase, 275

Vimentin, 524
Von Willebrand Factor, 241